International Academic Symposium of Social Science 2022

International Academic Symposium of Social Science 2022

Editors

Mohamad Rahimi Mohamad Rosman
Nor Erlissa Abd Aziz
Mohd Nasir Ismail
Ghazali Osman
Khalid Abdul Wahid

MDPI • Basel • Beijing • Wuhan • Barcelona • Belgrade

Editors

Mohamad Rahimi Mohamad Rosman
Universiti Teknologi MARA Kelantan Branch
Malaysia

Nor Erlissa Abd Aziz
Universiti Teknologi MARA
Kelantan Branch
Malaysia

Mohd Nasir Ismail
Universiti Teknologi MARA
Kelantan Branch
Malaysia

Ghazali Osman
Universiti Teknologi MARA
Kelantan Branch
Malaysia

Khalid Abdul Wahid
Universiti Teknologi MARA
Kelantan Branch
Malaysia

Editorial Office
MDPI
St. Alban-Anlage 66
4052 Basel, Switzerland

This is a reprint of Proceedings published online in the open access journal *Proceedings* (ISSN 2504-3900) in 2022 (available at: https://www.mdpi.com/2504-3900/82/1).

For citation purposes, cite each article independently as indicated on the article page online and as indicated below:

LastName, A.A.; LastName, B.B.; LastName, C.C. Article Title. *Journal Name* **Year**, *Article Number*, Page Range.

Volume 2
ISBN 978-3-0365-6095-3 (Hbk)
ISBN 978-3-0365-6096-0 (PDF)

Volume 1-2
ISBN 978-3-0365-6017-5 (Hbk)
ISBN 978-3-0365-6018-2 (PDF)

Contents

proceedings

Proceeding Paper

Bring-Your-Own-Device (BYOD) and Productivity: A Conceptual Framework †

Mohamad Rahimi Mohamad Rosman *, **Nurfatihah S Baharuddin**, **Noor Azreen Alimin**, **Nik Nur Izzati Nik Rosli**, **Amira Idayu Mohd Shukry** and **Noor Masliana Razlan**

Faculty of Information Management, Universiti Teknologi MARA Kelantan Branch, Machang 18500, Malaysia
* Correspondence: rahimimr@uitm.edu.my; Tel.: +60-199891306

† Presented at the International Academic Symposium of Social Science 2022, Kota Bharu, Malaysia, 3 July 2022.

Abstract: BYOD is defined as the act of bringing your own gadget, facilities, or device to the organization or institution. The concept of BYOD has spread almost to many sectors, especially in education, due to a shortage of financial resources in the aftermath of the Novel Coronavirus 2019 pandemic. BYOD is a helpful concept in face-to-face education by giving the needy access to adequate resources. However, most of the time, students especially in higher learning institutions, are having a problem accessing adequate resources and facilities standards that may influence their productivity, performance, and perceived benefits. Moreover, the inadequacy of standardized facilities and requirements may also deprive the students of necessary productivity standards. Furthermore, the pandemic of Novel Coronavirus 2019 has transformed the current workplace practices, changing the work-life environment and warranting further exploration. Therefore, the purpose of this study is twofold; first, to identify user behavior intention to adopt BYOD, and second, to propose a conceptual model of BYOD underlying the interrelationship between BYOD antecedents and productivity. A structured literature review methodology was adopted, and a conceptual model was developed for further exploration of the topic. The contribution of this paper is as follows; first, this study identifies the antecedents of behavioral intention to adopt BYOD in the aftermath of the Novel Coronavirus 2019. Second, this study proposes a conceptual model underlying the relationship between BYOD antecedent, behavioral intention to adopt BYOD, and its impact in terms of productivity.

Keywords: organizational behavior; user intention; information technology usage; organizational policy; digital information usage behavior; information security assurance

Citation: Rosman, M.R.M.; Baharuddin, N.S.; Alimin, N.A.; Rosli, N.N.I.N.; Shukry, A.I.M.; Razlan, N.M. Bring-Your-Own-Device (BYOD) and Productivity: A Conceptual Framework. *Proceedings* **2022**, *82*, 10. https://doi.org/10.3390/proceedings2022082010

Academic Editor: Mohd Nasir Ismail

Published: 7 September 2022

1. Introduction

Bring-your-own-device (BYOD) is defined as an act of bringing your own gadget, facilities, or device to the organization or institution. It's a concept that is fast-spreading, especially in information technology (IT) consortiums and educational institutions [1–4]. BYOD is simply a process of allowing employees or students to attach or utilize their own facilities or device to the organizational or institutional facilities and network. According to [2], BYOD allows the employee to utilize their privately-owned device to assist business purposes, however, not without the added pressure of increasing individual cost, benefits, and extra cyber security threats. Previous research shows that BYOD has the capability to increase student learning engagement [5,6] via mastery of their privately-owned device and facilities. BYOD also encourages cost reduction, better management of organizational assets, and management of IT security [2].

However, the Novel Coronavirus 2019 (COVID-19) pandemic has seen drastic changes in the IT infrastructure acquisition and policy. Many organizations and education institutions were hampered by a lack of financial ability to maintain, sustain, or improve their current IT investment and facilities. Most organizations were also faced with a budget shortage for IT investment due to the cause of the COVID-19 pandemic [7,8]. Due to these limitations, users were left with incomplete, unstandardized, and insufficient access to IT facilities that may influence their level of productivity [1,9–11]. However, knowledge of this relationship is scarce within the field of Information System (IS); thus, warranting further exploration of the topic. For example, the COVID-19 pandemic has transformed the current workplace practices, thus, the changing work-life environment. Therefore, the purpose of this study is twofold; first, to identify user behavior intention to adopt BYOD, and second, to propose a conceptual model of BYOD underlying the interrelationship between BYOD antecedents and productivity.

The rest of the paper is organized as follows: First, the literature of the study is presented, followed by the methodology of the study. Next, the conceptual model of the study is presented and discussed. Lastly, the conclusion of the study is outlined.

2. Literature Review

2.1. Concept of BYOD

The concept of bring-your-own-device or BYOD can be traced from multiple disciplines: Information systems (IS), and business and management. BYOD can be categorized into two explicit terms. First, BYOD is related to the tendency of employees or students in an organization to connect their privately-owned devices to the work network [1,2]. Second, it relates to the mobile phone capability that allows users to activate their current phone independently without the need to buy a new phone (provider independence). BYOD has been on an emerging trend since its inception in early 2012. Surveys in 2013 showed that respondents in emerging countries (including Malaysia) showed enormous interest in using their privately-owned device in their workplace [12]. These employees believe that having direct and constant access to their workplace network helps them to be competitive and sustainable. Interestingly, a recent survey conducted in 2020 indicated that more than 77% of Malaysians connect their privately-owned device to their organizational corporate network [13].

Even though BYOD is on the rise, much of the discussion is on the security of corporate data resides within the privately-owned devices [3,5,14,15]. Allowing privately-owned devices to access corporate information may lead to several problems: first, privately-owned devices may be susceptible to cybersecurity attacks due to low-security mechanisms on personal hardware. Second, un-erased corporate data residing in the privately-owned devices may be leaked to outsiders once employees discarded or sold their devices. Thus, implementing a BYOD concept requires an appropriate workflow and policy enforcement.

2.2. BYOD and Education

The concept of BYOD has shifted from a business-based environment to an educational-centric environment. As mentioned in [5,6], the concept of BYOD has the capability to enhance student engagement and subsequently improve productivity. Previous surveys conducted on the interrelationship between BYOD and education show a wider acceptance of the BYOD concept in the educational institution, and most students own their personal devices such as laptops and mobile phones. Unlike IT companies, educational institution deals with less sensitive information, thus ensuring the information can be accessed from outside of the institution [1].

On the other hand, recent studies show that a total of 42% of students were satisfied with the teaching and learning approach in the wake of BYOD [16]. Besides, it also encourages professional competence (by bringing their privately-owned device), inspires the learning process, increases the reliance on IT, and decreases the dependency on the conventional document [16]. On top of it, the BYOD concept also appeals to many schools as it was actively adopted concept over the years due to cost-saving and the comfort of using the privately-owned device [17].

3. Methodology

This study adopted the structured literature review (SLR) methodology of [18] and [19]. First, the literature search started with a leading journal that was relevant to the field of study. Second, a backward search was conducted to identify relevant articles. Third, a forward search was conducted to identify any news articles that cite the publication.

In the first phase, we utilize the EzAccess, online database tools from one of the biggest universities in Malaysia to identify and search relevant articles from leading journals. The keywords used are "BYOD", "Productivity", and "Factors" or *(TITLE-ABS-KEY(BYOD)) AND ((productivity)) AND (factor)*. To improve the accuracy of search results, we limit the article to research manuscripts, theses, and books. A total of 37 articles were found relevant to the topic of interest.

In the next stage, we evaluate the article based on the method suggested by [20]. First, we skimmed each of the papers to identify its topics and critical issues, Second, we focus on the introduction, abstract, discussion, and conclusion of the study. Third, we identify potential variables underlying the topic for the development of our proposed conceptual framework.

4. Conceptual Model

Based on the literature search, four (4) main themes or dimensions were identified, namely technological readiness [1,15,21–23], individual readiness [22,24–27], contextual readiness [1,6,28–30], and organizational readiness [31–33]. The following Figure 1 shows the proposed conceptual model of the study.

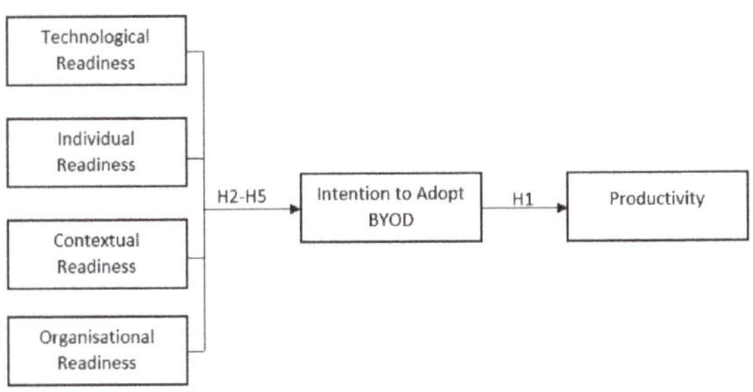

Figure 1. Conceptual Framework.

Behavioral intention to adopt BYOD is operationalized as the degree to which an individual has the urgency or tendency to adopt or refuse to adopt BYOD approaches. Behavior intention to adopt BYOD plays an important role in productivity improvement. Disengaging behavior or reluctance to adopt BYOD will lead to a negative impact on productivity improvement. As found out by [34], BYOD is no longer an option as many IT organizations turn to the concept of BYOD, thus, understanding the behavioral intention of users is very crucial to ensure the success of the transformation process. Similarly, [35,36] argue that BYOD implementation with related purposed has the capability to influence one's productivity. Furthermore, BYOD also has the capability to enhance morale and increase engagement—thus closely related to productivity improvement [1,30]. Therefore, it can be hypothesized that:

H1. *The intention to adopt BYOD has a positive and significant relationship with productivity.*

Technological readiness is operationalized as the extent of individual technological establishment in support of BYOD implementation. According to [37], it is predicted that more than 95% of organizations allow the use of privately-owned devices and more than 85% of organizations implemented BYOD due to the COVID-19 pandemic. Subsequently, more employees relied on their privately-owned devices (65% of employees disclosed their personal contact numbers), and the number of privately-owned devices increased to over

USD 11.96 billion in 2021. This recent statistic serves as evidence of the influence of the technological readiness towards user behavioral intention to implement or adapt the BYOD concept in the workplace. Therefore, it can be hypothesized that:

H2. *Technological readiness has a positive and significant relationship with the intention to adopt BYOD.*

Individual readiness is conceptualized as the extent of individual intrapersonal and interpersonal influence toward BYOD adoption. Individual readiness is believed to be one of the most relevant factors in IS success factor studies [38–41]. Among the factors that differentiate an individual is often related to demographic profiles such as age, education, income, etcetera [42]. These factors are crucial in the decision-making process that will shape user behavioral intention and either encourage or discourage the adoption of BYOD. Moreover, a lack of individual readiness usually discourages further interaction with the BYOD implementation. Therefore, it can be hypothesized that:

H3. *Individual readiness has a positive and significant relationship with the intention to adopt BYOD.*

Contextual readiness refers to external factors of an individual that shape their interaction and behavior towards BYOD implementation. Contextual factors can include multiple factors and antecedents, such as task resources and task complexity [6,43,44]. As argued by [45], the level of complexity influence user behavioral intention to adopt BYOD; for instance, BYOD increase cybersecurity threats and require more effort to ensure the safety of corporate data. Moreover, an individual has the preference to be familiar with their privately-owned devices, subsequently creating less pressure or a less intimidating environment [38]. Therefore, it can be hypothesized that:

H4. *Contextual readiness has a positive and significant relationship with the intention to adopt BYOD.*

Organizational readiness can be defined as the facilities, services, and support provided by the organization to the individual in an organization or institution. As mentioned by [31], organizational commitment is an important indicator of adoption success, as it may reduce resistance among the employee. Intriguingly, the number of adoption of BYOD among educational institutions is rather high with or without the organizational consent or policy [46]. According to [47], the adoption of BYOD may contribute to organizational benefits and performance; such as productivity enhancement. Therefore, it can be hypothesized that:

H5. *Organizational readiness has a positive and significant relationship with the intention to adopt BYOD.*

5. Conclusions

This study was conducted with two purposes; first, to identify user behavior intention to adopt BYOD, and second, to propose a conceptual model of BYOD underlying the interrelationship between BYOD antecedents and productivity. To proceed, a systematic literature review (SLR) methodology for [18,19] was adopted. Subsequently, four (4) antecedents underlying the adoption intention of BYOD were identified, technological readiness, individual readiness, contextual readiness, and organizational readiness. Next, the conceptual model of the study was proposed, and discussions were made on each hypothesis.

The next stage of this research is to validate the conceptual research model in a quantitative study. An instrument will be developed by adapting and adopting the previous instrument and it will be sent for expert validation. Following the expert validation, a pilot study will be conducted, and Cronbach's Alpha coefficient will be used to determine the reliability of the instrument. Upon completion of the reliability assessment, a full-scale data collection will be performed among the respondents selected via the convenience sampling method. The potential respondents are undergraduate students from one of the local universities in Malaysia, selected based on demographic information explored via the SLR. Next, data will be analyzed based on descriptive using Statistical Package for Social Sciences (SPSS) version 26 and inferential using SmartPLS version 3.3.3.

Author Contributions: Conceptualization, M.R.M.R. and N.S.B.; methodology, N.A.A. and N.N.I.N.R.; validation, M.R.M.R. and A.I.M.S.; formal analysis, N.M.R.; investigation, N.A.A., N.N.I.N.R. and A.I.M.S.; resources, N.S.B.; data curation, M.R.M.R.; writing—original draft preparation, M.R.M.R., N.S.B., N.A.A., N.N.I.N.R., N.M.R. and A.I.M.S.; writing—review and editing, M.R.M.R.; visualization, M.R.M.R.; supervision, M.R.M.R.; project administration, M.R.M.R. All authors have read and agreed to the published version of the manuscript.

Funding: This research received no external funding.

Institutional Review Board Statement: Not applicable.

Informed Consent Statement: Not applicable.

Data Availability Statement: Not applicable.

Acknowledgments: This research was partially supported by Universiti Teknologi MARA Kelantan Branch, Malaysia.

Conflicts of Interest: The authors declare no conflict of interest.

References

1. Afreen, R. Bring your own device (BYOD) in higher education: Opportunities and challenges. *Int. J. Emerg. Trends Technol. Comput. Sci.* **2014**, *3*, 233–236.
2. Loose, M.; Weeger, A.; Gewald, H. BYOD–the next big thing in recruiting? Examining the determinants of BYOD service adoption behavior from the perspective of future employees. In *AMCIS 2013 Proceedings, Chicago, IL, USA, 15–17 August 2013*; Association for Information Systems: Atlanta, GA, USA, 2013; p. 12.
3. Ratchford, M.; El-Gayar, O.; Noteboom, C.; Wang, Y. BYOD security issues: A systematic literature review. *Inf. Secur. J. A Glob. Perspect.* **2022**, *31*, 253–273. [CrossRef]
4. Blay, A. Factors Influencing Employees' Intention to Participate in a Bring Your Own Device Program in the Workplace: A Correlational Study in Ghana. Ph.D Thesis, Capella University, Minneapolis, MN, USA, 2022.
5. Cheng, G. Using the community of inquiry framework to support and analyse BYOD implementation in the blended EFL classroom. *Internet High. Educ.* **2022**, *54*, 100854. [CrossRef]
6. Rosman, M.R.M.; Arshad, I.H.; Alias, N.R.; Idris, Z.; Hamid, H.A. Managing Individual Online Learning Experience: The Roles of Perceived Engagement and Perceived Performance. In *Advances in Information, Communication and Cybersecurity—Proceedings of ICI2C'21*; Maleh, Y., Alazab, M., Gherabi, N., Tawalbeh, L., Abd El-Latif, A.A., Eds.; Lecture Notes in Networks and Systems; Springer: Cham, Switzerland, 2022; Volume 357, pp. 331–344.
7. Baharuddin, N.S.; Rosman, M.R.M. Factors affecting the usage of Library e-services in the aftermath of COVID-19 Pandemic. *Acad. J. Bus. Soc. Sci.* **2020**, *4*, 1–14.
8. Rosman, M.R.M.; Zawawi, M.Z.M.; Fadzil, F.H.; Abdullah, N.; Arshad, I.H.; Saleh, M.S.M. Does Information System Aids Decision Making Approach? The Case of Jury Management System. In Proceedings of the 2021 International Conference on Decision Aid Sciences and Application (DASA), Sakheer, Bahrain, 7–8 December 2021; pp. 85–90.
9. Singh, N. BYOD genie is out of the bottle—"Devil or angel". *J. Bus. Manag. Soc. Sci. Res.* **2012**, *1*, 1–12.
10. Alotaibi, B.; Almagwashi, H. A review of BYOD security challenges, solutions and policy best practices. In Proceedings of the 2018 1st International Conference on Computer Applications & Information Security (ICCAIS), Riyadh, Saudi Arabia, 4–6 April 2018; pp. 1–6.
11. Soubhagyalakshmi, P.; Satyanarayan Reddy, K. SWOT Analysis of BYOD (Bring Your Own Device). In *Emerging Technologies in Data Mining and Information Security*; Springer: Berlin/Heidelberg, Germany, 2021; pp. 681–688.
12. Cook, I. BYOD—Research Findings Released. Available online: https://cxounplugged.com/2012/11/ovum_byod_research-findings-released/ (accessed on 22 May 2022).
13. Murugiah, S. Majority of Malaysians Connect Corporate Laptops to Home Network—Survey. Available online: https://www.theedgemarkets.com/article/majority-malaysians-connect-corporate-laptops-home-network-%E2%80%94-survey (accessed on 22 May 2022).
14. Palanisamy, R.; Norman, A.A.; Mat Kiah, M.L. BYOD policy compliance: Risks and strategies in organizations. *J. Comput. Inf. Syst.* **2022**, *62*, 61–72. [CrossRef]
15. Downer, K.; Bhattacharya, M. BYOD Security: A Study of Human Dimensions. *Informatics* **2022**, *9*, 16. [CrossRef]
16. Jehma, H.; Punkhoom, W. How Thai Students Perceive BYOD as an Effective Strategy Applied in English Learning Environment. *Dhammathas Acad. J.* **2022**, *22*, 29–38.
17. Siyam, N.; Hussain, M.; Alqaryouti, O. Factors impacting teachers' acceptance and use of Bring Your Own Device (BYOD) in the classroom. *SN Soc. Sci.* **2022**, *2*, 1–30. [CrossRef]
18. Webster, J.; Watson, R.T. Analyzing the past to prepare for the future: Writing a literature review. *MIS Q.* **2002**, *26*, xiii–xxiii.
19. Rosman, M.R.M. Reviewing the Concept of Enterprise Content Management (ECM). *J. Digit. Inf. Manag.* **2020**, *18*, 125.

20. Glaser, B.G.; Strauss, A.L.; Strutzel, E. The discovery of grounded theory; strategies for qualitative research. *Nurs. Res.* **1968**, *17*, 364. [CrossRef]
21. Tsarapkina, J.M.; Dunaeva, N.; Kireicheva, A. Application of BYOD technology in education on the example of Lecture Racing mobile application. *Информатика Образование* **2019**, *9*, 56–64. [CrossRef]
22. Cristol, D.; Gimbert, B. Academic achievement in BYOD classrooms. In Proceedings of the 12th World Conference on Mobile and Contextual Learning (mLearn 2013), Doha, Qatar, 22–24 October 2013; p. 15.
23. Koh, E.B.; Oh, J.; Im, C. A study on security threats and dynamic access control technology for BYOD, smart-work environment. In Proceedings of the International MultiConference of Engineers and Computer Scientists, Hong Kong, China, 12–14 March 2014; pp. 1–6.
24. Leclercq-Vandelannoitte, A. Managing BYOD: How do organizations incorporate user-driven IT innovations? *Inf. Technol. People* **2015**, *28*, 2–23. [CrossRef]
25. Köffer, S.; Ortbach, K.; Junglas, I.; Niehaves, B.; Harris, J. Innovation through BYOD? *Bus. Inf. Syst. Eng.* **2015**, *57*, 363–375. [CrossRef]
26. Santos, I.M.; Bocheco, O. Exploring BYOD usage in the classroom and policies. *Int. J. Inf. Commun. Technol. Educ. (IJICTE)* **2016**, *12*, 51–61. [CrossRef]
27. Arpaci, I. Arpaci, I. A theoretical framework for IT consumerization: Factors influencing the adoption of BYOD. In *Handbook of Research on Technology Integration in the Global World*; IGI Global: Hershey, PA, USA, 2018; pp. 114–129. [CrossRef]
28. Alaskar, M.; Shen, K.N. Understanding bring your own device (BYOD) and employee information security behaviors from a work-life domain perspective. In *AMCIS 2016 Proceedings, San Diego, CA, USA, 11–14 August 2016*; Association for Information Systems: Atlanta, GA, USA, 2016; p. 38.
29. Dang, D.P.T.; Pittayachawan, S.; Nkhoma, M.Z. Contextual difference and intention to perform information security behaviours against malware in a BYOD environment: A protection motivation theory approach. In *ACIS 2013 Proceedings*; Association for Information Systems: Atlanta, GA, USA, 2013; p. 49. Available online: https://aisel.aisnet.org/acis2013/49 (accessed on 24 May 2022).
30. Yin, C.; Liu, L.; Liu, L. BYOD implementation: Understanding organizational performance through a gift perspective. In *PACIS 2014 Proceedings, Chengdu, China, 24–28 June 2014*; Association for Information Systems: Atlanta, GA, USA, 2014; p. 129. Available online: https://aisel.aisnet.org/pacis2014/129/ (accessed on 24 May 2022).
31. Doargajudhur, M.S.; Dell, P. Impact of BYOD on organizational commitment: An empirical investigation. *Inf. Technol. People* **2018**, *32*, 246–268. [CrossRef]
32. Hovav, A.; Putri, F.F. This is my device! Why should I follow your rules? Employees' compliance with BYOD security policy. *Pervasive Mob. Comput.* **2016**, *32*, 35–49. [CrossRef]
33. Putri, F.F.; Hovav, A. Employees' Compliance with BYOD Security Policy: Insights from Reactance, Organizational Justice, and Protection Motivation Theory. In Proceedings of the European Conference on Information Systems (ECIS) 2014, Tel Aviv, Israel, 9–11 June 2014; Association for Information Systems: Atlanta, GA, USA, 2014. Available online: https://aisel.aisnet.org/ecis2014/proceedings/track16/2/ (accessed on 24 May 2022).
34. Gupta, R.; Bhardwaj, G.; Singh, G. Employee Perception and Behavioral Intention to Adopt BYOD in the Organizations. In Proceedings of the 2019 International Conference on Automation, Computational and Technology Management (ICACTM), London, UK, 24–26 April 2019; pp. 73–78.
35. Hayes, B.; Kotwica, K. *Bring Your Own Device (BYOD) to Work: Trend Report*; Newnes: Amsterdam, The Netherlands, 2013.
36. Gaff, B.M. Byod? omg! *Computer* **2015**, *48*, 10–11. [CrossRef]
37. Chang, J. 44 Basic BYOD Statistics: 2022 Market Share Analysis & Data. Available online: https://financesonline.com/byod-statistics/ (accessed on 24 May 2022).
38. Mohamad Rosman, M.R.; Ismail, M.N.; Masrek, M.N. Investigating the predictors of digital library engagement: A structured literature analysis. *Pak. J. Inf. Manag. Libr.* **2021**, *22*, 60–82.
39. Hong, K.-K.; Kim, Y.-G. The critical success factors for ERP implementation: An organizational fit perspective. *Inf. Manag.* **2002**, *40*, 25–40. [CrossRef]
40. Hong, W.; Thong, J.Y.; Wong, W.-M.; Tam, K.-Y. Determinants of user acceptance of digital libraries: An empirical examination of individual differences and system characteristics. *J. Manag. Inf. Syst.* **2002**, *18*, 97–124. [CrossRef]
41. Thong, J.Y.; Hong, W.; Tam, K.-Y. Understanding user acceptance of digital libraries: What are the roles of interface characteristics, organizational context, and individual differences? *Int. J. Hum.-Comput. Stud.* **2002**, *57*, 215–242. [CrossRef]
42. Rosman, M.R.M.; Nik Rosli, N.N.I.; Shukry, A.I.M.; Razlan, N.M.; Alimin, N.A. Entangling the Interrelationship Between Demographics Profiles, Referencing Competencies and Individual Performance in the Digital Environments. *Int. J. Emerg. Technol. Learn.* **2022**, *17*, 125–136. [CrossRef]
43. Rosman, M.R.M.; Ismail, M.N.; Masrek, M.N. Development and validation of a tool for measuring digital library engagement. *Int. J. Electr. Comput. Eng. (IJECE)* **2022**, *12*, 4146–4154. [CrossRef]
44. Danner-Schröder, A.; Ostermann, S.M. Towards a Processual Understanding of Task Complexity: Constructing task complexity in practice. *Organ. Stud.* **2022**, *43*, 437–463. [CrossRef]
45. Disterer, G.; Kleiner, C. BYOD bring your own device. *Procedia Technol.* **2013**, *9*, 43–53. [CrossRef]

46. Vejayon, J.; Samy, G.N.; Maarop, N.; Megat, N.; Shanmugam, B.; Magalingam, P. Adopting Factors of Bring Your Own Device (BYOD) at the Selected Private Higher Learning Institution in Malaysia. *J. Adv. Res. Soc. Behav. Sci.* **2016**, *2*, 24–32.
47. Astani, M.; Ready, K.; Tessema, M. BYOD Issues and strategies in organizations. *Issues Inf. Syst.* **2013**, *14*, 195–201.

proceedings

MDPI

Proceeding Paper

Key Determinants of Continuance Usage Intention: An Empirical Study of Mobile Food Delivery Apps among Malaysians †

Emily H. T. Yapp [1,*] and Saraniya Kataraian [2]

[1] Labuan Faculty of International Finance, Universiti Malaysia Sabah, Labuan International Campus, Labuan 87000, Malaysia

[2] Accenture Technology Solution Sdn. Bhd. Sunway, Kuala Lumpur 47500, Malaysia

* Correspondence: emilyyht@ums.edu.my

† Presented at the International Academic Symposium of Social Science 2022, Kota Bharu, Malaysia, 3 July 2022.

Abstract: Mobile food delivery apps have been widely used, especially among Malaysians during the COVID-19 pandemic. When the government allowed restaurants to reopen, mobile food delivery apps slightly declined as Malaysians rushed to dine in restaurants. What determinants Malaysian users to continue to use mobile food delivery apps services after the COVID-19 pandemic remain unknown. Underlying the Extended Unified Theory of Acceptance and Use of Technology (UTAUT2) determinants, this study introduces two other relevant determinants: convenience and time saving. Partial least squares structural equation modeling techniques using SmartPLS were used to analyse the data.

Keywords: mobile food delivery apps; UTAUT2; COVID-19 pandemic

check for updates

Citation: Yapp, E.H.T.; Kataraian, S. Key Determinants of Continuance Usage Intention: An Empirical Study of Mobile Food Delivery Apps among Malaysians. *Proceedings* **2022**, *82*, 15. https://doi.org/10.3390/proceedings2022082015

Academic Editor: Mohamad Rahimi Mohamad Rosman

Published: 8 September 2022

1. Introduction

From 2018 to 2025, the Malaysian online meal delivery industry is expected to grow at 18.6% [1]. Mobile food delivery apps allow users to order their food fast and conveniently with real-time access, and users may either takeaway or obtain their food at their doorstep [2]. Users may need only a smartphone to download and visit restaurants, no physical contact is required to view menus and make orders and payments [3]. According to Ariel [4], food-related apps were users' second most frequently installed application. By using mobile food delivery apps, users may plan their meals from a broad variety of outlets more simply and reliably at times and locations that are convenient for them. These apps often provide more thorough, updated, and reliable information regarding food places and the meals they serve. Since Malaysia banned dine-in on March 16, 2019 as part of a government Movement Control Order, food delivery has become a lifeline for restaurateurs. Mobile food delivery apps have progressively increased as people want to purchase meals in a contactless way by utilising food delivery apps. However, as the Malaysian government allows restaurants and cafés to reopen, will the usage of mobile food delivery apps decrease? It is questionable if Malaysians would continue actively to utilise mobile food delivery apps as before. As a result, this research focuses on identifying the factors that may influence users' continued use of mobile food delivery apps through the lens of the Unified Theory of Acceptance and Use of Technology 2 (UTAUT2) by Venkatesh et al. [5]. In addition, this research also offers two more crucial factors to the model which is time saving and convenience.

2. Literature and Hypothesis

Table 1 provides a selected study on food delivery apps that used the UTAUT theory over the past 3 years. Although researchers have mainly used the UTAUT theory as their underlining theory, researchers such as Zhao and Bacao [6] have also merged the task–technology fit model and expectancy confirmation theory into the UTAUT theory. These researchers have mainly focused on multi-dimensional attributes of food delivery apps and concentrated on users' attitude and/or their behaviour. Although UTAU is the underlying theory, not every researcher retained the original independent variables [6,7] but also added new variables into their study [7–9]. In addition, given the importance of identifying what are the important aspects when evaluating food delivery apps, researchers such as Ray et al. [10]; Zhao and Bacao [6], and Fakfare [11] have urged for more research to be performed.

Table 1. Studies on food delivery apps using UTAUT theory.

Authors	Variables
Alalwan [8]	Performance expectancy, effort expectancy, social influences, facilitating conditions, hedonic motivation, price value, habit, online review, continued intention, online rating online tracking, and satisfaction
Gunden et al. [7]	Performance expectancy, congruity with self-image, habit, impulse buying tendency, mindfulness, and intention
Tam, Santos, and Oliveira [9]	Social influence, performance expectancy, effort expectancy, perceived fear, facilitating conditions, and continuous intention
Zhao and Bacao [6]	Performance expectancy, effort expectance, social influence, trust, perceived task–technology fit, confirmation, satisfaction, continuance intention
Surya, Sukresna, and Mardiyano [12]	Performance expectancy, effort expectancy, social influence, facilitating condition, behavioural intention
Agarwal and Sahu [13]	Delivery Experience, time saving orientation, societal pressure, search for restaurants, listing, performance expectancy, effort expectancy, facilitating conditions, price saving orientation, hedonic motivations, habit, e-satisfaction, usage intention, and repeat use intention bandwagon effect

2.1. *Performance Expectancy*

Performance expectation was defined by Venkatesh et al. [5] as "how much the use of technology in specific activities will benefit consumers". Individuals utilising mobile food delivery apps would help them accomplish a certain job [14]. According to Marinkovic, Dordevic, and Kalinic [15] performance expectation is equivalent to perceived usefulness. As a result, if the user considers mobile food delivery apps beneficial, they are more likely to continue to utilise the mobile food delivery services. Consequently, performance expectation is a major element in mobile food delivery apps continued use intention [5,6]. Hence, the following hypothesis was formulated:

H1. Performance expectancy will positively influence the continuous usage intention of the mobile food delivery apps.

2.2. *Effort Expectancy*

Effort expectancy is "the extent of ease connected with the use of a system" [13]. Based on the literature, one of the most significant reasons customers use mobile food delivery apps is the expectation of effort [14] and not to mention particularly relevant in food delivery apps [16]. In meal delivery applications, users order their food without assistance from restaurant or café workers. Thus, a customer's propensity to continue using mobile food delivery apps may be influenced by their perception of mobile food delivery apps simplicity and ease of use especially after they have adopted the apps for quite some time [15]. Therefore, effort expectations are anticipated to substantially impact customers' continued use intentions of mobile food delivery applications. The following hypothesis was formulated:

H2. The effort expectancy will positively influence the continuous usage intention of mobile food delivery applications.

2.3. Social Influence

One of the important factors in determining whether or not a user will continue to use an application is the opinions of their friends and family members [14]. For new technology, products, and services, social influence has been shown to affect customers' behaviour [16] positively. Customers are more likely to consult their social network to learn more about mobile food delivery apps or to obtain social support for their choice to continue using them. In addition, social influence may impact how individuals think and act [17]. These applications continue to be used by customers if customers receive societal acceptance. However, consumers' willingness to continue to use mobile food delivery apps decreases if they do not obtain social validation from their friends and family members. As a result, it was hypothesised that:

H3. The social influence will positively influence the continuous usage intention of mobile food delivery applications.

2.4. Facilitating Condition

Facilitating conditions are the customers' perceptions of available resources and support to perform the behaviour [5]. They also refer to consumers' perception of the system's access to instruction, guidance, and assistance [17]. As a result, if users are satisfied with the apps facilitating conditions, users are less likely to choose a new service and more likely to reuse food delivery applications. It may also be proposed that if consumers consider a sufficient degree of technological, operational, infrastructural, and human support by utilising food delivery applications, and more comfortable experience using them [18]. Therefore, it was hypothesised that:

H4. The facilitating conditions will positively influence the continuous usage intention of mobile food delivery applications.

2.5. Hedonic Motivation

Hedonic motivation relates to multisensory, fantasy, and emotive aspects when using the products or services [5]. It is connected to users' desire, entertainment, and pleasure after the experiences of using the products or services and, in this case, the mobile food delivery apps [19]. From a mobile food delivery app perspective, ordering food through an application enables consumers to enjoy their favourite food without going out of the house or workplace. As consumers find pleasure after using the food delivery application, this encourages them to continue using it. Therefore, it was hypothesised that:

H5. The hedonic motivation will positively influence the continuous usage intention of mobile food delivery applications.

2.6. Price Value

Price value exerts a favourable impact on intention to use as the advantage of technology use is considered more than currency benefits. The findings showed that the concept of price is essential in attracting consumers and that certain quality value has a beneficial impact on intentions to reuse mobile food delivery apps [20]. Mobile food delivery apps can save the expense of buying food from restaurants. Not to mention mobile food delivery apps collaborate with the restaurants by giving more promotions such as promotions as well as other attractive incentives for users if they use food delivery apps instead of visiting restaurants. Hence, the following hypothesis is proposed:

H6. The price value will positively influence the continuous usage intention of mobile food delivery applications.

2.7. Habit

Habit has been a crucial factor in the potential adoption of technologies. Research on habitual intentions has shown that the habit is a strong indicator of technology usage to facilitate behavioural improvements. Customer's inclination to act spontaneously as a result of their collected learning experience can be defined as habit [21]. In addition, the influence of habit influences consumers' continuous usage intention, especially when consumers are familiar with the food delivery apps [22]. This relationship has also been confirmed by research such as Hsu et al. [23]; and Zanetta et al. [24]. Thus, the following hypothesis proposes that:

H7. Habit will positively influence the continuous usage intention of mobile food delivery applications.

2.8. Time Saving

Society is busy with daily routines and experiencing time constraints. As a result, products or services that can provide them time saving are chosen. In the mobile food delivery apps context, time saving is a perceived advantage to consumers [25]. Using the food delivery, users do not need to drive to the restaurant or café for food. Furthermore, time savings may boost a user's feeling of self-control, which has been shown to increase the likelihood of continuing to use food delivery apps [26]. Thus, the following hypothesis proposes that:

H8. Time saving will positively influence the continuous usage intention in mobile food delivery applications.

2.9. Convenience

In previous research on routes and mobile applications, Ozturk et al. [27], Xu, Hung, and Li [26] and Shah, Yan, and Qayyum [2] discovered that users' continuing intention was substantially influenced by convenience. Convenience is known as perceived advantage associated with mobile apps [26]. In the mobile food delivery apps context, the application allows users to access the application anytime and anywhere if they want to order food. Not to mention, the food is then be delivered to their doorsteps. Due to these reasons, it encourages the user to reuse the application from time to time. As such, convenience positively influences continuance usage intention, as presented in the following hypothesis.

H9. Convenience will positively influence continuance usage intention in mobile food delivery applications.

3. Research Method

This study concentrated exclusively on Malaysians. In this study, a non-probability sampling method, specifically purposive sampling, was used. To be eligible for this study, respondents had to be Malaysian and have use mobile food delivery applications. The measurement items for performance expectancy, effort efficacy, social influence, facilities conditions, hedonic motivation, price value, and habit were adopted from Alalwan [8]. Time saving and convenience were adopted from Yeo, Goh, and Rezaei [25]. The questionnaire was constructed using Google Forms and was distributed online on Whatsapp and Facebook. The G-Power software [28] suggested a sample size of 114 respondents (minimum). In total 301 respondents who participated in the research. Descriptive statistics for a demographic profile were analysed using SPSS 26 package and Partial least square analysis was analysed using SmartPLS version 3.3.9 [29].

4. Result

4.1. Respondents' Profile

In total, there were 301 respondents, 165 are female, and 136 are male. A total of 53.5% are between 18 and 23 years of age, 35.5% between 24 and 39 years of age, 7.3% between 40 and 55 years of age, and 3.7% between 56 years of age and above. Malay (42.5%) comprise the majority of the respondents followed by the Indian (34.9%), Chinese (19.6%), and other ethnicities (3%).

4.2. Common Method Bias

The data for this research came from a single source (an experienced user), it is crucial to determine if common method bias has occurred [30]. To overcome this challenge, MacKenzie and Podsakoff [31] and Podsakoff, MacKenzie, Lee, and Podsakoff [32] suggested procedural control and statistical control strategies. Both procedural and statistical control were used since using a single control technique would not eliminate the possible influence of common method bias. For procedural control, a cover letter describing the foundation of the study was included in the surveys. A marker variable titled cognitive rigidity [33] was included as part of the questionnaire as statistical control. Cognitive rigidity is the cognitive process underpinning an individual's inclination to resist or avoid making changes; nonetheless, this measure has no theoretical relationship with the other factors in this research. The approach of a partially out marker variable [33] was applied. The R2 changes are negligible (0.041). This suggests that methodological common method bias was not present in this research.

4.3. Measurement Model

Construct validity of the measuring model was evaluated using convergent and discriminant validity. Refer to the table below (Table 2), except for CIN4, HBT1, PVE4, FCO4, and PEY3, all indicator loadings were more than 0.50 [34]. For each latent variable, the average variance extracted (AVE) values were more than 0.50 [35]; and the composite reliability scores were greater than 0.70 [34]. These data demonstrate the achievement of convergent validity.

Table 2. Measurement model.

Construct	Items	Loadings	CR	AVE
Continuance Intention	CIN1	0.786	0.754	0.507
	CIN2	0.622		
	CIN3	0.719		
Convenience	CON1	0.779	0.819	0.602
	CON2	0.737		
	CON3	0.809		
Effort Expectancy	EEY1	0.676	0.811	0.519
	EEY2	0.697		
	EEY3	0.722		
	EEY4	0.782		
Facilitating Conditions	FCO1	0.762	0.766	0.522
	FCO2	0.698		
	FCO3	0.705		
Habit	HBT2	0.754	0.768	0.525
	HBT3	0.685		
	HBT4	0.734		
Hedonic Motivation	HMN1	0.749	0.780	0.542

Table 2. *Cont.*

Construct	Items	Loadings	CR	AVE
	HMN2	0.746		
	HMN3	0.714		
Performance Expectancy	PEY1	0.759	0.796	0.567
	PEY2	0.800		
	PEY4	0.696		
Price Value	PVE1	0.714	0.784	0.548
	PVE2	0.722		
	PVE3	0.783		
Social Influence	SIF1	0.714	0.758	0.511
	SIF2	0.698		
	SI3	0.732		
Time Saving	TS1	0.773	0.851	0.588
	TS2	0.793		
	TS3	0.717		
	TS4	0.781		

Note: CON: convenience; EEY: effort expectancy; FCO: facilitating conditions; HBT: habit; HMN: hedonic motivation; PEY: performance expectancy; PVE: price value; SIF: social influence; TSG: time saving; CIN: continuance intention.

To ascertain the discriminant validity of the measurement model, the Fornell–Larcker Criterion was used. Table 3 demonstrates that each indication has a higher load on its constructs but a lower load on others. This implies that discriminant validity has been established.

Table 3. Discriminant validity.

	CIN	CON	EEY	FCO	HBT	HMN	PEY	PVE	SIF	TSG
CIN	0.712									
CON	0.528	0.776								
EEY	0.474	0.460	0.720							
FCO	0.471	0.464	0.562	0.722						
HBT	0.522	0.394	0.417	0.493	0.725					
HMN	0.492	0.437	0.482	0.480	0.518	0.736				
PEY	0.447	0.392	0.671	0.503	0.472	0.412	0.753			
PVE	0.513	0.440	0.487	0.482	0.558	0.463	0.400	0.740		
SIF	0.476	0.400	0.436	0.450	0.502	0.427	0.445	0.421	0.715	
TSG	0.554	0.732	0.542	0.506	0.441	0.450	0.491	0.414	0.416	0.767

Note: CON: convenience; EEY: effort expectancy; FCO: facilitating conditions; HBT: habit; HMN: hedonic motivation; PEY: performance expectancy; PVE: price value; SIF: social influence; TSG: time saving; CIN: continuance intention. The black background show that CIN and CON do not have value and so on for other variables.

4.4. Structural Model

Following the evaluation of the measurement model, the investigation moved on to the evaluation of the structural model. A bootstrapping approach with 5000 samples was utilised to generate the path coefficients and accompanying t-values. Table 4 reveals that all variables' variance inflation factor (VIF) ratings were less than 5 [35]. To be considered sufficient, the R2 must be greater than the value of 0.02. The model explained 48.1% of the variation in continuance intention (R^2 = 0.481) in this study. The model used in this study has strong explanatory power since the R is rather significant. Social influence (β = 0.119; t = 1.995); hedonic motivation (β = 0.111; t = 1.983); price value (β = 0.153; t = 2.619); habit (β = 0.140; t = 2.27); time saving (β = 0.177; t = 2.662) and convenience (β = 0.135; t = 1.968) are the factors that are significant and have positive influence on users' continuance intention towards food delivery apps except for performance expectancy (β = 0.045; t = 0.671); effort expectancy(β = 0.032; t = 0.476), and facilities conditions (β = 0.028; t = 0.426). Except for effort expectancy, facilities conditions, and performance expectancy, none of the variables crossed the '0' line at the lower and upper confidence levels. Furthermore, it is critical to determine if the exogenous constructions of convenience, habit, hedonic motivation, price value, social influence, and time saving have a significant influence on the endogenous constructs. Cohen's [36] criteria for effect sizes (f^2) were used to measure this, with 0.02 being a small impact, 0.15 representing an average effect, and 0.35 representing a large effect. According to this research, habit (f^2 = 0.020), price value (f^2 = 0.026) and time saving (f^2 = 0.023) have little effect. The Q^2 value for continuous intention (0.222) was more than 0, indicating that the model has sufficient predictive relevance.

Table 4. Structural model.

	Std. Beta	Std. Error	T-Value	*p* Values	Decision	LL	UL	f^2	VIF	R^2	Q^2
H1: PEY->CIN	0.045	0.067	0.671	0.251	Not supported	−0.071	0.15	0.002	2.068	0.481	0.222
H2: EEY->CIN	0.032	0.067	0.476	0.317	Not supported	−0.078	0.144	0.001	2.354		
H3: SIF->CIN	0.119	0.06	1.995	0.023	Supported	0.031	0.222	0.017	1.568		
H4: FCO->CIN	0.028	0.067	0.426	0.335	Not supported	−0.075	0.138	0.001	1.849		
H5: HMN->CIN	0.111	0.056	1.983	0.024	Supported	0.025	0.21	0.014	1.676		
H6: PVE->CIN	0.153	0.058	2.619	0.005	Supported	0.054	0.247	0.026	1.759		
H7: HBT->CIN	0.140	0.062	2.27	0.012	Supported	0.027	0.232	0.020	1.932		
H8: TSG->CIN	0.177	0.066	2.662	0.004	Supported	0.066	0.281	0.023	2.577		
H9: CON->CIN	0.135	0.068	1.968	0.025	Supported	0.023	0.248	0.015	2.325		

Note: CON: convenience; EEY: effort expectancy; FCO: facilitating conditions; HBT: habit; HMN: hedonic motivation; PEY: performance expectancy; PVE: price value; SIF: social influence; TSG: time saving; CIN: continuance intention.

5. Discussion

Data analysis results demonstrated that convenience, habit, hedonic motivation, price value, social influence, and time saving are the factors that are a positive influence on users' continuance intention towards food delivery applications. As a result, users' continued intention of using food delivery apps are based on their convenience, habit, hedonic motivation, price value, social influence, and time saving. However, users' continuous usage intention is not influenced by users' effort expectancy, facilities conditions, and performance expectancy. The results for facilities conditions and performance expectancy are consistent with Lee, Sung, and Jeon [16]. Thus, the result for effort expectancy is inconsistent with Lee, Sung, and Jeon [16]. The research respondents are mainly 18–39 years of age, which are the generations familiar with the technology and mobile applications. As a result, they do not need much effort expectancy and facilities conditions when they use the food delivery applications. As for performance expectancy, respondents do not perceive that mobile food delivery apps are helpful post-pandemic because users do not rely only on food delivery apps for food as dine-in is allowed.

6. Conclusions

This study contributes both to theoretical and practical research. From a theoretical perspective, this study added two additional determinants: time saving and convenience, into the UTAUT2 model. Furthermore, by adding these two determinants, it is also confirmed that both time saving and convenience contribute to continued usage intention. In addition, this study also contributes to mobile food delivery and the framework given in this model can be used as a foundation for future research on mobile food delivery apps.

From a practical perspective, organisations that offer mobile food delivery apps services need to understand the importance of convenience, habit, hedonic motivation, price value, social influence, and time saving to users, especially if they want to encourage users to continue to use mobile food delivery app services. The focus of this project is constrained in certain aspects: data were solely obtained among Malaysians; therefore, the findings' generalizability is constrained. As a result, this study may not be appropriate for other countries. Secondly, most of the respondents in this study are between 18 and 39 years of age, favouring the younger generation, particularly Millennials. Therefore, future research needs to focus or consider older generations and identify their determinant of continuance usage of mobile food delivery applications.

Author Contributions: Conceptualization, methodology, formal analysis, supervision, writing—review and editing, E.H.T.Y.; Writing—original draft, investigation, S.K. All authors have read and agreed to the published version of the manuscript.

Funding: This research received no external funding.

Institutional Review Board Statement: Not applicable.

Informed Consent Statement: Not applicable.

Data Availability Statement: Not applicable.

Acknowledgments: Paper presented during a panel—organized by Faculty of Information Management, Universiti Technologi Mara, Kelantan Branch, Malaysia—in the context of the International Academic Symposium of Social Science 2022 conference, organized and hosted by Universiti Technologi Mara, Kelantan Branch, during the IASSC 2022.

Conflicts of Interest: The authors declare no conflict of interest.

References

1. Adroit Market Research. Available online: https://www.adroitmarketresearch.com/industry-reports/malaysia-online-food-delivery-market (accessed on 2 February 2022).
2. Shah, A.; Yan, X.; Quayyum, A. Adoption of mobile food ordering apps for O2O food delivery services during the COVID-19 outbreak. *Br. Food J.* **2021**, *28*. ahead-of-print. [CrossRef]
3. Okumus, B.; Ali, F.; Bilgihan, A.; Ozturk, A. Psychological factors influencing customers acceptance of smartphone diet apps when ordering food at restaurants. *Int. J. Hosp. Manag.* **2018**, *72*, 67–77. [CrossRef]
4. Ariel, M. App Stores Growth Acceleratesin 2014. Available online: https://www.developernation.net/blog/app-stores-growth-accelerates-2014 (accessed on 2 February 2022).
5. Venkatesh, V.; Thong, J.Y.; Xu, X. Consumer acceptance and use of information technology: Extending the unified theory of acceptance and use of technology. *MIS Q.* **2012**, *36*, 157–178. [CrossRef]
6. Zhao, Y.; Bacao, F. What factors determining customer continuingly using food delivery apps during 2019 novel coronavirus pandemic period? *Int. J. Hosp. Manag.* **2020**, *92*, 102683. [CrossRef] [PubMed]
7. Gunden, N.; Morosan, C.; DeFranco, A. Consumers' intentions to use online food delivery systems in the USA. *Int. J. Contemp. Hosp. Manag.* **2020**, *32*, 1325–1345. [CrossRef]
8. Alalwan, A. Mobile food ordering apps: An empirical study of the factors affecting customer e-satisfaction and continued intention to reuse. *Int. J. Inf. Manag.* **2020**, *50*, 28–44. [CrossRef]
9. Tam, C.; Santos, D.; Oliveira, T. Exploring the influential factors of continuance intention to use mobile apps: Extending the expectation confirmation model. *Inf. Syst. Front.* **2020**, *22*, 243–257. [CrossRef]
10. Ray, A.; Dhir, A.; Bala, P.; Kaur, P. Why do people use food delivery apps (FDA)? A use of gratification theory perspective. *J. Retail. Consum. Serv.* **2019**, *51*, 221–230. [CrossRef]
11. Fakfare, P. Influence of service Attributes of food delivery application on customers satisfaction and their behavioural responses: The IPMA. *Int. J. Gastron. Food Sci.* **2021**, *25*, 100392. [CrossRef]
12. Surya, A.; Sukresna, I.; Mardiyona, A. Fctors affecting intention to use food order-delivery feature of ride-hailing applications: The UTAUT approach. *Int. J. Bus. Soc.* **2021**, *22*, 1363–1383. [CrossRef]
13. Agarwal, V.; Sahu, R. Predicting repeat usage intention towards O2O food delivery: Extending UTAUT2 with user gratifications and bandwagoning. *J. Foodserv. Bus. Res.* **2022**, *25*, 434–474. [CrossRef]
14. Venkatesh, V.; Morris, G.B.; Davis, F.D. User acceptance of information technology: Toward a unified view. *MIS Q.* **2013**, *27*, 425–478. [CrossRef]
15. Marinković, V.; Đorđević, A.; Kalinić, Z. The moderating effects of gender on customer satisfaction and continuance intention in mobile commerce: A UTAUT-based perspective. *Technol. Anal. Strateg. Manag.* **2019**, *32*, 306–318. [CrossRef]

16. Lee, S.W.; Sung, H.J.; Jeon, H.M. Determinants of continuous intention on food delivery apps: Extending UTAUT2 with information Quality. *Sustainability* **2019**, *11*, 3141. [CrossRef]
17. Chao, C.M. Factors Determining the Behavioral Intention to Use Mobile Learning: An Application and Extension of the UTAUT Model. *Front. Psychol.* **2019**, *10*, 1652. [CrossRef]
18. Al-Saedi, K.; Al-Emran, M.; Ramayah, T.; Abusham, E. Developing a general extended UTAUT model for M-payment adoption. *Technol. Soc.* **2020**, *6*, 101293. [CrossRef]
19. Akdim, K.; Casalo, L.V.; Flavian, C. The role of utilitarian and hedonic aspects in the continuance intention to use social mobile apps. *J. Retail. Consum. Serv.* **2022**, *66*, 102888. [CrossRef]
20. Dodds, W.B.; Monroe, K.B.; Grewal, D. Effects of price, brand, and store information on buyers. *J. Mark. Res.* **1991**, *28*, 307–319.
21. Limayem, M.; Hirt, S.G.; Cheung, M.K. How Habit Limits the Predictive Power of Intention: The Case of Information Systems Continuance. *MIS Q.* **2007**, *31*, 705–737. [CrossRef]
22. Nascimento, B.; Oliveira, T.; Tam, C. Wearable technology: What explains continuance intention in smartwatches? *J. Retail. Consum. Serv.* **2018**, *43*, 157–169. [CrossRef]
23. Hsu, M.-H.; Chang, C.M.; Chuang, L.W. Understanding the determinants of online repeat purchase intention and moderating role of habit: The case of online group-buying in Taiwan. *Int. J. Inf. Manag.* **2015**, *35*, 45–56. [CrossRef]
24. Zanetta, L.D.A.; Hakim, M.P.; Gastaldi, G.B.; Seabra, L.M.A.J.; Rolim, P.M.; Nascimento, L.G.P.; Medeiros, C.O.; Cunha, D.T.D. The use of food delivery apps during the COVID-19 pandemic in Brazil: The role of solidarity, perceived risk, and regional aspects. *Food Res. Int.* **2021**, *149*, 110671. [CrossRef] [PubMed]
25. Yeo, V.; Goh, S.K.; Rezaei, S. Consumer experiences, attitude and behavioral intention toward online food delivery (OFD) services. *J. Retail. Consum. Serv.* **2017**, *35*, 150–162. [CrossRef]
26. Xu, F.; Huang, S.; Li, S. Time, money, or convenience: What determines Chinese consumers' continuance usage intention and behavior of using tourism mobile apps? *Int. J. Cult. Tour. Hosp. Res.* **2019**, *13*, 299–322. [CrossRef]
27. Ozturk, A.B.; Nusair, K.; Okumus, F.; Hua, N. The role of utilitarian and hedonic values on users' continued usage intention in a mobile hotel booking environment. *Int. J. Hosp. Manag.* **2016**, *57*, 106–115. [CrossRef]
28. Faul, F.; Erdfelder, E.; Buchner, A.; Lang, A.G. Statistical power analyses using G*Power 3.1: Tests for correlation and regression analyses. *Behav. Res. Methods* **2009**, *41*, 1149–1160. [CrossRef]
29. Ringle, C.M.; Smith, W.R.; Becker, J. *SmartPLS 3*; SmartPLS GmbH: Boennigstedt, Germany, January 2015.
30. Tehseen, S.; Ramayah, T.; Sajilan, S. Testing and Controlling for Common Method Variance: A Review of Available Methods. *J. Manag. Sci.* **2017**, *4*, 142–168. [CrossRef]
31. MacKenzie, S.B.; Podsakoff, P.M. Common Method Bias: Nature, Causes, and Procedural Remedies. *J. Retail.* **2012**, *88*, 556–562. [CrossRef]
32. Podsakoff, P.M.; MacKenzie, S.B.; Lee, J.-Y.; Podsakoff, N.P. Common Method Biases in Behavioral Research: A Critical Review of the Literature and Recommended Remedies. *J. Appl. Psychol.* **2003**, *88*, 879–903. [CrossRef]
33. Oreg, S. Resistance to change: Developing an individual differences measure. *J. Appl. Psychol.* **2003**, *88*, 680–693. [CrossRef]
34. Byrne, B.M. *Structural Equation Modeling with AMOS: Basic Concepts, Appsand Programming*, 3rd ed.; Routledge: New York, NY, USA, 2016.
35. Hair, J.; Hult, G.T.; Ringle, C.; Sarstedt, M. *A Primer On Partial Least Squares Structural Equation Modeling (PLS-SEM)*, 2nd ed.; SAGE Publications: Thousand Oaks, CA, USA, 2017.
36. Cohen, J. *Statistical Power Analysis for the Behavioural Science*, 2nd ed.; Lawrance Elrbaum: Mahwah, NJ, USA, 1988.

proceedings

MDPI

Proceeding Paper

Visual Design as the Key Factor for Effective Environmental Analytics Dashboard †

Zada Zainuddin [1], Farashazillah Yahya [1,*] and Ahmad Fareez Yahya [2]

1 Faculty of Computing and Informatics, Universiti Malaysia Sabah, Kota Kinabalu 88400, Malaysia
2 Faculty of Hotel and Tourism Management, Universiti Teknologi MARA Sabah Branch, Kota Kinabalu Campus, Kota Kinabalu 88997, Malaysia
* Correspondence: fara.yahya@ums.edu.my
† Presented at the International Academic Symposium of Social Science 2022, Kota Bharu, Malaysia, 3 July 2022.

Abstract: This paper proposes visual design as the key factor for an effective environmental analytics dashboard. Organisations found it challenging to organise data before emerging visualisation techniques such as using dashboards and analytics. Environmental analytics dashboards support analysis of performance and progress of a task/situation for some time, i.e., weekly, monthly or yearly. However, there is a dilemma in choosing a suitable visual design for the data, especially for the environmental domain. The objective of the papers is to conduct a mini-review on the visual design factor and subfactors in the context of environmental dashboards. These factors can be the initial focal point when producing an effective dashboard framework. The methodology chosen for this review is a systematic literature review (SLR). Based on the SLR, 23 papers were reviewed, and visual design was deemed the key factor. Two subfactors, highlighted regions and values, and different visualisation techniques mainly support the key factor. Additionally, the current research shows increased research for effective dashboards from 2006 until 2022 that focuses on visual designs. The listed subfactors can be used as guidance later to build an effective environmental analytics dashboard.

Keywords: visual design; effective dashboard; factor; environmental

Citation: Zainuddin, Z.; Yahya, F.; Yahya, A.F. Visual Design as the Key Factor for Effective Environmental Analytics Dashboard. *Proceedings* **2022**, *82*, 21. https://doi.org/10.3390/proceedings2022082021

Academic Editor: Mohamad Rahimi Mohamad Rosman

Published: 9 September 2022

1. Introduction

Data analytics is growing in every field as it becomes critical to analyse all data. The dashboard is one of the tools that can help process the data. Dashboards have concise, precise, intuitive display mechanisms to fit on a single screen and the information in the dashboards is customisable, following the users' requirements. The additional characteristics that the users can consider are metrics or key performance indicators (KPIs) of an individual or organisation, real-time data display, and the availability to view on the web browser or other platforms. Dashboard content may be in the form of tables, graphics, or visual key performance indicators (KPIs) [1]. Some of the types of dashboards include strategic, operational, and analytical. The strategic dashboard monitors the key performance indicators. The operational dashboard displays the days-to-days immediate performance, and the analytical dashboard analyses a large amount of data to find trends and insights. Analytical information that employs visual information can draw user attention to critical situations, trends, and exceptions [2]. Analytical dashboard in the environmental sector mainly uses real-time data, acts as a decision-making tool, and monitors progress [3]. Then, an effective dashboard is helpful to the user where it combines the appropriate data with the proper visualisation technique [4] and interactions depending on their situational utility [1]. In the meantime, analytics analyses data and statistics in a computational mode. Nadj et al. [5] state that an analytical dashboard allows users to use an interactive dashboard with analytical features. The visual design emphasises the appeal of text, colour, and images to get users to engage with the platform. However, sometimes, it is used interchangeably with visualisation [4] or data visualisation [6–8]. Some researchers have used visualisation and then used data visualisation or visual design. However, there must be an audience and goals to figure out what makes data visualisations effective [1], as mentioned by Few [8]; a dashboard is a visual representation of the most relevant information needed to achieve one or more objectives. The data is integrated and organised on a single screen to monitor the information at a glance. The core of the dashboards is to be effective first [9,10], it should consist primarily of high-level summaries as it can quickly give an overview of activities to the users. Hence, this paper aims to discuss the key factor of visual design for an effective environmental analytics dashboard.

2. Methodology

Concurrently, design is abstract and does not seem rigid to a specific regulation, however, it can impact how people think and make a decision. That is why the role of choosing the right visual design in building a dashboard is important. It can manipulate the users based on the use of colours, size, type of chart, and others [1]. It is crucial to better understand these factors so that it is simple and straightforward to use for non-technical users [11,12]. Next, a dashboard's main value is that it reduces the number of metrics to a single visual display [13], which means synchronising between the space and the visual design. There were 23 papers shortlisted using a systematic literature review (SLR) and four stages in the Preferred Reporting Items for Systematic Reviews and Meta-Analysis (PRISMA): identification, screening, eligibility, and inclusion, to extract the subfactors. The entire stages taken with the PRISMA approach are depicted in Figure 1. When looking for the papers, the following keywords were used: dashboard design, effective dashboard, visual design factors, and analytics dashboard.

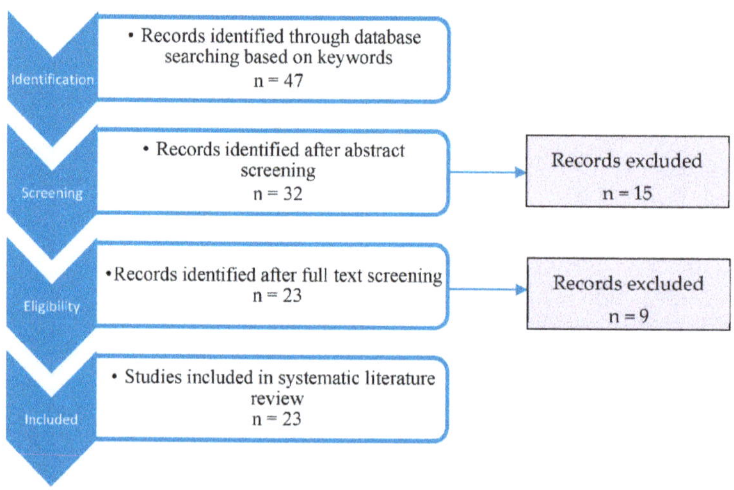

Figure 1. Systematic literature review flow diagram.

3. Visual Design as the Key Factor

The effectiveness of dashboards can be measured based on user customisation, knowledge discovery, security, information delivery, alerting, visual design, integration, and system connectivity [14]. Designing effective dashboards includes choosing an accurate data visualisation to display clear and concise information on a task. The accomplishment of the task can be used to support decision-making or monitoring. The visualisation should be easy to interpret without an explanation, so only important text (like graph titles, category labels, or data values) should be on the dashboard. In terms of visual features and interactivity, the dashboard also allows users to change the presentation of data in terms of construction, composition, and multipage by using a tab layout and interactive interface that allows the selection of the relevant items for views or analysis.

Seven factors use the concepts of effective dashboards introduced by Karami et al. [14]; the existing literature is synthesised into factors, subfactors and/or indicators. Nevertheless, the most mentioned factor by the researchers is visual design, or visualisation, as a vital part of a dashboard. There are 11 subfactors from [14] except for the last; the different visualisation techniques are from personal observations, as shown in Table 1. Visual design or visualisation intends to convey messages using aesthetic visual display techniques. The usability of a dashboard can be improved by highlighting areas and values [1,7,14–32] and displaying the table or graphs on the same page without scrolling [1,14,16,18–21,28]. Among other indicators are switching back and forth between tabular and chart views [1,14–16,18–20,24,30] and functionalities such as resizing, zooming in and out, and re-ordering zones [1,14,19]. Customisation flexibility, such as allowing different dashboard layouts, is also deemed significant based on existing research [1,14,15,18–21,24,28,30]. Too much data overloads a dashboard; prioritising important data through metrics can help display organisational performance measurements. The subfactors included for metrics includes linking metrics together [1,7,14,15,18,21,22,30], linking objectives with metrics [1,8,14] and displaying the metrics calculations [14]. Then, metadata and guidance, displaying instructions and user guide on the dashboards [14]. The last subfactors use different visualisation techniques such as bar charts, graphs, heatmaps and others [1,7,14–32].

Table 1. Factor and subfactors/indicators for a dashboard.

Factors	Subfactors/Indicators	Sources/References
Visual design	Regions and values that highlighted	[1,4,7,8,14–32]
	Table and graphs on the same page	[1,14,16,18–21,28]
	Changing the view from tabular to chart	[1,14–16,18–20,24,30]
	Resizing, maximising/minimising, re-ordering of zones	[1,14,19]
	Arrange in various layouts	[1,14,15,18–21,24,28,30]
	Definition and calculation of metrics included	[14]
	Metrics and aim linked	[1,8,14]
	Metrics linked	[1,7,14,15,18,21,22,30]
	Metadata and guidance	[4,14,21,28]
	No scrolling on a single screen	[1,8,14,16,18,28]
	Different visualisation techniques	[1,4,7,8,14–32]

Figure 2 shows the research pattern related to the visual design by year. Four phases are divided into five years, except the last phase, which is only for two years. From this graph, it can be seen that the topic has risen over the years. Only one researcher, [7], started the process in the first phase. Then, it increased to four, whereas there was one paper for each year except in 2015. In 2011; [1], 2012; [16], 2013; [4], and in 2014; [17]. After that, it increased to eight since in 2017; [14,18], 2018; [7,19], 2019; [15,20,28,29]. In the last phase, although only two years, there were ten papers, whereas 2021 had five; [21–23,30,31] and five in 2022; [24–27,32]. A total of 23 papers have discussed this key factor directly or indirectly.

Eleven subfactors influence visual design key factors for effective dashboards: regions and values that are highlighted, table and graphs on the same page, changing the view from tabular to chart, resizing, maximising/minimising, re-ordering of zones, arranged in various layouts, definition, and calculation of metrics including, metrics and aim linked, metrics linked, metadata and guidance, no scrolling on a single screen and using different visualisation technique as shown in Figure 3. Regions and highlighted values are the most highlighted subfactor by the researchers, followed by different visualisation techniques, while the definition and calculation of metrics included are the least. Regions and values are the most highlighted as it is the basis in the dashboard for showing the data, with the capacity to highlight the key regions and numbers on a dashboard in response to a user's cursor movement [14]. Arranged in various layouts is the second most mentioned by the researchers. The dashboard should be properly visual without overwhelming the users, allowing them to adjust the information layout.

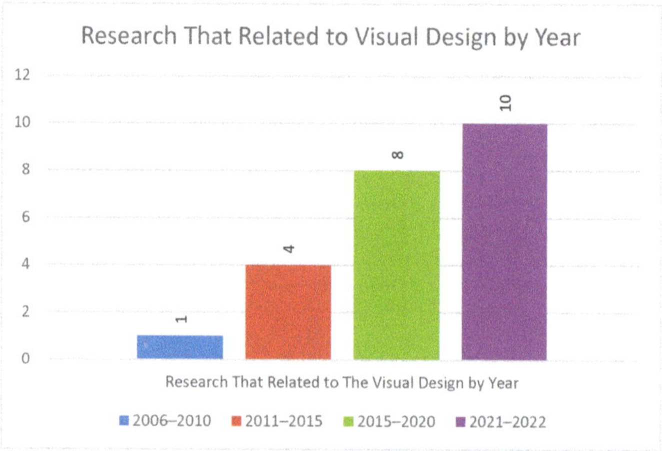

Figure 2. Research that related to visual design by year.

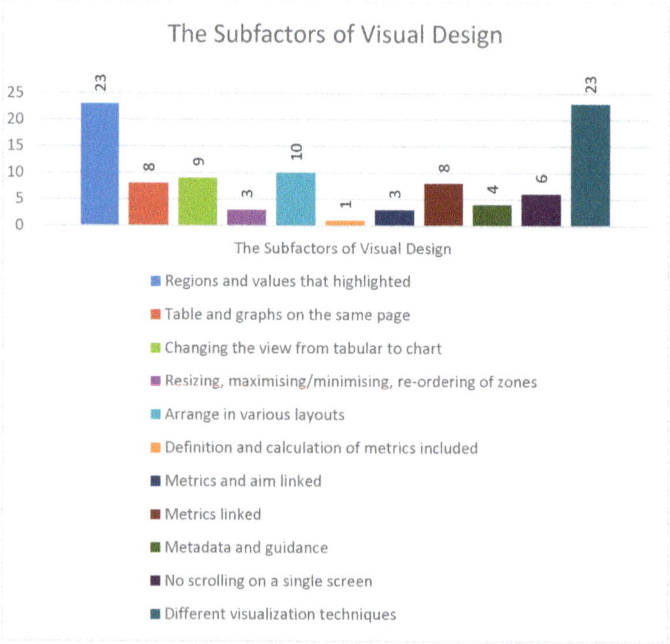

Figure 3. The Subfactors of Visual Design.

Figure 4 shows the division of subfactors by year. In 2021–2022, the research topic mainly focused on highlighted regions and values and different visualisation techniques. Next, in 2016–2020, the research harmoniously highlighted regions and values, different visualisation techniques, various layouts arranged, and tables and graphs on the same screen; it also changed the view from tabular to chart. There are only a few studies in both early phases, of 2006–2010 and 2011–2015.

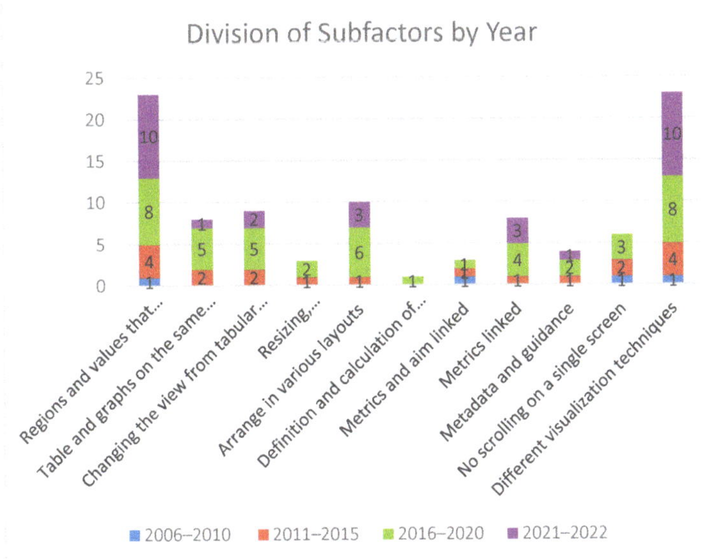

Figure 4. Division of Subfactors by Year.

4. Discussion

According to [28], from the result of a fair model kit, it is shown that the attraction of the visual design of the dashboard can change the user's understanding and behaviour. However, the case studies have not disproven that it can change the behaviours; instead, only the usability of the dashboard conveys the information. The dashboard becomes effective when it is frequently checked [24]. It is in synch with the top subfactors, highlighted regions and values and different visualisation techniques applied to build the effective dashboard. Regions and highlighted values are where the information is emphasised by the colour, font, or style intended to catch the reader's attention within the platform. It requires a good balance between information utility and visual complexity [22] to convey a story, eliminate noise from data, and emphasise essential details [31]. Then, different visualisation techniques mean a chart, graph, map, or mathematical visual. A suitable visualisation must be configured to deliver the right message due to multiple data types. For example, a bar chart represents categories on a two-way axis, and it would be convenient to show data on water usage by month. At the same time, the heatmap used a hue saturation for the area and was relevant to show water quality data in certain areas. Thus, minimalist and aesthetic design [19,26] can also be related to these subfactors. The second top subfactor is arranged in various as analytics dashboards need flexibility. As a result, users do not get bored quickly seeing it every day as the display is varied, and the information is clear and concise. However, past researchers discuss these factors in transportation; [30,32] and health [14,33]. There is less research on the environmental but limited such as [29]. Visual design has been discussed generally by past researchers in [1,4,15,27]. The researchers have indicated how to design different types of the dashboard with multiple types of visualisation, choose the right visualisation, fit all the data into the dashboard, and last, no matter the design pattern, each has its strength and weakness. Definition and metrics calculation are the least mentioned, since anyone who uses the dashboard already knows about them. Hence, it is not relevant to be an indicator for measuring the effectiveness of the dashboard. The trend from 2006 until now supports that visual design is essential when building a dashboard, as the total number of papers published has increased from year on year. Returning to the definition by Few [8]—that dashboards merge all the data and can be readable at a glance— he supports that key factor of visual design as effective to the dashboard.

5. Conclusions and Future Work

Nowadays, data analytics are crucial since much data needs control. Environmental operations generate data that needs to be visualised well in the dashboard for people to understand what is happening. For the time being, the analytics dashboard offers a package with guidelines to let the users build an effective environmental dashboard. This research emphasises eleven visual design subfactors for an effective environmental analytics dashboard, such as the need for visualisation to be arranged in various layouts, allowing the viewer to change the display and so on. However, researchers most mention highlighted regions and values, and different visualisation techniques. Both of the subfactors supported the visual design to be the key factor. Visual design is vital when managing the data into the layout, so no data are left alone without reason. In summary, the rise in visual design research shows a tendency for more technological research. Finally, since there are many subfactors, future work can explore and divide them into groups, followed by urgency level, and facilitate the development of an effective dashboard framework in the environmental domain.

Author Contributions: Conceptualisation, Z.Z. and F.Y.; methodology, Z.Z.; validation, Z.Z. and F.Y.; formal analysis, Z.Z. and F.Y.; investigation, Z.Z. and F.Y.; resources, Z.Z. and F.Y.; writing—original draft preparation, Z.Z. and F.Y.; writing—review and editing, F.Y. and A.F.Y.; visualisation, Z.Z. and F.Y.; supervision, F.Y. All authors have read and agreed to the published version of the manuscript.

Funding: This research was funded by the Ministry of Higher Education Malaysia, Grant No: RACER/1/2019/ICT04/UMS//1. The article processing charge was funded by Universiti Malaysia Sabah.

Institutional Review Board Statement: Not applicable.

Informed Consent Statement: Not applicable.

Data Availability Statement: Not applicable.

Conflicts of Interest: The authors declare no conflict of interest.

References

1. Pappas, L.; Whitman, L. Riding the Technology Wave: Effective Dashboard Data Visualization. In *Human Interface and the Management of Information. Interacting with Information*; Springer: Berlin/Heidelberg, Germany, 2011; pp. 249–258. [CrossRef]
2. Negash, S.; Gray, P. Business Intelligence. Handbook on Decision Support Systems. *Int. Handbooks Inf. Syst.* **2008**, *2*, 19–29.
3. Yahya, F.; Zaki, A.F.A.; Moung, E.G.; Sallehudin, H.; Bakar, N.A.A.; Utomo, R.G. An IoT-Based Coastal Recreational Suitability System Using Effective Messaging Protocol. *Int. J. Adv. Comput. Sci. Appl.* **2021**, *12*, 558–567. [CrossRef]
4. Janes, A.; Sillitti, A.; Succi, G. Effective Dashboard Design. *Cut. IT J.* **2013**, *26*, 17–24.
5. Nadj, M.; Maedche, A.; Schieder, C. The Effect of Interactive Analytical Dashboard Features on Situation Awareness and Task Performance. *Decis. Support Syst.* **2020**, *135*, 113322. [CrossRef]
6. Raheem, N. Big Data Visualization. In *Big DataA Tutorial-Based Approach*; Chapman and Hall/CRC: Boca Raton, FL, USA, 2019; pp. 75–102. [CrossRef]
7. Bikakis, N. Big Data Visualization Tools. *arXiv* **2018**, arXiv:1801.08336.
8. Few, S. *Information Dashboard Design*; Eff. Vis. Commun. Data; O'Reilly Media, Inc.: Sebastopol, CA, USA, 2006; p. 37.
9. Few, S. Dashboard Confusion. Available online: https://www.perceptualedge.com/articles/ie/dashboard_confusion.pdf (accessed on 21 June 2021).
10. Baharum, A.; Wan, L.Y.; Yahya, F.; Nazlan, N.H.; Nor, N.A.M.; Ismail, I.; Noor, N.A.M. Mobile Learning Application: Flipped Classroom. *Indones. J. Electr. Eng. Comput. Sci.* **2020**, *17*, 1084–1090. [CrossRef]
11. Natalia, F.; Eko, Y.; Ferdinand, F.V.; Murwantara, I.M.; Ko, C.S. Interactive Dashboard of Flood Patterns Using Clustering Algorithms. *ICIC Express Lett. Part B Appl.* **2019**, *10*, 413–418. [CrossRef]

12. Lee, D.; Felix, J.R.A.; He, S.; Offenhuber, D.; Ratti, C. CityEye: Real-Time Visual Dashboard for Managing Urban Services and Citizen Feedback Loops. In Proceedings of the 14th International Conference on Computing in Urban Planning and Urban Management (CUPUM), Cambridge, MA, USA, 7–10 July 2015.
13. Malik, S.; Sharma, P. Dashboard Necessary in Every Field. *Int. J. Comput. Sci. Inf. Technol.* **2014**, *5*, 3511–3513.
14. Karami, M.; Langarizadeh, M.; Fatehi, M. Evaluation of Effective Dashboards: Key Concepts and Criteria. *Open Med. Inform. J.* **2017**, *11*, 52–57. [CrossRef]
15. Sarikaya, A.; Correll, M.; Bartram, L.; Tory, M.; Fisher, D. What Do We Talk about When We Talk about Dashboards? *IEEE Trans. Vis. Comput. Graph.* **2019**, *25*, 682–692. [CrossRef]
16. Lempinen, H. Constructing a Design Framework for Performance Dashboards. In *Nordic Contributions in IS Research*; Springer: Berlin/Heidelberg, Germany, 2012; pp. 109–130.
17. Abd-Elfattah, M.; Alghamdi, T.; Amer, E. Dashboard Technology Based Solution To Decision Making. *Int. J. Comput. Sci. Eng. Inf. Technol. Res.* **2014**, *4*, 59–70.
18. Mayer, B.; Weinreich, R. A Dashboard for Microservice Monitoring and Management. In Proceedings of the 2017 IEEE International Conference on Software Architecture Workshops (ICSAW), Gothenburg, Sweden, 5–7 April 2017; pp. 66–69. [CrossRef]
19. Dowding, D.; Merrill, J.A. The Development of Heuristics for Evaluation of Dashboard Visualizations. *Appl. Clin. Inform.* **2018**, *9*, 511–518. [CrossRef]
20. Kamulegeya, G.; Mugwanya, R.; Hebig, R. Requirements for Measurement Dashboards and Their Benefits: A Study of Start-Ups in an Emerging Ecosystem. In Proceedings of the 2019 45th Euromicro Conference on Software Engineering and Advanced Applications (SEAA), Kallithea, Greece, 28–30 August 2019; pp. 300–308. [CrossRef]
21. Harding, C. A User Analysis of the Environmental Finance Center's Water and Wastewater Dashboard Tool. Master's Thesis, University of North Carolina at Chapel Hill, Chapel Hill, NC, USA, 2021; 48p.
22. Hehman, E.; Xie, S.Y. Doing Better Data Visualization. *Adv. Methods Pract. Psychol. Sci.* **2021**, *4*, 1–18. [CrossRef]
23. Zhwan, M.; Khalid, S.R.M.Z. Big Data Analysis for Data Visualization: A Review. *Int. J. Sci. Bus.* **2021**, *5*, 64–75. [CrossRef]
24. Bloemen, S. Designing a Dashboard Development Methodology for Heterogeneous Stakeholders in Dutch Nursing Homes. Master's Thesis, University of Twente, Enschede, The Netherlands, 2022; pp. 1–40.
25. Antonini, A.S.; Ganuza, M.L.; Castro, S.M. VISUEL—A Web Dynamic Dashboard for Data Visualization. *J. Comput. Sci. Technol.* **2022**, *22*, 42–57. [CrossRef]
26. Khanbhai, M.; Symons, J.; Flott, K.; Harrison-White, S.; Spofforth, J.; Klaber, R.; Manton, D.; Darzi, A.; Mayer, E. Enriching the Value of Patient Experience Feedback: Web-Based Dashboard Development Using Co-Design and Heuristic Evaluation. *JMIR Hum. Factors* **2022**, *9*, 1–14. [CrossRef]
27. Bach, B.; Freeman, E.; Abdul-Rahman, A.; Turkay, C.; Khan, S.; Fan, Y.; Chen, M. Dashboard Design Patterns. *arXiv* **2022**, arXiv:2205.00757.
28. Park, Y.; Jo, I.H. Factors That Affect the Success of Learning Analytics Dashboards. *Educ. Technol. Res. Dev.* **2019**, *67*, 1547–1571. [CrossRef]
29. Jing, C.; Du, M.; Li, S.; Liu, S. Geospatial Dashboards for Monitoring Smart City Performance. *Sustainability* **2019**, *11*, 5648. [CrossRef]
30. Deee, G.; Pack, M.L.; Young, S.; Ph, D.; Franz, M.L.; Pack, M.L.; Young, S. Transportation Energy Analytics Dashboard (TEAD). (No. DEEE0008465). 2021. Available online: https://www.osti.gov/biblio/1784082 (accessed on 21 June 2021).
31. Wagh, S.J.; Bhende, M.S.; Thakare, A.D. Data Visualization Using Tableau. In *Fundamentals of Data Science*; Chapman and Hall/CRC: Boca Raton, FL, USA, 2021; pp. 249–267. [CrossRef]
32. Bachechi, C.; Po, L.; Rollo, F. Big Data Analytics and Visualization in Traffic Monitoring. *Big Data Res.* **2022**, *27*, 100292. [CrossRef]
33. Wimba, P.M.; Bazeboso, J.A.; Katchunga, P.B.; Tshilolo, L.; Longo-Mbenza, B.; Rabilloud, M.; Vanhems, P.; Iwaz, J.; Étard, J.F.; Écochard, R. A Dashboard for Monitoring Preventive Measures in Response to COVID-19 Outbreak in the Democratic Republic of Congo. *Trop. Med. Health* **2020**, *48*, 74. [CrossRef] [PubMed]

proceedings

MDPI

Proceeding Paper

Developing an E-College Monitoring System as a Web-Based Monitoring Tool Application †

Nur Ainatul Mardiah Mat Nawi [1,*], Nur Shaliza Sapiai [1], Siti Aswani Mohd Ghazali [1], Nur Hazelen Mat Rusok [2], Fara Zuliana Zulkifli [1] and Fatin Masturina Mazlan [1]

1 Faculty of Information Management, Universiti Teknologi MARA Kelantan Branch, Machang 18500, Malaysia
2 Faculty of Business and Management, Universiti Teknologi MARA Kelantan Branch, Machang 18500, Malaysia
* Correspondence: ainatulmardiah@uitm.edu.my; Tel.: +60-145142921

† Presented at the International Academic Symposium of Social Science 2022, Kota Bharu, Malaysia, 3 July 2022.

Abstract: The E-College Monitoring System is an online system that can manage students' outings in UiTM and to study and understand the system application for students who live in UiTM. In addition, it will be used to test the e-college monitoring application system functioning. The methodology that we chose is by using the system development life cycle (SDLC). The qualitative findings showed that the current manual system in UiTM Machang is not reliable. This project contributes to the technological value and the impacts of the system that have been developed. Different from the prior system, this study is an exploratory attempt to investigate the manual system that has been used by some universities and uncovered that the mechanisms with technology influence the university's performance.

Keywords: E-College Monitoring; SDLC; mobile applications

check for updates

Citation: Nawi, N.A.M.M.; Sapiai, N.S.; Ghazali, S.A.M.; Rusok, N.H.M.; Zulkifli, F.Z.; Mazlan, F.M. Developing an E-College Monitoring System as a Web-Based Monitoring Tool Application. *Proceedings* **2022**, *82*, 25. https://doi.org/10.3390/proceedings2022082025

Academic Editor: Mohamad Rahimi Mohamad Rosman

Published: 13 September 2022

1. Introduction

Every university will provide accommodation for students, especially those who are far from their homeland. UiTM Machang Kelantan agrees. The E-college Monitoring System, which was designed, makes it simpler for hostel or college residents to keep track of how many students are entering and exiting. This approach will benefit the University of Technology MARA (UiTM) Machang Campus hostel or college and many others.

The E-College Monitoring System is a method for UiTM college students to help with the problem of resident staff approval of outings. With this method, students will not need to ask the resident staff for permission. Compared to a manual method, the system makes it possible to track outing activities more effectively. Students who use the E-college Monitoring System must complete forms for their guardians or parents so that residents or staff can get in touch with them in an emergency. When a student scans their outing activity, the resident staff and the student's parents are notified. The resident staff can also look over the report to see if any students failed to return to the dorm or college.

Residents can more easily keep track of where their students are using the E-College Monitoring System. It can also act as a reminder to students that they have until 11 p.m. to enter the UiTM property. There are thousands of students at UiTM, and it is challenging to keep track of them all, so using a manual system, like filling out a form, is a burden. Using the E-College Monitoring System, parents may keep an eye on their children without worrying about their whereabouts or other circumstances.

Tracking the whereabouts of the students is an important step in the hostel management process, as mentioned by [1]. The authors indicate that a lot of data, including students' in and out times, disciplinary issues, details about their parents or guardians, merits, and so on, must be kept when needed to conduct the outing system for students at every boarding school. In most public and private institutions nowadays, manual registration is employed to track students' movements, and students are required to record every aspect of their activities in a log book. Other than that, in accordance with [2], the creation of the smart outing and attendance system in this project as a suggested solution to the mentioned issues is anticipated to improve the safety and security of boarding school students.

2. Backgrounds of the Study

2.1. Problems Statements

According to certain views, some public and private institutions continue to use manual systems, such as the requirement that students carry their outing cards with them everywhere they go. The drawback of the manual system is that students must carry the card at all times, and if it is lost, they must purchase a new card, which can cost anywhere from MYR 5 to MYR 10 for one card. The student monitoring process is now carried out manually in UiTM Machang utilizing a paper-based system. They rely on manual procedures, such as coming to the office to ask the resident staff for permission to go on an outing, because they lack an E-College Monitoring Application System.

As a consequence, it will be challenging for the College Resident Staff (CRS) to keep track of the students who do not arrive on time and figure out how many students are currently absent. To find out the status of the students, the CRS must manually count the number of outing cards at the guardhouse. It consumes a lot of effort and work to complete this process. In addition, the possibility of data loss is particularly high, since data is stored in multiple files at one location. Due to the high number of students enrolled, it is possible that human mistakes will cause all of the manually kept data to be destroyed and that there is no backup data.

With the development of the technology and the advancement of the Internet, researchers decided to develop a web-based system that can be used to assist their management, especially the CSR in hostel management. According to [3], in today's environment, information systems are critically needed to change the manual paper-based work, which, at the same time, helps to improve the efficiency and optimize the time in delivering the maximum service. In line with this, a study by [4] discussed that computerized information systems benefit people in many ways, especially in helping the organization to reduce the use of space to store data and the reduction of costs, which leads to efficient time management. The design of this web-based computer experiment management system was carried out within three aspects: overall design, database design, and function module design. The system design should be based on the principle of security, and different users should be able to validly acquire a corresponding authority [5].

2.2. Objectives of the Systems Development

This system will work to improve the outing system for UiTM students. This system will take the place of the current manual system. Hence, the objectives of this system include:

OBJ 1: To identify appropriate data from the college administration;
OBJ 2: To organize the data gathered from the college administration;
OBJ 3: To develop a system that can manage the student's outings at UiTM.

2.3. Scopes of the Systems Development

This system will work to improve the outing system for UiTM students. This system will take the place of the current manual system. The system will include a feature that allows parents to receive notifications through email. The resident staff, who is in charge of updating the student data, will provide the initial data gathering. The mobility of students is a major area, as is delivering reports (statistics) after each month. The initial data will be collected from the resident staff members who are in charge of updating the student information. The E-College Application System is a mobile application that caters to two types of users: residents and students. The system consists of three main modules.

The first module is the registration module. This module offers a conceptual framework for entering information about those students in a way that: facilitates data entry and accuracy by matching the entry to the data source, connects registers to data by easily tying back to individual student records, and gathers data elements to enable better student supervision. The second module of the system is the monitoring module. This module is used to monitor the location of the students. Through this module, the CRS can identify the students who are absent from the campus or college. The next module is the reporting module. This module involves the process of generating reports and statistics for the use of management.

3. Methodology

Referring to [6], the E-College Monitoring System is developed following several empirical stages. First, we conduct a review of the existing forms, procedures, and reports. Second, interview sessions are conducted with the stakeholders, which are the College Resident Staff (CRS) and students. Finally, to have a structured process, the System Development Life Cycle (SDLC) is used as the methodology in developing the E-Outing System. The SDLC comprises five (5) stages, which include planning, analysis, design, implementation, and maintenance [7].

Planning is the first stage of the SDLC. In an attempt to optimize a new system that is user-friendly and helps to streamline the monitoring process, we evaluated manual operations throughout the planning phase. The management's policies and practices, as well as reports, are then analyzed during the analysis phase to fully understand the outing monitoring activity process. Based on a review of the present systems, we could determine the benefits and drawbacks of the current practices and could see the issue much more clearly [8]. Additionally, we included any hardware and software that is best-suited for use in developing the E-College Monitoring System.

In the design phase, once the requirements are gathered and understood, a detailed design is constructed, and the hardware and software are identified. The design phase starts with the data flow of the system, which will describe more about the process of our system. Our system design is continued by designing and creating an Entity Relational Relationship Diagram (ERD) and Data Dictionary (Metadata) that will explain the structure of the database system that will be developed. Apart from that, in the design phase, the user interface is designed, which will act as a guideline in the development process and, at the same time, will help to endure that the system is user-friendly and easy to use [9].

The next phase is the implementation stage. This implementation stage is about how the system will be used by the user and implement the function. This phase is essential to ensure that this system will be implemented correctly and efficiently on the device involved. Coding or programming is the main activity in this implementation process, and the most important thing is that it fulfills the users' expectations and avoids critical errors in the program to ensure that the system runs effectively. Additionally, during this stage, the process of hardware and application installation occurred. Furthermore, the implementation phase is important to identify the system function, as well as to verify whether it works as expected and as documented in the requirements analysis phase. In this phase, the system is tested to ensure no bugs or errors.

After the product is fully operational, the SDLC's maintenance phase begins. As stated by [7], software maintenance might involve software upgrades, repairs, and software patches if the software breaks. In addition, as discussed by [10], once the product or application is launched or officially in full operation, the maintenance phase of the SDLC occurs. The author of [11] stated that the process of maintaining the software or application may include software upgrades, application repairs, and bug fixing. In this phase, some final alterations and bug fixing are done. Moreover, the system is also presented and trained for the users to ensure that the system is fully operational and developed according to their needs.

Figure 1 shows the context diagram of the E-College Monitoring System, where we can see the external entities in this system, which are a student and the College Resident Staff.

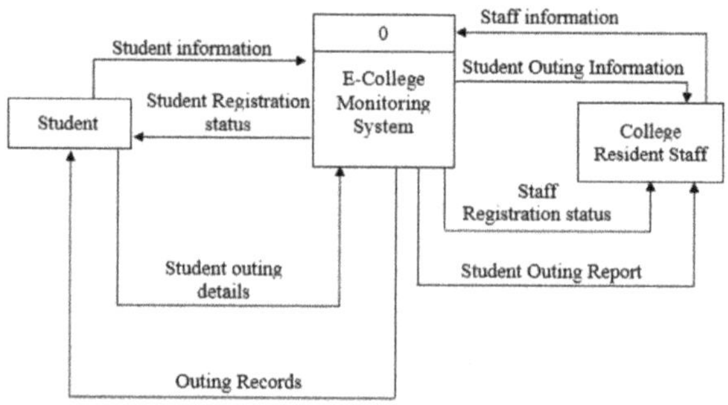

Figure 1. Context diagram.

The figure specifies the system's boundaries. It identifies the information flows between the system and the external entities. The E-College Monitoring System was developed for college students to assist them in keeping track of their time spent on and off the campus of UiTM. To log into the system, students must complete a form through the E-College Monitoring System. Once users have successfully registered with the system, they can use the system's QR code scanning feature. The system will display the outing records. As for the College Resident Staff, they can log in with their staff IDs. A student's scan results are reflected in the system, and their check-in and check-out times are immediately displayed.

In Figure 2, the data flow diagram of the E-College Monitoring System identifies the flow of the information, which consists of inputs, outputs, data stores, and subprocesses. Students create an account with the E-College Monitoring System. Once the students successfully register with the system, their information will be stored and displayed. Students that scan their QR codes are likewise tracked by the E-College Monitoring System. The administrator then receives information about the students who scan QR codes when entering and exiting the UiTM campus. Finally, the system will generate a report for the administrator to keep track of how many students are entering and exiting the UiTM.

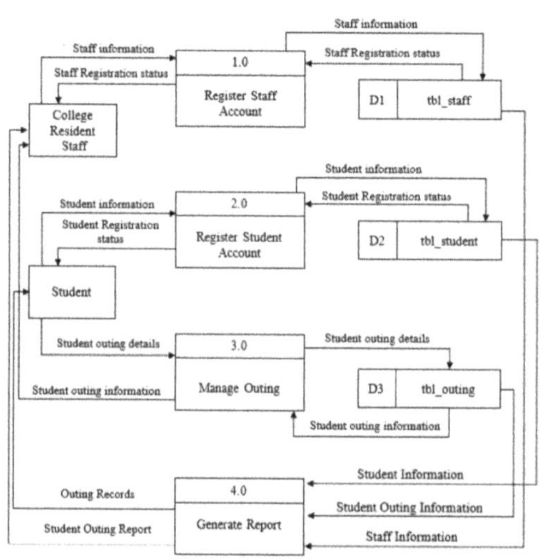

Figure 2. Data Flow Diagram (DFD).

4. Main Interfaces of the E-College Monitoring System

The application was developed using PHP programming language, adopting the Laragon framework. The following figures show print screens of the user interfaces for the E-College Monitoring System. The following Figure 3 illustrates the login page of the system.

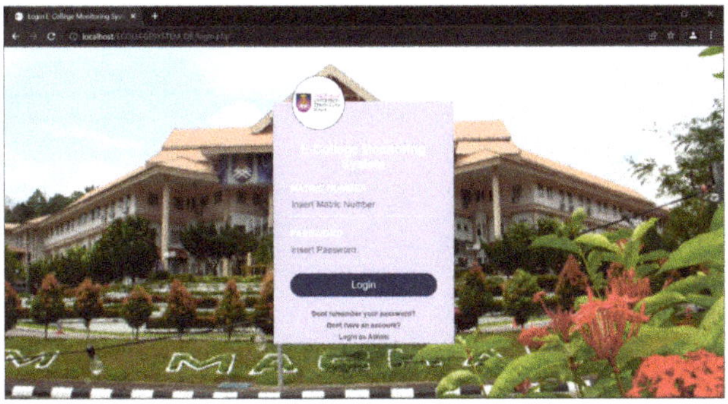

Figure 3. Login page of the E-College Monitoring System.

Figure 4 shows the registration page of the system. Each user are required to register an account to use the system. Once the registration process is complete, users' detail will be displayed as in Figure 5.

Figure 4. Students' registration page.

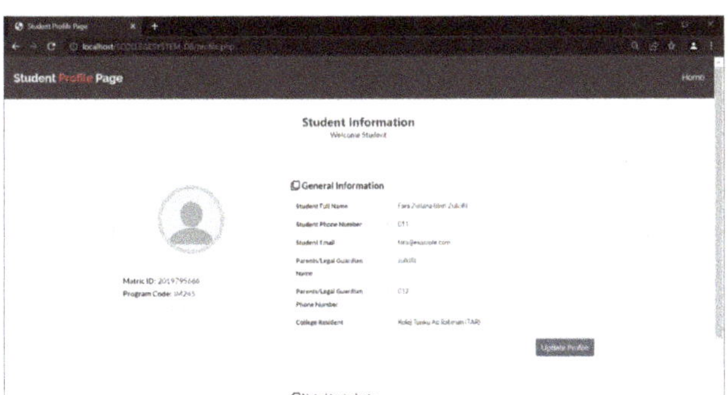

Figure 5. Profile page.

The following Figure 6 indicates students' outing history. Students can view their outing history in this page. This page will display their in and our time. The check in and check out function is shown in Figure 7. Through this module, staffs are able to monitor the students.

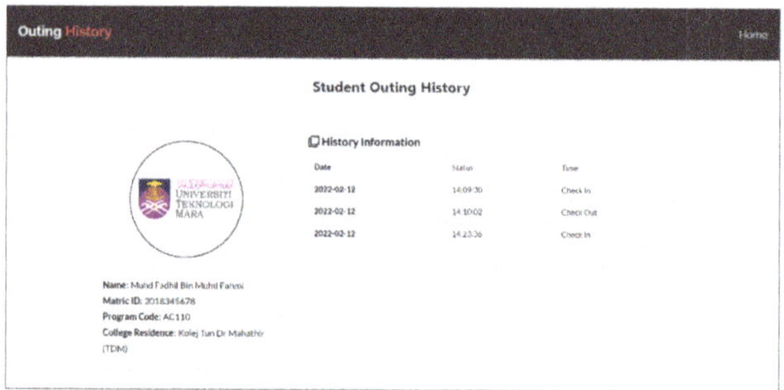

Figure 6. Students' outing history page.

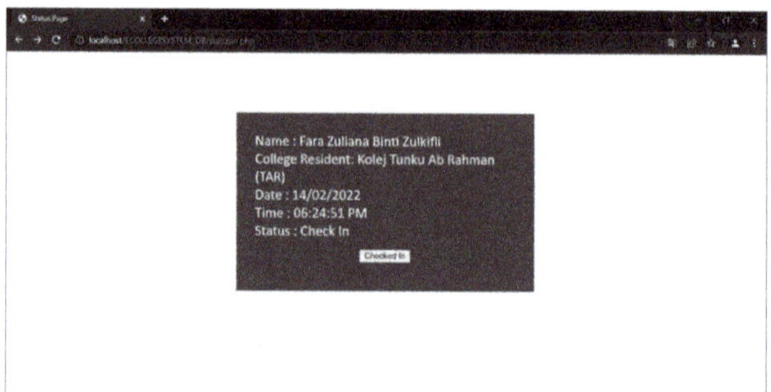

Figure 7. Status check-in page.

This system is not only beneficial and useful to the students but also to the management. Through this system, management can monitor the students as shown in Figure 8. Apart from that, by using this e-College Monitoring System, the managements especially the College Residents Staff (CSR) can view the details of the students including their contact number as illustrated in Figure 9.

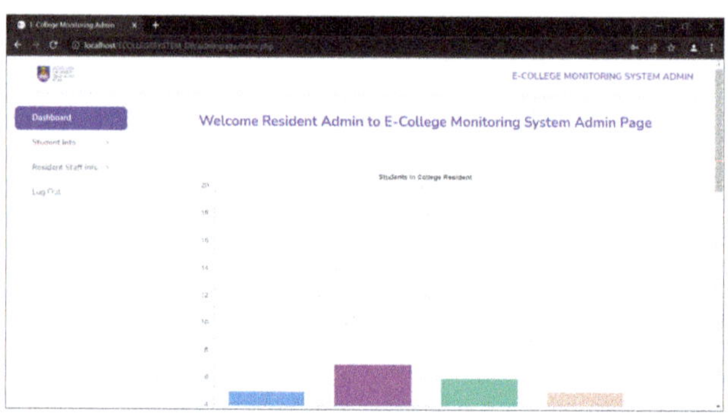

Figure 8. Administrator's dashboard.

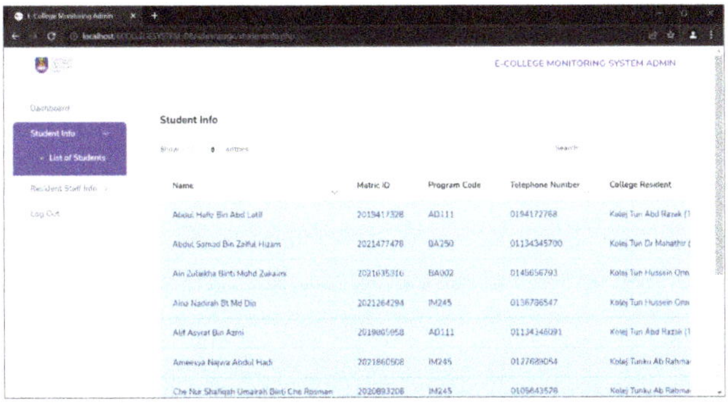

Figure 9. Students' information page for the College Resident Staff (CSR).

5. Conclusions

The E-College Monitoring System will provide students and resident staff with numerous benefits. The usage of technology can help to strengthen and secure the systems. Through the installation of the new system, any issues that developed in the prior manual system will be fixed. When students embark on outings, the system will ensure that both the resident staff and students are aware. It will be a more efficient approach to keep track of the students' information. The QR code scanner features in this system can track the student entering and leaving information with specific dates and times. Residents will be informed that the students will be absent from the UiTM premises on specific dates. This skill will aid in keeping staff up-to-date on the student data information from their history of outings information.

More services, such as a resident activity monitor, are also being added to strengthen the system. The parental activity monitor can assist in tracking the students going out or entering UiTM Machang. This feature will make the system more functional for all users participating in the application management.

When fully established and following a successful implementation at UiTM Machang, it is our hope and vision that the system will be adopted by other UiTM universities, along with other universities in Malaysia, whether public or private. As a result, the E-College Monitoring System may be a viable option for all students and resident staff. There will be significant cost savings, and student outing administration will be much more simplified.

Author Contributions: Conceptualization, N.A.M.M.N. and N.S.S.; methodology, S.A.M.G. and N.H.M.R.; validation, N.A.M.M.N.; formal analysis, F.Z.Z.; investigation, F.Z.Z., N.S.S. and F.M.M.; resources, S.A.M.G. and N.H.M.R.; software, F.Z.Z. and F.M.M.; writing—original draft preparation, N.A.M.M.N., N.S.S., S.A.M.G., N.H.M.R., F.Z.Z. and F.M.M.; writing—review and editing, N.A.M.M.N.; visualization, N.S.S.; supervision, N.A.M.M.N. and project administration, N.A.M.M.N. All authors have read and agreed to the published version of the manuscript.

Funding: This research received no external funding.

Institutional Review Board Statement: Not applicable.

Informed Consent Statement: Not applicable.

Data Availability Statement: Not applicable.

Acknowledgments: This research was partially supported by the Universiti Teknologi MARA Kelantan Branch, Malaysia.

Conflicts of Interest: The authors declare no conflict of interest.

References

1. Mansor, M. Sistem Keluar Masuk Asrama Berkomputer Sek. Men. Sains Pokok Sena. Doctoral dissertation, Universiti Teknologi Petronas, Tronoh, Malaysia, 2003.
2. Wen, M.K.K.; Ahmad, N.; Ruslan, S.H. Arduino based outing and attendance system for boarding school students. *Indones. J. Electr. Eng. Comput. Sci.* **2020**, *20*, 1053.
3. Kuo, S.E.; Lai, H.S.; Hsu, J.M.; Yu, Y.C.; Zheng, D.Z.; Hou, T.W. A clinical nutritional information system with personalized nutrition assessment. *Comput. Methods Programs Biomed.* **2018**, *155*, 209–216. [CrossRef] [PubMed]
4. Azza, A.M. *An Empirical Study of MIS Impacts and Organizational Efforts on Improving Business Performance*; Taylor & Francis Group: London, UK, 2017.
5. Yu, J.; Xu, J. Design and Implementation of Warehouse Management System Based on Web Technology. In Proceedings of the 2018 3rd International Conference on Mechanical, Control and Computer Engineering (ICMCCE), Huhhot, China, 14–16 September 2018. [CrossRef]
6. Rosman, M.R.M.; Abdullah, R.Y.R.; Salleh, M.I.M. Development of Vehicle Observation System as security mechanism towards electronic records conversion. In Proceedings of the 2010 International Conference on Science and Social Research (CSSR 2010), Kuala Lumpur, Malaysia, 5–7 December 2010; pp. 699–704.
7. Hoffer, J.A.; George, J.; Valacich, J.S. *Modern Systems Analysis and Design*; Pearson: Boston, MA, USA, 2017.
8. Chassiakos, A.P.; Sakellaropoulus, S.P. A Web-based System for managing Construction Information. *Adv. Eng. Softw.* **2008**, *39*, 865–876. [CrossRef]
9. Taylor, M. A General Understanding of Microsoft Excel, Visual Studio, and WebMatrix. Available online: http://533507.weebly.com/notepad.html (accessed on 21 May 2022).
10. Maintenance Phase in SDLC. Available online: https://study.com/academy/lesson/maintenance-phase-in-sdlc.html (accessed on 21 May 2022).
11. Mohamad, S.M.S. Development and implementation of a web-based system to study children with malnutrition. *Comput. Methods Programs Biomed.* **2009**, *93*, 83–92. [CrossRef] [PubMed]

proceedings

Proceeding Paper

Developing a COVID-19 Tool as a Web-Based Means of Assessing Gold (iGold) †

Khadijah Abdul Rahman [1,*], Nur Ainatul Mardiah Mat Nawi [1], Asiah Said [2], Fadhilah Mohd Ishak@ Zainudin [3], Wan Masnieza Wan Mustapha [3] and Asma Shazwani Shaari [3]

1 Faculty of Information Management, Universiti Teknologi MARA Kelantan Branch, Bukit Ilmu, Machang 18500, Kelantan, Malaysia
2 Academy of Language Studies, Universiti Teknologi MARA Kelantan Branch, Bukit Ilmu, Machang 18500, Kelantan, Malaysia
3 Faculty of Business and Management, Universiti Teknologi MARA Kelantan Branch, Bukit Ilmu, Machang 18500, Kelantan, Malaysia
* Correspondence: khadijah10@uitm.edu.my; Tel.: +60-19-386-3710
† Presented at the International Academic Symposium of Social Science 2022, Kota Bharu, Malaysia, 3 July 2022.

Abstract: The focus of this study was the design and development of a web-based application for the assessment of gold during the COVID-19 pandemic known as Cov_iGold. The pandemic had clear and serious consequences for the global economy due to the stringent quarantine guidelines introduced which led to economic activities becoming dormant. Despite the extreme restrictions on operations and face-to-face transactions, there is still a demand for gold. As the global market has been widely affected, this platform is expected to provide a general tool to meet the demands of gold collectors and merchants. It is a self-exploratory application, through which users will be able to gain daily access to gold-related information, e.g., current prices, gold market trends and the selling and purchasing of gold with added features. This tool was developed according to five stages of the Systems Development Life Cycle (SLDC), namely, planning, analysis, design, implementation and maintenance. Gold merchants can easily track their buyers, receive payment hassle-free and effectively manage large numbers of buyer-related data in one repository. With these significant functions, inaccuracies in data will be reduced or eliminated. Having these resources at their fingertips, Cov_iGold users will be able to fully optimize its features from the comfort of their own homes.

Keywords: gold; web-based tool; COVID-19; iGold

Citation: Rahman, K.A.; Nawi, N.A.M.M.; Said, A.; Zainudin, F.M.I.; Mustapha, W.M.W.; Shaari, A.S. Developing a COVID-19 Tool as a Web-Based Means of Assessing Gold (iGold). *Proceedings* **2022**, *82*, 27. https://doi.org/10.3390/proceedings2022082027

Academic Editor: Mohamad Rahimi Mohamad Rosman

Published: 13 September 2022

1. Introduction

Since the emergence of the COVID-19 crisis in 2021, small enterprise and business owners have become a focus of attention in the research on small and medium enterprises (SMEs) in the Malaysian economy. According to the SME Annual Report (2016), a micro-enterprise is defined as an organization with fewer than five employees and yearly sales of less than MYR 300,000. Such organizations account for about two-thirds of Malaysia's SMEs. Utilizing a web-based tool, an online approach has been developed with the aim of identifying the biggest challenges facing SMEs after the COVID-19 outbreak and charting entrepreneurial performance and trends over the course of the pandemic. Three research objectives are highlighted, namely, to (1) provide access to the current price of gold, (2) allow comparison of the prices of gold among different sellers and (3) reduce the numbers of people going out to survey the price of gold. This study employed a web-based tool to obtain materials from authoritative sources related to the aims of the study. Researchers have studied the contents of journals, papers and books to explicate the strategies used by successful entrepreneurs in managing their businesses during the COVID-19 pandemic.

Their findings have allowed them to provide support for the sustainability of micro-enterprise entrepreneurs during this crisis. This study was carried out in accordance with the objectives of this research.

The COVID-19 pandemic started in the year 2020 and has tremendously affected micro-enterprise operators. However, the implementation of the business continuity strategy for micro-enterprises as a means of coping with the crisis has not been fully explored. Furthermore, gold merchants in Malaysia will likely face new challenges as the pandemic continues. They might be sued or forced to shut down their operations if they fail to comply with standard procedures [1]. Fortunately, with the rapid growth of online platforms in the market, there is an open-source opportunity for gold merchants to expand and conduct their businesses via e-commerce platforms. Therefore, the primary goals of this study were to provide a means of accessing the current price of gold and comparing the price of gold among sellers, thereby reducing the numbers of people going out to survey the price of gold.

2. Literature Review

Gold as jewelry is a good investment and has been dominant in world markets. In fact, gold sets the benchmark in terms of prestige due to its ability to withstand any crisis. According to one study [2], gold is a great asset with increased liquidity and represents a solid investment that is stable both short-term and long-term. However, due to massive demand, the price of gold has greatly increased over time. Before making an investment, it is vital for gold buyers to grasp the nature of gold investment. A considerable amount of research has attempted to ascertain the variables that affect customers' decisions to purchase gold and the significance of these factors. The uncertain economic situation during the COVID-19 pandemic and the gold market's persistence are encouragements to people to explore the feasibility of gold as a financial instrument that is resistant to depreciation. Another study [3] concurrently supported the potential of gold as a safe financial asset. This finding was strongly supported by the findings of previous research [4–6]. According to these authors, central banks have kept significant amounts of gold in their reserves for a variety of reasons, including economic security, because gold keeps its purchasing power, is suitable for portfolio diversification, is physically secure since it is a highly liquid asset, has guaranteed value in times of economic complexity and represents a relatively risk-free investment in economic crises. The authors of one study [7] examined several aspects of gold as a financial asset and discovered that gold acts as an inflationary hedge. Their research revealed that gold meets the criteria for zero beta assets and market risk. From a Malaysian perspective [8], it was reported that during times of economic uncertainty, particularly when the stock market has collapsed, gold investments have made important contributions to economic stabilization. In addition to the above studies, the authors of [9] discovered through their research that a significant proportion of the population bought gold for religious festivals, anniversaries, birthdays and family weddings. According to [10], the leading causes of gold jewelry purchases are cost, purity, design, the wide range of choices, image and pressure from family and friends. Customers need to have sufficient knowledge in order to make decisions when buying gold. The knowledge required by consumers to decide on a gold purchase process is liable to vary between individuals [11]. The authors of the previously mentioned study also stated, based on their research, that product knowledge affects consumers' online purchasing decisions. It is human nature to reflect on social engagement, which is a very common factor in people's daily activities. Various cyber activities can be performed via numerous digital platforms which enable users to exchange information and share ideas online [12].

Today, the evolution of the Internet propels the evolution of online web-based systems as essential communication media, resulting in the emergence of virtual communities. The online world has presented a new perspective on social communication, connecting people to a variety of evolving online communities over the past decade. Vendors of products and customers may never meet in the real world, yet vendors are able to affect customer behavior, including purchasing decisions [13]. The Internet is a social place where information is created and shared among consumers [14]. Web pages are considered highly involving advertising spaces due to the ease of their access by consumers [15]. Furthermore, web sites with interactive links provide ample and unlimited spaces for users to craft their information. In fact, the dynamic character [16] of web-page communication is due to the interactivity which allows individuals the freedom to choose and consult suppliers and vendors.

In today's setting, especially during the COVID-19 era, people do not mind paying extra charges to ease their online shopping experiences, as these processes are convenient and less time-consuming than actual shopping [17]. One study [18] revealed that only people who already trusted the Internet and had prior online experience would want to buy gold online. Consumer trust levels are likely to be influenced by a website's quality, technological reliability and online shopping capabilities. Another study [19] identified several factors which affect consumer behavior in purchasing goods online. The factors prioritized by consumers were found to be price, convenience, security, information quality and access to a website.

3. Background of the Study

3.1. Problem Statement

From prehistoric times, people across the globe have displayed a special interest in gold either in the form of jewelry or gold bars. Gold is prized for ownership, prestige and status. The passion for jewelry is widespread. However, due to the COVID-19 pandemic, global markets were severely affected and most stores selling gold were prohibited from operating. As many nations adjust to increasingly rigorous quarantine regulations and find their economic operations severely constrained, the epidemic can be seen to have had an obvious and significant impact on economies all over the world [20]. Despite this, there is still a demand for gold. Based on survey data, it was found that gold shops were inundated with visitors seeking to purchase the coveted gold collections. Most of the customers ignored the physical social distancing procedures due to their excitement. Some stores were relatively small and crowded with visitors and the owners did not impose standard operating procedures (SOPs) or ask visitors to enter in stages [21]. Such practices are particularly unsafe, as customers are prone to COVID-19 infection. Meanwhile, according to [22], it is evident that demand for gold persists even though prices are high, with increments in the price of gold expected in the future. This is due to the expectation of higher returns upon resale. Facing unprecedented risks in the markets, people feel an increasing need to make secure investments. The study reported in [23] presented evidence that gold was a safe investment in times of post-political and financial crises. Gold is preferable to other types of assets generally considered secure due to the behavioral prejudices associated with gold's history as a commodity—a product of constant value and a secure investment. Remarkably, people will go from one store to another to find the cheapest prices. However, due to the risk of being infected by COVID-19 and the imposition of MCOs, it is difficult for people to complete surveys. Due to the fact that gold remains popular, even during the pandemic, and can be hedged against crises, researchers are keen to develop a web-based tool that can be employed to assess and easily compare the price of gold among different sellers. Such a web-based tool would save time and reduce the numbers of people leaving their homes during the Movement Control Order (MCO) period.

3.2. Objectives of the System's Development

This paper will examine the motivation to develop a web-based tool to assess the current price of gold during the COVID-19 pandemic. The development of such a web-based tool is imperative, since gold is still in demand during the pandemic despite the strict MCOs. A web-based tool would be beneficial and save the time of people wanting to compare prices and services offered by different stores. Hence, the research objectives of this study were:

1. To provide access to the current price of gold and a means of comparing the price of gold among different sellers;
2. To facilitate the process of surveying and purchasing gold;
3. To design and develop a web-based tool for assessing gold and evaluating performance.

4. Materials and Methods

Five stages of the System Development Life Cycle (SLDC), namely, planning, analysis, design, implementation and maintenance, were deployed as a methodology in order to gain a clearer view for each phase involved [24]. A few existing systems with similar functions were evaluated to discover problems with these systems in order to develop a new system that would be much more user-friendly. Moreover, for the Cov_iGold system, we included any software and hardware that was deemed most suitable for use in building the system. Next, we determined user requirements by several methods, such as interviews and observation. Individual and group interviews were conducted, with gold merchants especially, to obtain information about operations, issues and the need for an online platform to help sustain trade in gold. Furthermore, we also observed, at selected times, how data were handled and what information people needed so that we could develop a system that would be better for the user [25].

Regarding the design phase, the detailed process of the proposed system's construction is illustrated in a data flow diagram. Our system design was expanded by designing and creating an entity relationship diagram (erd) and data dictionary (metadata) which explained the structure of the database system. We also created an interface for the system, namely, an input, output and dialog box, to make it easier for users to utilize and comprehend its functions. In addition, during this phase, we created an interface to ensure that the system was user-friendly to facilitate engagement [26].

This phase entailed the determination of the project's outcome, and the implementation of the project become clearer, as coding and testing were executed at this stage. During this period, the public became aware of the project and may have believed that it had already been initiated. This was a crucial stage, where all of the actions taken were crucial to sustaining momentum throughout the rest of the process. The PHP programming language was used to code the system. According to [27], the process of coding uses just a few functions, for example, an insert and update function to make sure that the buyer and seller can update the system and insert the necessary data. Furthermore, we also incorporated buttons, such as buyer profile updates, so that buyers could register and update their information, enabling them to compare fluctuations in prices and choose the merchants they prefer. The payment method features also provided updates on their payments after the approval of transactions. Additionally, the interactive buttons for the "Contact Us" and "About Us" pages navigated users to information about the Cov_iGold system and future correspondences.

SDLC's maintenance phase begins after a product is fully functioning and includes upgrades, repairs and software patches if the software has broken down [24,28]. During this phase, we made the final alterations to the system, fixed a few problems and display errors and organized the design of the system. Finally, we presented the system to users, informing them about how the system operates and the functions that can be performed.

5. Main Interfaces of the Cov_iGold System

The following figures display the main interfaces of Cov_iGold. Figure 1 indicates the main interface of the system. The figure shows the landing page of the Cov_iGold system. Figure 2 illustrates the login page of the system. Users are required to register an account with the system to facilitate the process of browsing and purchasing items. Next, Figure 3 shows the search page of the system. Users can search by keywords and price.

Figure 4 indicates the promotion page of the system. The page provides information on items, including price, a description and images.

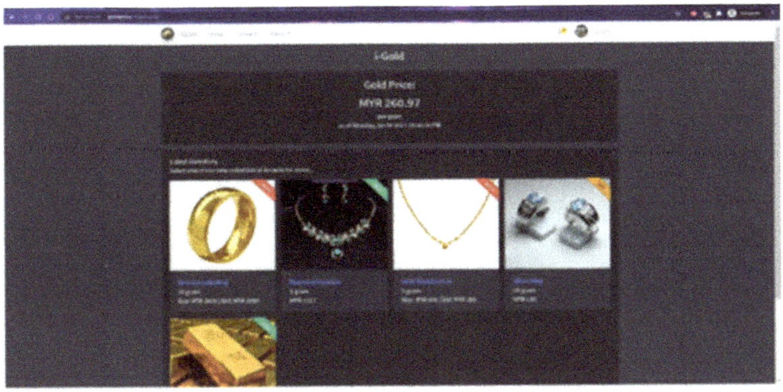

Figure 1. Main page of Cov_iGold.

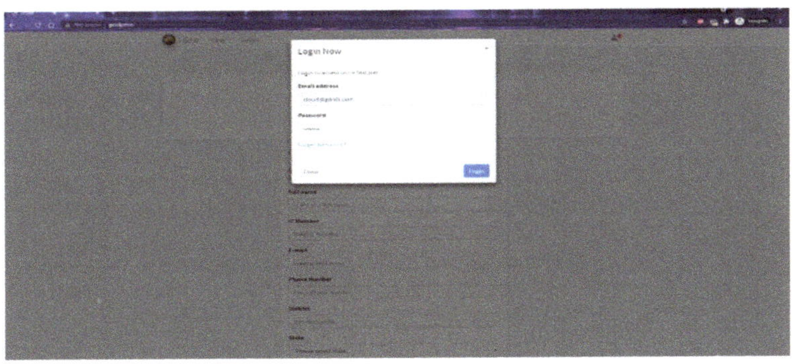

Figure 2. Login page.

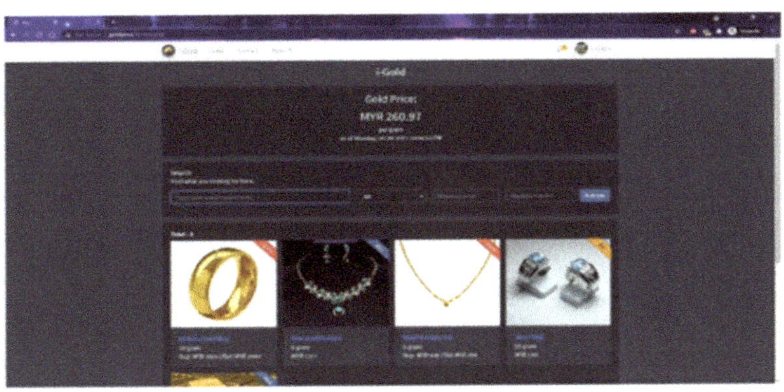

Figure 3. Search page.

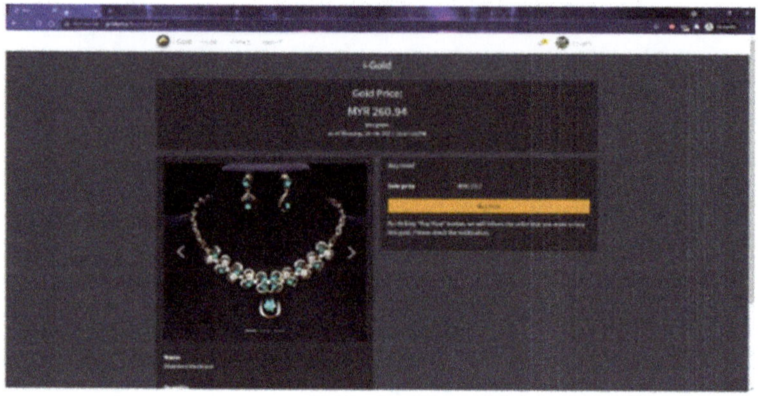

Figure 4. Promo page.

6. Conclusions

We would like to express our deep thanks to our research members for their continuous support and encouragement, as well as our esteemed institution, University Teknologi MARA, Machang Campus, Kelantan. This appreciation also extends to the individuals involved directly or indirectly throughout this valuable journey for the opportunities they have all given wholeheartedly. The completion of this project could not have been accomplished without the support of our colleagues and family members, whom we thank for allowing us the precious time away from them in order to carry out the research and write the present paper. They all deserve a special appreciation. The generosity of the time allowed us, despite our hectic schedules, will not be forgotten.

Author Contributions: Conceptualization, K.A.R., W.M.W.M. and F.M.I.Z.; methodology, N.A.M.M.N., A.S.S. and K.A.R.; formal analysis, A.S. and A.S.S.; investigation, N.A.M.M.N.; resources, W.M.W.M. and F.M.I.Z.; data preparation, N.A.M.M.N.; writing—original draft preparation, A.S., W.M.W.M. and A.S.S.; writing—review and editing, K.A.R., N.A.M.M.N., A.S., F.M.I.Z., A.S.S. and W.M.W.M.; visualization, F.M.I.Z.; supervision, K.A.R. and A.S.; project administration, N.A.M.M.N. All authors have read and agreed to the published version of the manuscript.

Funding: This research received no external funding.

Institutional Review Board Statement: Not applicable.

Informed Consent Statement: Not applicable.

Data Availability Statement: Not applicable.

Acknowledgments: This research was partially supported by Universiti Teknologi MARA Kelantan Branch, Malaysia. We would also like to express our gratitude to all of the individuals who contributed to the development of the system.

Conflicts of Interest: The authors declare no conflict of interest.

References

1. Nor, F.A.R. Kedai emas gagal patuh SOP mungkin diarah tutup. *Berita Harian*. 29 November 2020. Available online: https://www.bharian.com.my/berita/nasional/2020/11/759919/kedai-emas-gagal-patuh-sop-mungkin-diarah-tutup (accessed on 22 May 2022).
2. Sukri, M.K.A.; Mohd, Z.N.H.; Zainal, A.N.S. The Relationship Between Selected Macroeconomic Factors and Gold Price in Malaysia. *Int. J. Bus.* **2015**, *8*, 82–193.
3. Baur, D.G.; Lucey, B.M. Is Gold a Hedge or a Safe Haven? An Analysis of Stocks, Bonds and Gold. Available online: https://papers.ssrn.com/sol3/papers.cfm?abstract_id=952289 (accessed on 22 May 2022). [CrossRef]
4. Kaufmann, T.; Winters, R. The price of gold: A simple model. *Resour. Policy* **1989**, *19*, 309–318. [CrossRef]
5. Faugere, C.; Van Erlach, J. The price of gold: A global required yield theory. *J. Invest.* **2016**, *14*, 99–111. [CrossRef]

6. Lucey, B.; Tully, E. Seasonality, risk and return in daily comex gold and silver 1980–2002. *Appl. Financ. Econ.* **2006**, *16*, 319–333. [CrossRef]
7. McCown, J.; Zimmerman, J. Is Gold a Zero-Beta Asset? Analysis of the Investment Potential of Precious Metals. Available online: https://papers.ssrn.com/sol3/papers.cfm?abstract_id=920496 (accessed on 1 June 2022). [CrossRef]
8. Baharom, A.H.; Mansor, H.I. The Role of gold in financial investment: A Malaysian perspective. *Econ. Comput. Econ. Cybern. Stud. Res.* **2011**, *45*, 227–238.
9. Asha, K.; Christopher, S.E. A study on buying behaviour of customers towards branded and non-branded gold jewellery with preference to Kanyakumari district. *Int. J. Manag.* **2014**, *5*, 105–114.
10. Raju, K.V.; Kumar, D.P. A Study on Consumer Preference on Branded Jewellery in Hyderabad. *Int. J. Sales Mark. Manag.* **2013**, *2*, 23–34.
11. Hölscher, C.; Strube, G. Web search behaviour of Internet experts and newbies. *Comput. Netw.* **2000**, *33*, 337–346. [CrossRef]
12. Heinonen, K. Consumer activity in social media: Managerial approaches to consumers' social media behaviour. *J. Consum. Behav.* **2011**, *10*, 356–364. [CrossRef]
13. Evans, M.; Jamal, A.; Foxall, G. *Consumer Behaviour*, 2nd ed.; John Wiley & Sons Ltd.: Hoboken, NJ, USA, 2009.
14. Impact of Social Media on Purchasing Behavior. Available online: http://174.133.170.120/files/DEIStudyEngaging%20ConsumersOnlineSummary.pdf (accessed on 24 May 2022).
15. José, G.S.M.; Gutiérrez, A.A.M.; San, J.C.R. Determinants of internet use in the purchasing process. *J. Bus. Ind. Mark.* **2006**, *21*, 164–174. [CrossRef]
16. Hoffman, D.; Novak, T.; Chatterjee, P. Commercial scenarios for the web: Opportunities and challenges. *J. Comput.-Mediat. Commun.* **1995**, *1*, JCMC136. [CrossRef]
17. Bhoomika, K.M.; Neetha, M. Consumer buying behaviour and trust towards online jewellery shopping. *J. Bus. Manag.* **2020**, *22*, 9.
18. Corbitt, B.; Thanasankit, T.; Han, Y. Trust on the World Wide Web: A study of consumer perceptions. *Sch. Work. Pap. Ser.* **2017**, *2*, 34–37.
19. Kalia, P.; Kaur, N.; Singh, T. A review of factors affecting online buying behaviour. *Apeejay J. Manag. Technol.* **2016**, *1*, 58–73.
20. Mohammed, M.E.; Walid, M.A.; Abdullah, A. Price and volatility spillovers between global equity, gold, and energy markets prior to and during the COVID-19 pandemic. *Resour. Policy* **2021**, *74*, 102–334.
21. Nor, F.A.R. Orang ramai serbu kedai emas di Kota Bharu. *Berita Harian*. 29 November 2020. Available online: https://www.bharian.com.my/berita/nasional/2020/11/759706/orang-ramai-serbu-kedai-emas-di-kota-bharu (accessed on 22 May 2022).
22. Yoganandham, G.; Govindaraj, A.; Kumaran, P.S. Knowledge, awareness, attitude and purchasing decision making behaviour among women towards gold jewellery in vellore city–an assessment. *Int. Rev. Bus. Econ. Spec. Issue* **2020**, *4*, 300–311.
23. Baur, D.G.; McDermott, T.K.J. Why is Gold a Safe Haven? *J. Behav. Exp. Financ.* **2016**, *10*, 63–71. [CrossRef]
24. Hoffer, J.A.; George, J.; Valacich, J.S. *Modern Systems Analysis and Design*; Pearson: Boston, MA, USA, 2017.
25. Rosman, M.R.M.; Abdullah, R.Y.R.; Salleh, M.I.M. Development of Vehicle Observation System as security mechanism towards electronic records conversion. In Proceedings of the 2010 International Conference on Science and Social Research (CSSR 2010), Kuala Lumpur, Malaysia, 5–7 December 2010; pp. 699–704.
26. Taylor, M. A General Understanding of Microsoft Excel, Visual Studio and Webmatrix. Available online: http://533507.weebly.com/notepad.html (accessed on 25 May 2022).
27. Chen, Y.; You, H. Selection and Research for Online Registration System's. *J. Softw. Eng. Appl.* **2017**, *6*, 33. [CrossRef]
28. Maintenance Phase in SDLC. Available online: https://study.com/academy/lesson/maintenance-phase-in-sdlc.html (accessed on 22 May 2022).

proceedings

MDPI

Proceeding Paper

Conscious or Unconscious: The Intention of Hate Speech in Cyberworld—A Conceptual Paper †

Noramira Fatehah Azman and Norena Abdul Karim Zamri *

Faculty of Communication and Media Studies, Universiti Teknologi MARA (UiTM), Cawangan Melaka Kampus Alor Gajah, KM 26 Jalan Lendu, Alor Gajah, Melaka 78000, Malaysia

* Correspondence: norena@uitm.edu.my

† Presented at the International Academic Symposium of Social Science 2022, Kota Bharu, Malaysia, 3 July 2022.

Abstract: Exponential growth in social media has led to the increasing popularity of hate speech and hate-based propaganda. Hate speech or malicious expression refers to the use of offensive, violent, or offensive language and its religious conduct with a specific group of people who share a common property, such as gender, ethnicity, race, or beliefs. Online hate diffusion has now become a serious problem as it creates a series of international initiatives aimed at defining problems and developing effective countermeasures; this study delves into the exploration of the intention of hate speech posting on social media, especially on Twitter. Both dramaturgical models of social interaction and cultivation theory were used to explain the hate speech culture phenomenon. A qualitative method is proposed for this study as part of the exploration. Results revealed that most of the previous studies on hate speech focused on the field of computer science but rarely on the communication field. The paper presents the results of past studies and shows the new proposed framework. The investigation suggests future directions for the problem and possible solutions; it starts with the background of the research, the explanation of the problem, the meaning of the research, and pursuing the research questions and goals of the research before finally explaining the limits.

Keywords: hate speech; cyberworld; social media; Twitter; offensive language; malicious comment

Citation: Azman, N.F.; Zamri, N.A.K. Conscious or Unconscious: The Intention of Hate Speech in Cyberworld—A Conceptual Paper. *Proceedings* **2022**, *82*, 29. https://doi.org/10.3390/proceedings2022082029

Academic Editor: Mohamad Rahimi Mohamad Rosman

Published: 13 September 2022

1. Introduction

There is no denying that Malaysia is one of the countries that is not left behind in pursuing technological modernity. Today, technology is evolving with its variety of sophistication. The traditional details lead to more modern technical pieces of information in line with the current of modernity over time. One of the latest technologies is social media, where we can socialize and communicate well with people worldwide.

Twitter is one of the most popular social media platforms, with almost every adolescent and young adult having an account. Between 2005 and 2010, the popularity of Internet social networking devices such as Twitter skyrocketed. Twitter was founded in March 2006 by Jack Dorsey, Noah Glass, Biz Stone, and Evan Williams and was launched in July of the same year [1]. According to [2], as of early 2019, Twitter has over 330 million active monthly users. Twitter is unique because of the brevity in which users interact with other users; Twitter messages, or tweets, are limited to 280 characters, a gif, an emoji, a tag location and a poll, we can "tweet" anything we want with our own freedom of speech. Twitter offers three different ways for users to interact with each tweet: a like button, a retweet button, and a reply button. The like button is to show that we love the thing by clicking the love-shape button and our mutual friends on Twitter will get a notification by the apps. Other than that, the Retweet button is a sharing button, so our followers see what we share. The reply button is where users can comment on the tweet to give feedback on some tweets.

According to [3], the numbers posted on Twitter's advertising resources showed that Twitter had 4.4 million users in Malaysia in early 2022; this number indicates that 13.3% of

Malaysia's population was reached via Twitter advertising at the time. However, because Twitter limits the use of the platform to people over the age of 13, it may be helpful to know that 16.7% of Malaysia's "qualified" audiences will use Twitter in 2022. Furthermore, Twitter's Malaysian advertising reach earlier this year accounted for 14.9% of the local Internet user base (regardless of age). Twitter's popularity as a new medium for online communication has caused language issues that require this research, as it clearly imposes word restrictions on each update; this limitation also created another issue with messages that tend to be ambiguous when communicating over Twitter. With only 280 characters, a photo or video, a gif, an emoji, a tag location, and a poll, we can "tweet" anything we want with our own freedom of speech.

Malicious posting or hate speech has frequently been examined in relation to the communication of (extreme) right-wing populist leaders [4]; these players are also becoming increasingly linked to the unchecked spread of misinformation [5]. As it is used by millions of people or we call netizens, some negative things like dissemination of false information, identity fraud and mostly hate speech will occur uncontrollably as social media like Twitter has no gatekeeper; these platforms partially guarantee the user's freedom of expression, so this means it can be used to attack or insult others. Safety concerns result from this circumstance. Online attacks and abuse have been shown to cause not only psychological and physiological health problems for victims, but also self-harm and even suicide [6]. Therefore, our focus is to analyze the intention of hate speech in controlling hate speech in the cyberworld to avoid more negative impacts occurring to users.

In accordance with [7], malicious language or hate speech is used to identify messages that violate existing legal norms and require government regulation or publicly shared statement that incites, encourages, justifies, or are based on a particular group of hatred, discrimination, or hostility. For instance, race or ethnicity, beliefs, abilities or disabilities, gender, age, sexual orientation, or gender identity. Other than that, ref. [8] defined hate speech as showing disgust as it spreads, incites, and promotes hatred, violence, and discrimination against individuals or groups because of their protected nature; these include "race", ethnicity, religion, gender, sexual orientation, and disability, among other social distinctions.

Davidson et al. [9] said that the definition of hate speech is a language of expression that describes the use of words that are hatred or blame, offensive or insulting to the subject. Hate speech as a jargon is no longer an individual-dominated territory in the legal world, but within reach of amateurs. In the context of the Internet and social networks, hate speech does not only create tensions between groups of people, but its impact can affect businesses and create serious conflicts in real life. For this reason, sites such as Facebook, YouTube, and Twitter prohibit the use of malicious expressions as they can endanger the lives of users.

The definition of hate speech is highly controversial and is often used in place of various forms of group-related hostilities such as anti-Semitism, racism, and anti-gypsy [10]. In fact, as [11] pointed out, hate speech is a seemingly complex term, with expressions that can be grouped into such general terms using a variety of derivative and competing concepts. Cyber world hate speech can also change the first intention of social media, which is to communicate and socialize well with other people in the whole world. The "Like" button is made to make people happy, but nowadays, youth can be depressed if they do not get one; it deviates from the original purpose of social media. As a result, the issue that requires this research is to explore the intention of hate speech on Twitter from Malaysians perspectives. Simultaneously, the interaction among human beings should be included as part of this study too.

In an online interaction, people can say what they want without any filter or gatekeeper, either their own posts, tweets, or their comments on some other individual posts. In this way, the whole world can react to each other either in a good or bad way; it is not a problem if good things spread all over the internet, but we are very concerned that bad things, such as hate speeches, will spread, with negative consequences for other people who believe what they see is 100 percent true and real. In sum, throughout this research, the study wanted to delve into the intention of hate speech on social media, on Twitter. Other than that, this study wanted to explore how hate speech affected other users or victims of hate speech itself.

2. Problem Statement

Watanabe H et al. [12] defined malicious expression or hate speech as the use of offensive, violent, or abusive language; it targets a specific group of people who have something in common, such as gender, ethnicity, race, beliefs or faith, or skin color. Hate speech does not only create tensions between groups of people on the Internet and social networks, but its impact can influence businesses and cause serious conflicts in real life. For this reason, sites such as Facebook, YouTube, and Twitter prohibit the use of malicious language. However, controlling, monitoring, or filtering all content is always difficult. Hate speech and malicious expression is a particular form of offensive language in which the person using it expresses an opinion based on either their biases, personality segregation, racist or radical background, stereotype, or in the worst scenario, just hating the other users.

Oxford Constitutional Law [13] described hate speech as "verbal or non-verbal communication with hostility towards certain social groups, most often racial and ethnicity (racism, exclusion of foreigners, anti-semithism, etc.)" based on gender (sexism), misogyny), sexual orientation (homophobia, transphobia), age (agism), and disability. Therefore, hate speech is considered a global issue in which many countries and organizations are fighting against each other.

Past studies have reported many cases of hate speech all over the world, which can be cancer to the cyberworld society. King and Sutton [14] reported that in the year following 9/11, 481 hate crimes with specific anti-Islamic motivations were committed, 58% of which were committed two weeks after the incident (4% of the risk period). Such evidence shows that unfavorably motivated crimes often occur in close time to electric shock events such as terrorist attacks. As Malaysians and Asians, we are no exception to the pitfalls of hate speech in the cyber world.

Murad D [15] stated that Malaysians living abroad were also victims of racist attacks on Asian communities. There are numerous explanations for this behavior; this is primarily due to misinformation and stigma, alienation, and anti-immigrant sentiments that link the COVID-19 pandemic to China and Asia; these experiences are not limited to the United States. Newcastle University student Sze Qi, 21, was also a target of anti-Asian hatred while studying in the UK. She recalls the incident when she passed a group of teens and one of them coughed fake and said "coronavirus." Last April, a Malaysian student and her Singaporean friend were violently attacked while shopping for groceries in Australia's central business district, Melbourne. The attackers reportedly beat, kicked, and yelled at the "coronavirus" and threatened to kill the two women. In response to this hatred bullying, the #StopAsianHate movement is gaining momentum online by holding physical rallies in the United States, Canada, and Taiwan to raise awareness. UN Secretary-General Antonio Guterres also expressed concern about the increase in violence against Asians during the COVID-19 pandemic.

Meanwhile in Malaysia, some act or policy is needed as transgender people were recently killed in Bukit Tinggi, Klang in 2019, and hate speech to ethnicity, religion and even gender is on the rise [16]. Recent stories at Seafield Temple about land conflicts and their escalation to racial and religious issues show how easily and violently the issues can be transformed into problems of race and religion. In Malaysia, there are no specific laws or policies that can be imposed on those who spread and promote hate speech, especially for the race, ethnic or faith groups, and even for gender.

Article 10 of the Federal Constitution states that all citizens have the right to free speech, but this norm is not absolute. As in other countries, this freedom is subject to some restrictions to protect minorities and people of various ethnic, religious, and sexual orientations, as well as those suffering from other forms of discrimination and aggression. Words, ironically, are sometimes considered a weapon. When archaic incitement is used against people who disagree or hold opposing views on the government at the time, the law's intended spirit is "to fall into hatred, cause disdain or dissatisfaction with the government." It incites hostility, violence, and discrimination against groups and causes "malicious or interracial hostility." Hate speech, which was created by some problematic groups and has recently been distributed through many communication channels such as Facebook, Twitter, Instagram, and Snapchat, helps to eliminate structural violence, racism, and cyber hatred.

Due to the openness of the Internet and the spread of social media, hate speech has become a prominent topic on the Internet; this will result in experiments with increasing needs and automatic hate speech detection [17,18]; they also stated that in addition to the lack of problems with the traditional terms mentioned above, the lack of a common dataset for conducting research is a difficult obstacle to progress in this area [19].

Based on the past studies, clearly hate speech on social media, especially in Twitter needs some more concern and more studies. Therefore, this study wanted to delve into the intention of hate speech on cyberworld culture based on race, religion, and ethnicity from the Malaysian perspective; this research is relevant to Malaysian social media culture. Azreen H [20] stated that the Malaysian Communications and Multimedia Commission (MCMC) should promote awareness of self-regulation and set a guideline on what constitutes hate speech for the public, as said by a lawmaker, Petaling Jaya (PJ) MP, Maria Chin Abdullah and civil liberties lawyer, Syahredzan Johan, after the institution drop a hotline initiative, where the commission welcomes the public's reports on social media comments that touch on the 3R–Race, Religion and Royal Institution; they also said that education is key because if a person does not have knowledge on what hate speech is, most likely this person would bombard MCMC with various reports just because the content is not up to his liking in the same article. By this research, we can analyze and inform MCMC on the current state of our society nowadays and educate netizens more about hate speech. Based on The Federal Constitution in Article 10, it declares that Malaysians have the right to free expression, but the standard is not absolute. Therefore, this research can make it easier for the government to rebuild or revise an act or article according to hate speech, as it is becoming one of the most harmful things on the Internet after fake news. On the other hand, this research can also contribute a lot to our ministry to conduct an awareness campaign among the netizens to be more polite online. The impactful-ness of the pre-campaign can be measured by this intention of hate speech in the first place.

3. Literature Review

3.1. *Hate Speech*

"Hate speech," according to [19], is communication or messages that insult an individual or group based on religion, race, caste, sexual orientation, gender, country of origin, or other characteristics. In accordance with [7], malicious language, hate speech or flaming comments is used to identify messages that violate existing legal norms and require government regulation like a publicly shared statement that incites, promotes, or justifies hatred, discrimination, or hostility towards a particular group such as race or ethnicity, beliefs or faith, abilities or disabilities, age, sexual orientation, and gender identity.

Aside from that, Kilvington [8] stated that hate speech is disgusting because it spreads, inflames, and stimulates hatred, violence, and discrimination against individuals or groups due to their protected status; these include "race", ethnicity, religion, gender, sexual orientation, and disability, among other social distinctions. According to the Oxford English dictionary, the term hatred means someone has a strong disgust or dislike, evoking a disliked temperament [2]. Malicious expression or hate speech is a special form of offensive language, where the individual using it is expressing their opinion based on either race, radical background, or stereotype [13].

Thus, by these definitions in previous studies, we can conclude that hate speech is a type of vulgar, obscene, abusive, and abusive language, often expressing hatred and including blasphemous language. Hate messages posted online can harm victims by developing post-traumatic-stress-disorder such as psychological and pathophysiological symptoms (PTSD)—panic, fear, anxiety, nightmares, threatening intrusive thoughts, contempt [6]. Before this, a lot of researchers focused on how to recognize, detect, and analyze hate speech on social media, especially in the data mining field or how hate speech affected victims in psychology. To differentiate this study and the others, researchers want to investigate more on intentions and motives behind hate speech. The researchers believe that all sentiments and phenomena in previous research are so helpful as they can be proof of the need to do the study.

3.2. *Types of Hate Speech*

In the untrained eye, hate speech consists of two very common terms: hatred and speech; it refers to emotions and words as well as meanings that can evoke emotions to someone who hates a particular object; it is also usually defined in a broader sense and includes insults, discrimination, dehumanization, demonization, and incitement to violence. According to [21], there are eight types of hate speech as follows: race, behavior, body, class, gender, ethnicity, disability, and religion. The motive of hate speech in [22] study, determine hate speech by Nnamdi Kanu, the leader of the independent people of Biafra; they consider Nnamdi Kanu's speech has five intents or motives which are declaration, commitment, judgment, direction, and assertion. Most of the intent of Nnamdi Kanu's speech is assertive.

In sum, hatred messages or hate speech can be everything or anything that gives negative feedback to other users. Insults or offensive language can be "innocently" negative or sarcasm. In Malaysian cultures, words such as "babi" or "anjing" are so offensive but not "pig" or "dog" in American society. Other than that, calling other individuals using some tribe names such as "batak", "sakai" and "keling" can be offensive and negatively affect other users.

3.3. *Intention*

Intention is the mental state in which an agent act. One example is the plan to go to the museum tomorrow. An action plan is the content of an intention, and a commitment is an attitude toward that content. Other mental states may be accompanied by the following action plans. As there is no practical obligation to carry out a plan, one can still unintentionally make it. However, only successful intent leads to the intended course of action, whereas unsuccessful intent does not.

Matczak M [23] discovered three types of intention. The three types of intent correspond to the three types of speech: locutionary, illocutionary, and perlocutionary. The initial positional intent is nearly identical to the semantic intent. The intention, for example, is to say a word with a specific meaning and reference. The intent to take specific actions by uttering these words, such as giving orders or making promises, is referred to as spoken intent. Perlocutionary intent is to alter reality by speaking words that influence the behavior of others. The dominant approach in legal theory holds that the desire to legislate is primarily a semantic (linguistic) desire.

3.4. *Hate Speech Crime World and Malaysia*

Other countries, such as the United States of America [14], reported that in the year following 9/11, 481 hate crimes with anti-Islamic motivations were committed, with 58% of them occurring within two weeks of the incident (4 percent of the risk period). Such evidence indicates that unfavorably motivated crimes frequently occur in close proximity to electric shock events such as terrorist attacks. Other than that, [24] currently in India from 26 May 2022, an inclusive, incredible and colossal India has been in the news for all the worst and wrong reasons in which two apparatchiks of the country's governing party, the Bharatiya Janata Party, or BJP, had made inexcusable incendiary and insulting remarks that have offended more than 10% of the country's minority and the adherents of one of the world's largest practised religious faiths, Islam, by committing an insult towards Prophet Muhammad (SAW).

According to [25] on 25 March, a video went viral as a Chinese woman falsely accused Malacca Hospital of not treating patients, according to an article in The Center. In addition, a series of online attacks and malicious comments about the woman's ethnicity, such as claiming she left her brain in Wuhan, hoarding, and violations of the Movement Control Order (MCO) have occurred. Between 27 February and 25 March, a total of 706 negative references to Malays/Muslims were registered in the Tweet. "Malas" was used most often, followed by "bodoh". The pinnacle of anti-Malay Muslim mention was on 12 March, when the first few COVID-19 cases were reported by the Tabligh Cluster. Except for anti-Chinese tweets, 506 negative references were recorded during the same period. The highlight of such coverage occurred on March 20, the day after a Chinese cardiologist was discovered jogging in the park against the MCO. The case caused racism on Twitter after his video, discussing with law enforcement officials, became viral. In comparison to other ethnic groups, there were 156 repugnant mentions in anti-Indian tweets during the same time. The highest peak took place on March 22nd, the day a mechanic of Indian descent was charged in court for public intoxication and refusing to cooperate with the police.

There is no specific law in Malaysia that regulates or criminalizes hate speech, especially against ethnicities, religious groups, or skin color. Malaysians have the right to free expression under Article 10 of the Federal Constitution, but this standard is not absolute. As in other countries, some limitations apply to this freedom in terms of protecting minorities, people of different ethnicities, religions, sexual orientations, and others who suffer persecution and aggression.

4. Underpinning Theory

The study is conducted within various theoretical frameworks, one of them is based on the work of [26]. His theory of human engagement and behavior was based solely on an analytical understanding of face-to-face situations in which actors physically coexisted. [27] pioneering work is crucial in understanding how human behavior relates to communication. He was the first scholar to use the dramaturgy metaphor to help him understand himself in social interactions.

Goffman's theoretical model contributed to an understanding of why and how individual and group performance differ in social contexts, particularly when trying to express hate speech. According to [26], individuals acting in a specific social context make "moral demands" on the audience and encourage them to respond in the expected or appropriate manner.

Cultivation is a sociocultural theory of television's part in shaping viewers' notions, beliefs, attitudes, and values [28]. Back then, humans used television as their main tool to get information, news, or public service announcements from the government. Nowadays, when watching football games or concerts, people usually have their mobile phones, tablets, or laptops at arm's length or in the palm of their hands. Often, these devices are used to stream this content. People tend to believe, use ideas and some may follow the things they see on the social media. If hatred messages or hate speech are happening throughout their timeline in social media, it is not impossible if they believe what they watch and see.

According to [8] researchers should try to understand why hatred messages are uploaded online, rather than simply applying Goffman's model to virtual communication. Because the virtual frontstage and virtual backstage are blurred, Goffman's work necessitates extensive rethinking, and redesigning. As a result, researchers combine this model with Gerbner and Gross's cultivation theory from 1976.

In accordance with [29] cultivation is a sociocultural theory that includes three components: media institutions, message production, and message consequences on viewers; however, the effects on viewers are the most popular subject that has been researched. The cultivation effect, in its most basic form, is the correlation between the amount of time people spend watching television and their worldview. Traditionally, the more people who watch television, the more their worldviews reflect the dominant narrative messages transmitted by television. Consistent with this, the more people who see and read hate speech on social media, the more their worldview reflects the message of the story conveyed by social media. Cultivation theory believes that the potential for storytelling on social media is enormous.

Both models and theories come from different fields as The Dramaturgical Model of Social Interaction comes from psychology and the Cultivation theory is a classic theory from the communication field; it may seem to not be related to each other but in this study, hate speech is in the form of social interaction online and what media has been telling us. In online interaction, people can say what they want without any filter or gatekeeper, either way their own posts or tweets or their comments on some other individual posts. In this way, the whole world can actually react to each other either in a good or bad way; it is not an issue if good things spread all over the Internet, but we are concerned that bad things, such as hate speeches, will have a negative impact on other people who presume what they see is 100 percent true and real; they will always be impacted by what they see on social media, spreading pessimism without questioning whether it is true or not; this could be a viral disease or cancer for the virtual world. Figure 1 shows the relation between The Dramaturgical Model of Social Interaction [26] and the Cultivation theory [28].

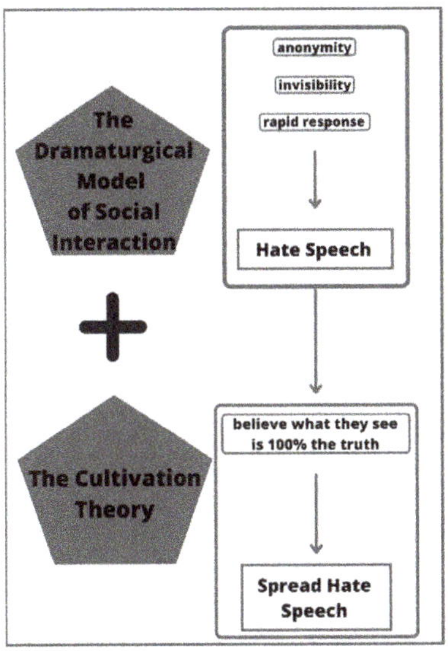

Figure 1. This figure explains how both theories are related to hate speech in the cyberworld.

5. Methodology

The proposed research aims to identify the intentions of hate speech on Twitter from Malaysians perspectives. Through this, the focus of this research is to know the structure of how hate speech affects Malaysians' Twitter lives. The proposed method for this study is to use an interview, which is the qualitative method. Since we want to reach the root of hate speech intent, qualitative research methods are the most appropriate because they are designed to help clarify the behavior and perceptions of the target audience associated with a particular problem [30]. The researcher will interview participants as it is a purely conversational method and encourages participants to receive detailed information [31]. In terms of experiencing using social media, the participants will share how they watch or read hate speech and handle it, and mainly to know the factors that contribute to the occurrence of hate speech.

As we know in the endemic era, we are advised to lessen human-to-human interaction as it can increase the spread of COVID-19. Our target is Malaysians that use social media such as Twitter so all our informants should know how to interact online. We will use video conference platforms such as Zoom Meeting applications or Google Meet applications; it should be easier for everyone, including researchers, to get to interact with everyone without the need to travel around Malaysia.

The Zoom Meeting application will allow you to communicate, chat and discuss things online with other people all around the world. The researchers can connect to those informants that are far from us and it makes it easier in this endemic era; they will also encourage the participants to open their cameras and microphone to make sure that the opinions, as well as facial expressions that the participants want to express, can be recognized in this focus group discussion.

As the target is wide, the kind of informants that should be interviewed or joined the focus group discussion need to be decided. After some meetings with other researchers, it has been decided that the informants, who use Twitter, should come from various backgrounds, such as different races, ethnic groups and academic backgrounds.

This study aims to enhance the validity and reliability of this research paradigm. According to [32], validity and reliability are important because it will help to improve the accuracy of a research work's evaluation and assessment. One of them is to try to make our informants understand why the instruments were used and what our research goals are; this can cause informants to respond to questions in their best minds, increasing the number of ideas that will be used in research findings. Other than that, to make sure our informants give their best opinions, we need to provide some examples or proof of hate speeches that arise on social media, such as Twitter. By this strategy, informants can put their shoes on victims of hate speech as it will increase their empathy.

6. Conclusions

Previous studies paid more attention to how to track and trace hate speech or detect hatred messages in social media. For example, Paschalides D et al. [33] proposed an integrated architecture for detecting online hate speech using multilingual datasets in English, Italian, and German. Corazza M et al. [34] developed a three deep learning method that monitors, detects, and visualizes the occurrence of hate speech using Twitter messages. Essentially, there is little research or article that investigate the motive or intention of hatred message in the cyberworld. In addition, according to recent statistics, Malaysia comprises approximately 20.1 million internet users and 16.8 million social media accounts [3] and cyberbullying was brought to attention when an opinion poll of 28 countries discovered that Malaysia was sixth in the world's cyberbullying rankings, trailing only India among Asian countries [35]; this is very much concerning because it can exacerbate other users and victims of hateful words psychologically and develop post-traumatic stress disorder-like psychological and pathophysiological symptoms (PTSD). Negative consequences include intrusive thoughts of distress, fear, anxiety, nightmares, threats, and injuries [6]. The researchers believe that understanding such extreme negative effects will have both theoretical and practical implications into real-life social interaction.

Malicious expression refers to public statements that express hatred or incite violence towards race, ethnicity, faith, gender or sexual orientation. Online hate diffusion and culture have now become serious problems, prompting several international initiatives aimed at defining the issue and creating effective countermeasures and policies.

Because the goal of this study is to examine the intent of hate speech, researchers believe that human interaction should be studied. A combination of motivation and intention research, the functionalist theory of psychology, the dramaturgical model of social interaction, Erving Goffman [26] and the cultivation theory [28] are used in this research to meet the objective; it is hoped that this study can provide incremental contributions to the hate speech research and provide motives and intentions for other cyberworld users.

Author Contributions: Conceptualization, N.F.A. and N.A.K.Z.; methodology, N.F.A. and N.A.K.Z.; validation N.F.A. and N.A.K.Z.; formal analysis N.F.A. and N.A.K.Z.; investigation, N.F.A. and N.A.K.Z.; resources, N.F.A. and N.A.K.Z.; data curation, N.F.A. and N.A.K.Z.; writing—original draft preparation, N.F.A. and N.A.K.Z.; writing—review and editing, N.F.A. and N.A.K.Z.; visualization, N.F.A. and N.A.K.Z.; supervision, N.F.A. and N.A.K.Z.; project administration, N.F.A. and N.A.K.Z.; funding acquisition, N.F.A. and N.A.K.Z. All authors have read and agreed to the published version of the manuscript.

Funding: This research received funds under Fundamental Research Grant Scheme (600-RMC/FRGS 5/3 (041/2021)).

Institutional Review Board Statement: Not applicable.

Informed Consent Statement: Not applicable.

Data Availability Statement: Not applicable.

Acknowledgments: I would like to express my sincere gratitude to everyone who gave us the opportunity to complete this report. I would also like to express our sincere gratitude to the seniors who cooperated in assembling the parts and made proposals for this research. Finally, many thanks go to the Ministry of Higher Education (FRGS/1/2021/SS0/UITM/02/15) and Universiti Teknologi MARA (UiTM) who have invested funding in order to achieve the goal.

Conflicts of Interest: The authors declare no conflict of interest.

References

1. Twitter | History, Description, Elon Musk, and Uses. Available online: https://www.britannica.com/topic/Twitter (accessed on 8 July 2022).
2. Jay, A. 85 Twitter Statistics You Must Know: 2021/2022 Market Share Analysis & Data. Available online: https://financesonline.com/twitter-statistics/ (accessed on 14 January 2022).
3. Kemp, S. Digital 2022: Malaysia. DataReportal—Global Digital Insights. Available online: https://datareportal.com/reports/digital-2022-malaysia (accessed on 15 February 2022).
4. Pattersson, K. The discursive denial of racism by Finnish populist radical right politicians accused of anti-Muslim hate-speech. In *Nostalgia and Hope: Intersections between Politics of Culture, Welfare, and Migration in Europe*; Springer: Cham, Switzerland, 2020; pp. 35–50.
5. Bennett, W.L.; Livingston, S. The disinformation order: Disruptive communication and the decline of democratic institutions. *Eur. J. Commun.* **2018**, *33*, 122–139. [CrossRef]
6. Phoey, L.T.; Chi-Bin, C. Profanity and Hate Speech Detection. *Int. J. Inf. Manag. Sci.* **2020**, *31*, 227–246.
7. Brown, A. What is hate speech? Part 1: The myth of hate. *Law Philos* **2017**, *36*, 419–468. [CrossRef]
8. Kilvington, D. *The Virtual Stages of Hate: Using Goffman's Work to Conceptualize The Motivations For Online Hate*; Leeds Beckett University: Leeds, UK, 2020.
9. Davidson, T.; Warmsley, D.; Macy, M.; Weber, I. Automated Hate Speech Detection and the Problem of Offensive Language. *Proc. Int. AAAI Conf. Web Soc. Media* **2017**, *11*, 512–515.
10. Zick, A.; Krause, D.; Berghan, W.; Küpper, B. Gruppenbezogene Menschenfein-dlichkeit in Deutschland 2002–2016. In *Gespaltene Mitte—Feindselige Zustände: Rechtsextreme Einstellungen in Deutschland 2016*; Dietz: Bonn, Germany, 2016; pp. 33–81.
11. Thurston, N. Back to front truths. Hate library. In *Post-Digital Cultures of the Far Right: Online Actions and of INE Consequences in Europe and the US*; Fielitz, M., Thurston, N., Eds.; The Deutsche Nationalbibliothek: Bielefeld, Germany, 2019; pp. 193–204.
12. Watanabe, H.; Bouazizi, M.; Ohtsuki, T. Hate speech on twitter: A pragmatic approach to collect hateful and offensive expressions and perform hate speech detection. *IEEE Access* **2018**, *6*, 13825–13835. [CrossRef]
13. Oxford Constitutional Law. Max Planck Ensyclopedia of Comparative Constitutional Law. Available online: https://oxcon.ouplaw.com/view/10.1093/law-mpeccol/law-mpeccol-e130?prd=MPECCOL (accessed on 20 June 2022).
14. King, R.D.; Sutton, G.M. High Times for Hate Crime: Explaining the Temporal Clustering of Hate Motivated Offending. *Criminology* **2013**, *51*, 871–894. [CrossRef]
15. Murad, D. Caught in the Hate-Demic. *The Star*, 18 April 2021.
16. Lih, Y.B. Death of Transgender Woman in Malaysia Sparks Fears of Rising Hate Crime. U.S. Available online: https://www.reuters.com/article/malaysia-lgbt-crime-idUSL8N1YM12C (accessed on 20 June 2022).
17. Kiela, D.; Firooz, H.; Mohan, A.; Goswami, V.; Singh, A.; Ringshia, P.; Testuggine, D. The hateful memes challenge: Detecting hate speech in multimodal memes. *Adv. Neural Inf. Processing Syst.* **2020**, *33*, 2611–2624.
18. Malmasi, S.; Zampieri, M. Detecting hate speech in social media. In Proceedings of the International Conference Recent Advances in Natural Language Processing, Varna, Bulgaria, 2–8 September 2017; pp. 467–472.

19. Schmidt, A.; Wiegand, M. A survey on hate speech detection using natural language processing. In Proceedings of the Fifth International Workshop on Natural Language Processing for Social Media, Valencia, Spain, 3 April 2017; Association for Computational Linguistics: Valencia, Spain, 2017; pp. 1–10.
20. Azreen, H. MCMC Urged to Promote Awareness On Self-Regulation, Hate Speech Online. The Malaysian Reverse. Available online: https://themalaysianreserve.com/2019/08/19/mcmc-urged-to-promote-awareness-on-self-regulation-hate-speech-online/ (accessed on 20 June 2022).
21. Mondal, M.; Silva, L.A.; Benevenuto, F. A measurement study of hate speech in social media. In Proceedings of the 28th ACM Conference on Hypertext and Social Media, Prague, Czech Republic, 4–7 July 2017; pp. 85–94.
22. Alabi, T.O.; Ayeloja, A.K. Hate speech and security challenges: A pragmatic study of Nnamdi Kanu's speeches in the south-eastern Nigeria. *Int. J. Engl. Res.* **2019**, *5*, 1–9.
23. Matczak, M. Three Kinds of Intention in Law making. *Law Philos.* **2017**, *36*, 651–674. [CrossRef]
24. Santhananaban, M. Hate speech has no place anywhere. *The Sundaily*, 10 June 2022.
25. Jia, V.T.; Omar, N. Like a Virus: How Racial Hate Speech Looks Like in Malaysia During the COVID-19 Pandemic. *The Centre*, 2 April 2020.
26. Goffman, E. *The Presentation of Self in Everyday Life*; Penguin Books: London, UK, 1959.
27. Goffman, E. *Frame Analysis: An Essay on the Organization of Experience*; Harvard University Press: Harvard, MA, USA, 1974.
28. Gerbner, G.; Gross, L. Living with television; The violence profile. *J. Commun.* **1976**, *26*, 182–190. [CrossRef] [PubMed]
29. Shrum, L.J. Cultivation theory: Effects and underlying processes. In *The International Encyclopedia of Media Effects*; Roessle, P., Ed.; John Wiley and Sons, Inc.: Hoboken, NJ, USA, 2017.
30. Creswell, J.W. *Qualitative Inquiry and Research Design: Choosing among Five Approaches*, 3rd ed.; Sage: Thousand Oaks, CA, USA, 2013.
31. Mathers, N.; Fox, N.; Hunn, A. *Using Interviews in a Research Project*; Trent Focus Group: Sheffield, UK, 2002.
32. Mohajan, H.K. Two criteria for good measurements in research: Validity and reliability. *Ann. Spiru Haret Univ. Econ. Ser.* **2017**, *17*, 59–82. [CrossRef]
33. Paschalides, D.; Stephanidis, D.; Andreou, A.; Orphanou, K.; Pallis, G.; Dikaiakos, M.D.; Markatos, E. Mandola: A big-data processing and visualization platform for monitoring and detecting online hate speech. *ACM Trans. Internet Technol. (TOIT)* **2020**, *20*, 1–21. [CrossRef]
34. Corazza, M.; Menini, S.; Cabrio, E.; Tonelli, S.; Villata, S. A multilingual evaluation for online hate speech detection. *ACM Trans. Internet Technol. (TOIT)* **2020**, *20*, 1–22. [CrossRef]
35. Rosli, J. Malaysia Sixth-Worst in Global Cyber-Bullying Ladder, Survey Shows. Available online: https://www.malaymail.com/news/malaysia/2018/10/27/malaysia-sixth-worst-inglobal-cyber-bullying-ladder/1687181 (accessed on 15 June 2022).

proceedings MDPI

Proceeding Paper

Software Reuse Practices among Malaysian Freelance Developers: A Conceptual Framework †

Mohd Akmal Faiz Osman *, Mohamad Noorman Masrek and Khalid Abdul Wahid

Faculty of Information Management, UiTM Cawangan Selangor Kampus Puncak Perdana, Seksyen U10, Shah Alam 40150, Malaysia
* Correspondence: akmalfaiz@uitm.edu.my
† Presented at the International Academic Symposium of Social Science 2022, Kota Bharu, Malaysia, 3 July 2022.

Abstract: Software reuse development practices have been proven to benefit software development in terms of a quicker time to market and quality. Although being aware of the benefits, software developers tend to overlook this method due to various reasons. Impacted by the advancement of technology, the software development scenario has been dramatically altered as the popularity of software development organizations hiring third party developers acutely rises. Freelance developers enable software development organizations to obtain high expertise in development specific areas on a project-by-project basis that gives development cost flexibility. However, information regarding freelance developer development practice is found to be limited in the related literature. By applying the systematic literature review approach, this study's main objective is to develop a conceptual framework to investigate software reuse practices among Malaysian freelance developers. The framework consists of individual characteristics, project characteristics, technological characteristics, software reuse development practices, quality, and achievement goals as constructs. The framework would be a basis for future empirical studies.

Keywords: software reuse; development practice; freelance developer

Citation: Osman, M.A.F.; Masrek, M.N.; Wahid, K.A. Software Reuse Practices among Malaysian Freelance Developers: A Conceptual Framework. *Proceedings* **2022**, *82*, 30. https://doi.org/10.3390/proceedings2022082030

Academic Editor: Mohamad Rahimi Mohamad Rosman

Published: 13 September 2022

1. Introduction

Software reuse practices entails capitalizing on existing software and systems to create new products [1]. Instead of developing new software from scratch, developers use reusable software assets, such as components, information resources, source codes, conception, frameworks, and documentation to create new software systems [2]. The idea behind using reusable assets during the development of new software applications is to envisage an efficient way of development, rather than rewriting the modules that already exist. Software reuse practices can be applied in any of the software development phases such as requirement gathering, design, implement, testing, or maintenance. For instance, the development of software using reusable assets that have been tested rigorously by other developers will result in the testing phase to be shortened [3]. However, despite the benefits of adopting software reuse during software development, professional developers tend to overlook this method due to various reasons [4]. This has captured the attention of researchers, scholars, and the software development community [2,4] to investigate influencing factors of software reuse development practice. In general, among the prominent factors identified in literature are individual characteristics [2,4] and technological characteristics [5]. More thorough investigation is needed to support the informal view on the relationship by statistically presenting a thorough analysis of the relationship. Moreover, there are limited studies in the context of freelance software developers. Therefore, by applying the systematic literature review (SLR) approach by [6], the objective of this study is to propose a conceptual framework to investigate software reuse practices among Malaysian freelance software developers.

2. Freelance Software Developer

Freelance software developers are self-employed individuals who work on short term, project-based associations with hiring employers [7]. The term third-party developer explains the situation where software development organizations outsource software development tasks to a freelancer via freelancing websites or contacts. This is proven as [8] highlights the rapid increase in both the number of software development projects as well as tasks posted on online freelancing platforms, and this freelance industry is worth billions of dollars [9]. In 2018, it was reported that more than 9 million freelancers were registered in 110 freelance services websites, where the IT and programming section is one of the most important fields [9]. These freelance developers are available in various online marketplaces that provides flexibility for hiring organizations [10,11]. Many organizations have redefined their structures to incorporate freelance workers to development tasks [12]. Freelance developers possess highly professional skills in a specific area of software development, but there is no fixed monthly salary, insurance, pension, or any other benefits that are usually enjoyed by traditional developers [11]. The rapid advancement of IT technology has led to the software lifecycle being dramatically shortened where frequent updates were needed to cater to human software needs. This situation has prompted development organizations to find development alternatives by hiring these freelancers. To this end, freelance developers are required to develop the software of high quality more quickly.

3. Related Studies

Factors that influenced software reuse among 35 software developers were investigated by [5] and technological factors were found out to be the most significant predictors of software reuse, as reliability and availability of software assets are among the determinants. Similarly, findings by [13] found out that despite the awareness of software reuse benefits, developers tend to write new codes instead of reusing existing codes due to availability, accessibility and acceptability constraints of the reusable assets. In the research of [2] have presented similar findings with [5], in which technological factors and individual factors influenced intention to reuse software assets. In other research, software reuse practices have positive impacts on quality and efficiency [14]. In order to investigate the effect of software reuse development practice among freelance developer, [15] categorized two types of software reuse development practice, namely replicative usage and innovative usage. Although this study is still in its early stage, where the author only pilot tested the conceptual framework, this study is worth expanding into an empirical analysis as the study promotes innovative developer performance, which is crucial to freelance developers [7–10]. As shown in [2,5,13–15] studies, software reuse contributes to product quality and development quality, while individual characteristics and technological characteristics contribute to software reuse practice. However, the samples used from the studies was a traditional developer and not the freelance developer. Traditional developer works permanently in organization, in a team of other developers, and with considerable resources, such as funding, platform, and manpower [16]. Conversely, a freelance developer is self-employed and works on a project-by-project basis with limited resources [7].

4. Literature Review

4.1. *Quality*

Software that is developed using reused components results in better quality [1,17]. The reusable components used in the development have already been tested rigorously, with prior bug removal and defect corrections done. Findings from the SLR approach confirmed that product quality and development quality ranked highest in terms of the number of studies that have been published. To accommodate the quality of software development using freelancers, this study proposes two variables to measure the quality dimension, which are product quality and development quality.

4.1.1. Product Quality

Product quality is the extent of which a product or service meets or exceeds a client's expectations [15]. Software product quality includes having low defects, and being maintainable, portable, reliable, upgradable, and safe [14,15,17]. Direct relationship between the amount of reuse and level of software quality have been found in [17–19], while the quality of software is better when the software was developed using reusable components [20]. The consistency of the findings from the early 2000s until the 2020s have provided strong evidence of the effect of reuse towards software quality, and those freelance developers must be able to either replicate or innovate software reuse assets to develop quality software with efficiency [20].

4.1.2. Development Quality

Development quality is key performance indicator in software development [1,2,15]. Previous literature often pairs product quality and development quality as a measure of development success and quality [1,2,17,19,21]. Development quality captures the extent to which the software development process was managed with efficiency in terms of whether the software was completed on time and within budget [15]. The life span of software has been significantly shortened as human software needs are ever more complex, and more efficient development is needed. Therefore, it is assumed that software reuse practices would influence reduced software time to market and cost [1,2,17,19].

4.2. Achievement Goals

The difference between freelance developers and traditional developers are well noted in [11]. In terms of measuring freelance developer performance, achievement goals need to be included; these are defined as proficiency-relevant goals that an individual would make in relation to their competency [22]. Achievement goals are important in determining the career success and growth of a freelance developer, as they are self-employed individuals, compared to traditional developers who usually have a predetermined key performance indicator set by the attached organization. Hence, freelance developers would need to set their own achievement goals for career advancement and survival. Motivational factor of freelance developer was explored in [20]. The authors highlighted achievement goals that involves personal mastery and performance as significant predictors to motivation.

4.2.1. Satisfaction

Satisfaction is defined as a feeling of pleasure or displeasure that results from aggregating all the benefits that a developer hopes to receive from software reuse development practice [23]. Satisfaction sometimes is the best measure of success in terms of software development practices [24]. Satisfaction of software developers is closely related to motivation, where satisfied software developers were found to be more focused, committed, and hardworking [25]. Moreover, satisfaction of developers was found to influenced job productivity [26,27]. Instead of committing to repetitive tasks, reuse practices allow the expertise of freelance developers to be used more efficiently by focusing on developing higher quality software. Software reuse enables developers to complete pending projects quickly, sometimes ahead of schedule. With that said, therefore, satisfaction is critical to development practice success.

4.2.2. Personal Sense of Accomplishment

Personal sense of accomplishment is defined as the developer's feelings of self-esteem resulting from practicing software reuse [28]. In IS research, personal sense of accomplishment has been used to measure the success of certain practice effectiveness [29]. The performance of certain practices can only be considered successful if it gives a positive impact to the developer in enhancing their personal growth and productivity. Software reuse practices are considered to be development practices and their effectiveness can be measured through the developer's personal sense of accomplishment. A freelance developer's personal growth and productivity are of the utmost importance to their career, if not their survival.

4.3. *Software Reuse Development Practices*

In regard to the knowledge reuse theory, a conceptual framework has been developed by [15] to measure the impact of software reuse usage on software quality and development quality. According to [15], that software development jobs require a high level of innovation among developers to solve complex problems, and categorized the software reuse usage into two, namely replicative usage and innovative usage. Reusable assets, such as components, source codes, frameworks, and documentation, can be used accordingly with only slight modifications, or integrated and manipulated by combining them into novel components to develop a new software. In the freelance development world, with fierce competitors and an uncertain market, having the ability to replicate or innovate reusable software assets is considered a competitive advantage. The freelance developer could find the source and reusable assets and integrate them into a new software that satisfies the client's needs. Following and extending the study of [15], this study classifies software reuse practice into three forms: replicative usage, innovative usage, and extent of usage. Therefore, it is hypothesized that,

H1: *Software reuse development practices have a positive effect on Quality.*

H2: *Software reuse development practices have a positive effect on Achievement Goals.*

4.3.1. Replicative Usage

Replicative usage occurs when a developer reuses available modules to solve a common technical problem that involves a little adaptation or integration of the modules [15]. When there is a particular development problem that needs solving, the developer reuses previously created assets to solve the problem. As there is little adaptation or integration needed in this process, no novel results are generated. Replicative usage refers to best practices. This reusable module usually resides in the developer community with support from other developers in the form of development guides, templates, and tutorials.

4.3.2. Innovative Usage

Innovative usage occurs when the developer reuses available modules to explore a challenging problem that involves deliberate adaptation or integration of the modules [15]. Contrary to replicative usage, innovative usage refers to the reuse of software assets in novel ways such as combining and integrating existing software assets to develop a novel software asset that can be used again for a similar problem. Instead of using best practices that are available in the software community, the developer opts for the trial-and-error learning process, which leads to the creation of new assets. The developer starts with evaluating the reuse module to identify the intended function. By going through this process, developers are able to implement new functionalities into the software development process and attract more clients.

4.3.3. Extent of Use

Software reuse practice can be considered as a class of innovation [30]. Consequently, the diffusion of the innovation theory is adopted as a reference theory that implies to IS implementation and software reuse adoption. Extent of use captures the extent of which software reuse features and functionality are used in a complete and comprehensive manner in the development process [31]. Referring to [30], this study has modified the original extent of use of the CASE tool into extent of use of reusable software artefacts to suit the objective of the study. Previous studies have empirically shown that extent of use of software reuse has a positive impact on quality [14,17,31].

5. Individual Characteristics

Human factor was often found to contribute towards success or failure of software. Software developers differ from other developers in terms of their experience, expertise, and self-efficacy. From the literature, it is found that experience, expertise, and self-efficacy positively affect developer development practice [2,5].

5.1. Experience

Experience in the software engineering context can be defined as the extent of professional experience an individual possesses of being involved in systems and software development in years [32]. Experience enhances the capability and maturity of software developers, which is crucial in software reuse. Experience is listed as one of the variables to measure individual factor influence towards software reuse [5]. Similarly, experience was found to influenced software reuse practices adoption [4]. The accumulated experience of software development and reuse practice enables the developer to understand the value of the assets. Therefore, it is hypothesized that,

H3: *Experience positively influences software reuse development practice among freelance developers.*

5.2. Expertise

Expertise is defined as the degree of proficiency in a specific software development job. Recent studies have demonstrated that as organizational environments have evolved, work-related factors need to be revised through the perspective of professionals and the reality of modern software development companies [33], given, e.g., the emergence of developer expertise in front-end developers, back-end developers, full-stack developers, database architecture, and mobile apps developers. Domain knowledge and capability of developers are important in software reuse [5]. Moreover, expertise of a developer would determine innovative capabilities [34]. Therefore, it is hypothesized,

H4: *Expertise positively influences software reuse development practice.*

5.3. Self-Efficacy

In software development context, self-efficacy can be defined as individual self-judgment of their ability to use technology or innovation to accomplish a software development task [2]. Findings from [2] revealed that when developers are perceived to possess necessary skills to develop and reuse software assets, they are likely to practice software reuse. The practical reason of self-efficacy terms incorporated into software development is the assumption that an individual would most likely adopt a technology or innovation if they perceived themselves to have the ability to incorporate the practices in completing a task. Based on this assumption, it is hypothesized,

H5: *Self-efficacy positively influences software reuse development practice.*

6. Project Characteristics

Project characteristics concerns the characteristics of a software terms of the cost, duration and complexity [35]. Cost, duration, and complexity was found to influence the developer adoption of project management software to achieve higher productivity [35,36]. In the IS research context, project characteristics have a strong effect on system utilization [37]. Factors, such as increased number of modules in development, larger amount of information required, the needs of being able to track and reusing artefacts without developing a new one, will initiate the developer to embark upon software reuse practice.

6.1. Cost

Cost concerns with the amount of monetary budget allocated for development project [35]. Software development includes costs, such as purchase of hardware, license, components, templates, modules, knowledge, and training. While there is little known about the effect development cost on the adoption of software reuse, the cost of a software project could provide positive influence on the software reuse practice. When the development cost is high, developers thus need some innovation to reduce the cost. The total cost of the development project would encourage freelance developers to adopt software reuse practices. Therefore, it is hypothesized that,

H6: *Cost positively influences software reuse development practice.*

6.2. Duration

Duration concerns with the time consumed during development from start to market. Duration of software to be developed has influence the method used in the development [37]. Consider this example. Software project A took 12 months to complete that included 20,000 lines of code while software project B took 3 months to complete that consists of 5000 lines of code. Both projects very much differed in terms of duration to complete. It is interesting to see how freelance developers perceived software reuse regarding two different kinds of projects, as there is no information regarding this issue in the literature. Therefore, it is hypothesized that,

H7: *The duration of software projects influences software reuse development practice.*

6.3. Complexity

Complexity is the degree to which the software project is perceived difficult and challenging to complete [38]. Project complexity has been shown to determine the use of a variety of technologies, such as support systems [39]. From this proposition, it is possible that when freelance developers face a very complex task, they are more likely to perceive that software reuse will aid them to complete the task. Therefore, it is hypothesized that,

H8: *The complexity of software projects influences software reuse development practice.*

7. Technological Characteristics

In this study, technological characteristics pertain to technical characteristics of the reusable software assets. This construct is based on the diffusion of innovation theory by Rogers which posits the technological characteristic of innovation, such as compatibility, relative advantage, trialability, and result demonstration, that would influence the individual adoption of innovation. Studies have shown that these technological characteristics have influenced individuals to adopt specific innovation [31,40]. Although this theory proved to provide strong determinants in multiple contexts, this study would generalize and extend further into the context of freelance developers.

7.1. Compatibility

Compatibility is defined as the degree to which software reuse adoption is perceived as being consistent with the existing values, needs, and past experiences of the developer [40]. Freelance developers engaged in software development activities, such as software requirement gatherings, design specifications, source code writing, and software testing. Perhaps this kind of activity can be reused for new development [2,4]. For any developer, the compatibility and acceptance of development practice is important before they use it [5]. This study posits that,

H9: *Compatibility of reused assets influences software reuse development practices.*

7.2. Relative Advantage

Relative advantage is defined as the degree to which using software reuse is perceived as being better than developing from scratch [40]. Among important items in relative advantage is economic profitability, low initial cost, saved time, decrease in discomfort, and immediacy of rewards. Relative advantage has been found to be a consistent and strong predictor of individual adoption in various fields of IS studies such as software development practice [31,40]. However, in software reuse context, it is postulated by [31] that developers who believe that a new technology can enhance their task and job performance are more likely to use the new technology compared to those developers who do not hold this belief. Therefore, it is hypothesized that,

H10: *Relative advantage of reuse assets influences software reuse development practice.*

7.3. Trialability

Trialability is the degree to which an innovation may be experimented with on a limited basis [41]. Trialability plays a prominent role during the persuasion stage of the innovation process, as many reusable software artefacts usually are free to try and be experimented by developers. In the literature, trialability is often positively related to innovation adoption [31]. Trialability is an important trend towards adoption in the real world today, for instance, many software vendors usually provide users with a 30-day trial access. This study tends to widen the context to encompass freelance developers by positing,

H11: *Trialability of reuse assets influences software reuse development practice.*

7.4. Result Demonstration

Result demonstration is tangibility of the results of using innovation, consists of observability and communicability [41]. A clearly defined result from observation might result in adoption among developers. Freelance developers might find software reuse is convincing during development. Result demonstration proved to be an important determinant in IS adoption research [40]. Therefore,

H12: *Result demonstration influences software reuse development practices.*

8. Proposed Framework

Figure 1 shows the proposed framework that was developed based on previous relationships determined from previous literature. Experience of software developers was found to affect reuse development practice in the study of [2,4,5], where previous experience in systems development and adoption of software reuse would reduce the uncertainty of developers. Self-efficacy was found to influence reuse practices in [2,4], where the perceived competency of developers regarding reusable assets influenced their decision. Although project characteristics were not thoroughly investigated in software reuse literature, the relationship between software process improvement and adoption was discovered in the findings of [42]. Relationship between task complexity and knowledge reuse was discovered in findings of [39]. The same could be applied to software reuse practice as software reuse practice is considered as an improvement process using previous knowledge. Technological characteristics proved to be the strongest predictor of reuse practice, as showcased in the study of [30,31]. Software reuse practices were found to be strongly influenced both quality of developed software and development quality [1,2,14,17,19–21]. Satisfaction and a personal sense of accomplishment have been found to correlate with other development practices in the context of software development [26,27].

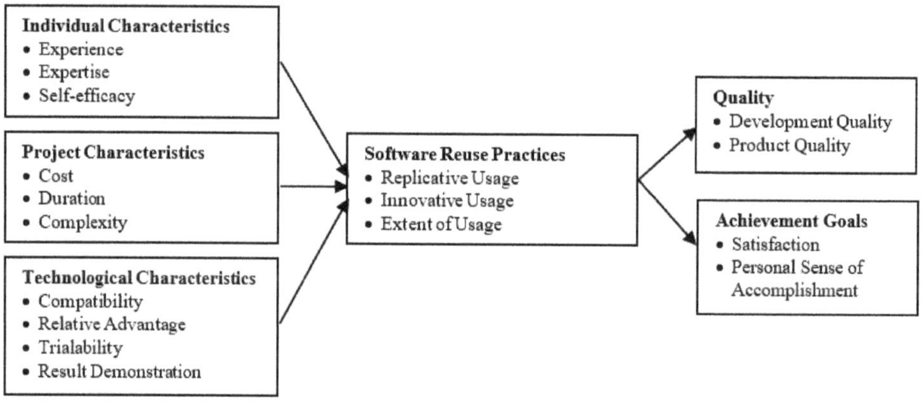

Figure 1. The proposed framework.

9. Discussion and Conclusions

This study proposed a conceptual framework to investigate the software reuse development practice among Malaysian freelance software developers, as there is scarce research in the literature discussing this topic. As the software crisis is well known, software development organizations have redefined their strategies to incorporate freelance developers to remedy the organization's lack of expertise in software development. However, comprehensive studies regarding freelance developer development practice are limited. Therefore, identifying important variables for further study would minimize this limitation. Individual characteristics of developers are varied. Some developers could be ten times more skilled than others. Software development projects differ from other projects, as some projects consist of 1000 lines of codes while other projects consist of 20,000 lines of codes. Information regarding development projects of freelance developers in the literature is unknown. By discovering the project characteristics of freelance developers, projects can then be profiled, determining the areas for improvement. Previous literature shows that many developers opt to develop software from scratch rather than use reusable assets due to their being incomplete, outdated, and lacking documentation. By investigating the technological characteristics of reusable assets used by freelancers, the templates, modules, functions, source codes, and documentation of reusable assets could perhaps be improved to encourage the increased usage of these assets.

Author Contributions: Conceptualization, M.N.M. and M.A.F.O.; validation, M.N.M.; formal analysis, M.A.F.O.; investigation, M.A.F.O.; resources, M.A.F.O.; data curation, M.A.F.O.; writing—original draft preparation, M.A.F.O.; writing—review and editing, M.N.M. and K.A.W.; visualization, M.A.F.O.; supervision, M.N.M. and K.A.W.; project administration, M.A.F.O. All authors have read and agreed to the published version of the manuscript.

Funding: This research has not received any external funding.

Institutional Review Board Statement: Not Applicable.

Informed Consent Statement: Not Applicable.

Data Availability Statement: Not Applicable.

Acknowledgments: This research was partially supported by Universiti Teknologi MARA Kelantan Branch, Malaysia.

Conflicts of Interest: The authors declare no conflict of interest.

References

1. Barros, J.L.; Pinciroli, F.; Matalonga, S.; Martínez-Araujo, S. What software reuse benefits have been transferred to the industry? A systematic mapping studies. *Inf. Softw. Technol.* **2018**, *103*, 1–21. [CrossRef]
2. Mellarkod, V.; Appan, R.; Jones, D.R.; Sherif, K. A multi-level analysis of factors affecting software developers' intention to reuse software assets: An empirical investigation. *Inf. Manag.* **2007**, *44*, 613–625. [CrossRef]
3. Mohagheghi, P.; Conradi, R. An empirical study of software reuse defect-density. In Proceedings of the 26th International Conference on Software Engineering, Edinburgh, UK, 23–28 May 2004; pp. 282–291. [CrossRef]
4. Bakar, N.H.; Kasirun, Z.M. Exploring Practitioners Requirements Reuse Empirical Study. *Int. J. Softw. Eng. Technol.* **2014**, *1*, 33–42.
5. Tung, Y.H.; Chuang, C.J.; Shan, H.L. A framework of code reuse in open source software. In Proceedings of the 16th Asia-Pacific Networking Operating Management Symposium, Hsinchu, Taiwan, 17–19 September 2014.
6. Webster, J.; Watson, R.T. Analysing the past for prepare the future: Writing a review. *MIS Q.* **2002**, *26*, xiii–xxiii.
7. Gupta, V.; Fernandez-Crehuet, J.M.; Hanne, T. Freelancers in the development process: A Systematic Mapping Study. *Processes* **2020**, *8*, 1215. [CrossRef]
8. Sison, R.; Lavilles, R. Grounded theory of online development freelancing. In Proceedings of the International Conference on Information System, San Francisco, CA, USA, 13–16 December 2018; p. 26. Available online: https://aisel.aisnet.org/icis2018/behavior/Presentations/26 (accessed on 20 May 2022).
9. Haq, N.U.; Raja, A.A.; Nosheen, S.; Sajjad, M.F. Determinants of client satisfaction in web development projects from freelance marketplaces. *Int. J. Manag. Proj. Bus.* **2018**, *11*, 583–607. [CrossRef]
10. Mullins, N. Career-Related Attitudes, Competencies of Freelance Workers. Master's Thesis, University of Pretoria, Pretoria, South Africa, 2019.
11. Kazi, A.G.; Yusoff, R.; Khan, A.; Kazi, S. Freelancer: A Conceptual Review. *Sains Hum.* **2014**, *2*, 3. [CrossRef]
12. Hsieh, J.K.; Hsieh, Y.C. Internet-based freelance developers in app marketplaces. *Int. J. Inf. Manag.* **2013**, 33, 308–317. [CrossRef]
13. Agresti, W.W. Software Reuse: Developers' Experiences and Perceptions. *J. Softw. Eng. Appl.* **2011**, *4*, 48–58. [CrossRef]
14. Ha, W.; Sun, H.; Xie, M. Reuse of embedded software in small and medium enterprises. In Proceedings of the 2012 IEEE International Conference on Management of Innovation & Technology (ICMIT), Bali, Indonesia, 11–13 June 2012; pp. 394–399.
15. Zhou, J. Application developer's innovation performance on mobile platforms-Investigating the effect of module reuse. In Proceedings of the 24th Pacific Asia Conference of Information System, Dubai, United Arab Emirates, 22–24 June 2020.
16. Sherif, K.; Vinze, A. Barriers to adoption of software reuse A qualitative study. *Inf. Manag.* **2003**, *41*, 159–175. [CrossRef]
17. Deniz, B.; Bilgen, S. Empirical study of software reuse and quality. *Lect. Notes Comput. Sci.* **2014**, *8583*, 508–523.
18. Frakes, W.B.; Succi, G. An industrial study of reuse, quality, and productivity. *J. Syst. Softw.* **2001**, *57*, 99–106. [CrossRef]
19. Rine, D.C.; Nada, N. An empirical study of a software reuse reference model. *Inf. Softw. Technol.* **2000**, *42*, 47–65. [CrossRef]
20. Slyngstad, P.; Gupta, A.; Landre, E. An empirical study on software reuse. *Int. Conf. Comput. Sci. Softw. Eng.* **2008**, *6*, 509–512.
21. Bauer, V.; Vetro, A. Comparing reuse practices in two large software-producing companies. *J. Syst. Softw.* **2016**, *117*, 545–582. [CrossRef]
22. Wulandari, A.; Qamara, T.; Bawazir, M. ELancing Motivation on Sribulancer. *J. World Conf.* **2019**, *1*, 193–202.
23. Masrek, M.N.; Jamaludin, A.; Mukhtar, S.A. Evaluating academic library portal effectiveness. *Libr. Rev.* **2010**, *59*, 198–212. [CrossRef]
24. Green, G.C.; Hevner, A.R. Successful doi: Guidance for software development organizations. *IEEE Softw* **2000**, *17*, 96–103.
25. Franca, C.; Da Silva, F.; Sharp, H. Motivation and Satisfaction of Software Engineers. *IEEE Trans. Softw. Eng.* **2020**, *46*, 118–140. [CrossRef]
26. Mannaro, K.; Melis, M.; Marchesi, M. Empirical analysis on satisfaction of IT employees. *Lect. Notes Comput. Sci.* **2004**, 166–174. [CrossRef]
27. Graziotin, D. Happiness and the Productivity of Software Engineers. *Rethinking Prod. in Soft. Eng.* **2019**, 109–124. [CrossRef]
28. Samadi, I.; Masrek, M.N.; Yatin, S.F. The effect of individual characteristics and digital library characteristics on digital library effectiveness: A survey at university of Tehran. *World Appl. Sci. J.* **2014**, *30*, 214–220.
29. Masrek, M.N.; Karim, N.S.A.; Hussein, R. Measuring corp. *Intranet effectiveness. Info. Manag. Comput. Secur.* **2006**, 89–112. [CrossRef]
30. Iivari, J. Why are case tools not used? *Commun. ACM* **1996**, *39*, 94–103. [CrossRef]
31. Kishore, R.; McLean, E.R. Reconceptualizing innovation compatibility as organizational alignment in secondary IT adoption contexts: An investigation of software reuse infusion. *IEEE Trans. Eng. Manag.* **2007**, *54*, 756–775. [CrossRef]
32. Ahmad, M.A.; Ubaidullah, N.H.; Lakulu, M. Current Practices in Monitoring Software Development Process in Malaysia. *World Comput. Sci. Inf. Technol. J.* **2014**, *4*, 62–67.
33. Magalhaes, C.; Santos, R. The Role of Job Specialization in the Software Industry. In Proceedings of the International Conference on Information Technology & Systems, Libertad City, Ecuador, 4–6 February 2021.
34. Fulk, H.; Nagy, D. Expertise and Information Technologies: A Multidisciplinary Review. In Proceedings of the Twenty-Sixth Americas Conference on Information Systems, Salt Lake City, UT, USA, 15–17 August 2020.
35. Pellerin, R.; Perrier, N.; Guillot, X.; Léger, P.M. Project characteristics, project management software utilization and project performance: An impact analysis based on real project data. *Int. J. Inf. Syst. Proj. Manag.* **2013**, *1*, 5–26. [CrossRef]

36. Ottow, J. The Individual Adoption of New IT System within Organizations. Master's Thesis, Tilburg University, Tilburg, The Netherlands, 2016.
37. Xu, J.; Quaddus, M. Examining a model of KM systems adoption and diffusion: A PLS approach. *Knowl. Based Syst.* **2012**, *27*, 18–28. [CrossRef]
38. Morgeson, F.; Humphrey, S. Job design: Toward integrative conceptualization. In *Research in Personnel and Human Resources Management*; Emerald Group Publishing: Bingley, UK, 2008; Volume 27, pp. 39–91. [CrossRef]
39. Iyer, G.; Ravindran, S. Do task complexity and knowledge recency affect knowledge reuse? Implications for knowledge management efforts. In Proceedings of the 13th Americas Conference of Information Systems, Keystone, CO, USA, 10–12 August 2007; p. 493.
40. Kügler, M.; Smolnik, S.; Raeth, P. Why don't you use it? Assessing the determinants of enterprise software usage: A conceptual model integrating innovation diffusion and social capital theories. *Int. Conf. Inf. Syst.* **2012**, *5*, 3672.
41. Rogers, E.M. Innovation diffusion at the implementation stage of a construction project: A case study of information communication technology. *Constr. Manag. Econ.* **2006**, *24*, 321–332.
42. Mangalaraj, G.; Mahapatra, R.; Nerur, S. Acceptance of SPI: The case of extreme programming. *Eur. J. Inf. Syst.* **2009**, *18*, 344–354. [CrossRef]

proceedings

Proceeding Paper

Enhancing ESL Writing Instruction Using Headgram †

Norhaslinda Hassan *, **Norhafizah Abdul Halil**, **Maizatul Akmal Mohd Mohzan** and **Hanani Ahmad Zubir**

Academy of Language Studies, Universiti Teknologi MARA Cawangan Pulau Pinang, Bukit Mertajam 13500, Malaysia
* Correspondence: haslinda.hassan@uitm.edu.my
† Presented at the International Academic Symposium of Social Science 2022, Kota Bharu, Malaysia, 3 July 2022.

Abstract: Language instructors have long focused on improving students' competency. The greatest methods for involving students in the classroom are constantly evolving to meet the demands of the students. To accomplish the goal, new strategies and methods have been developed. As a result, some form of motivation is required to make sure that students are inspired and involved in the class so that it will be more meaningful. In light of this context, the researchers suggest the use of a board game called Headgram. Headgram is a board game developed by the researchers of the present study. A total of 62 students enrolled in the Integrated Language Skills III at Universiti Teknologi MARA, Cawangan Pulau Pinang were chosen as participants of the present study, and they played Headgram. This preliminary study is set out with the purpose of examining the students' perception of Headgram and to learn whether Headgram has helped the students perform better with their writing skills. In doing so, an action research study was conducted, and information was acquired using a combination of methods, including semi-structured individual face-to-face interviews and 15 questions on pre- and post-tests. The results of this study indicate that the students enjoy Headgram, though it is a challenging game, and Headgram effectively encourages a sense of enjoyment while enhancing students' writing abilities—specifically, sentence structure. Further, it was revealed that Headgram encourages students to work well as a team. Future studies might concentrate on the opinions of the teachers in order to have a balanced viewpoint, whereas the current study concentrated only on student responses using Headgram. Additionally, since technology is a cutting-edge tool, it would be amazing if Headgram became accessible online in the future.

Keywords: ESL; writing; board game; news headlines; UiTM

Citation: Hassan, N.; Halil, N.A.; Mohzan, M.A.M.; Zubir, H.A. Enhancing ESL Writing Instruction Using Headgram. *Proceedings* **2022**, *82*, 41. https://doi.org/10.3390/proceedings2022082041

Academic Editor: Mohamad Rahimi Mohamad Rosman

Published: 14 September 2022

1. Introduction

By concentrating on student learning outcomes, the Malaysian Education Blueprint 2015–2025 underlined the significance of English language proficiency among higher education students. It is thought to be important for educators to change the nature of their teaching and learning process, particularly in the context of ESL. ESL consists of four skills: speaking, reading, writing, and listening. Of the four skills, mastering writing is crucial for college students because it is a skill that is required in all occupations [1]. Employees must specifically possess strong writing abilities because the workplace necessitates documentation. According to Zhu, having strong writing abilities is crucial in the corporate sector because they are valued as a competence. Therefore, it is crucial for students to have strong writing skills in order to boost the employment rate for students in higher education. The ability to communicate effectively through writing is crucial in today's globalized environment.

The most difficult skill to learn and teach in an ESL classroom, according to [2], p. 470, is writing. Additionally, students should develop writing literacy as it is an essential ability, particularly at the higher education level [3]. Despite the range of writing strategies that teachers have created and used in their classrooms, a unifying goal is to ensure that students understand that they write in order to carry out specific intentional purposes. This has resulted in a difficult task for both students and teachers: teaching and learning writing.

Due to the difficulty of writing for both teachers and students, students eventually responded negatively. One of the factors contributing to students' lack of English language proficiency is their low motivation to learn the language. Notably, social interactions in the classroom are related to students' motivation for and disengagement from their English learning [4]. Social interactions allow teachers and students to communicate ideas in ESL classes, which may enhance student learning. In order to improve student learning, it is important to provide a joyful and relaxing classroom environment.

Reiterating this, writing has traditionally been seen as intimidating by both students and teachers, which has led to low motivation to improve writing skills, particularly among students pursuing higher education. According to [5], ESL writing courses need to make thoughtful use of effective instructional approaches that emphasize writing for meaningful goals, mastery of pertinent genres, and growing students' skill. Given this context, the authors of the current study believe that an intervention and creativity are required in the teaching of writing skills. In order to escape the passive learning process, the researchers of the present study created a board game.

Newspaper headlines from the local newspapers were included into the creation of the board game. It should be emphasized that news headlines can grab readers' attention by summarizing the crucial information [6]. Ref [7] also noted that news headlines use distinct grammar rules compared to standard English, which will help students master the different elements of speech and writing grammar. In addition to textbooks, there are a variety of other important and authentic teaching resources, and newspapers are one of them. Ref [2] asserts that reading newspapers improves language abilities as well as knowledge and adds to the fun and interaction of learning. Additionally, a blend of interactive and didactic aspects must be used in order to successfully teach [8]. In light of recent game aficionados, games are infused and included into the teaching and learning process to promote enjoyment and joy regardless of the nature: conventional or modern, offline or online. Furthermore, teaching languages is not an exception.

The board game, called Headgram, was created to test players' ability to accurately rearrange newspaper headlines and to promote cooperation among participants. Teamwork is becoming a key component of gameplay, according to [9]. Each player may have a more positive and joyful experience as a result of the group's active participation. They will learn from one another as they play the game, working together to defeat the other groups. The researchers are of the opinion that this board game could serve as an engaging and stimulating ESL classroom exercise while also helping students improve their writing and grammar.

The goal of the current study is to learn how students view Headgram as a tool to improve their writing abilities. The goals of this study are to find out whether Headgram has helped students enhance their writing skills and to emphasize students' perceptions of Headgram. There are two research questions that correspond to the objectives of the study:

(1) What is the students' perception of Headgram? and
(2) Does Headgram help improve the students' writing performance?

2. Methodology

Action research is used in the current study by means of a mixed-methods approach. Teachers frequently use action research to examine their instructional strategies or investigate an issue and its solutions. Additionally, because it follows a step-by-step approach, it is a methodical and organized research method. According to Johnson, an action research project should include the following five components: (1) defining the question, problem, and area of interest; (2) planning the data collection; (3) gathering and analyzing the data; (4) describing the potential uses and applications of the findings; and (5) reporting the results and formulating a plan for action with others. This study examines the efficacy of Headgram in teaching ESL writing as a preliminary action research study. The fact that action research projects range in complexity from straightforward and informal to intricate and formal is interesting to observe. This served as the catalyst for the researchers of the current study to conduct modest, informal action research as a starting point. The qualitative instrument is individual face-to-face semi-structured interviews, which are set to answer the first research question, i.e., the students' perception on Headgram. The quantitative instrument is used to address the second research question of the current study. Pre-test and post-test were the instruments utilized for the quantitative component. There are two distinct writing-related test items for both the pre-test and post-test.

2.1. Headgram

In order to teach sentence construction through newspaper headlines, the researchers of the current study developed a board game called Headgram (see Figure 1). The use of educational games in second language writing instruction is beneficial. Headgram is designed to encourage kids to learn grammar in an entertaining way by fusing words, headlines, and syntax. To ensure that real resources are used in teaching and learning, the headlines in this board game were drawn from headlines in a local newspaper. Headgram may be played in a group of four because it was designed to foster teamwork among pupils. Each of the three levels, basic, intermediate, and advanced, have five news headlines. Each set of words will be placed on a magnetic board, and each group will have one minute to arrange them. Teachers should explore the answers and how they differ from the original headline as a follow-up activity, as well as the significance of using the right parts of speech, because different word placements convey varied emphasis.

Figure 1. Headgram set.

2.2. Participants

Students taking Integrated Language Skills III at UiTM, Cawangan Pulau Pinang are the participants of the present study. There are 62 students from three distinct Integrated Language Skills III classrooms. It was necessary to obtain the students' consent prior to treating them as study participants, and only then was that consent granted. Five of the students agreed to participate in the semi-structured face-to-face interview. The students were coded using pseudonyms to guarantee their confidentiality. As the researchers are the instructors of the three classes, they had a good view of what was going on during the exercise.

2.3. Data Collection and Data Analysis

To restate, this study employed both qualitative and quantitative approach. Face-to-face semi-structured interviews with students who have participated in Headgram are part of the qualitative approach. Questions on their experiences playing Headgram and the usefulness of the game in teaching grammar and writing skills are part of the interview protocol. Five students in total consented to having their responses videotaped. The videos were verbatim transcribed. Thematic analysis was utilized to analyze qualitative data using Nvivo12, and the results were displayed in diagrams for easy reference.

Pre- and post-tests were employed for the quantitative technique. Notably, pre- and post-tests are frequently employed in behavioral research, especially to gauge how an intervention or course of treatment has changed subjects [10]. The pre-test and post-test were used to evaluate the pupils' progress after playing Headgram. Prior to playing Headgram, the students took the pre-test, and the post-test was given to them after they played Headgram. A total of 15 questions were taken from newspaper headlines for the pre- and post-test. A Google Form was used to administer the pre- and post-tests. Students' marks for both the pre- and post-tests were totaled up and the mean score was computed. The mean scores of the pre- and post-tests for the respondents were compared, and the results were translated into percentages. The respondents' pre- and post-test mean scores were computed using Microsoft Excel 365. The mean scores were compared to determine whether any improvements had been made.

3. Findings and Discussion

The findings of this study are shown in accordance with the research questions. The first research question looks into how informants feel about Headgram. The second study question assesses if there have been any improvements by comparing the mean score on the respondents' pre- and post-tests.

3.1. *Informants' Perception on Headgram*

Regarding the students' perception, the five informants in the current study provided excellent feedback on Headgram. When asked about their perception playing Headgram, informants A and C stated that it was challenging, whereas informants D and E thought it was exciting. Figure 2 below displays this information:

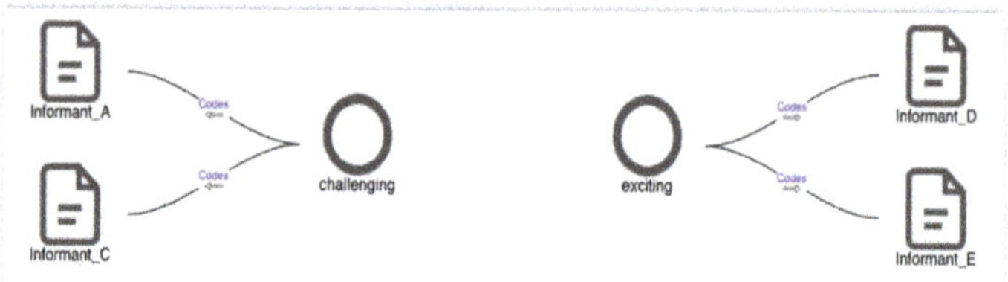

Figure 2. Informants' experience.

Headgram is challenging because it "tests...how to handle our sentences...like simple sentences to the expert one," according to informant A. Furthermore, informant C said that the game is challenging because "we need to finish the sentence in the time allotted". Informants D and E praised Headgram's excitement despite the game's difficulty. Informant D distinctly exclaimed, "Yes! The game is fun for me because it's so exciting". These findings support [11]'s observation that newspapers may not only increase knowledge, but also make learning fun and interactive. In order to make learning more dynamic and engaging, it is crucial to include some enjoyable and interactive features in classrooms. In terms of Headgram's usefulness, the informants stressed that it has assisted them in both grammatical and sentence construction improvement, as shown in the following Figure 3:

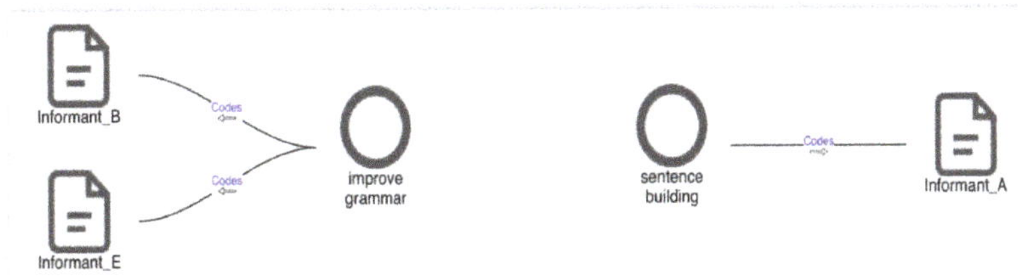

Figure 3. The effectiveness of Headgram.

Informant E added that "(this game) is wonderful and I wish we can play it every time since it can assist us in learning, advances, and so on,". Informant B underscored "I think this game improve our language." Informant A, however, made a point of underlining how Headgram improves sentence structure when she stated, "it improves our... how to organise our phrases... like simple sentences to the expert one." This implies that Headgram can improve pupils' writing skills, namely sentence structure and grammar. Notably, it has been demonstrated that using newspaper headlines to teach grammar and parts of speech to students is effective. It is also noteworthy that informant B said Headgram has, to some extent, encouraged the students to cooperate while playing. He claimed, "Headgram forces us to collaborate as a group." Teamwork is essential for learning in groups since peer learning also occurs, making it more than just a crucial component of a game.

3.2. Respondents' Pre- and Post-Test Mean Score Comparison

Figure 4 displays the mean score for the respondents' pre- and post-tests. Since the mean score before the test and the mean score after the test differ by 35%, it can be said that the respondents' mean scores have improved overall. Pre-test scores on average were 17.6%, and post-test scores on average were 52.6%. It is safe to say that Headgram has, in some manner, enhanced the responders' writing abilities.

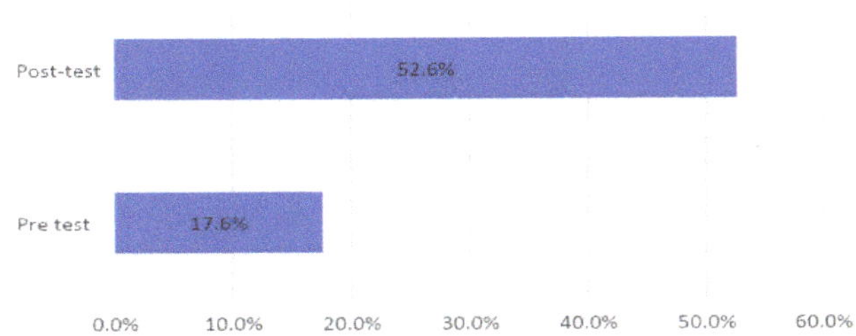

Figure 4. Respondents' mean score comparison.

4. Conclusions

It is important to highlight that the current study is a pilot project to evaluate the efficacy of utilizing Headgram in an ESL writing classroom. The preliminary feedback obtained shows that Headgram has effectively addressed the objectives described in the present study. This study also demonstrates how Headgram has improved students' writing skills, particularly in the construction of sentences and the application of grammar from authentic materials When compared to traditional grammar courses, which usually emphasize practice and rote memorization, Headgram provides students with an interesting way to understand how different sentence construction elements, from punctuation to word order, work. As a result, it gives students the opportunity to study, as well as comprehend, the delivered lesson in a style that may inspire them immensely.

Undoubtedly, the thrill of educational games, such as Headgram, instils in students a spirit of competition and challenge that could result in a more engaging and supportive learning environment. The researchers expect that academicians would recognize Headgram's potential for improving students' linguistic skills in an amusing way. The game components included in Headgram do raise the students' level of involvement while they participate in the activity. Additionally, Headgram is said to be able to foster extraordinary collaboration that encourages collaborative tasks in order to accomplish a shared objective [12].

Future research may focus on the instructors' input in order to have a balanced perspective, as the current study focused only on students utilizing Headgram's feedback. Additionally, it would be fantastic for Headgram to be accessible online in the future because technology is a modern tool.

Author Contributions: Conceptualization, N.H., N.A.H., M.A.M.M., and H.A.Z.; methodology, N.H.; software, H.A.Z.; validation, N.A.H.; formal analysis, M.A.M.M.; investigation, N.H., N.A.H., M.A.M.M., and H.A.Z.; resources, N.A.H. and H.A.Z.; data curation, M.A.M.M.; writing—original draft preparation, N.H., N.A.H., M.A.M.M., and H.A.Z.; writing—review and editing, N.H., N.A.H., M.A.M.M., and H.A.Z.; visualization, N.H; supervision, N.H; project administration, N.A.H. and M.A.M.M.; funding acquisition, H.A.Z. All authors have read and agreed to the published version of the manuscript.

Funding: This research received no external funding.

Institutional Review Board Statement: Not applicable.

Informed Consent Statement: Informed consent was obtained from all subjects involved in the study.

Data Availability Statement: Not applicable.

Acknowledgments: The researchers of the present study would like to thank the participants of this study and UiTM for granting us monetary sponsorship.

Conflicts of Interest: The authors declare no conflict of interest.

References

1. Selvaraj, M.; Aziz, A.A. Systematic review: Approaches in teaching writing skill in ESL classrooms. *Int. J. Acad. Res. Progress. Educ. Dev.* **2019**, *8*, 450–473. [CrossRef]
2. Jusun, K.D.; Yunus, M.M. The effectiveness of using sentence makers in improving writing performance among pupils in Lubok Antu rural schools. In Proceedings of the International Conference on Education (ICE2) 2018: Education and Innovation in Science in the Digital Era, Surabaya, Indonesia, 28 July 2018; pp. 469–475.
3. Jamil, M.; Hassan, N.; Abdul Halil, N.; Mohd Mohzan, M.A.; Ahmad Zubir, H. Headgram: Transcending grammar through news headlines. In Proceedings of the Innovation in Teaching and Learning Competition (INTELEC2020), Cawangan Perak, Malaysia, 29 October 2020; pp. 83–86.
4. Tanduklangi, A.; Lio, A. Classroom action research in teaching English for senior high school students through blended learning in Kendari of Indonesia. *J. E-Learn. Knowl. Soc.* **2019**, *15*. [CrossRef]
5. Warschauer, M. New tools for teaching writing. *Lang. Learn. Technol.* **2010**, *14*, 3–8.
6. Tuncer, H. Newspaper headlines in EFL classes: Development of reading and writing skills. *Read. Matrix Int. Online J.* **2019**, *19*, 58–76.
7. Williams, P. 8 Grammar Rules for Writing Newspaper Headlines. Available online: http://www.englishlessonsbrighton.co.uk/8-grammar-rules-writing-newspaper-headlines/ (accessed on 19 November 2013).
8. Taspinar, B.; Schmidt, W.; Schuhbauer, H. Gamification in education: A board game approach to knowledge acquisition. *Procedia Comput. Sci.* **2016**, *99*, 101–116. [CrossRef]
9. Harteveld, C.; Kleinman, E.; Rizzo, P.; Schouten, D.; Nguyen, T.H.; Liberty, S.; Kimbrough, W.; Fombelle, P.; El-Nasr, M.S. Teamwork and adaptation in games (TAG) a survey to gauge teamwork. In Proceedings of the 14th International Conference on the Foundations of Digital Games 2019, San Luis Obispo, CA, USA, 26–30 August 2019; pp. 1–12.
10. Dimitrov, D.M.; Rumrill, P.D., Jr. Pretest-posttest designs and measurement of change. *Work* **2003**, *20*, 159–165. [PubMed]

11. Tafida, A.G.; Dhalhatu, B.M. Using Newspapers in Teaching English as a Second Language. *J. Educ. Res. Rev.* **2014**, *2*, 61–65.
12. Talak-Kiryk, A. Using Games in a Foreign Language Classroom. Master's Thesis, SIT Graduate Institute, Brattleboro, VT, USA, 2010; p. 484.

proceedings

Proceeding Paper

The Impact of the Malaysia ICT Volunteers (MIV) Programme: The Volunteers' Perspective †

Mazni Omar [1,*], Huda Ibrahim [1,*], Mazida Ahmad [1,*], Hapini Awang [1], Azman Yasin [1] and Azizah Ahmad [2]

[1] Institute for Advanced and Smart Digital Opportunities (IASDO), School of Computing, Universiti Utara Malaysia, Sintok 06010, Kedah, Malaysia
[2] School of Computing and Informatics, Albukhary International University (AIU), Alor Setar 05200, Kedah, Malaysia
* Correspondence: mazni@uum.edu.my (M.O.); huda753@uum.edu.my (H.I.); mazida@uum.edu.my (M.A.)
† Presented at the International Academic Symposium of Social Science 2022, Kota Bharu, Malaysia, 3 July 2022.

Abstract: The Malaysia ICT Volunteers (MIV) is an ICT volunteer programme that aims to empower Malaysians towards becoming digital citizens of the Digital Smart Nation. As its effectiveness in attaining this goal has never been measured using rigorous scientific methods; the present study aimed to fill this gap. The aim of this study is to investigate the effectiveness of the MIV programme in terms of its impact on the MIV volunteers' behaviour. This study is guided by the Theory of Planned Behaviour (TPB), and a set of questionnaires was developed for the online survey. During the data collection period from August to October 2021, 329 valid cases from the MIV volunteers were gathered. Structural Equation Modelling (SEM) was used to analyse the date. Statistical results confirmed that Attitude towards MIV and Perceived Behavioural Control exert a significant influence on Intention to Volunteer, which in turn significantly influences the ICT Volunteering Behaviour of MIV volunteers. On the other hand, Subjective Norms do not play a significant role in the participants' attitudes toward the MIV programme. Another important finding that emerged from this study is that the MIV programme is effective in promoting ICT Volunteering Behaviour by exerting a positive influence on the participants' willingness and attitude by increasing their confidence in the MIV programme advantages. However, its benefits can be further enhanced by strengthening partnerships with schools, higher education institutions, private/government agencies, and non-governmental organisations (NGOs), as well as through more active social media utilisation for the promotion of ICT knowledge sharing and engagement.

Keywords: MIV volunteers; community; ICT; Theory of Planned Behaviour; volunteering behaviour; digital literacy

Citation: Omar, M.; Ibrahim, H.; Ahmad, M.; Awang, H.; Yasin, A.; Ahmad, A. The Impact of the Malaysia ICT Volunteers (MIV) Programme: The Volunteers' Perspective. *Proceedings* **2022**, *82*, 52. https://doi.org/10.3390/proceedings2022082052

Academic Editor: Mohamad Rahimi Mohamad Rosman

Published: 15 September 2022

1. Introduction

The Malaysian Communication and Multimedia Commission (MCMC) has taken many initiatives to raise awareness and educate the community. One of the initiatives is the Malaysia ICT Volunteer (MIV) programme that was initiated in 2016. The MIV aims to empower Malaysians to become digital citizens besides supporting the 'Smart Digital Nation' initiative. It focuses on four focus groups, namely MIV with community, MIV with school, MIV with Institutions of Higher Education (IHE), and MIV with international ICT volunteers (IIVs).

The objective of the MIV programme is to impart the positive use of ICT knowledge and encourage digital adoption among communities through MIV volunteers. In a nutshell, the MIV programme promotes and encourages digital and media literacy among citizens. Nevertheless, the effectiveness of the MIV programme in promoting positive behaviour use among the MIV volunteers has yet to be assessed. It is imperative to review and assess how effective it is in promoting ICT volunteering behaviour and, in turn, helping the community to improve their digital literacy awareness. Thus, this study aims to measure the effectiveness of the MIV programme in terms of its impact on the MIV volunteers' behaviour. This study was carried out based on the ICT Volunteering Behaviour model to gauge the effectiveness of the MIV programme based on the Theory of Planned Behaviour (TPB).

2. Literature Review

This section will review related works on the relation of TPB and volunteerism. According to the TPB, attitudes towards something can be operationalised using different labels depending on the context [1]. While Harrison [2] failed to establish a link between attitudes and intention to volunteer, Okun and Sloane [3] reported contrasting results. This view is supported by Maes, Leroy, and Sels [4], who argued that, according to the TPB postulates, entrepreneurial intention and behaviour depend on three factors, one of which is attitude. Similarly, Lee, Reisinger, Kim, and Yoon [5] and Chua, Meng, Ryu, and Han [6] found that positive attitudes towards volunteering enhance the likelihood of working as a volunteer even in the long term. This study therefore hypothesised:

Hypothesis 1 (H1). *Attitude Towards MIV has an influence on Intention to Volunteer.*

Subjective Norms influence the Intention to Volunteer via social pressure, as individual behaviours are often guided by the opinions of significant others [7]. According to Brayley et al. [8], Subjective Norms are essential for a person to decide whether to be a volunteer. Based on their study, one will continue to volunteer in the mentoring programme if they can feel the support of significant others towards them. Even regarding intention towards purchasing green food, it was found that Subjective Norms have a significant correlation with the intention to purchase [9]. It is because people tend to consider other people's perspectives and expectations on their behaviour and are concerned about how people will behave. Thus, when making a decision Subjective Norms have an influence on them. Thus, this study hypothesised:

Hypothesis 2 (H2). *Subjective Norms have an influence on Intention to Volunteer.*

According to the TPB, human behaviour is guided by behavioural, normative, and control beliefs, which will in turn give rise to perceived behavioural control [10]. Randle and Dolnicar [11] found that Perceived Behavioural Control was among the most important influential factors on volunteering intentions. France et al. [12] similarly noted that Perceived Behavioural Control was strongly related to one's intention to donate blood. Perceived Behavioural Control was identified as a reliable predictor of Perceived Behavioural Control by several authors, including Warburton and Terry [1], Lee, Won, and Bang [13], and Reuveni and Werner [14]. Thus, this study hypothesised:

Hypothesis 3 (H3). *Perceived Behavioural Control has an influence on Intention to Volunteer.*

ICT Volunteering Behaviour is a special form of prosocial behaviour geared toward helping others, often through non-profit organisations [15]. According to Ajzen [10], stronger functional beliefs related to volunteering will lead to a stronger Intention to Volunteer and thus the actual prosocial behaviour. In an earlier study, Ajzen [16] found that intention is a reliable predictor of behaviour and these findings were later confirmed by Hagger, Chatzisarantis, and Biddle [17], and Sheeran [18]. Teo and Lee [19] also found that pre-service teachers' intention to use technology was influenced by their behaviour toward computers, while Ayob, Sheau-Ting, Jalil, and Chin [20] established that intention leads to better waste management. Therefore, this study hypothesised:

Hypothesis 4 (H4). *Intention to Volunteer has an influence on ICT Volunteering Behaviour.*

Based on the reviews, the following Figure 1 illustrates the research model that comprises four predictors (Attitude Towards MIV, Subjective Norms, Perceived Behavioural Control, and Intention to Volunteer) and a dependent variable (ICT Volunteering Behaviour).

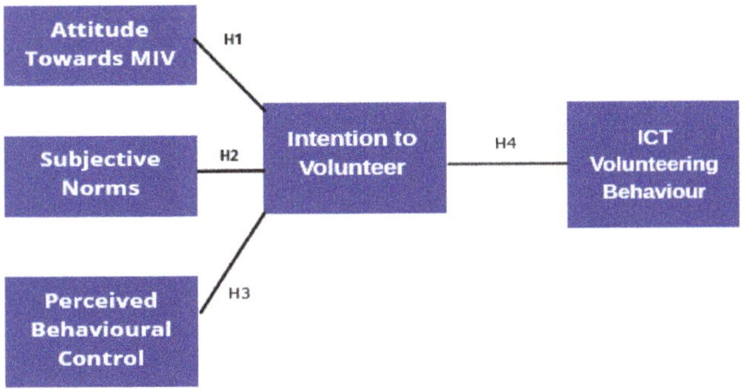

Figure 1. Research model.

In general, this proposed research model depicted in Figure 1 describes how the MIV programme would influence the ICT volunteering behaviour among MIV volunteer participants. It posits that three antecedent constructs (Attitude Towards MIV, Subjective Norms, and Perceived Behavioural Control) will positively impact the Intention to Volunteer. Finally, the increase in Intention to Volunteer should improve the ICT Volunteering Behaviour of the participants.

3. Methodology

This section describes the steps involved in achieving the study's objectives.

Figure 2 shows the research design used in this study. The quantitative approach was applied in this study based on the research objectives. To examine the relationships between variables, this study proposed four hypotheses as the research model. Thus, the use of such an approach is appropriate since quantitative studies usually investigate the relationships between variables and occasionally explain the causes of those relationships [21].

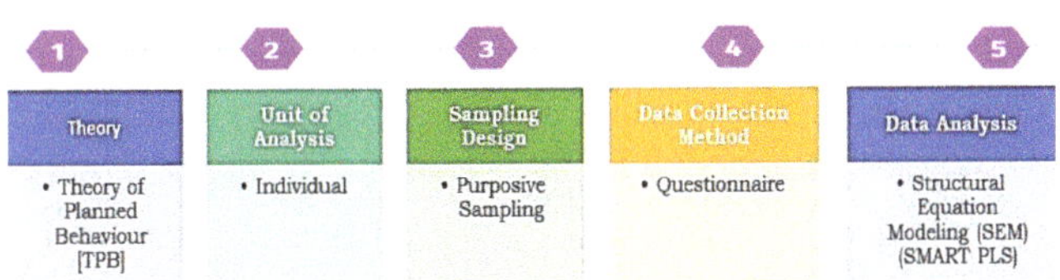

Figure 2. Research design.

A survey is considered the most appropriate method because of its accuracy in gathering information and enabling researchers to generalise findings, from a sample to a population [22]. It is also appropriate for studies with large sample sizes, as it is expedient, cost-effective, and administratively efficient [23]. Finally, a survey is also relevant when querying respondents about their perceptions, opinions, and feelings [24], which fits the nature of this study. The potential respondents were invited via email invitations, and an online survey is used to capture responses from respondents during the data collection period. This method allows this study to gain a faster and higher response rate while permitting real-time data monitoring [25].

3.1. Research Instrument

The outcome of the initial theoretical investigation was later used to develop the instruments to measure the effectiveness of the MIV programme. The instruments consist of a set of statements that were adapted from several sources and reformulated to suit the perspective of this study. Additionally, face validation, content validation, and pilot study were employed during the instrument development stage to produce a valid and reliable instrument. The instrument comprises two main parts: (i) Part A: Respondent Demographics; and (ii) Part B: MIV programme Evaluation.

To facilitate the understanding among respondents, the items are presented in two languages, Malay and English. The measurement scale for every construct is a seven-point numerical scale, which ranges from 1 to 7 ('1' extremely disagree to '7' extremely agree). The scale is applied because it provides a wider range and prevents respondents from selecting the neutral value. Thus, the chances of bias will be decreased [26].

3.2. Data Collection

Once a valid and reliable instrument was successfully produced, the study proceeded with the data collection procedures. The main data collection was undertaken from August to October 2021. The respondents were MIV volunteers that were selected from the list of MIV volunteers given by the MCMC. The MIV volunteers answered the online survey through the URL link provided: https://bit.ly/MIV-VOLUNTEER (accessed on 31 October 2021).

3.3. Data Analysis

The gathered data was analysed using statistical techniques to validate the research model that was developed based on the TPB model. Accordingly, the descriptive statistics and PLS-SEM were used to discover participants' change in behaviours prior to the MIV programme. This data analysis phase was completed in two stages. Both datasets underwent data cleaning and preparation procedures: (i) outliers, (ii) multicollinearity, and (iii) normality. Next, the main analysis began with two types of descriptive statistics for respondents and constructs of this study. Finally, the PLS-SEM analysis was performed in two stages: the Measurement Model and Structural Model. The research model based on the TPB Model was validated using PLS-SEM during the implementation phase. Finally, the results of the survey were reported and presented to the MCMC.

4. Results

This study aimed to investigate the behavioural changes triggered by MIV programmes among MIV volunteers. In order to make sense of the statistical findings produced in the prior stage, the interpretation and discussion were carefully and critically completed during the last stage of this study. This stage also highlighted the strengths and weaknesses of the current implementation of MIV programmes, along with suggestions for future improvements. These were completed based on empirical findings yielded from the descriptive and inferential statistical analyses. At the end of the data collection period, this study managed to gather 415 responses. During the cleaning procedures, 86 deletions were made, thus, producing 329 valid cases for the final PLS-SEM analysis.

This section presents a summary of demographic information of the respondents based on the returned questionnaires. The information obtained from the survey included the phone number, name, email, age, gender, race, highest education, and working status. From the survey, 329 respondents of Malaysia ICT Volunteers (MIV) aged between 18 years old and above 50 years old answered the questionnaires. Out of 329 respondents, 263 respondents are female volunteers (79.94%) and another 66 respondents are male volunteers (20.06%). From the data collected, the majority of the respondents were aged between 31 and 40 years old (150 respondents, 45.59%), 96 respondents were aged between 26 and 30 years old (29.18%), 64 respondents were aged between 18 and 25 years old (19.45%), 18 respondents were aged between 40 and 50 years old (5.47%), and 1 respondent was aged > 50 years old (0.30%).

Moreover, the respondents are of different races, where most of the respondents are Malay (231 individuals, 70.22%), while 97 respondents (29.48%) are of other races (Siam, Bajau, Banjar, Bidayuh, Bugis, Brunei, Cocos, Dusun, Iban, Iranun, Jawa, Kadazan,

Kadazan Dusun, Kenyah, Kimaragang, Kedayan, Kayan, Kagayan, Lunbawang, Murut, Rungus, Salako, Sino, Siam, Sungai, Suluk, and Tidung), and only 1 respondent (0.30%) is Chinese. Regarding highest education, 183 respondents (55.62%) have a degree, 79 respondents (24.01%) have a diploma, 52 respondents (15.81%) SPM/SPMV/STPM/Foundation, 12 respondents (3.65%) have a Master's degree, 1 respondent (0.30%) Sijil Kemahiran Malaysia, 1 (0.30%) SRP, and another 1 UPSR (0.30%). Lastly, regarding the working status, most of the volunteers work full-time in private sectors, which was 220 respondents (66.87%). Secondly, 26 volunteers are students (7.90%), followed by 25 volunteers (7.60%) working in a full-time government/agency position. Then, 14 respondents (4.26%) were self-employed, 12 (3.64%) housewives, 11 respondents (3.34%) were unemployed, 2 respondents (0.61%) were self-employed and housewives, 2 respondents (0.61%) were students and had full-time government/agency positions, and another 2 respondents (0.61%) were in part-time government employment, while the rest of the 15 respondents (4.56%) had another working status or double working status. Table 1 indicates a summary of the demographic information of the respondents.

Table 1. Summary of demographic information of the respondents.

Characteristics	Categories (%)	
Gender	**Male (20.1)**	**Female (79.9)**
Age	≤30 (65.6)	>30 (34.4)
Race	Malay (70)	Non-Malay (30)
Education	Graduate (79.6)	Non-Graduate (20.4)
Working Status	Working (85.1)	Unemployed (14.9)

Structural Equation Modelling

The measurement model was examined for construct reliability and validity. First, the construct reliability was assessed using Cronbach's Alpha (CA) and composite reliability (CR), with the recommended value of 0.700 [27]. The result shows that Perceived Behavioural Control has the lowest construct reliability (CA = 0.906, CR = 0.925) while the highest belongs to Intention to Volunteer (CA = 0.970, CR = 0.977). Thus, the results illustrate that the indicators used to present the construct have desirable internal consistency reliability. Second, the constructs' validity was evaluated through convergent and discriminant validity. In this study, the Average Variance Extracted (AVE) was tested to examine the convergent validity. The results indicate that the AVE values range between 0.640 and 0.894, which were all higher than the suggested value of 0.500. Given these results, the convergent validity is ascertained. This study employed the HTMT criterion for assessing the discriminant validity. The HTMT criterion is considered to be satisfied as all the values were below the accepted value of 0.900. Therefore, the discriminant validity was established.

After the CFA that was completed during the measurement model analysis, the structural model was examined. The conceptual model of this study consists of four hypotheses. The R2 values were 0.618 and 0.497 for Intention to Volunteer and the ICT Volunteering Behaviour, respectively. It is suggested that 61.8% of the variance in MIV Awareness can be explained by Attitude towards MIV, Subjective Norms, and Perceived Behavioural Control. Furthermore, 49.7% of the variance in Digital Literacy Behaviour can be explained by MIV Awareness. Table 2 summarises the assessment of four hypotheses in the structural model.

Table 2. Hypotheses in the structural model.

Hypothesis		Beta	*p* Value	T Value	Decision
H1	ATT → ITV	0.255	0.001 ***	4.409	Supported
H2	SN → ITV	0.029	0.709 NS	1.400	Returned
H3	PBC → ITV	0.574	0.000 ***	7.664	Supported
H4	ITV → VB	0.732	0.000 ***	17.924	Supported

Note: 1.65 ($p < 0.10$), 1.96 ($p < 0.05$), 2.58 (*** $p < 0.01$); NS = not significant; ATT = Attitude towards MIV; SN = Subjective Norms; PBC = Perceived Behavioural Control; ITV = Intention to Volunteer; VB = Volunteering Behaviour.

Hypothesis H1 postulated that ATT significantly influences the ITV among MIV volunteers. From the analysis, it was discovered that this relationship is significant (β = 0.255, t = 4.409, $p < 0.01$). Therefore, this finding supports hypothesis H1. However, the result shows that SN did not significantly affect ITV (β = 0.029, t = 1.400, $p = 0.709$), which returned the hypothesis H2. Meanwhile, hypothesis H3 posited that PBC would significantly influence the ITV. The analysis shows that this hypothesis is supported (β = 0.574, t = 7.664, $p < 0.01$), thus, supporting the proposed hypothesis. Similarly, hypothesis H4 which postulated a significant relationship between ITV and VB is also supported (β = 0.732, t = 17.924, $p < 0.01$) in this study. Figure 3 illustrates the bootstrapping analysis of the structural model for MIV volunteers.

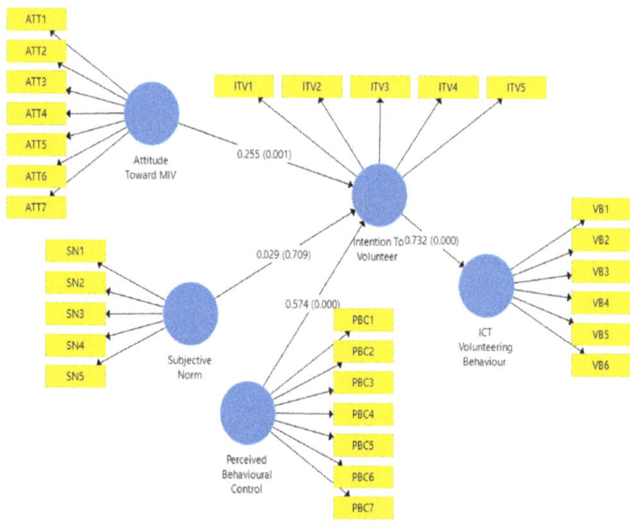

Figure 3. The structural model for MIV volunteers.

5. Discussion

The study findings indicate that the MIV volunteers are committed to volunteering in their community, school, MIV, and institutions of higher education because they are aware of the importance of promoting digital and media literacy, thus, supporting H1 and concurring with the results reported by Edelman [28]. ICT volunteering is increasingly recognised as the effective means of promoting digital literacy at the global level, as this allows the local talents, ideas, insights, and resources to be leveraged to enhance innovation and influence community attitudes toward digital literacy. However, as volunteers' subjective norms (i.e., opinions of friends and family and other influential individuals) were found not to affect their intention to volunteer, H2 is not supported. These results are also countered by the findings reported by other authors [29–31], possibly due to the differences in the research aims and the methodology adopted, or the participants' characteristics.

As noted by Knabe [32], the decision to volunteer in MIV programmes is guided by individual judgment. This is consistent with the current study finding that Perceived Behavioural Control Influences Intention to Volunteer, thus, supporting H3. Nonetheless, to sustain volunteers' motivation, they should be offered appropriate financial or non-financial awards [33], and their efforts should be recognised more explicitly, via appreciation letters from MCMC superiors or similar initiatives. In terms of the MIV programme's ability to impact the MIV volunteering behaviour, the analyses revealed that this influence was exerted via Intention to Volunteer, thus, supporting H4 and concurring with available evidence [34,35]. Hence, to ensure that volunteers remain engaged in this and other ICT initiatives in Malaysia, strategies for improving their willingness, effort, readiness, and confidence should be investigated.

6. Recommendations

There is a need to identify the ways to or factors that attract more volunteers to be involved in the MIV programme in terms of improved, better rewards, global recognition, and a proper ICT volunteerism policy. To sustain volunteer motivation, it is necessary to adopt a suitable award system, while continuing to improve the ICT programme quality by partnering with diverse stakeholders such as schools, higher institutions, private and public agencies, and non-governmental organisations (NGOs).

The MIV volunteers would also benefit from the adoption of a Digital Volunteer Champion, which would allow more experienced individuals to teach new recruits. They should also be encouraged to share their ideas with their peers and promote ICT to their family and friends while actively engaging in the MIV programme recruitment. For this purpose, social media platforms such as Facebook, Instagram, etc., can be used alongside other marketing strategies, including hashtag projects, online interactive quizzes, and online gamification. The MCMC should also consider customising the MIV digital literacy module to different age and aptitude levels, starting with children, teenagers, and the elderly, as this will increase interest in participation as well as the benefits derived from the programme.

7. Conclusions

As MIV programmes rely on volunteers, it is essential to ensure their continued commitment, which can be achieved by a better-designed reward and recognition system, as well as opportunities to network and collaborate with ICT volunteers in other parts of the world. Moreover, even though digital literacy is important for the social growth of communities and the economic prosperity of nations, the effectiveness of such ICT programmes has not been extensively studied. This gap in extant knowledge was addressed in the present study. The findings indicate that attitude towards MIV, subjective norms, perceived behaviour control, intention to volunteer, and ICT volunteering behaviour can potentially influence digital and media literacy at both individual and community levels. Thus, it is essential to develop new ways of promoting the MIV programme to the wider community to increase the number of potential participants as well as MIV volunteers. It is also important to forge relationships with relevant entities that can provide knowledge and resources for the MIV programmes.

Author Contributions: Conceptualisation, M.O. and H.I.; methodology, M.A. and H.A.; validation, A.Y. and A.A.; formal analysis, M.O. and H.I.; investigation, M.A and H.A.; resources, A.Y. and A.A; writing—original draft preparation, M.O., H.I., M.A., H.A., A.Y. and A.A.; writing—review and editing, M.O., H.I., M.A. and H.A.; project administration, M.O.; funding acquisition, M.O., H.I., M.A., H.A., A.Y. and A.A. All authors have read and agreed to the published version of the manuscript.

Funding: This work was supported by the Malaysian Communications and Multimedia Commission (MCMC) under the Digital Society Research Grant (DSRG) programme [S/O Code: 14972].

Institutional Review Board Statement: Not applicable.

Informed Consent Statement: Not applicable.

Data Availability Statement: Not applicable.

Acknowledgments: This research is funded by the Malaysian Communications and Multimedia Commission through the Digital Society Research Grant.

Conflicts of Interest: The authors declare no conflict of interest.

References

1. Warburton, J.; Terry, D.J. Volunteer decision making by older people: A test of a revised theory of planned behavior. *Basic Appl. Soc. Psychol.* **2000**, *22*, 245–257. [CrossRef]
2. Harrison, D.A. Volunteer motivation and attendance decisions: Competitive theory testing in multiple samples from a homeless shelter. *J. Appl. Psychol.* **1995**, *80*, 371–385. [CrossRef]
3. Okun, E.S.; Sloane, M.A. Application of planned behavior theory to predicting volunteer enrollment by college students in a campus-based program. *Soc. Behav. Personal. Int. J.* **2002**, *30*, 243–250. [CrossRef]
4. Maes, L.; Leroy, J.; Sels, H. Gender differences in entrepreneurial intentions: A TPB multi-group analysis at factor and indicator level. *Eur. Manag. J.* **2014**, *32*, 784–794. [CrossRef]
5. Lee, M.C.; Reisinger, K.; Kim, Y.; Yoon, M.J. The influence of volunteer motivation on satisfaction, attitudes, and support for a mega-event. *Int. J. Hosp. Manag.* **2014**, *40*, 37–48. [CrossRef]
6. Chua, H.; Meng, B.L.; Ryu, B.; Han, H.B. Participate in volunteer tourism again? Effect of volunteering value on temporal re-participation intention. *J. Hosp. Tour. Manag.* **2021**, *46*, 193–204. [CrossRef]
7. Rhodes, K.S.; Courneya, R.E. Investigating multiple components of attitude, subjective norm, and perceived control: An examination of the theory of planned behavior in the exercise domain. *Br. J. Soc. Psychol.* **2003**, *42*, 129–146. [CrossRef]
8. Brayley, N.M.N.; Obst, P.L.; White, K.M.; Lewis, I.M.; Warburton, J.; Spencer. Examining the predictive value of combining the theory of planned behaviour and the volunteer functions inventory. *Aust. J. Psychol.* **2015**, *67*, 149–156. [CrossRef]
9. Ham, A.F.; Jeger, M.; Ivković, M. The role of subjective norms in forming the intention to purchase green food. *Econ. Res. -Ekon. Istraživanja* **2015**, *28*, 738–748. [CrossRef]
10. Ajzen, I. Perceived behavioral control, self-efficacy, locus of control, and the theory of planned behavior. *J. Appl. Soc. Psychol.* **2002**, *32*, 665–683. [CrossRef]
11. Randle, S.; Dolnicar, M. Does cultural background affect volunteering behavior? *J. Nonprofit Public Sect. Mark.* **2009**, *21*, 225–247. [CrossRef]
12. France, H.J.; Kowalsky, L.; France, J.M.; McGlone, C.R.; Himawan, S.T.; Kessler, L.K.; Shaz, D.A. Development of common metrics for donation attitude, subjective norm, perceived behavioral control, and intention for the blood donation context. *Transfusion* **2014**, *54*, 839–847. [CrossRef]
13. Lee, H.; Won, Y.J.; Bang, D. Why do event volunteers return? Theory of planned behavior. *Int. Rev. Public Nonprofit Mark.* **2014**, *11*, 229–241. [CrossRef]
14. Reuveni, P.; Werner, Y. Factors associated with teenagers' willingness to volunteer with elderly persons: Application of the theory of planned behavior (TPB). *Educ. Gerontol.* **2015**, *41*, 623–634. [CrossRef]
15. Law, D.T.; Shek, B.M. Beliefs about volunteerism, volunteering intention, volunteering behavior, and purpose in life among Chinese adolescents in Hong Kong. *Sci. World J.* **2009**, *9*, 855–865. [CrossRef]
16. Ajzen, I. The theory of planned behavior. *Organ. Behav. Hum. Decis. Processes* **1991**, *50*, 179–211. [CrossRef]
17. Hagger, S.J.; Chatzisarantis, M.S.; Biddle, N.L. The influence of autonomous and controlling motives on physical activity intentions within the theory of planned behaviour. *Br. J. Health Psychol.* **2002**, *7*, 283–297. [CrossRef]
18. Sheeran, P. Intention—behavior relations: A conceptual and empirical review. *Eur. Rev. Soc. Psychol.* **2002**, *12*, 1–36. [CrossRef]
19. Teo, C.B.; Lee, T. Explaining the intention to use technology among student teachers: An application of the Theory of Planned Behavior (TPB). *Campus-Wide Inf. Syst.* **2010**, *27*, 60–67. [CrossRef]
20. Ayob, H.C.; Sheau-Ting, S.F.; Jalil, L.; Chin, R.A. Key Determinants of Waste Separation Intention: Empirical Application of TPB. *Facilities* **2017**, *35*, 696–708. [CrossRef]
21. Fraenkel, N.E.; Wallen, J.R. *The nature of qualitative research: How to Design and Evaluate Research in Education*, 7th ed.; McGraw-Hill: Boston, MA, USA, 2009.
22. Creswell, J.W. *Educational Research: Planning, Conducting, and Evaluating Quantitative and Qualitative Research*, 4th ed.; Pearson Education Limited: Boston, MA, USA, 2012.
23. Sekaran, R.; Bougie, U. *Research Methods for Business: A Skill-Building Approach*, 5th ed.; John Wiley and Sons Inc.: West Sussex, UK, 2009.
24. Shaughnessy, J.J.; Zechmeister, E.B.; Zechmeister, J.S. *Research Methods in Psychology*, 9th ed.; McGraw-Hill: New York, NY, USA, 2012.
25. Cooper, P.; Schindler, D. *Business Research Methods*, 10th ed.; McGraw-Hill/Irwin: New York, NY, USA, 2008.
26. Dwivedi, B.; Papazafeiropoulou, Y.K.; Brinkman, A.; Lal, W.P. Examining the influence of service quality and secondary influence on the behavioural intention to change internet service provider. *Inf. Syst. Front.* **2010**, *12*, 207–217. [CrossRef]

27. Hair, M.; Hult, J.F.; Ringle, G.T.M.; Sarstedt, C.M. *A Primer on Partial Least Squares Structural Equation Modelling*; Sage Publications, Inc.: Thousand Oaks, CA, USA, 2014.
28. Edelman, B. *GE Global Innovation Barometer 2014 Edition—Global Report*; GE: Boston, MA, USA, 2014.
29. Ajzen, I. The theory of planned behavior: Frequently asked questions. *Hum. Behav. Emerg. Technol.* **2020**, *2*, 314–324. [CrossRef]
30. Battista, I.V.; Curmi, F.; Said, E. Influencing Factors Affecting Young People's Attitude Towards Online Advertising: A Systematic Literature Review. *Int. Rev. Manag. Mark.* **2021**, *11*, 58–72. [CrossRef]
31. Wang, J.; Shahzad, F.; Ahmad, Z.; Abdullah, M.; Hassan, N.M. Trust and Consumers' Purchase Intention in a Social Commerce Platform: A Meta-Analytic Approach. *SAGE Open* **2022**, *12*, 21582440221091262. [CrossRef]
32. Knabe, A. Applying Ajzen's Theory of Planned Behavior to a Study of Online Course Adoption in Public Relations Education. Ph.D. Thesis, Marquette University, Milwaukee, WI, USA, 2012.
33. Sander, L.; Nummenmaa, D. Reward and emotion: An affective neuroscience approach. In current opinion in behavioral sciences. *Curr. Opin. Behav. Sci.* **2021**, *39*, 161–167. [CrossRef]
34. Yi, G. From green entrepreneurial intentions to green entrepreneurial behaviors: The role of university entrepreneurial support and external institutional support. *Int. Entrep. Manag. J.* **2021**, *17*, 963–979. [CrossRef]
35. Peltokorpi, V.; Allen, D.G.; Shipp, A.J. Time to leave? The interaction of temporal focus and turnover intentions in explaining voluntary turnover behaviour. *Appl. Psychol.* **2022**, 1–20. [CrossRef]

proceedings

Proceeding Paper

Conceptual Framework: Factors Affecting Learner Performance and Cognitive Load in Educational Multimedia Learning †

Khadijah Abdul Rahman [1,*], Siti Aishah Mokhtar [1,*], Noor Arina Md Arifin [1,*], Mohd Nasir Ismail [1] and Tengku Mohd Farid Tengku Abdul Aziz [2]

1 Faculty of Information Management, Universiti Teknologi MARA Kelantan Branch, Bukit Ilmu, Machang 18500, Kelantan, Malaysia
2 Academy of Languange Studies, Universiti Teknologi MARA Kelantan Branch, Bukit Ilmu, Machang 18500, Kelantan, Malaysia
* Correspondence: khadijah10@uitm.edu.my (K.A.R.); aishah835@uitm.edu.my (S.A.M.); arina848@uitm.edu.my (N.A.M.A.)
† Presented at the International Academic Symposium of Social Science 2022, Kota Bharu, Malaysia, 3 July 2022.

Abstract: The spread of COVID-19 has compelled educational institutions to make the transition to online communication. It has limited teaching and learning to utilize a fully online learning mode. However, the problem with online learning is that it is entirely dependent on technological devices and the internet: teachers and students with poor internet connections face the problem of access to online learning. These challenges will affect the effectiveness of educational multimedia in the learning process and how students will be able to grasp the educational lesson to remain focused throughout the lesson. Therefore, this study aims to investigate the factors affecting cognitive load and learner performance and proposes a conceptual framework of factors affected in educational multimedia learning underlying the interrelationship between cognitive load and learner performance.

Keywords: learner performance; cognitive load; educational multimedia learning

Citation: Rahman, K.A.; Mokhtar, S.A.; Arifin, N.A.M.; Ismail, M.N.; Aziz, T.M.F.T.A. Conceptual Framework: Factors Affecting Learner Performance and Cognitive Load in Educational Multimedia Learning. *Proceedings* **2022**, *82*, 56. https://doi.org/10.3390/proceedings2022082056

Academic Editor: Mohamad Rahimi Mohamad Rosman

Published: 16 September 2022

1. Introduction

The COVID-19 pandemic has forced the teaching and learning environment to transition to a fully online learning mode. The non-face-to-face teaching made deliverables of academic syllabuses for technical subjects in universities has been a daunting task for educators. Thus, the advancement of internet technology has created opportunities for educators to create educational videos for lesson delivery. However, questions remain on how effective the multimedia learning is. Is the student able to grasp the educational lesson? Can they remain focused throughout the lesson? However, the challenges with online learning is that it is entirely dependent on technological devices and the internet. Both teachers and students with poor internet connections are facing problems with access to online learning. Numerous students, particularly those living in rural areas, lack the high-speed internet connection required for online learning instruction. They had difficulty going live for virtual learning as a result. Some of them had difficulty with technology, as they lack expertise in computers and other forms of technology. There is also a possibility that they did not have a strong internet connection, causing them to have problems downloading some information on the topic. Some students have difficulty managing their time online learning since it is unfamiliar and takes a substantial amount of work. They need to figure out how to schedule their time wisely. Online learning is more flexible than traditional learning; however, some students experience difficulty adjusting to the time required to study online. This study intends to investigate the factors that affect cognitive load and learner performance in educational multimedia learning. The method of the study is a structured literature review, and the expected outcomes of this study is the conceptual framework that educators can use in developing educational videos by applying the principles of multimedia learning and reducing the cognitive load in the video.

2. Literature Review

2.1. Modality Principle

One of the principles of multimedia learning is the modality principle. According to Mayer's cognitive theory of multimedia learning, people learn better from visuals and spoken words than visuals and written words. This is because learners can become overwhelmed if there is too much text. According to the modality principle, cognitive load is likely to occur when combined with written–verbal and visual information. Therefore, providing written–verbal information through an auditory channel is preferable when visual information is also available [1]. Ref. [2] stated in their 2019 study that narration is preferable to text according to the modality principle since individuals may focus on visuals while listening to audio. This indicated the modality's effect on cognitive load and learning performance.

The principles of modality suggest that learning from visuals with spoken text is effective compared to learning from displays with written text [3]. Ref. [4] proved that learning from visuals completed by spoken words is more effective than learning from visuals supplemented by written words. They also argued that learning with visuals alone is not as effective when compared to learning with visuals and spoken words. When learning from visuals alone, it might overload the visual channel; meanwhile, when learning from visuals and narration, both channels are be used to share the load. The modality principles stated learning is at its best when using dual-modal (e.g., text with image) compared to single-modal (text only or image only).

2.2. Spatial Contiguity Principle

The Spatial Contiguity Principle highlighted the concerns with the placement or position of both image and text in an educational video. The image should be placed close to the text explaining the image, rather than being placed separately. An example of spatial contiguity principle in a video is when a caption is explaining an image, its caption should be placed close to the image rather than it being distant. Their experiment showed how this spatial contiguity principle affects cognitive load and learning outcome by presenting he multimedia lesson with printed text at the bottom of each image serving as a caption, and placing the text next to the corresponding part of each image. Spatial contiguity is highlighted when students correspond with the words and visuals are placed close together on the page or screen rather than far apart [5]. As a result, learners are less likely to be able to maintain both in working memory simultaneously. Ref. [3] states that spatial contiguity is the degree to which similar pieces of information are arranged in close physical proximity to simplify processing. It is also about learners learning best when relevant information and visual are physically close together on the screen.

2.3. Segmenting Principle

According to the segmenting principle, multimedia instructions should be delivered in segments rather than as continuous units to maximize learning. For instance, the durations of the chapter 1 video lesson is originally one hour. However, after applying the segmenting principle, the video is divided into five subtopics of 20 min duration each. Ref. [2] study revealed that their experiment groups with segmented clips had higher learning performance than the non-segmented groups. The segmentation principle would allow learners to adapt to the pace of the lesson with cognitive needs. Both studies found that using the segmenting principle also increased knowledge retention and knowledge transfer test marks compared to the lesson presented in the continuous unit. Segmenting is being used as one of the techniques to manage essential processing, which is breaking down a complex lesson into advanced parts under the leaner's control [6]. Segmenting is a technique that gives the learner control over the pacing of an online lesson, such as instructing the learner to go from one slide to the next of a slideshow by pressing a key [7]. According to [8], there are two main benefits of segmentation that have impact on learning facilitation: a decrease in cognitive overload and the learner better configuring the learning materials.

2.4. Interpolated Assessment

Apart from using these principles, the current trend of measuring effective educational videos is to include the interactivity of the video. In recent years, interactivity of videos has been a subject of research interest in which, as technology evolves, it is now possible for videos to interact with learners. For instance, while a student watches the video lesson, questions that require answers could suddenly pop out. Interpolating assessment in an educational video enables the leaner to interact with the video. Ref. [9] studied the effects on embedded questions in educational videos where there were two versions of the video, one with embedded questions and the other without questions. The author used a comprehension test and self-efficacy as an indicator of effectiveness. He found that the group with embedded questions performed better in assessment scores.

2.5. Cognitive Load

The researchers measured self-efficacy based on Cognitive Load Theory (CLT), and embedded questions also were found to reduce cognitive load. They stated that embedded questions made individuals feel better about themselves, improved confidence, added new information to what they already knew, exercised their memories, and that they learned more. In contrast, the majority of affected variables utilized by previous studies were derived from Sweller's (1988) CLT. There are three main variables: Intrinsic Load (IL), Extrinsic Load (EL), and Germane Load (GL). Extraneous load (EL) is a load that is not required due to unclear instructions. Intrinsic load (ILL) demonstrates the difficulty or complexity of learning materials. As for Germane load (GL), it refers to the effort required to construct knowledge. When a researcher wants to determine how effectively a student learns from a video, they use subjective measurement. In contrast, objective measuring is based on a learner's test scores or final grades. This is significantly different from subjective measuring, based on a leaner's word, ratings, or what he or she saw or heard while studying.

2.6. Learner Performance

According to [10] study, the online self-regulated learning questionnaire can measure self-regulation and motivation in the student characteristics and learning outcome categories. Other custom-made instruments were employed to assess attitudes, computer skills, workload management, social and family support, satisfaction, knowledge construction, and technology quality, interactions, tools and resources for learning management systems, and face-to-face support. The rubrics can be used to evaluate a student's performance. Rubrics provide students with a description of the performance levels for each criterion dimension that demonstrates what is expected of them in a given task. Also available for self-evaluation are self-assessment papers, checklist, and portfolios. Ref [11] adds that the process evaluating teachers should focus on the following significant indicators: identifying indicators of teacher practice that lead to maximum learner performance; establishing current performance levels; identifying alternative ways to improve performance, and providing feedback on the efficacy of the new strategies.

3. Methodology

This study adopted the structured literature review (SLR) methodology. First, the literature search started with a leading journal relevant to the field of study. Second, a backward search was conducted to identify relevant articles. Third, a forward-search was conducted to identify any news articles that citing the publication.

In the first phase, the study uses online databases to identify and search relevant articles from leading journals. The keywords used are "Learner Performance", "Cognitive Load", and "Principles of Multimedia Learning". In the next stage, the study evaluates the article base and browsed each paper to identify its topics and critical issues. Next, the focus is on the study's introduction, abstract, discussion, and conclusion of the study. Finally, the study identifies potential variables underlying the topic for the development of our proposed conceptual framework.

4. Conceptual Framework

Based on the literature search, this study proposes an integrated framework to illustrate these relationships of identified variables: Modality principle, segmenting principle, spatial contiguity, interpolated assessment cognitive load, and learner performances depicted in Figure 1 below:

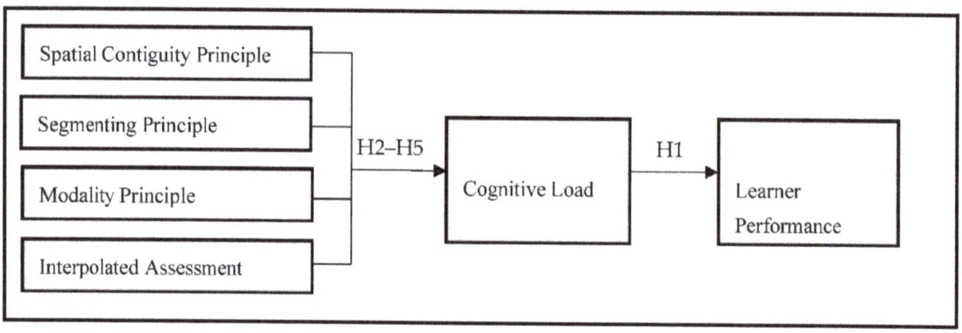

Figure 1. The proposed framework.

H1: *Cognitive load influences learner performance.*

Student performance in university usually measured using test. The test includes a knowledge retention test and a knowledge transfer test. Knowledge Retention tests are designed to test the learner's ability to recall the lesson, while knowledge transfer tests are designed to test the learner's ability to apply the lesson learnt. Using this knowledge retention and knowledge transfer as variables to measure student performance, substantial empirical evidence was found that cognitive load significantly affects student performance. Ref. [2] found a significant positive relationship by conducting pre-tests and post-tests.

H2: *Spatial contiguity principle influences cognitive load in educational multimedia learning.*

Ref. [3] stated that in multimedia learning research, the spatial contiguity effect enhances learning when written text and visuals are spatially integrated rather than physically separated in multimedia learning studies. In their research, Ref. [12] proposed that the spatial contiguity principle is the most effective way for slow learners. During complex lectures, words are necessary to understand the image, whereas, during interactive multimedia, learners move text physically closer to the diagram. Thus, it shows that spatial contiguity positively impacts learner performance. In a recent meta-analysis, Ref. [13] found strong evidence that increasing spatial contiguity can lead to significant learning improvements. As a result, physical closeness has the potential to impact unconscious as well as cognitive thinking. Therefore, the spatial layout of information representations must be considered, particularly in a learning setting. When related representations are spatially integrated by the designer and nearby to one other, it promotes information processing, integration, and learning [14]. The spatial continuity principle, where learners learn better when text and graphics are close to each other, was supported in the previous research.

Physical closeness can impact learners' performance, where they learn and understand faster than before.

H3: *Segmenting Principle influences cognitive load in educational multimedia learning.*

Segmenting can help to enhance the performance of the learner. A study by [6] indicated that it is more effective to split up a complicated slide into portions that are shown one at a time, as opposed to providing all of the information in a single self-paced slideshow lesson. According to [8], segmenting permits the learner to construct a mental image of one section of the content before going on to the next. It has been demonstrated in previous research (e.g., Mautone and Mayer, 2007, Mayer and Chandler, 2001, Mayer, Dow, and Mayer, 2003, and Sung and Mayer, 2013) that segmenting a slideshow in a variety of ways may improve transfer test performance compared to a continuous presentation.

H4: *Modality Principle influences cognitive load in educational multimedia learning.*

The modality principle also related to the leaners' performance. According to [15] the modality effect can be replicated using extended multimedia instructions on a non-technical toxic such as instructional design. This is demonstrated by the fact that students in the audio group perform just as well as those in the visual group on both the retention and transfer tests. This has been carried out by requiring less mental effort to achieve the same score on the transfer test. The effectiveness of multimedia teachings will only increase if the student cannot determine the pace of the instructions and the narrator sets the pace [15]. According to the study, using a mixed modality to offer the learner materials that can be processed as both text and images simultaneously improves learner performance [3]. When text is replaced with narration, verbal and nonverbal visual inputs that were previously competing are eliminated. Intriguingly, this benefit of storytelling over text remained even when presentations were arranged in chronological sequence.

H5: *Interpolated Assessment influences cognitive load in educational multimedia learning.*

Ref. [16] assert that interpolated assessment can facilitate the learning of a videotaped lecture in several ways. For instance, it can alter the types of ideas people have throughout the lecture, making it easier to understand the lecture's content and making it simpler to connect the lecture's session. Highly driven students may benefit from interpolated evaluations in video-based learning, although the gains are shown in the transfer of knowledge rather than its retention (Ref. [17]). Based on the previous study's findings, it can be deduced that assessment influences cognitive load in educational multimedia learning.

5. Conclusions

This study was conducted with two purposes: first, to investigate the factors affecting cognitive load and learner performance, and second, to propose a conceptual framework of factors affecting educational multimedia learning underlying the interrelationship between cognitive load and learner performance. In order to answer both purposes, a systematic literature review (SLR) methodology was adopted. Subsequently, four multimedia principle factors that will affect the cognitive load and leaner performance were identified: spatial contiguity principle, segmenting principle, modality principle, and interpolated assessment. Next, the study's conceptual model was proposed, and discussions were made on each hypothesis. Future studies to empirically test this framework are encouraged to generalize the findings and further support the theories. The next stage of this research is to validate the conceptual research model in a quantitative study. An instrument will be developed by adapting and adopting the previous instrument. The potential respondents are students from one of the local universities in Malaysia, selected based on demographic information explored via the SLR.

Author Contributions: Conceptualization, N.A.M.A. and K.A.R.; methodology, S.A.M. and K.A.R.; resources, N.A.M.A.; writing—original draft preparation, N.A.M.A., K.A.R., S.A.M., M.N.I. and T.M.F.T.A.A.; writing—review and editing, N.A.M.A., K.A.R. and S.A.M.; project administration, M.N.I.; funding acquisition, T.M.F.T.A.A. All authors have read and agreed to the published version of the manuscript.

Funding: This research as funded by Research management Centre (RMC) of Universiti Teknologi MARA Kelantan Branch 600-TNCPI 5/3/DDN (03) (006/2021).

Institutional Review Board Statement: Not applicable.

Informed Consent Statement: Not applicable.

Data Availability Statement: Not applicable.

Conflicts of Interest: The authors declare no conflict of interest.

References

1. Kozan, K.; Erçetin, G.; Richardson, J.C. Input modality and working memory: Effects on second language text comprehension in a multimedia learning environment. *System* **2015**, *55*, 63–73. [CrossRef]
2. Chen, C.Y.; Yen, P.-R. Learner control, segmenting, and modality effects in animated demonstrations used as the before-class instructions in the flipped classroom. *Interact. Learn. Environ.* **2019**, *29*, 1–15. [CrossRef]
3. Moreno, R.; Mayer, R.E. Cognitive Principles of Multimedia Learning: The Role of Modality and Contiguity. *J. Educ. Psychol.* **1999**, *91*, 358–368. [CrossRef]
4. Castro, J.C.; Sweller, J. The Modality Effect of Cognitive Load Theory. *AHFE 2019* **2020**, *963*, 75–84. [CrossRef]
5. Mayer, R. *Multimedia Learning*, 3rd ed.; Cambridge University Press: Cambridge, UK, 2020.
6. Mayer, R.E. Evidence-based principles for how to design effective instructional videos. *J. Appl. Res. Mem. Cogn.* **2021**, *10*, 229–240. [CrossRef]
7. Mayer, R.E.; Wells, A.; Parong, J.; Howarth, J. Learner control of the pacing of an online slideshow: Does segmenting help? *Appl. Cogn. Psychol.* **2021**, *33*, 930–935. [CrossRef]
8. Spanjers, I.A.E.; Gog, T.V.; Merrienboer, J.J.G.V. A theoretical analysis of how segmentation of dynamic visualizations optimizes students' learning. *Educ. Psychol. Rev.* **2010**, *22*, 411–423. [CrossRef]
9. Tweissi, A. The Effects of Embedded Questions Strategy in Video Among Graduate Students at a Middle Eastern University. Ph.D. Thesis, Ohio University, Athens, OH, USA, 2016.
10. Barnard, L.; Lan, W.Y.; To, Y.M.; Paton, V.O.; Lai, S. Measuring self-regulation in online and blended learning environments. *Internet High. Educ.* **2009**, *12*, 1–6. [CrossRef]
11. Rockwood School District Committee. *Teacher Evaluation Instrument*; Rockwood School District Committee: Rockwood, TX, USA, 2014.
12. Mayer, R.E.; Fiorella, L. *Principles for Reducing Extraneous Processing in Multimedia Learning: Coherence, Signaling, Redundancy, Spatial Contiguity, and Temporal Contiguity Principles*; Florida Atlantic University: New York, NY, USA, 2014.
13. Ginns, P. Integrating Information: Meta-Analyses of the Spatial Contiguity and Temporal Contiguity Effects. *Learn. Instr.* **2006**, *16*, 511–525. [CrossRef]
14. Clark, R.C.; Mayer, R.E. *E-Learning and the Science of Instruction: Proven Guidelines for Consumers and Designers of Multimedia Learning*, 2nd ed.; John Wiley & Sons: Pfeiffer, Germany, 2008.
15. Tabbers, H.K.; Martens, R.L.; Van Merrienboer, J.J.G. The modality effect in multimedia instructions. *Proc. Annu. Meet. Cogn. Sci. Soc.* **2001**, *23*. Available online: https://conferences.inf.ed.ac.uk/cogsci2001/pdf-files/1024.pdf (accessed on 16 March 2022).
16. Jing, H.G.; Szpunar, K.K.; Schacter, D.L. Interpolated Testing Influences Focused Attention and Improves Integration of Information During a Video-Recorded lecture. *J. Exp. Psychol. Appl.* **2016**, *22*, 305–318. [CrossRef] [PubMed]
17. Yang, J.; Zhang, Y.; Pi, Z.; Xie, Y. Students' achievement motivation moderates the effects of interpolated pre-questions on attention and learning from video lectures. *Learn. Individ. Differ.* **2021**, *91*, 102055. [CrossRef]

proceedings

MDPI

Proceeding Paper

Success Factors for Using E-Court in Indonesian Courts †

Muhamad Romdoni [1,*], Assed Lussak [2] and Ibnu Darmawan [3,*]

1 Faculty of Law, Primagraha University, Griya Gemilang Sakti Complex, Trip Jamaksari St. No. 1A, Serang 42111, Indonesia
2 Baldwin Boyle Group, GridAKL Building, Wynyard Quarter, 12 Madden St., Auckland 1010, New Zealand
3 Communication Department, Islamic University of Indonesia, Kaliurang St. Km 14.5, Sleman 55584, Indonesia
* Correspondence: muhamadromdoni@primagraha.ac.id (M.R.); 213210501@uii.ac.id (I.D.)

† Presented at the International Academic Symposium of Social Science 2022, Kota Bharu, Malaysia, 3 July 2022.

Abstract: The goal of this research was to investigate whether or not electronic court petitions are beneficial in Indonesia. The proposed research model for analyzing the success criteria for using the case management system was tested in the courts of Greater Jakarta. The model's explanatory relevance was revealed by using structural equation modeling to examine survey data from 30 attorneys who utilize the system. Furthermore, the significant contribution of system and information quality characteristics is demonstrated. The investigation's findings have far-reaching implications for both theory and practice. System quality, information quality, and system utilization all have a positive influence on user performance. However, the influence of information quality on usage was not statistically significant. This study adds to the body of research on court management by seeking to implement the success model in Indonesian courts. Furthermore, the study focuses on which parts of the court administration procedure have the most explanatory power. It is recommended that additional research be conducted on the role that technology plays in determining judicial system performance, with a special emphasis on determining the relations between ICT and judicial system performance.

Keywords: e-court success; IS success; judicial system; case management system

Citation: Romdoni, M.; Lussak, A.; Darmawan, I. Success Factors for Using E-Court in Indonesian Courts. *Proceedings* **2022**, *82*, 58. https://doi.org/10.3390/proceedings2022082058

Academic Editor: Mohamad Rahimi Mohamad Rosman

Published: 16 September 2022

1. Introduction

Electronic government (e-government) initiatives have grown considerably in the past two decades. These projects help governments deliver information and communication technology (ICT)-based information and services. It gives residents, businesses, and the government additional chances to provide better services [1]. ICTs in public administration also help modernize governance and have huge administrative possibilities [2]. This could improve government relationships with communities, businesses, and workers through decentralization, openness, and internal/external responsibility [3,4]. Many sorts of e-government programs are utilized globally in healthcare, education, and transportation. When the judicial system (JS) uses ICTs to improve its efficacy and efficiency, it can ease communication among JS parties (ministry, courts, and prisons) and help improve access to justice and the efficiency of legal proceedings [5]. E-court itself is a set of hardware, software, and networking solutions for the JS to aid court personnel and judges enhance case filing, hearing scheduling, and efficiency. Daily responsibilities include workflow, filing, and hearing scheduling. This research project refers to courtroom and client-courtroom technology as e-court. E-court refers to courtroom technology used globally. According to research on court management, there are two types of court technologies: those used in the courtroom and those used to improve court, party, and public communication [6,7]. This category covers automated registers, case management systems, and financial management systems. First, we have court administration software that helps administrative staff and/or judges (such as jurisprudence databases and sentencing support systems). Second, we have apps that let parties share data and communicate (such as e-filing and provision through court websites). A legal system case registry is automated and helps improve court cases and data access [8].

ICT helps implement the JS's administrative theory and improve judicial efficiency [9]. Despite significant investments, technical progress is questionable [10,11]. Multiple studies find it to be inconsistent in terms of value for money [12,13]. Previously, court management specialists decided to explore the relationship between technology and individual and organizational performance, mostly using descriptive research [14,15]. This method covers judicial technology, processes, and systems. Unfortunately, no quantitative research has been done on technology and court performance, notwithstanding the court management literature. Moreover, no research has studied how court technology affects lawyers' IT use (IS). Thus, this study examines how court administrative personnel use e-court applications. With the Indonesian JS's efforts to digitize all records, judges may be resistant to court technology [16]. This also threatens Indonesia's justice system. Several countries have used DeLone and McLean's implementation strategy to analyze IS effects [17], which was also used in this study. Despite academics' attention to information security, no research has tested this technology in courts. Understanding the efficacy of electronic court applications and the factors that determine their performance is crucial for addressing research gaps and resolving future challenges.

2. Literature Review

The ability to effectively manage or use information systems is the cornerstone of determining the value and utility of information system management and investments. Many academics recognize the need for analyzing the success (or efficacy) of information systems [18]. End-user satisfaction is the most widely used metric for measuring IT performance [19]. Information quality refers to the measurement of the information system's output, system quality refers to the measurement of the system itself, information use refers to the recipients' consumption of the output, user satisfaction refers to the recipients' responses to using the output, and organ quality refers to the overall quality of the organ [20]. System quality, information quality, and information use are all terms that refer to the quality of the system, the information provided by the information system, and how efficiently the information is used. The consistency of the user interface comes first, followed by the quality of the documentation, and last by the system's failures. Some academics have assessed the quality of systems based on how simple the system is perceived to be or how easy it is to use [19]. When analyzing the quality of information produced by a system, it is vital to take into account the system's precision, timeliness, and relevance [21]. User satisfaction is described as an individual's reactions to a set of criteria used to assess a system's effectiveness [17]. The word user satisfaction refers to a broader sense of contentment with a system's use than simply using an information system. The usage of information systems can affect user satisfaction at the individual, organizational, and user levels [21]. A successful variable indicates that an information system enhanced a user's understanding of the decision context, increased the user's decision-making productivity, influenced the user's behavior, or influenced the decision maker's evaluation of the information system's significance or utility [15]. An effective information system assists users in better understanding the context in which decisions are made.

There are several approaches for quantifying the impact of information on a business. Higher income, increased profitability, and increased return on investment are common measures of organizational effectiveness [15]. The strategy is divided into three stages [17]. The ability of an item to positively influence individuals is referred to as effectiveness. The term technical relates to the accuracy and effectiveness with which information is generated inside the communication system. The ability of information to convey the intended meaning is referred to as its semantic value. The quality dimension assesses the system's success in terms of its technical aspects, the information quality dimension assesses the system's success in terms of its semantic aspects, and the effectiveness dimension assesses the system's success in terms of its efficacy in multiple areas, including utilization, user satisfaction, and individual and organizational benefits. In summary, this is an evaluation of efficacy, quality, and information gathering quality.

The technical levels are determined by the information's quality, usefulness, and efficiency. The success variables are interconnected rather than independent in this model since it is based on process and cause considerations [16]. This is due to the fact that the model is built on considerations. The model's creators state that it studies the knowledge of system operations and the effects of those processes, as well as the system's growth, implementation, and the consequences of those actions. The information system success model has served as the foundation for both academic and empirical research on the efficacy of information and communication technology [20]. DeLone and McLean revised the initial IS success model in response to the findings of studies that called for revisions. Individual and organizational impact have been replaced with net benefits to take into account benefits at any level of analysis, and the quality of service offered is the new criterion for determining whether or not an information system is successful. The information systems success model can be used in the process of evaluating cutting-edge internet-based apps [22].

DeLone and McLean's success model for information systems has been evaluated in a range of contexts, including commercial, public, voluntary, and mandatory settings [23]. Despite its broad use in assessing the usefulness of ICTs in other public administration (PA) sectors, few studies have looked at the effectiveness of e-government apps using this technique [24]. There have not been many studies looking at e-government from a business perspective. All of these studies begin with the assumption that the success of e-government is more dependent on public participation than on technological improvement. Because user behavior is so important to the success of e-government systems, understanding the elements that influence it is an intriguing issue [25]. The majority of court management research has focused on evaluations of local justice systems, e-justice methodologies, court software development, and user experiences with court technology [26], while user acceptance of intergovernmental services is also critical to the success of an e-government operation [27]. This applies to all e-government rollouts when user participation is critical to e-government success [28].

3. Research Model and Hypotheses

The success of information technology in the judicial system is investigated from the point of view of court administration professionals in this study. These professionals are responsible for utilizing apps in order to carry out their duties when evaluating the technical success, semantic success, and effectiveness of CMS applications, and the effectiveness of CMS applications in general [18], in which key variables apply, such as system quality, information quality, information system use, user satisfaction, and individual impact. These are all considered to be key variables in the assessment. Both the quality of the system and the content are major predictors of user satisfaction and the use of information systems, which are two factors that have an effect on an individual's level of performance [18]. The majority of research focused on information systems holds the belief that the quality of the system and the output have an effect on user behavior [29], while CMS systems incorporate automated court dockets and other court register processes into the JS. This makes it possible for different court activities, procedures, and practices to be consolidated into a single platform. Employees that have a positive perception of the ease of use, dependability, and timeliness of CMS applications are more likely to utilize it. The combination of previous studies results the construct of this study, as shown in Figure 1.

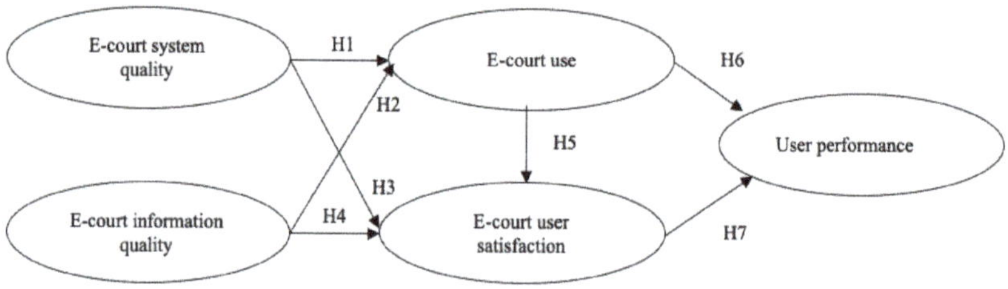

Figure 1. Research framework.

According to research conducted on information systems, the quality of both the system and the information is an important factor in determining the level of user satisfaction. When clients have the perception that a system is reliable, precise, and punctual, it leads to an increase in user enjoyment and delight, as demonstrated by the following: CMSs are used to automate the activities of administrators and judges within the JS. This gives them the ability to quickly access data and information important to particular cases that have been submitted from any location, at any time, and with no restrictions whatsoever. These systems, as a result of their deployment, have also resulted in the complete description of organizational processes, which includes reading the procedure codes and other applicable legislation, and have also resulted in an improvement in the standardized application of regulations by each court office [17]. It has also resulted in the deployment of these systems, which has resulted in the complete description of organizational processes. As a consequence of this, court employers who place a high value on the quality of e-court applications and output will feel more satisfied with the systems that are now being considered.

As seen in the following explanation, utilization and user happiness are intricately related [20]. In the sense of a process, usage must come before user happiness; however, in the sense of a causal relationship, having a pleasant usage experience will lead to increased user pleasure [20]. In the end, a person's degree of satisfaction with a system is determined not by how satisfied they are with the system itself, but by how satisfied they are with the results or impacts of their use of the system [23]. Additionally, the impact of use components to system performance has been researched in a range of contexts, such as voluntary and forced settings. For instance, while its contribution may be negligible in a context where participation is a choice, it may be utterly unimportant in a system where participation is mandated. When system utilization is necessary, it is possible that removing use as a success variable is erroneous [21]. This is due to the fact that system use can fluctuate greatly; hence, the variable use must be kept. As a consequence of this, and in agreement with the findings of a number of scholars [21], use is then defined as the degree to which the information system has been integrated into the daily activities of each individual [29]. The contentment level of those who work in court administration rises in proportion to the amount of time spent using e-court applications.

A previous study examined the links between semantic features such information system usage and user happiness, and effectiveness-related variables like an individual's impact [20]. An empirical study showed that using information systems benefits the individual. Infrequent usage of a system has little effect on an individual's performance, but consistent use does. Another study found a correlation between user satisfaction and performance in various conditions [21]. User satisfaction is the idea that task requirements and information system capabilities match. There may be a link between user satisfaction and individual influence. Judges and courtroom staff can obtain timely information via a CMS. This improves individual and court performance by shortening the case resolution time. The individual's performance is likely to be favorably related to their best estimate of the CMS software's requirements and capabilities.

4. Methodology

The quantitative study was carried out on Indonesian court users, i.e., lawyers, in the Greater Jakarta area. It covered all Jakarta cities and regencies, including Bogor Regency and City, Depok, South Tangerang, Tangerang Regency and City, and Bekasi Regency and City. Questionnaires were collected from 30 attorneys who are able to carry out their duties with the support of e-court content management systems, which are more popularly known as e-court. During a legal procedure, this application is a content management system (CMS) that was developed to help automate court operations and provide assistance. It provides support for case management tasks such as case planning and tracking, and the scheduling of hearings and other court appearances in general. In particular, CMS assists in the operation of the court through the utilization of capabilities that include calendaring and scheduling, docketing, case information management, ticklers, notes, and case linkages.

The comprehensive questionnaire was sent out in February 2022 in order to collect information. In addition to gathering information on user profiles and the e-court software that is being utilized, the questionnaire was designed to investigate the elements that are related with the success of e-courts. In the beginning, comments from both court users and information systems specialists were solicited. This was to undertake preliminary and pilot testing to analyze and validate the approved procedures. Using data from the IS made publicly available in the past, the questionnaire was first crafted, and after that, it underwent testing and analysis. The components had some minor alterations so that they would function better within the e-court program. The purpose of the 15 questions in the survey was to investigate five primary components of the system. These were the quality of the system, the quality of the information, the usage of the system, the happiness of the users, and the individual impact. The ease of use of a system was used as one of the criteria for measuring system quality in the research. The scale consisted of two items. The quality of the information that is produced by e-court applications can be evaluated along three dimensions: the material's substance, its correctness, and how it is presented.

To evaluate a user's level of dependence on the information system that was available to them, a reliance measure was applied. The utilization of IS was evaluated based on a single item of this metric. User satisfaction is a measurement of how satisfied users are with a system, and it was computed by modifying a single item based on the work done by Rai and colleagues to capture the level of enjoyment provided by the system [27]. Individual impact, on the other hand, is defined as the extent to which application software was successful in improving the quality of the user's work, making the end user's job easier, reducing the amount of time spent on the end user's job, and assisting the end user in meeting the end user's job needs and requirements. The four-item user-performance metric was modified in order to arrive at the result. On a scale from one to five, the allegations were rated, with a score of one indicating major disagreement and a score of five indicating strong agreement.

5. Analysis and Results

Based on the Table 1 below, the hypothesis is accepted, i.e., there is an effect of the e-court system quality variable on e-court use. While Table 2 concludes that the hypothesis is not accepted, i.e., there is no effect of the e-court information quality variable on the e-court use. The Sig value is $0.003 < 0.05$, as shown in Table 3, the hypothesis is accepted, i.e., there is an influence of the e-court system quality variable on e-court user satisfaction. Moreover, Table 4 shows that the hypothesis is accepted, i.e., there is an effect of the e-court information quality variable on e-court user satisfaction. Table 5 later concludes that the hypothesis is accepted, i.e., there is an influence of the e-court use variable on the e-court use satisfaction. Based on the Sig value of $0.002 < 0.05$ in Table 6, it can be concluded that the hypothesis is accepted, i.e., there is an influence of the e-court use variable on user performance.

Table 1. E-court system quality and e-court use.

Model		Sum of Sq.	df	Mean Sq.	F	Sig.
	Regr.	5.106	1	5.106	6.281	0.018 b
1	Residual	22.761	28	813		
	Total	27.867	29			

Dependent Variable: EU. b. Predictors: (Constant), ESQ.

Table 2. E-court information quality and e-court ese.

Model		Sum of Sq.	df	Mean Sq.	F	Sig.
	Regr.	3.085	1	3.085	3.486	0.072 b
1	Residual	24.782	28	885		
	Total	27.867	29			

Dependent Variable: EU. b. Predictors: (Constant), EIQ.

Table 3. E-court system quality and e-court user satisfaction.

Model		Sum of Sq.	df	Mean Sq.	F	Sig.
	Regr.	4.255	1	4.255	10.626	0.003 b
1	Residual	11.212	28	400		
	Total	15.467	29			

Dependent Variable: EUS. b. Predictors: (Constant), ESQ.

Table 4. E-court information quality and e-court user satisfaction.

Model		Sum of Sq.	df	Mean Sq.	F	Sig.
	Regr.	3.452	1	3.452	8.046	0.008 b
1	Residual	12.014	28	429		
	Total	15.467	29			

Dependent Variable: EUS. b. Predictors: (Constant), EIQ.

Table 5. E-court use and e-court user satisfaction.

Model		Sum of Sq.	df	Mean Sq.	F	Sig.
	Regr.	2.864	1	2.864	6.363	0.018 b
1	Residual	12.603	28	450		
	Total	15.467	29			

Dependent Variable: EUS. b. Predictors: (Constant), EU.

Table 6. E-court use and user performance.

Model		Sum of Sq.	df	Mean Sq.	F	Sig.
	Regr.	54.023	1	54.023	11.232	0.002 b
1	Residual	134.677	28	4.810		
	Total	188.700	29			

Dependent Variable: UP. b. Predictors: (Constant), EU.

Lastly, Table 7 shows that the hypothesis is accepted, i.e., there is an influence of the e-court user satisfaction variable on user performance.

Table 7. E-court user satisfaction and user performance.

Model		Sum of Squares	df	Mean Square	F	Sig.
1	Regr.	157.782	1	157.782	142.890	0.000 b
	Residual	30.918	28	1.104		
	Total	188.700	29			

Dependent Variable: UP. b. Predictors: (Constant), EUS.

6. Discussion

According to this study, the model captures the multimodal and interrelated nature of CMS performance [18], making it a potential court prediction model. The majority of the hypothesized links in the e-court success model were proven, except for the quality-consumption connection. The data show that system quality affects its effectiveness. Previous studies evaluated system quality based on usability [21]. This study found that the time spent using a system increases with its quality and convenience of use. Due to these findings, e-court applications require a high-quality system. Many court clerks lack computer skills and IT experience. Despite CMS being required in courts, administrative officers employ e-court applications since they are easy to use. The content management system (CMS) interface provides straightforward and visible access to content, eliminating the need for organization-wide training and making the system more simply understood and utilized. IT competence strengthens this idea. By using CMS, more court administration officers will be available to complete tasks.

According to the study, the impact of information quality on consumption is not statistically significant. Consider these examples to evaluate CMS content: real-time data access gives consumers fast, accurate information from a database, less erroneous data input, and more consistent data entry over time. Real-time data access also allows users to provide data consistently throughout time. CMS is more significant for judges than court personnel since its output helps judges settle cases more quickly and effectively and plan and organize their operations. CMS output is largely used by courts. Since judges are not required to use CMS, many prefer paper-based paperwork to best utilize employees. Research shows that system quality, information quality, and system use all affect user performance. These findings are compatible with a model [21] for determining an information system's success. The quality of the information affects user happiness more than the system and how it is used, both of which are important. CMS users in the justice system believe the output is reliable, accurate, and timely. The output of an electronic court application allows administrative employees to swiftly and simply investigate court-docket books and court registers from any location and at any time. Many courts still use paper-based documentation. This study found that the total time spent using a system increases with its quality and convenience.

7. Research Limitations and Conclusions

The primary goal of this study was to investigate the elements that contributed to the effective adoption of CMS as a first step toward a better understanding of the utility of e-courts in Indonesian courts. Because this research is exploratory in nature, there are a number of limitations that must be addressed and taken into account in any future research endeavors. To begin, variable measurement scales are frequently self-report measures, which means they can be skewed, skewing, distorting, and exaggerating the causal relationship between independent and dependent variables. Second, because variable measuring scales are dependent on self-report assessments, they may contain biases. Furthermore, because the study is primarily concerned with analyzing the e-court success model in Greater Jakarta courts, which use a specific type of technology (CMS) and are located in a specific metropolitan area, it is critical to proceed with extreme caution when attempting to generalize the findings to other locations. Additional study, including an assessment of alternative technologies, is required to evaluate the suggested model across a nationwide sample of court users that is representative of the population as a whole and in a range of e-court scenarios, including ones in which participation is voluntary. Finally, one of the research model's flaws is that it fails to account for the impact of CMS on the daily operations of the judicial system as a whole.

In order to better comprehend the utility of e-court technologies from the perspective of court administration staff, the study focuses on the effects of CMS on individual user performance. Furthermore, because the research model was only employed in two unique courts and no attempt was made to conduct a longitudinal analysis, it cannot assess court efficiency in terms of organizational structure. The first step, however, in acquiring a better understanding of the overall effectiveness of the court system is to conduct an evaluation of the performance of court administration staff. The study's goal was to obtain a better understanding of information technology's success in courts and to add to a long-running debate among e-justice authorities, scholars, and practitioners. According to the study's findings, the ease of use and output of the deployed systems are important factors in determining the efficacy of CMS adoption in courts. This finding has far-reaching implications for court administration theory and practice. It adds to the body of studies on court management by aiming to implement the success model in Indonesian courts more frequently. Furthermore, the study focuses on which parts of the court administration procedure have the most explanatory power. It is recommended that additional research be conducted on the role that technology plays in determining judicial system performance, with a particular emphasis on determining the relationship between information and communications technology (ICT) and judicial system performance.

Author Contributions: Conceptualization, A.L.; methodology, A.L.; software, I.D.; validation, I.D.; formal analysis, I.D.; investigation, M.R.; resources, M.R.; data curation, M.R.; writing—original draft preparation, A.L. and I.D.; writing—review and editing, A.L. and I.D.; visualization, I.D.; supervision, A.L.; project administration, A.L.; funding acquisition, Not Applicable. All authors have read and agreed to the published version of the manuscript.

Funding: This research received no external funding.

Institutional Review Board Statement: Not applicable.

Informed Consent Statement: Informed consent was obtained from all subjects involved in the study.

Data Availability Statement: Not applicable.

Conflicts of Interest: The authors declare no conflict of interest.

References

1. Bernroider, E.W.N.; Koch, S.; Stix, V. A comprehensive framework approach using content, context, process views to combine methods from operations research for IT assessments. *Inf. Syst. Manag.* **2013**, *30*, 75–88. [CrossRef]
2. Yildiz, M. E-government research: Reviewing the literature, limitations, and ways forward. *Gov. Inf. Q.* **2017**, *24*, 646–665. [CrossRef]
3. Currie, W.; Guah, M.W. IT-enabled healthcare delivery: The U.K. National Health Service. *Inf. Syst. Manag.* **2006**, *23*, 7–22. [CrossRef]
4. Lussak, A.; Abdurachman, E.; So, I.G.; Setiowati, R. The effect of operational improvement on business survival with moderating role of continuous improvement (Case study: Food and beverage small business in Bali). *Int. J. Sci. Tech. Res.* **2019**, *8*, 617–621.
5. Commission of the European Communities. *Towards a European E-Justice Strategy*; The European Parliament and the European Economic and Social Committee: Brussels, Belgium, 2018.
6. Reiling, D. Technology in courts in Europe: Opinions, practices and innovations. *Int. J. Court Adm.* **2012**, *4*, 1–10. [CrossRef]
7. Velicogna, M. *Use of Information and Communication Technologies (ICT) in European Judicial Systems*; Commission for the Efficiency of Justice: Brussels, Belgium, 2018.
8. Kiškis, M.; Petrauskas, R. ICT adoption in the judiciary: Classifying of judicial information. *Int. Rev. Law, Comput. Technol.* **2004**, *18*, 37–45. [CrossRef]
9. Fabri, M.; Langbroek, P.M. *The Challenge of Change for Judicial Systems. Developing a Public Administration Perspective*; IOS Press: Amsterdam, The Netherlands, 2020.

10. Hamin, Z.; Othman, M.B.; Mohamad, A.M. Benefits and achievements of ICT adoption by the high courts of Malaysia. *IEEE Symp. Humanit. Sci. Eng. Res.* **2012**, *2*, 1233–1238. [CrossRef]
11. Palumbo, G.; Giupponi, G.; Nunziata, L.; Mora-Sanguinetti, J. *Judicial Performance and Its Determinants: A Cross-Country Perspective*; OECD: Paris, France, 2018.
12. Commission for the Efficiency of Justice (CEPEJ). *European Judicial Systems. Efficiency and Quality of Justice*; Commission for the Efficiency of Justice: Brussels, Belgium, 2020.
13. World Bank. *Doing Business*; World Bank: Washington, DC, USA, 2018.
14. Azmi, Z.T. Using technology to improve court performance: Malaysia's experience. *Asia Pac. Judic. Reform Forum Conf.* **2020**, *37*, 195–220.
15. Contini, F.; Cordella, A. Information system and information infrastructure deployment: The challenge of the Indonesian e-justice approach. *Electron. J. E-Gov.* **2017**, *5*, 43–52.
16. DeLone, W.H.; McLean, E.R. Information systems success: The quest for the dependent variable. *Inf. Syst. Res.* **1992**, *3*, 60–95. [CrossRef]
17. Wang, Y.S.; Liao, Y.W. Assessing eGovernment systems success: A validation of the DeLone and McLean model of information systems success. *Gov. Inf. Q.* **2008**, *25*, 717–733. [CrossRef]
18. Petter, S.; DeLone, W.H.; McLean, E.R. Measuring information systems success: Models, dimensions, measures, and interrelationships. *Eur. J. Inf. Syst.* **2008**, *17*, 236–263. [CrossRef]
19. Davis, F.D. Perceived usefulness, perceived ease of use, and user acceptance of information technology. *MIS Q.* **2019**, *13*, 319–340. [CrossRef]
20. Lussak, A. Triple A Strategy to Improve Supply Chain Performance in Semarang City SMEs. *Int. J. Sci. Technol. Res.* **2020**, *9*, 218–222.
21. Petter, S.; McLean, E.R. A meta-analytic assessment of the DeLone and McLean IS success model: An examination of IS success at the individual level. *Inf. Manag.* **2009**, *46*, 159–166. [CrossRef]
22. Lussak, A.; Abdurachman, E.; So, I.G.; Setiowati, R. The effect of financial performance, service and product innovation on the survival of small businesses mediated by TAM. *Int. J. Innov. Creat. Change* **2019**, *7*, 259–275.
23. Prybutok, V.; Zhang, X.; Ryan, S. Evaluating leadership, IT quality, and net benefits in an e-government environment. *Inf. Manag.* **2008**, *45*, 143–152. [CrossRef]
24. Teo, T.S.H.; Srivastava, S.C.; Jiang, L. Trust and electronic government success: An empirical study. *J. Manag. Inf. Syst.* **2008**, *25*, 99–131. [CrossRef]
25. Casanovas, P.; Sartor, G.; Casellas, N.; Rubino, R. *Computable Models of the Law*; Springer: Berlin, Germany, 2018.
26. Cerrillo, M.; Fabra, P. *E-Justice: ICT in the Court System*; IGI Global: Hershey, PA, USA, 2018.
27. Contini, F.; Lanzara, G.F. *ICT and Innovation in the Ublic Secor*; Pelgrave: London, UK, 2018.
28. Hung, S.Y.; Tang, K.Z.; Chang, C.M.; Ke, C.D. User acceptance of intergovernmental services: An example of electronic document management system. *Gov. Inf. Q.* **2009**, *26*, 387–397. [CrossRef]
29. Doll, W.J.; Torkzadeh, G. The measurement of end-user computing satisfaction. *MIS Q.* **1988**, *12*, 259–274. [CrossRef]

proceedings

MDPI

Proceeding Paper

Improving IT Self-Efficacy, Experience and Training, and Technological Anxiety's Impact on Remote Work Quality †

Ibnu Darmawan [1,*] and Assed Lussak [2]

[1] Communication Department, Islamic University of Indonesia, Kaliurang St. Km 14.5, Sleman 55584, Indonesia
[2] Baldwin Boyle Group, GridAKL Building, Wynyard Quarter, 12 Madden St., Auckland 1010, New Zealand
* Correspondence: 213210501@uii.ac.id

† Presented at the International Academic Symposium of Social Science 2022, Kota Bharu, Malaysia, 3 July 2022.

Abstract: The COVID-19 pandemic that happened in 2020 has forced people worldwide to practice remote work with little or no prior experience, working for companies and organizations that are most likely unprepared for this change. The quality of remote work then becomes an ultimate question, whether people can adapt or not, and what determinants are influencing it. Earlier, a remote work self-efficacy model was developed to accommodate such situations. However, the development was meant to assess virtual companies that have reliable ICT and enough training for the employees. The research tries to dig deeper into its antecedents' components amid unpreparedness. There were 46 respondents in the Jakarta, Central Java, and Yogyakarta provinces participating in the study conducted in May 2021, when increasing virus transmission reinforced companies to close their premises. The study illustrates how two-way conversations that generate social persuasion, physiological and emotional states, and self-efficacy affect remote job quality, which differs from previous research.

Keywords: communication; self-efficacy; remote work; social persuasion; physiological; emotion

Citation: Darmawan, I.; Lussak, A. Improving IT Self-Efficacy, Experience and Training, and Technological Anxiety's Impact on Remote Work Quality. *Proceedings* **2022**, *82*, 59. https://doi.org/10.3390/proceedings2022082059

Academic Editor: Mohamad Rahimi Mohamad Rosman

Published: 16 September 2022

1. Introduction

The research topic of distant work is significant to a wide range of domains, including information systems, communication, and psychology, and it regularly contradicts findings [1]. Managing remote workers or running a virtual company is critical and needs to be well understood. Advances in information and communication technology (ICT) provide the infrastructure needed to guide the development of new organizational forms [2]. Furthermore, in 2020, COVID-19 pandemic conditions drove millions of people all over the world to practice remote work with little or no prior experience, working for firms and organizations that are most likely unprepared for this transformation [3,4]. To handle such conditions, a self-efficacy model for remote working has been developed. Firms that learn how to increase employees' self-efficacy judgments and to conduct remote working activities will have greater performance [5,6].

To be significantly implemented in a remote working environment [7], self-efficacy theory can include various aspects. However, because almost none of these studies were carried out at a time when far-flung operations were being carried out on such an unprecedented scale as they were during the pandemic, some of the previously gathered knowledge on far-flung operations may have little contextual relevance in the context of the pandemic. The study investigated workers' self-efficacy impact on remote working quality, as well as their experience, training, social relationships, and emotional and psychological states.

2. Literature Review

Among the fundamental models of technical communication, Shannon and Weaver identified the transmission model as one of the fundamental models that defines how messages are conveyed from an encoding sender to a decoding receiver [8]. What matters is whether or not the recipient understands what was meant by the sender. The phrase two-way communication refers to a procedure in which the receiver offers feedback and engages with the sender on both sides of the conversation. The sender strives to explain what he intended and double-checks if the recipient understood what he was saying. In addition, the sender can compare different views and supply extra information, which helps to avoid misinterpretation [9]. To communicate effectively, one must be able to ask questions, receive replies, and trade information [10], among other things. It is only through cooperation and teamwork that effective communication could be accomplished between management and employees [9]. To integrate the self-efficacy theory into major remote management challenges highlighted in the literature, a research model was built. The following discussion has been broken down into three sections to make further examination of this model: (1) self-efficacy antecedents, (2) self-efficacy judgments, and (3) self-efficacy outcomes.

Most workers had little remote work experience prior to the pandemic. Almost none of the organizations were also prepared to support this practice [4,11]. Whereas, individuals can learn about their past performance accomplishments through previous experience and instruction. The research showed that the longer someone has worked remotely, the easier they deal with the situation. Relevant training will also provide knowledge regarding performance achievement to people's self-efficacy views [5]. Information and technology (IT) are a significant enabler that can be a critical responsibility for successful remote work of the remote work environment [7,12]. This accelerated shift to digital communication impelled new IT knowledge and skills. As a result, formal and informal training sessions should be held [13].

In line with those studies, Staples found that people's experience and training using remote-access technology have an impact on individual self-efficacy assessment. Self-efficacy indicates that the more training people have in terms of available IT, the more successful they should be able to use it [5]. Giving an example or modeling can increase perceived efficacy. It teaches the spectators effective methods to deal with frightening or difficult situations and allows an observer to believe that potential problems can be managed more than before [6]. Effective remote work or management practiced by managers represents a source of information modeling that can affect workers' own judgment of their ability in doing effective remote work or tasks [5]. Management should provide consistent feedback based on an evaluation of the remote worker's achievement of the goals [14]. Individuals and organizations cannot evolve in ways that meet the standards of others without evaluative feedback [15]. Prior studies found that feedback is critical because it identifies significant communication barriers such as differences in background, interpretations of words, and reactions of emotion [16].

Feedback relates to a response to a person's behavior. It has an impact on whether that behavior will be continued [17]. It is, however, evaluative when it provides an assessment of behavior in relation to performance criteria [18]. Based on communication theory, evaluative feedback is a type of asymmetrical communication task, it has been proven to improve the performance of feedback receivers [19]. Technological anxiety, or well known as technostress, is defined as stress which users experience because of application multitasking, constant connectivity, information overload, frequent system upgrades, and consequent uncertainty, continual relearning and consequent job-related insecurities, and technical problems associated with the organizational use of information and technology [20]. Employee productivity can be influenced by the physical work environment [20,21]. Recent studies suggest the importance of creating a separate, suitable work environment by maintaining a clear boundary between work and home life. Physical working conditions could influence assessments of self-effectiveness. Thus, it is a serious issue for a remote worker [7].

However, it is critical to provide timely and forthright feedback in order to address potential issues before they become a source of resentment in remote workers. Advances in data communication technology have made it possible for any employee, regardless of their location, to join a company's virtual network [17]. Further, the perceived accessibility and ease of collaboration and information sharing via the chosen communication medium has an impact on team interactions and cohesiveness. As a result, connectivity level becomes a key indicator for people when deciding whether they can perform effectively [5]. Information and communication technologies (ICTs) can be used to improve communication, as well as assist in the maintenance of positive relationships in workgroups, particularly to foster the trust and cooperation required for virtual teamwork.

The ability to use technology is a critical part of employees' ability to effectively perform in the environment of remote management [5]. Later, Marques found that remote work requires specific abilities, skills, and knowledge of IT. High levels of information and technology self-efficacy could boost remote self-efficiency and the ability of remotely managed employees to work efficiently. Prior studies found a strong link between self-efficacy and task performance [6]. In the present context, employees with a high level of remote self-effectiveness believe they can accomplish tasks that permit remote work more efficiently.

Thus, in general, they become more effective remote workers. There is a linear, positive relationship between remote work and job satisfaction, implying that employees who work remotely are more frequently satisfied [14]. Furthermore, employee satisfaction perception in a virtual environment varies depending on management support and activities, also its remote skills [7]. These results together suggested that positive assessments of the ability to execute have a positive influence on job satisfaction. Another research also discovered a significant link between self-efficacy and coping ability. It was proposed as a valuable personal resource that can help the abilities to improve. Self-efficacy is the belief in individual ability to complete difficult or novel tasks and to deal with adversity in challenging situations [16].

Feelings of isolation were found to lower an individual's organizational commitment [7]. Remote workers are more willing to be engaged when they see connections between their own values and the values of their company. Employee engagement is critical for any organization, but it is especially critical for companies with remote employees. Employee engagement can be measured at a high level by an employee's commitment to an organization and motivation to achieve better. Self-efficacy to cope with stress was defined as the belief in one's personal resources to handle stressful conditions in an effective and competent approach. Earlier, studies underlined the link between self-efficacy and stress in the workplace. An increase in self-efficacy has been negatively associated with stress.

Self-efficacy judgments themselves are influenced by four factors, described in detail above [6,20]: performance accomplishments in which previous experiences of success result in boosting self-efficacy and in failure lowering it, the vicarious experience of observing and copying others' behavior who have completed successful projects, social persuasion in which they receive coaching or evaluative assessment, and physiological and emotional states as a reaction to a certain job that causes a lack of discernment or anxiety. Self-efficacy itself is translated as people's belief in their ability to summon the cognitive, motivational, and behavioral resources required to compete in a certain situation. People can have a strong or weak belief. Therefore, people with a strong self-efficacy belief are putting more effort into their work so that they can manage the challenge, while those who have a weak one put less effort into the work and tend to give up and quit [5]. Thus, the construct of the research is shown in Figure 1, below.

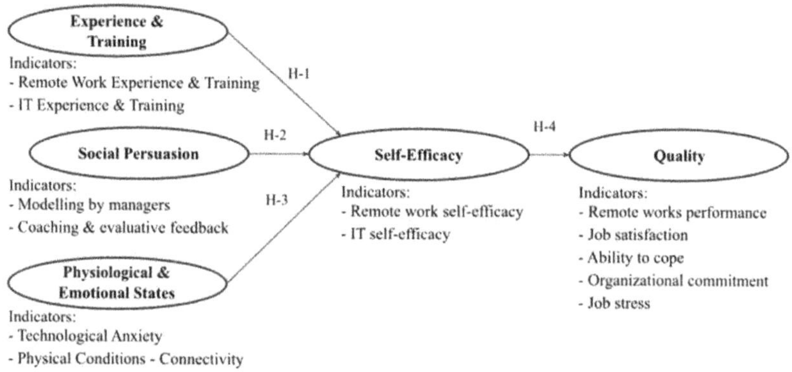

Figure 1. Construct.

3. Research Method

The study used a quantitative research approach, chosen to investigate planned relationships among constructs within the model. It provided a form for respondents to facilitate an assortment of knowledge from an oversized and geographically spread sample. Quantitative methods are also used to provide information for prediction, correlation, causation, and generalizability. Data were collected during May 2021 using Google Form, which has been sent through personal messages.

The respondents were selected from several locations in the Jakarta, Central Java, and Yogyakarta provinces with the random sampling method, and 46 people participated in the survey. The participants represented work function in the following sectors: Human Resources, Sales, Marketing, Operations, Inventory, Finance/Accounting/AP/AR, Administration, System/Technical Support, Creative/Design, Client Relations, Training/Teaching/Coaching, Research and Development, Legal, Health Care, and others. The questionnaires were close-ended questions containing single-choice response questions and using a five-point Likert scale, where 1 means strongly disagree, 2 means disagree, 3 indicates neutral, 4 reflects agreement, and 5 states strongly agree. All data gathered were then analyzed with SPSS Amos 23.0 (Armonk, NY, USA) using descriptive statistics.

4. Methodology

The quantitative study was carried out to Indonesian courts' users in the Greater Jakarta area, lawyers respectively. It covered all Jakarta cities and regencies, Bogor Regency and City, Depok, South Tangerang, Tangerang Regency and City, and Bekasi Regency and City. Questionnaires were collected from 30 attorneys who are able to carry out their duties with the support of e-court content management systems, which are more popularly known as e-Court. During a legal procedure, this application is a content management system (CMS) that was developed to help automate court operations and provide assistance. It provides support for case management tasks such as case planning and tracking, as well as the scheduling of hearings and other court appearances in general. In particular, CMS assists in the operation of the court through the utilization of capabilities that include calendaring and scheduling, docketing, case information management, ticklers, notes, and case linkages.

The comprehensive questionnaire was out in February 2022 in order to collect information. In addition to gathering information on user profiles and the e-court software that is being utilized, the questionnaire was designed to investigate the elements that are related with the success of e-courts. In the beginning, comments from both court users and information systems specialists were solicited. This was to undertake preliminary and pilot testing to analyze and validate the approved procedures. Using data from the IS made publicly available in the past, the questionnaire was first crafted, and after that, it underwent testing and analysis. The components had some minor alterations so that they would function better within the e-court program. The purpose of the 15 questions in the survey was to investigate five primary components of the system. These were the quality of the system, the quality of the information, the usage of the system, the happiness of the users, and the individual impact. The ease of use of a system was used as one of the criteria for measuring system quality in the research. The scale consisted of two items. The quality of the information that is produced by e-court applications can be evaluated along three dimensions: the material's substance, its correctness, and how it is presented.

To evaluate a user's level of dependence on the information system that was available to them, a reliance measure was applied. The utilization of IS was evaluated based on a single item of this metric. User satisfaction is a measurement of how satisfied users are with a system, and it was computed by modifying a single item based on the work done by Rai and colleagues to capture the level of enjoyment provided by the system [21]. Individual impact, on the other hand, is defined as the extent to which application software has been successful in improving the quality of the user's work, making the end user's job easier, reducing the amount of time spent on the end user's job, and assisting the end user in meeting the end user's job needs and requirements. The four-item user-performance metric was modified in order to arrive at the result. On a scale from one to five, the allegations were rated, with a score of one indicating major disagreement and a score of five indicating strong agreement.

5. Results and Discussion

5.1. Demographic Profile of Respondents

Table 1 below shows that based on respondents' age, 89.1% of them are millennials with an age range of 25–40 years.

Table 1. Age of Respondent.

		Freq.	Percent	Valid Percent	Cumulative Percent
Valid	<25 years	5	10.9	10.9	10.9
	25–40 years	41	89.1	89.1	100.0
	Total	46	100.0	100.0	

5.2. Results of Research Instrument

Based on Table 2, the experience and training variable obtained a minimum value of 11, a maximum value of 45, and an average value (mean) of 29.65.

Table 2. Descriptive.

	N	Min	Max	Mean	Std. Deviation
Experience and Training	46	11	45	29.65	7.109
Social Persuasion	46	18	70	50.52	12.976
Physiological and Emotional States	46	14	45	29.80	6.131
Self-efficacy	46	14	60	45.20	9.404
Quality	46	28	91	65.83	14.821
Valid N (listwise)	46				

Table 3 presents how Predictors: (Constant), Physiological and Emotional States, Experience and Training, and Social Persuasion simultaneously affect the dependent variable by 69.0%. Moreover, Table 4 below shows how the effect is statistically significant. Table 5 then provides calculations on how Social Persuasion and Physiological and Emotional States significantly affect while Experience and Training does not.

Table 3. Physiological and Emotional States.

Model	R	R Sq.	Adjst. R Sq.	Std. Error of the Estimate
1	0.843	0.710	0.690	5.237

Table 4. ANOVA.

Model		Sum of Sq.	df	Mean Sq.	F	Sig.
1	Regression	2827.146	3	942.382	34.355	0.000
	Residual	1152.093	42	27.431		
	Total	3979.239	45			

Dependent Variable: self-efficacy, Predictors: (Constant), Physiological and Emotional States, Experience and Training, Social Persuasion.

Table 5. Coefficients.

Model		Unstandardized Coefficients		Std. Coeff.	t	Sig.
		B	Std. Error	Beta		
1	(Constant)	5.888	4.181		1.408	0.166
	Experience and Training	0.131	0.142	0.099	0.927	0.359
	Social Persuasion	0.332	0.085	0.458	3.920	0.000
	Physiological and Emotional States	0.626	0.165	0.408	3.802	0.000

Dependent Variable: self-efficacy, The F test results in 34,355 in its calculated F and the probability is 0.000. As sig F_{count} < 5% (0.000 < 0.05).

Figure 2, as below, shows how the regression coefficient value of experience training on quality is 0.203, while the coefficient value of experience and training on quality through self-efficacy as a mediating variable is 0.066.

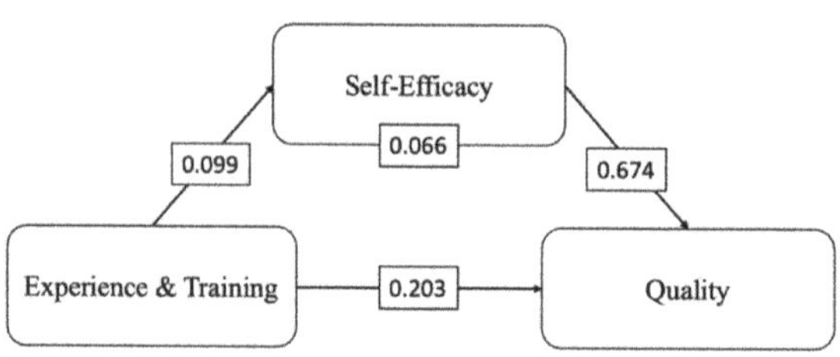

Figure 2. Path analysis of experience and training's effect on quality.

It follows that self-efficacy is unable to moderate the effects of experience and training on quality since the value coefficient is lower than the direct influence of experience and training.

Self-efficacy, on the other hand, as shown in Figure 3, can act as a mediator between the impacts of social persuasion on quality. The coefficient value is even bigger than the direct effect, as seen by the data presented above.

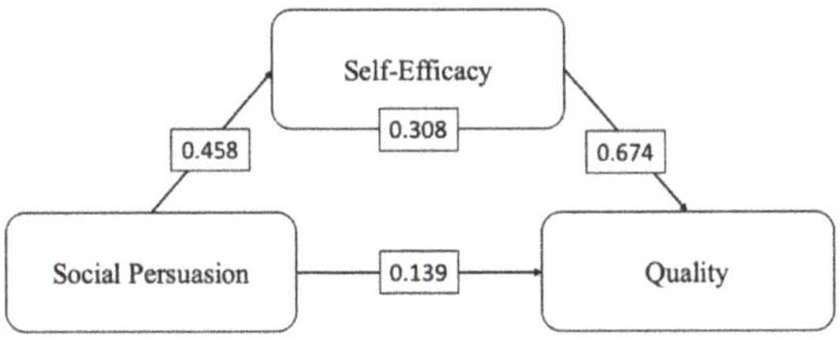

Figure 3. Path analysis of social persuasion's effect on quality.

6. Conclusions

Virtual organizations have become increasingly popular in recent years, particularly during pandemics and disasters. Employees in a virtual organization operate primarily in areas that are far from their peers and managers, posing numerous management and communication issues to the firm. The awareness of the key issues and critical drivers must be strengthened in order for firms to make effective transitions to this new method of conducting business. Using the self-efficacy theory, the current study contributes to this goal by predicting links between the remote work self-efficacy and the effects of self-efficacy. The validity of these interactions was investigated through a survey of varied individuals who worked distant from their colleagues and bosses in a variety of companies.

Overall, the findings demonstrate that remote employees' self-efficacy has a significant impact on their ability to perform well at their remote jobs. Because the independent variables of experience and training do not have an effect on remote work quality, self-efficacy plays a significant role in mediating the relationship between the two variables. As a result, the findings can also be interpreted as a positive outlook during a pandemic when people are not ready and well-prepared to deal with such a crisis, so long as there is strong two-way communication among the company members.

Author Contributions: Conceptualization, A.L.; methodology, A.L.; software, I.D.; validation, I.D.; formal analysis, I.D.; investigation, I.D.; resources, A.L. and I.D.; data curation, I.D.; writing—original draft preparation, A.L. and I.D.; writing—review and editing, A.L. and I.D.; visualization, I.D.; supervision, A.L.; project administration, A.L.; funding acquisition, not applicable. All authors have read and agreed to the published version of the manuscript.

Funding: This research received no external funding.

Institutional Review Board Statement: Not applicable.

Informed Consent Statement: Informed consent was obtained from all subjects involved in the study.

Data Availability Statement: Not applicable.

Acknowledgments: Not applicable.

Conflicts of Interest: The authors declare no conflict of interest.

References

1. Ale Ebrahim, N.; Ahmed, S.; Taha, Z. Virtual Teams: A Literature Review. *Aust. J. Basic Appl. Sci.* **2019**, *3*, 2653–2669.
2. Allen, T.D.; Golden, T.D.; Shockley, K.M. How Effective Is Telecommuting? Assessing the status of our scientific findings. *Psychol. Sci. Public Interest* **2015**, *16*, 40–68. [CrossRef] [PubMed]
3. Kniffin, K.M. *COVID-19 and The Workplace: Implications, Issues, and Insights for Future Research and Action*; Cornell University: New York, NY, USA, 2021.
4. Neeley, T. 15 Questions About Remote Work, Answered. *Harv. Bus. Rev.* **2020**, *8*, 3–7.
5. Staples, D.; Hulland, J.S.; Higgins, C.A. A Self-Efficacy Theory Explanation for the Management of Remote Workers in Virtual Organizations. *Organ. Sci.* **1999**, *10*, 758–776. [CrossRef]
6. Bandura, A. Reflections on Self-efficacy. *Adv. Behav. Res. Ther.* **1978**, *1*, 237–269. [CrossRef]
7. Staples, D.S. An Investigation of Some Key Information Technology-Enabled Remote Management and Remote Work Issues. *Australas. J. Inf. Syst.* **2017**, *5*, 1–4.
8. Shannon, C.E.; Weaver, W. *The Mathematical Theory of Communication*; University of Illinois Press: Champaign, IL, USA, 1949.

9. Lussak, A.; Romdoni, M. An exploratory study of youth peace movement effectiveness in yogyakarta. *Int. J. Sci. Tech. Res.* **2019**, *8*, 3579–3581.
10. Lussak, A. Triple A Strategy to Improve Supply Chain Performance in Semarang City SMEs. *Int. J. Sci. Tech. Res.* **2020**, *9*, 218–222.
11. Wang, B.; Liu, Y.; Qian, J.; Parker, S.K. Achieving Effective Remote Work During The Covid-19 Pandemic: A Work Design Perspective. *Appl. Psychol.* **2021**, *70*, 16–59. [CrossRef] [PubMed]
12. Weimann, P.; Pollack, M.; Scott, E.; Brown, I. Enhancing Team Performance Through Tool Use: How Critical Technology-related Issues Influence the Performance of Virtual Project Teams. *IEEE Trans. Prof. Commun.* **2013**, *56*, 332–353. [CrossRef]
13. Gigauri, I. Remote Working Concerns During the COVID-19 Pandemic. *Int. J. Soc. Sci. Econ. Res.* **2020**, *5*, 2803–2818. [CrossRef]
14. Staples, D. *Making Remote Workers Effective in Telecommuting and Virtual Offices: Issues and Opportunities*; Idea Group Publishing: Hershey, PA, USA, 2021.
15. Lussak, A.; Abdurachman, E.; So, I.G.; Setiowati, R. The effect of operational improvement on business survival with moderating role of continuous improvement (Case study: Food and beverage small business in Bali). *Int. J. Sci. Tech. Res.* **2019**, *8*, 617–621.
16. Bovee, C.L.; Thill, J.V. *Business Communication Today*; McGraw-Hill: New York, NY, USA, 2002.
17. Lussak, A.; Abdurachman, E.; So, I.G.; Setiowati, R. The effect of financial performance, service and product innovation on the survival of small businesses mediated by TAM. *Int. J. Innov. Creat. Change* **2019**, *7*, 259–275.
18. Kluger, A.N.; DeNisi, A. The effects of feedback interventions on performance: A historical review, a meta-analysis, and a preliminary feedback intervention theory. *Psychol. Bull.* **1996**, *119*, 254–284. [CrossRef]
19. Tarafdar, M.; Tu, Q.; Ragu-Nathan, B.; Ragu-Nathan, T.S. The Impact of Technostress on Role Stress and Productivity. *J. Manag. Inf. Syst.* **2007**, *24*, 301–328. [CrossRef]
20. Gist, M.E.; Mitchell, T.R. Self-efficacy: A Theoretical Analysis of Its Determinants and Malleability. *Acad. Manag. Rev.* **2022**, *17*, 183–211. [CrossRef]
21. Otterbring, T.; Pareigis, J.; Wästlund, E.; Makrygiannis, A.; Lindström, A. The Relationship Between Office Type and Job Satisfaction: Testing a Multiple Mediation Model Through Ease of Interaction and Wellbeing. *Scand. J. Work. Environ. Health* **2018**, *44*, 334–338. [CrossRef] [PubMed]

proceedings

Proceeding Paper

The Preliminary Exploration of Multimedia Effects in Sculpture Creation Inspired by Maslow's Hierarchy of Needs †

Xing Yu Liu * and Kin Meng Cheng

De Institute of Creative Arts and Design, UCSI University, Cheras, Kuala Lumpur 56000, Malaysia
* Correspondence: 1002161225@ucsiuniversity.edu.my

† Presented at the International Academic Symposium of Social Science 2022, Kota Bharu, Malaysia, 3 July 2022.

Abstract: Due to the limits of conventional sculpting, some artists may be unable to express themselves creatively. The research question is therefore posed: What are the factors that influence the use of multimedia effects in sculpture creation and Maslow's hierarchy of needs? The research objective of was to identify the factors that influence the use of multimedia effects in sculpture creation and Maslow's hierarchy of needs and to develop new creative methods for sculptors. The study used the purposive sampling method to invite six traditional sculpture artists and six multimedia sculpture artists to participate in interviews in an online format. Six factors were obtained from the analysis.

Keywords: multimedia effects; sculpture creation; Maslow's hierarchy of needs

check for updates

Citation: Liu, X.Y.; Cheng, K.M. The Preliminary Exploration of Multimedia Effects in Sculpture Creation Inspired by Maslow's Hierarchy of Needs. *Proceedings* **2022**, *82*, 62. https://doi.org/10.3390/proceedings2022082062

Academic Editor: Mohamad Rahimi Mohamad Rosman

Published: 19 September 2022

1. Introduction

Sculpture art forms often employ three methods of creation: carving, engraving, and modelling, according to the conventional notion. Various soft (such as plaster, resin, clay, etc.) or hard materials (such as wood, stone, metal, jade, etc.) can be carved and engraved to create a space of visible and tactile art images that mirror social life and communicate the artist's aesthetic feelings, emotions, and aspirations. With the advancement of science, technology, and changes in the aesthetic notion of the audience, the modern sculpture is increasingly employing new materials, concepts, and scientific and technical tools [1]. On the other hand, the classic sculptural notion is increasingly being superseded by new conceptions in contemporary circumstances. Because the new sculptural concepts and principles are still developing, individuals are not yet entirely removed from their thinking. The value of these thought concepts such as truth, goodness, and beauty has caused a significant distortion, and the traditional way of creating sculpture has failed to meet the contemporary sculpture art sought, and sculpture has also become blurred by the impact of various forms in this general environment [2]. As a result, many sculptors' goals and needs have shifted considerably throughout their work.

Multimedia exploration has offered artists a modern, personalised language to work within the creation of art. It has shattered the barrier between technology and sculpture art, allowing for the production of new methods and forms of sculpture [3]. Sculpture artists are no longer limited to using traditional materials such as clay, wood, stone, and metal to create their works; instead, they are increasingly incorporating multimedia effects such as audio, graphics, video, text, and light to create and communicate their emotional responses, resulting in multimedia sculptures [4]. The artists continue to experiment with new sculptural materials to communicate their emotions via their art, satisfying both their aesthetic and self-actualisation needs.

The study used a version of Maslow's eight levels of needs. In 1943, Maslow first proposed a theory of five levels of human needs in *A Theory of Human Motivation*, with self-actualisation needs introduced as the highest need [5]. The theory was expanded in 1970 to include cognitive and aesthetic needs [6], and later, transcendental needs [7]. Aesthetic and self-actualisation needs are located at levels six and seven in Maslow's hierarchy of needs theory (Figure 1) and are second only to transcendental needs, which are the highest level as higher-level spiritual needs [6]. In the process of creating art, artists are continually seeking these two spiritual requirements [8].

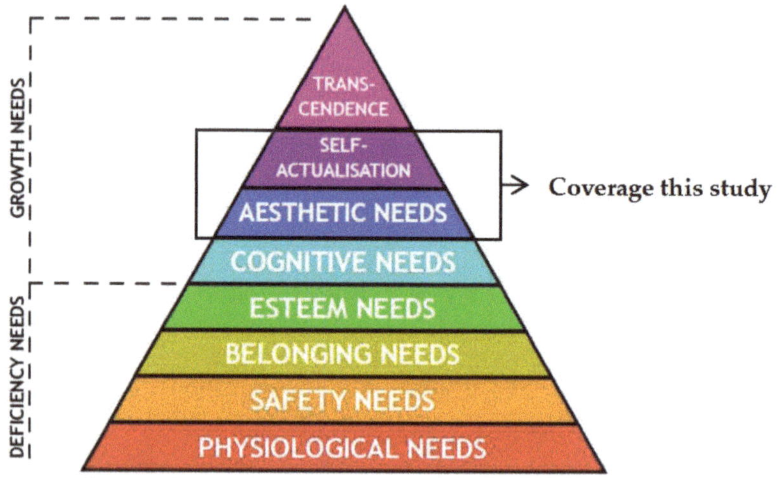

Figure 1. Maslow's hierarchy of needs model.

The research questions are: What are the factors that influence the use of multimedia effects in sculpture creation and Maslow's hierarchy of needs? What are the multimedia effects being utilised in sculpture creation inspired by Maslow's hierarchy of needs?

This research aimed to explore the multimedia effects being utilised in sculpture creation inspired by Maslow's hierarchy of needs and identify the factors that influence the use of multimedia effects in sculpture creation and Maslow's hierarchy of needs.

The basic descriptions of "sculpture creation in artists" and the development of sculpture-creation materials is given first, followed by a more detailed explanation of "multimedia, multimedia technology" and "art, effect", as well as a discussion of the possibilities and methods of combining the two in sculpture. In Maslow's hierarchy of needs, the relationship between aesthetic needs, self-actualisation needs, and artistic creation is discussed. Finally, the factors that influence the use of multimedia effects in sculpture creation and Maslow's hierarchy of needs are examined.

Significance

The significance of the exploration of multimedia effects in sculpture creation inspired by Maslow's hierarchy of needs has three main aspects.

Firstly, it is an extension of the way and form in which sculpture is created and promotes the development of sculptural art. The use of multimedia effects in sculpture creation can break down the boundaries between art, science, and technology, adding more artistic interest and visual tension to sculpture, thus injecting new connotations, expanding the creative space of sculpture art, and promoting its multi-dimensional development.

Secondly, it has an impact on the way sculptors create and think, creating more from different perspectives of needs and achieving a positive interaction between themselves and sculpture in a spiritual sense. Meanwhile, it can also bring diversified forms of expression to sculpture works, and the combination of media effects and art adds different colours to sculpture art, enriching the spirit of the times and the. cultural connotation of sculpture.

Finally, the realisation of multiple mediums in sculpture art allows the public to see more forms of sculpture art, satisfying the artistic needs of visitors and playing a positive role in the art education of the next generation.

Overall, research on the use of multimedia effects in sculpture creation, especially at the level of human psychological needs, is necessary and has a certain forward-looking and practical significance.

2. Research Conceptual Framework

The research focuses on the factors that influence the application of multimedia effects in sculpture creation and Maslow's hierarchy of needs. The relationship between the application of multimedia effects in sculpture creation and Maslow's hierarchy of needs is set out first. The research framework (Figure 2) consisted of three elements: the multimedia effects on sculpture creation, the influence factors, and Maslow's hierarchy of needs. The relationship between these influences is that the use of multimedia effects (audio, graphic, video, light) in the creation of a sculpture by the artist can satisfy the artist's own aesthetic and self-actualisation needs, and the factors that shape this relationship are an increase in the artist's aesthetic understanding, a greater emphasis on aesthetic enjoyment, and an increase in innovative ability and innovative thinking.

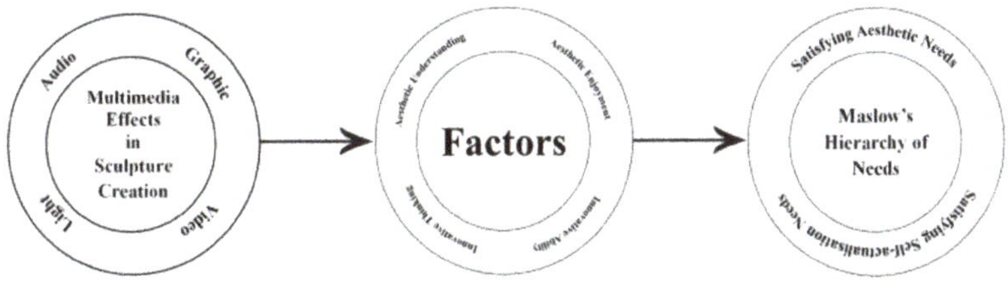

Figure 2. The preliminary framework of multimedia effects in sculpture creation with Maslow's hierarchy of needs.

2.1. *Sculpture Creation Artists and Sculpture Materials*

Sculpture creation is an entirely spatial embodiment of the artist's personality, and the sculptor expresses his thoughts and feelings via his work. Contemporary artists use various media materials to display and expand a large amount of artistic information, arrange and combine materials of various forms and structures in a specific way to form spatial symbols, and express their ideas and emotions in various forms such as realistic, abstract, decorative, and installation during the creation process [9].

Because of the constant growth of colonial times and science and technology, sculpture materials have steadily evolved from traditional materials such as clay, wood, bone, ivory, jade, and stone to current materials that can generate multimedia effects such as video, video, and graphics [10], to support modern artists in their pursuit of a distinctive style of expression and practical knowledge of life [11].

2.2. *Multimedia, Technology, Arts, and Effects*

Multimedia is a term that refers to a collection of different media (or mediums) [12]. Multimedia technology is the systematic combination of animation, audio, graphics, images, video, and other media, as well as their corresponding and reasonable processing, such as animation production, audio processing, graphics processing, video processing, and so on, to create a logical and reasonable connection [13]. The use of multimedia or multimedia technology in art creation innovates and expands the notion of art, enriching the creative form of art and giving the artist a new space of expression to convey his or her artistic point of view using new methods and materials.

Multimedia artworks combine a variety of mediums and artistic expressions to create a powerful aesthetic impression on viewers. Multimedia art may also combine visual, aural, temporal, and spatial art, giving multimedia works a multi-sensory influence on viewers and heightening the tension in the artworks [14]. Multimedia impacts in sculpture art may help it evolve and develop more effectively and help artists to more directly express emotions, vent inner desires, and realise their demands.

Audio on sculpture, graphics on sculpture, video on sculpture, and light on the sculpture are examples of multimedia effects employed in sculpture. These emerging forms of sculpture expression can help sculpture creators use diversified creation methods, compelling their audience through visual and auditory means to realise the communication of the content of the work, becoming a bridge of emotional communication between sculptors and visitors, and realising the Aesthetic needs and self-actualisation needs of sculptors.

2.3. Aesthetic Needs and Self-Actualisation Needs in Maslow's Hierarchy of Needs and Art

Aesthetic and self-actualisation requirements are higher-level spiritual desires in Maslow's hierarchy of needs, which are second only to transcendental needs in the 1970 edition of Maslow's hierarchy of needs [15]. When art is made, it must first fulfil the artist's spiritual needs as a derivative of spiritual needs. Aesthetics are an interior sense, a sensation of objects experienced during mental activity. The sense of objects in the course of life is the aesthetic necessity stated here for human beings. It is also the most fundamental characteristic of artistic creativity [16]. People's desire to develop their abilities and their potential in society is known as self-actualisation. Self-actualisation, or the desire for self-actualisation, aids in the development of the desire to utilise all of one's latent skills. According to Onah (2015), the urge for self-actualisation allows people to be more creative in society [17].

The use of multimedia elements in sculpture provides artists with new approaches and tools to achieve their aesthetic and self-needs. The varied multimedia effects make it possible to create sculptures with more possibilities, while allowing artists to show their creativity and granting full play to their artistic talents, allowing them to greatly satisfy their aesthetic and self-actualisation needs when creating their works.

3. Research Method

This study uses qualitative research methods. Qualitative research methods focus on the collection and analysis of non-digital data (e.g., text, video, or audio) used to understand the concepts, opinions, or experiences of the research participants [6]. A qualitative research method was used due to the exploration of human emotions and feelings towards multimedia effects in sculpture creation.

As one of the primary methods in social science, a qualitative method was utilised as a bridge between the research and design process, especially in the exploration of multimedia effects in sculpture creation inspired by Maslow's hierarchy of needs.

A case study is an in-depth, detailed examination of a specific case (or cases) in the real world [18]. This type of research allows for the gathering of extensive data on artists' perspectives on multimedia effects in sculpture creation, as well as the comprehension, organisation, and detailed analysis of the research topic "Multimedia Effects on Sculpture Creation Inspired by Maslow's Hierarchy of Needs" as a research topic.

In qualitative research, interviews are the most popular method of data collecting, with the goal of conducting in-depth personal interviews with a limited number of respondents to learn about their perspectives on certain ideas, plans, or circumstances [19].

This study mainly used semi-structured interviews. The research question is about the changing needs of art creators brought about by the use of multimedia effects in sculpture creation. The scope of the study is relatively broad, involving both sculpture art and Maslow's theory of needs, and it is also a topic that favours feelings, so it was not possible to set standardised answers and responses. The semi-structured interview, with its semi-open and guided communication, made it easier for the interviewees to communicate freely within a certain range and to obtain research information and data effectively [20].

4. Sampling

This study used purposive sampling because this study had a specific target study population, and the characteristics and advantages of a purposive sampling approach were more likely to help the study obtain valid data. The 12 sculpture artists interviewed for this study were between the ages of 25 and 30 and had been working in the sculpture industry for at least 5 years, including six artists who create sculptures using only traditional materials (henceforth referred to as traditional sculpture artists) and six artists who have already used multimedia effects for their work (henceforth referred to as multimedia sculpture artists). According to the age structure redefined by WHO in 2015, people 25 to 30 years old are considered young people [21], an indispensable part of the artistic development of contemporary young artists who are more willing to try new things, catch up with avant-garde art trends, have a spirit of experimentation, and have great plasticity and future creative potential [22]. Sculptural creation needs young people with a creative and experimental spirit in order to move forward. Young people are also quicker to accept and explore multimedia effects. Therefore, the age range of the sample was set between 25 and 30 years old.

The research sites were located in Hebei Academy of Fine Arts and Tianjin Academy of Fine Arts. The sculpture major of Hebei Academy of Fine Arts is one of the key construction majors of the school; the teachers and students are involved in the design of campus landscape sculpture, mostly using traditional sculpture materials [23]. Tianjin Academy of Fine Arts is one of the eight outstanding art and art institutions in fine arts, as rated by the China Academic Degrees & Graduate Education Development Center (CDGDC) of China [24], of which the School of Experimental Art responds to the current social development and artistic requirements, emphasizing the integration of values across fields, industries, disciplines, and cultures [25]. Thus, the Hebei Academy of Fine Arts was the sampling site for traditional sculpture artists, and the Tianjin Academy of Fine Arts was the sampling site for multimedia sculpture artists.

5. Analysis and Results

The research used the data analysis method of thematic analysis, which was often used to analyse the information from this paper and to help researchers digest the data. Braun and Clarke (2006) developed a six-step process: familiarisation of data, initial coding, generating themes, reviewing themes, defining and naming themes, interpretation, and reporting, for the inductive analysis of the data [26].

After analysing the pre-test data, the results of the analysis were obtained to answer the two research questions respectively.

In order to answer the first research question, "What are the factor that influence use the Multimedia Effects on Sculpture Creation and Maslow's Hierarchy of Needs?", interview questions were designed for traditional and multimedia sculptors as follows (Table 1).

Table 1. Pre-test interview questions.

	For Traditional Sculpture Artists:	For Multimedia Sculpture Artists:
1	Do you think that the use of multimedia effects in sculpture creation can satisfy the **aesthetic needs** of the creator even more?	Do you feel that your **aesthetic needs** are more satisfied when using multimedia effects to create sculptures?
2	Do you think that the use of multimedia effects in sculpture creation can satisfy the **self-actualisation needs** of the creator even more?	Do you feel that your **self-actualisation needs** are more satisfied when using multimedia effects to create sculptures?

Ten of the twelve artists interviewed agreed that the use of multimedia effects in sculpture can satisfy their own aesthetic and self-actualisation needs, accounting for 83.3% of all artists interviewed.

In the interview responses about aesthetic needs, traditional sculpture artist A mentioned that "the use of multimedia in sculpture can meet some of my aesthetic needs in sculpture .... I think the use of multimedia effects is more capable of achieving the subjective feelings and subjective aesthetics that I want to express in my work. ...... It will focus more on the artist's own ideas, as well as his own creative philosophy. ... Multimedia effects, materials, etc. ...... to help me realise certain creative ideas".

The traditional sculpture artist A clearly expressed the aesthetic needs that can be met by the artist. The reason for this is that it enables them to realise their aesthetic expression and creative ideas, and the artist's own subjective aesthetic can also be better expressed, allowing the aesthetic concept and space to be changed and expanded. Traditional sculpture artist C also expressed this view clearly, believing that there are more channels and ways of creative expression and that these lead to a better interpretation of her own aesthetic concepts and emotions.

Multimedia sculpture artist D expressed the same view as above, saying "The use of multimedia effects in sculpture allows me to be more in tune with what I want to express, ... It also inspires more formal or conceptual ideas, it's a mutually reinforcing process ...... It is a new form ...... Creates a new aesthetic and way of seeing ...... opens up a new aesthetic space ...... The richness of one's own life experience can be released more".

In the interview responses about self-actualisation needs, according to traditional sculpture artist D: "I think it should keep my mind active and I can often come up with new ideas to make new work better......make me more open-minded, because ideas may change and I will think about how to combine with these new technologies to make more interesting and 'story-like' works". Multimedia sculpture artist E expressed a similar view: "Multimedia effects have made this kind of coding richer and more diverse for me, allowing for many more combinations of codes, adding more possibilities for creation and inspiring me to be very creative and to actively explore more ways of creating".

It is clear from this that the use of multimedia effects in sculpture can be more satisfying to the artist's self-actualisation needs, mainly because it makes their creative thinking more active, increases their creative ability, and inspires them to experiment with new possibilities.

The analysis of the interview data obtained a total of six themes, three of which were aesthetic needs and three of which were self-actualisation needs. The specific codes and definitions of the themes are presented in the Tables 2 and 3 below.

Table 2. Aesthetic needs, coding themes, and defining results.

Themes	Coding	Defining
Aesthetic Expression	Satisfying creative expression, Diversified forms of expression, Interactivity, Interestingness	Artists can unleash more of their aesthetic expression
Aesthetic Concept	The artist's creative philosophy, The artist's self-aesthetic, Spiritual core, Self-subjective aesthetic awareness, Realisation of creative philosophy	The artist develops their own aesthetic
Aesthetic Space	Aesthetic space, Aesthetic diversity	Artists have more aesthetic options

Table 3. Self-actualisation needs, coding themes, and defining results.

Themes	Coding	Defining
Innovative Thinking	Inspiring creative mindset, Active minds, Dispersing the mind, Inspiration	Artists are able to think outside the box and find more innovative ways of doing their work
Creative Ability	Creativity, Releasing, Originality	Improving the creative capacity of artists to create more new sculptural forms
Innovative Drive	Sense of achievement, Motivating	Giving artists more incentive to produce innovative sculptures

A preliminary framework for the relationship between the use of multimedia effects in sculpture and Maslow's hierarchy of needs was devised in the above section.

Through the interviews and the analysis of the interview data, the preliminary conclusion is that artists indicate that the use of multimedia effects in sculpture creation can satisfy their personal aesthetic expression, form their own subjective and creative aesthetic concepts, and expand the diversity of the aesthetic space, thus satisfying the aesthetic needs of artists; they indicate that the use of multimedia effects in sculpture creation can inspire innovative thinking in the process, stimulate creative ability, and enhance the innovative drive, thus satisfying the artist's need for self-actualisation.

Therefore, in the framework, the "multimedia effects on sculpture creation can satisfy the artist's own aesthetic and self-actualisation needs" is valid. However, the factors that influence the application of multimedia effects in sculpture creation and Maslow's hierarchy of needs are different. The factors that were collected and analysed after the interviews are Aesthetic Expression, Aesthetic Concept, Aesthetic Space, Innovative Thinking, Creative Ability, and Innovative Drive.

The study framework will be redrawn after analysis in conjunction with the post-test.

The second question, "What are the Multimedia Effects to be utilised on Sculpture Creation Inspired by Maslow's Hierarchy of Needs?" was extracted and coded from the collected interview responses to answer.

The two tables below (Tables 4 and 5) code the multimedia effects that traditional sculpture artists are aware of using in their sculpture creation and the multimedia effects that multimedia sculpture artists have used in their sculpture creation. After comparing these two sets of codes, it is clear that four main forms of multimedia effects are currently used in the creation of sculpture: light, audio, video and graphics.

Table 4. The use of multimedia effects in sculpture as understood by traditional sculpture artists.

Artists	Multimedia Effects			
	Light	**Audio**	**Video**	**Graphics**
Artist A	√		√	√
Artist B	√		√	
Artist C	√	√		
Artist D	√	√	√	
Artist E	√	√	√	√
Artist F	√	√	√	

Table 5. The use of multimedia effects in sculpture as understood by multimedia sculpture artists.

Artists	Multimedia Effects			
	Light	**Audio**	**Video**	**Graphics**
Artist A		√	√	√
Artist B	√	√	√	√
Artist C	√			
Artist D	√			√
Artist E				√
Artist F		√	√	

6. Discussion and Conclusions

After a preliminary study of the topic, it can be found that the use of multimedia effects in sculpture has a positive impact on the satisfaction of the artists' aesthetic and self-actualisation needs.

In the interview, 10 out of the 12 artists interviewed fully agreed that the use of multimedia effects in sculpture creation can satisfy their own aesthetic and self-actualisation needs, while the other two artists held different opinions. Four out of the six traditional sculpture artists agreed that the use of multimedia effects in sculpture creation can satisfy the artist's aesthetic and self-actualisation needs; one artist was neutral, believing that the use of materials depends on the content the artist is seeking; one artist agreed that it can satisfy the self-actualisation needs, but not aesthetic needs because he was not interested in multimedia effects himself; and all six multimedia artists agreed that the use of multimedia effects in sculpture can satisfy the aesthetic and self-actualisation needs of the artist.

Both the traditional and multimedia artists interviewed generally felt that the use of multimedia effects in sculpture could satisfy their own aesthetic and self-actualisation needs. The main factors are: satisfying their aesthetic expression, forming their own subjective and creative aesthetic concepts, and expanding the diversity of the aesthetic space, thus satisfying the aesthetic needs of artists; they indicate that the use of multimedia effects in sculpture creation can inspire innovative thinking in the process, stimulate creative ability, and enhance the innovative drive, thus satisfying the artist's need for self-actualisation. The results of the analysis were summarised in six thematic groups: Aesthetic Expression, Aesthetic Concept, Aesthetic Space, Innovative Thinking, Creative Ability, and Innovative Drive. In the interviews, the artists generally identified light, audio, video and graphic as multimedia effects that could be used in sculpture. Therefore, the use of multimedia effects in sculpture can help artists improve their aesthetic potential and innovative creative ability, prompting them to create more developmental artworks and promote the positive development of sculpture art.

In the future, I will continue to complete the collection and analysis of the post-test and compare it with the pre-test to draw conclusions. The analysis of the post-test will help me to find the positive impact of the use of multimedia effects in sculpture creation and to in-depth explore the relationship between the use of multimedia effects in sculpture and Maslow's hierarchy of needs.

Author Contributions: Conceptualization, X.Y.L. and K.M.C.; methodology, X.Y.L. and K.M.C.; formal analysis, X.Y.L.; investigation, X.Y.L.; resources, X.Y.L.; data curation, X.Y.L.; writing—original draft preparation, X.Y.L.; writing—review and editing, K.M.C. and X.Y.L.; visualization, X.Y.L.; supervision, K.M.C. All authors have read and agreed to the published version of the manuscript.

Funding: This research received no external funding.

Institutional Review Board Statement: Not applicable.

Informed Consent Statement: Not applicable.

Data Availability Statement: The data presented in this study are available on request from the corresponding author.

Acknowledgments: This study was completed under my supervisor's kind care and careful guidance, Cheng Kin Meng. I want to express my sincere gratitude and respect to Cheng Kin Meng for his serious scientific attitude, rigorous academic spirit, and dedication to excellence. Cheng Kin Meng provided me with careful academic guidance and gave me great encouragement and support in spirit. I would also like to thank the artists who participated in my research, as it was thanks to your help and support that I was able to overcome one difficulty or doubt after another. Finally, I would like to thank my family for being my most reliable support and supporting my research work without asking for anything in return.

Conflicts of Interest: The authors declare no conflict of interest.

References

1. Chen, C. Between Virtual and Real-when Interactive Media Is involved in Contemporary Sculpture. Master's Thesis, Sichuan Fine Arts Institute, Chongqing, China, 2019.
2. Liu, X. Research on the Crossover Forms of Contemporary Sculpture under the Influence of Science and Technology. Master's Thesis, Dalian Polytechnic University, Liaoning, China, 2020.

3. Wu, N.N. From Art to Multi-Media Art—A Preliminary Study of the Expressions in Multi-Media Art Works. *Art Panor.* **2009**, 64–65. (In Chinese)
4. Askinglot. What's Multimedia Sculpture? Available online: https://askinglot.com/what-is-multi-media-sculpture (accessed on 5 June 2020).
5. Maslow, A.H. A theory of human motivation. *Psychol. Rev.* **1943**, *50*, 370–396. [CrossRef]
6. Maslow, A.H. *Motivation and Personality*; Harper & Row: New York, NY, USA, 1970.
7. Maslow, A.H. *Religions, Values, and Peak Experiences*; Penguin: New York, NY, USA, 1970.
8. Abudurousuli · Keqikeahong. Art appreciation as a need of spiritual life. *J. Xinjiang Univ. Philos. Soc. Sci. Uyghur Ed.* **2013**, 88–92, (In Uighur Language).
9. Chen, L. A Brief Introduction to Sculpture Creation—An Analysis of the Process from Nothing to Something. *Moon Over West River* **2012**, *21*, 133. (In Chinese)
10. Bqdiaosu. Available online: http://www.bqdiaosu.com/bbs/boke660/ (accessed on 20 February 2018).
11. Zhang, S. Analysis of the correspondence between traditional sculpture language and modern sculpture language. *Sci. Educ. Lit. Exch. (Midterm)* **2009**, 269. (In Chinese)
12. Liao, J. Multimedia Art—The Trend of Art Creation. Master's Thesis, Xiamen University, Fujian, China, 2019.
13. Lv, L.R. The use and prospect of multimedia technology in urban sculpture. *Art Panor.* **2015**, 114–115. (In Chinese)
14. Tian, Y. Research on Multimedia Art Forms in Visual Communication Design. *Art Des.* **2016**, *2*, 39–41. (In Chinese) [CrossRef]
15. Maslow's Eight Basic Needs and the Eight Stage Developmental Model. Available online: https://the-mouse-trap.com/2007/12/14/maslows-eight-basic-needs-and-the-eight-stage-devlopmental-model/ (accessed on 14 December 2007).
16. Onah, F.O. *Human Resource Management*, 4th ed.; John Jacob's Classic Publishers Ltd.: Enugu, Nigeria, 2015.
17. Bhandari. What Is Qualitative Research? | Methods & Examples. Available online: https://www.scribbr.com/methodology/qualitative-research/ (accessed on 19 June 2020).
18. Bromley, D.B. *The Case-Study Method in Psychology and Related Disciplines*; Wiley: Chichester, UK, 1986.
19. Boyce, C.; Neale, P. *Conducting In-Depth Interviews: A Guide for Designing and Conducting In-Depth Interviews for Evaluation Input*; Pathfinder International: Watertown, MA, USA, 2006.
20. Kiunchiku. What Is a Semi-Structured Interview? Available online: https://www.zhihu.com/question/403522406 (accessed on 4 September 2021).
21. Dyussenbayev, A. Age Periods of Human Life. *Adv. Soc. Sci. Res. J.* **2017**, *4*, 258–263. [CrossRef]
22. Zhang, H.; Ma, S. The Construction of Community-The Dual Aspects of the Young Artist Residency Program of Sichuan Fine Arts Institute. *Contemp. Artist.* **2021**, 58–61. (In Chinese)
23. Zxxy–hbafa. Sculpture. Available online: http://zxxy--hbafa--com--dd3c4720c.zy.ipv6transform.cmecloud.cn/info/1168/1873.htm (accessed on 9 August 2020).
24. Cdgdc. National Fourth Round Subject Assessment-Fine Arts. Available online: https://www.cdgdc.edu.cn/dslxkpgjggb/ (accessed on 28 December 2017).
25. Tjarts. School Philosophy. Available online: http://syxy.tjarts.edu.cn/xygk/bxsx.htm (accessed on 3 January 2022).
26. Braun, V.; Clarke, V. Using thematic analysis in psychology. *Qual. Res. Psychol.* **2006**, *3*, 77–101. [CrossRef]

proceedings

MDPI

Proceeding Paper

Pilot Study on Cyberbullying among Visually Impaired Youths †

Wen Li Ku [1], Pei Boon Ooi [1,*] and Nee Nee Chan [2]

1 Department of Medical Science, School of Medical and Life Sciences, Sunway University, Bandar Sunway, Selangor 47500, Malaysia
2 Department of Education, Faculty of Social Sciences, UCSI University, UCSI Heights, Taman Connaught, Cheras, Kuala Lumpur 56000, Malaysia
* Correspondence: peiboono@sunway.edu.my

† Presented at the International Academic Symposium of Social Science 2022, Kota Bharu, Malaysia, 3 July 2022.

Abstract: Social networking has dominated the modern technological era, with youths being the main users. However, technological advancement has instilled a form of online violence called cyberbullying. Visually impaired Malaysian youths are no exception to cyberbullying incidents; however, this aspect has been understudied in Malaysia. This study examined the associated factors of cyberbullying among visually impaired Malaysian youths. A total of 30 participants were included in the Braille/pen-and-paper and voice-over online survey. We found that only aggressive behavior was positively associated with cyberbullying (ß = 0.544, p = 0.002). The distinctiveness of the study can help to draw public attention to the issue of cyberbullying.

Keywords: aggressive behavior; cyberbullying; technological exposure; visually impaired youths

Citation: Ku, W.L.; Ooi, P.B.; Chan, N.N. Pilot Study on Cyberbullying among Visually Impaired Youths. *Proceedings* **2022**, *82*, 67. https://doi.org/10.3390/proceedings2022082067

Academic Editor: Mohamad Rahimi Mohamad Rosman

Published: 20 September 2022

1. Introduction

Social networking, or social media, has dominated the modern technological era and is especially popular among youths [1,2]. In the past, the tools of communication were limited and slower, especially over long distances. However, globalization has significantly revolutionized the nature of communication to fill in the void and meet the needs of individuals via the advancement of information technologies [3]. Wok and colleagues have stated that globalization has led Malaysians to change their internet usage patterns to adapt to and fit into the Information age [2]. Today, the advancement of information technologies has facilitated opportunities for effective communication and has afforded more convenience with the accessibility of the Internet [4,5]. Moreover, according to Hruška & Maresova, social media outlets have become important tools for acquiring and spreading information in different domains (e.g., entertainment, politics, social science, etc.) [6]. Given the accessibility of the internet, previous studies have shown that the youth population are the main users and subscribers of social media [1,7]. Recently, The Star reported that there were approximately 28 million social media users (86% of the total Malaysian population) as of January 2021 [8]. Additionally, the outbreak of the COVID-19 pandemic has increased people's dependency on the Internet and social networking platforms to stay connected for work and education [9]. Several studies have found that visually impaired youths are also active Internet users, and they have been observed as spending more time accessing the Internet than non-visually impaired individuals [10–12].

Undeniably, the Internet and social media have made life easier for people by providing different platforms for effective communication and learning [4]. However, these technological advancements have also instilled terror in some populations who use the Internet via an activity known as "cyberbullying" [1,5,13]. Various studies have defined "cyberbullying" as a form of online violence with the intention to harass, threaten, or harm a victim repeatedly over time via the Internet [5,14,15]. According to [16], cyberbullying has been found to be more threatening due to three factors: the anonymity of the cyber perpetrators, the breadth of potential targets, and the difficulty of escaping from it. Chan and colleagues' recent study proposed that cyberbullying is an extension of traditional bullying, with the exception that cyberbullying occurs in cyberspace via access to the Internet [5]. However, Wang and colleagues argued that cyberbullying is an overlap of traditional bullying, emphasizing aggressive behavior through indirect and verbal aggression [17]. The definition of "cyberbullying" is still open for scholarly debate and continues to evolve.

Cyberbullying is an ongoing phenomenon which has been a concern of youth advocates and which has dominated the headlines as a serious issue facing youths today [18]. Several studies have been conducted on the Malaysian youth population. A study by the Global Youth Online Behavior Survey conducted with 7600 Malaysians found that 1 out of every 3 respondents reported having been a cyber victim [19]. A 2016 MCMC study reported that there had been 1524 cyberbullying cases in Malaysia over five years [20]. Based on these statistics, Malaysia was counted among the countries with the highest number of cyberbullying cases recorded in 2018, ranking 6th among 28 other countries. In 2020, Malaysia ranked as second among all Asian countries in the number of cyberbullying cases reported among youths [21]. However, it is unfortunate that many people view this issue lightly and perceive its effect to be less severe [22]. Cyberbullying is associated with many negative effects on mental health (depression, low self-worth, hopelessness, and loneliness) and suicidal ideation [18,22–24]. Yusuf and colleagues found that youths are more vulnerable to the involvement of cyberbullying [20].

Visually impaired Malaysian youths are no exception to the trend in cyberbullying incidents. International studies have reported that visually impaired youths are at a higher risk of cyberbullying involvement due to their disabilities; however, this aspect has been understudied in Malaysia. Heiman & Olenik-Shemesh found that visually impaired youths reported twice as much cyber victimization as compared to sighted individuals [14]. Many of the assumptions about visually impaired youths involve a lack of access to the Internet, especially when information is presented in a visual form [25,26], and therefore they are assumed to be immune to cyberbullying. However, what is not known is that visually impaired youths are equally engaged in internet use [11]. Inventions have been made available for the visually impaired to access the Internet and improve the readability of webpages, for instance, computer-based assistive devices, such as screen-reading software [27]. Thus, their access to the Internet is undeniable and prominent. They, too, have been introduced to playing audio games as well, thus making them susceptible to online behaviors such as cyberbullying. Moreover, cyberbullying behaviors can occur in the form of sending audio messages or making disturbing calls [28], as the study reported hate speech as a negative online behavior. This would include, for instance, calling someone "stupid" and others continuing to comment hatefully. Therefore, to address the gap, the objectives of this study are:

1. To examine the relationship between technological exposure and cyberbullying among visually impaired Malaysian youths, and
2. To examine the relationship between aggressive behavior and cyberbullying among visually impaired Malaysian youths.

2. Materials and Methods

Approval from the institutional ethics board for the study has been obtained (Ethics Approval No.: PGSUREC2021/028). A quantitative approach has been employed as the research methodology for this study so as to test the following research hypotheses:

Hypothesis 1. *There is a significant positive relationship between technological exposure and cyberbullying.*

Hypothesis 2. *There is a significant positive relationship between aggressive behavior and cyberbullying.*

The proposed research model is shown in Figure 1.

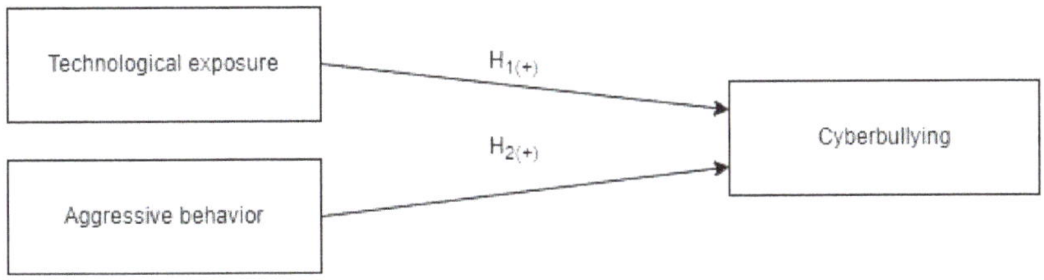

Figure 1. Proposed research model.

Using a stratified sampling technique, 30 visually impaired participants from Malaysia were chosen to participate in this study via both Braille/ pen-and-paper and online voice-over survey questionnaires. The participants used NonVisual Desktop Access (NVDA) (NV Access Limited, St Lucia, Australia) as a screen-reader tool to assist them in completing the online survey.

A four-section questionnaire was developed in both English and Bahasa Malaysia, utilizing three existing surveys and including demographic questions. The Social Networking Usage Questionnaire, the Aggression Questionnaire, and the Cyber Bullying Behaviors and Victimization Experiences measure were used to examine the potential factors (technological exposure and aggressive behavior) contributing to cyberbullying. Section A assessed participants' social networking usage in four dimensions—academic, socialization, entertainment, and informativeness, using the Social Networking Usage Questionnaire of 19 items using a Cronbach's alpha coefficient of 0.83 [29]. The Social Networking Usage Questionnaire measured responses on a 5-point Likert Scale: 1 = never, 2 = rarely, 3 = sometimes, 4 = often, and 5 = always. Section B assessed how aggressive participants were based on four aggression factors—physical aggression, verbal aggression, anger, and hostility—using the Aggression Questionnaire of 29 items with a Cronbach's alpha coefficient of 0.89 [30]. The anchors in the Aggression Questionnaire were 1 = extremely uncharacteristic of me, 2 = somewhat uncharacteristic of me, 3 = neither uncharacteristic nor characteristic of me, 4 = somewhat characteristic of me, and 5 = extremely characteristic of me. Section C assessed participants' frequency of involvement in cyberbullying and victimization experiences using the Cyber Bullying Behaviors and Victimization Experiences measure of 8 items with a Cronbach's alpha coefficient of 0.86 [31]. The anchors of the Cyber Bullying Behaviors and Victimization Experiences measure were 0 = never, 1 = once or twice, 2 = a few times, 3 = many times, and 4 = every day. Section D included demographic questions (e.g., age, gender, frequency of Internet usage, etc.).

Data were analyzed using Statistical Package for the Social Sciences (SPSS) version 26 (IBM, Armonk, NY, USA) to conduct descriptive, reliability, correlation, and regression analyses. The total score for all scales was used for analysis.

3. Results

3.1. Demographics

A total of 18 males and 9 females took part in this study, with a mean age of 17.85 (SD = 11.80). Some participants preferred to remain completely anonymous by not providing any personal information. Therefore, the cases with missing values were excluded, and the demographics of the participants have been reported in a list (Table 1). The majority of the participants (93.3%) actively used the Internet. Only 5 participants (16.7%) identified themselves as cyber victims, and only one participant (3.3%) reported himself/herself to be a cyber perpetrator, while only 11 participants (36.7%) identified themselves as cyber witnesses.

Table 1. Demographic information.

		n (%)
Age (n = 27)	15–18	13 (43.3%)
	19–24	14 (46.5%)
	Missing value	3 (10.0%)
Gender (n = 27)	Male	18 (60.0%)
	Female	9 (30.0%)
	Missing value	3 (10.0%)
Ethnicity (n = 28)	Malay	11 (36.7%)
	Chinese	14 (46.7%)
	Indian	2 (6.7%)
	Other	1 (3.3%)
	Missing value	2 (6.7%)
Internet usage (n = 28)	Yes	28 (93.3%)
	No	-
	Missing value	3 (6.7%)
Frequency of Internet usage (n = 28)	Almost every day	27 (90.0%)
	3–4 times a week	1 (3.3%)
	Missing value	2 (6.7%)
Have you experienced cyberbullying? (n = 20)	Yes	5 (16.7%)
	No	9 (30.0%)
	Maybe	6 (20.0%)
	Missing value	10 (33.3%)
Are you a cyberbully? (n = 20)	Yes	1 (3.3%)
	No	16 (53.3%)
	Maybe	3 (10.0%)
	Missing value	10 (33.3%)
Have you witnessed a cyberbullying incident? (n = 20)	Yes	11 (36.7%)
	No	7 (23.3%)
	Maybe	2 (6.7%)
	Missing value	10 (33.3%)

3.2. Reliability Analysis

Reliability analysis was performed to assess the consistency of the results based on each variable presented, and all variables have achieved a Cronbach's alpha score of 0.7 or above, which indicates high reliability (Table 2).

Table 2. Reliability analysis.

Variables	Items	Cronbach's Alpha (α)
Technological exposure	19	0.905
Aggressive behavior	29	0.913
Cyberbullying	8	0.737

3.3. Descriptive Statistics

A total of 30 participants participated in this pilot study. The mean, standard deviation, skewness, and kurtosis of the variables are shown in Table 3.

Table 3. Information on the descriptive statistics.

Variables	Mean	Std. Deviation	Skewness	Kurtosis
Technological exposure	60.07	14.65	−0.836	0.428
Aggressive behavior	81.00	20.84	0.006	−0.370
Cyberbullying	6.83	4.32	0.174	−0.888

3.4. Correlational Analysis

A preliminary analysis was run prior to the final analysis. Pearson correlational analysis was performed to determine the variables' relationships based on the empirical research model in this study. Table 4 shows that there was no significant relationship between technological exposure and cyberbullying, and therefore H_1 was not supported. There was a significant positive relationship between aggressive behavior and cyberbullying ($r = 0.583$, $p < 0.01$), and thus H_2 was supported.

Table 4. Correlation analysis.

Variables	1	2	3
Cyberbullying	-		
Technological exposure	0.293	-	
Aggressive behavior	0.583 **	0.262	-

** $p < 0.01$.

3.5. Regression Analysis

Regression analysis was performed to identify the association between cyberbullying and technological exposure and aggressive behavior. The results are presented in Table 5. We found that cyberbullying was positively associated with aggressive behavior ($\beta = 0.544$, $p = 0.002$), but it was not associated with technological exposure ($\beta = 0.150$, $p > 0.05$). In our model, aggressive behavior and technological exposure explain a 31.4% variance in cyberbullying, which achieves a small-effect size [32].

Table 5. Results from the regression analysis.

Variables	Cyberbullying					Adjusted R Square
	β	P	t	SE	95% CI	
Technological exposure	0.150	0.354	0.942	0.047	(0.052, 0.141)	0.314
Aggressive behavior	0.544	0.002	3.411	0.033	(0.045, 0.180)	

Note. Predictors: (Constant); Aggressive behavior, Technological exposure.

4. Discussion

The evolution of new technologies has made life easier for people and has yielded various benefits, for instance, convenience in acquiring information [6,33]. The dependency of people on the Internet has been reflected in the findings of this study, revealing that a majority of the participants (93.3%) actively engaged in Internet and social media use. Previous studies have found that increased time spent on the Internet and social media can potentially increase the risk of cyberbullying in cyberspace [7,15,22]. The findings of this study show that cyberbullying is present in the Malaysian visually impaired youth community, with 16.7% of participants reporting experiences as cyber victims, 3.3% of participants as cyber perpetrators, and 36.7% of participants as cyber witnesses. Although the findings help us to ascertain the presence of cyberbullying, the findings still showed that the majority of the participants did not report being cyberbullied nor as having cyberbullied others, in comparison to only being cyber witnesses. This is consistent with [14], who found that more people with low vision reported being a cyber witness; in the same study, they also found that people with low vision were more likely to know others who had suffered from being cyberbullied compared with their sighted counterparts. Therefore, we suggest that visually impaired Malaysian youth are highly aware of cyberbullying, while they are also well-versed in preventing themselves from being cyberbullied.

The focus of the present study is to examine the relationship between technological exposure and cyberbullying, as well as the relationship between aggressive behavior and cyberbullying. our findings reveal that there is no significant relationship between technological exposure and cyberbullying, and therefore H_1 was not supported. Parenting styles regarding internet and social media usage may be a factor to take into account.

This is supported by [34], who found that Chinese parents' restrictions on children's screentime on technological devices can significantly predict children's social skills. Asian parents or guardians are more likely to restrict their children from the Internet because they fear that their children will be exposed to potential hazards on the Internet, for instance, pornography sites, scam sites, and money-laundering sites. Visually impaired youths are especially vulnerable in this regard as compared with sighted individuals. Moreover, in a short meeting with the guardians of a home for the disabled, they revealed that the visually impaired youths there were not exposed to computers because the guardians are concerned that the children might browse websites that could potentially cause harm to them. Moreover, the parents of visually impaired may face financial difficulties as they may spend more money on treatments for their visually impaired children, and this may cause them to be unable to provide and constrain access to the Internet.

Our study found that there is a positive relationship between aggressive behavior and cyberbullying, and therefore H_2 was supported. This finding is supported by several studies by [20,35]. Runions & Bak's study further suggests that online settings may facilitate cyber aggression and cyberbullying [36]. The anonymity afforded to Internet users provides a chance for individuals to carry out aggressive behaviors without the need to face any consequences [37], while some of them use it to take revenge on others with hostile intent [38]. Another study by [23] also supports the relationships by explaining the nature and extent of cyberbullying involving direct and indirect aggression. Cyberbullying, according to [15], is a variant of traditional bullying with similar aggression, as intentional and repeated acts of doing harm, except that cyberbullying occurs using electronic devices (e.g., smartphones and laptops) and online media (e.g., Facebook, Instagram, and WhatsApp). Several unique cyberbullying behaviors have been described by [39], including bombing, flaming, and happy slapping. "Bombing" occurs when an aggressor uses an automated program to collapse a victim's e-mail with simultaneous messages and block the e-mail account. "Flaming" occurs when an aggressor sends electronic messages with hostile and vulgar language. "Happy slapping" occurs when an aggressor attacks a person with images or videos [39]. Wrzesinska and colleagues also found that individuals who are visually impaired were at a higher risk of engaging in electronic aggression [40]. This might be caused by the type of discrimination that happens so often on the Internet, in which a minority group (e.g., visually impaired individuals) feel the need to defend themselves while enlarging their feelings of anger and aggression towards others.

Implications

This study reveals the presence of cyberbullying among visually impaired Malaysian youths, which draws the attention of researchers and the public to cyberbullying issues among individuals with disabilities in Malaysia. Although there are works of literature showing the negative effects and presence of cyberbullying among young adults, the subject of cyberbullying among visually impaired youths remains understudied. This problem is further reinforced by the fact that society fails to learn about how and to what extent visually impaired youth become cyber victims or cyber perpetrators. Therefore, the present study can provide a new understanding of the prevalence of cyberbullying among the community of visually impaired youths in Malaysia. The findings can further contribute to understanding the theoretical aspects of technological exposure and aggressive behaviors in cyberbullying.

Based on our findings, aggressive behaviors appear to be an important factor contributing to cyberbullying among visually impaired youths in Malaysia. It can serve as a basis for awareness among counseling practitioners and schoolteachers of the presence of aggression while intervening in cyberbullying cases. Therefore, those professionals are encouraged to identify the aspects of aggression among cyberbullies and counter them with appropriate methods. Additionally, schools should implement programs and interventions which aim to reduce aggressive behaviors in their curriculum. This will help to raise the awareness of the visually impaired youth to refrain from using violence and aggression to approach problems, but rather to resort to healthier coping strategies.

5. Conclusions

The findings of the study give light to the presence of cyberbullying among the visually impaired Malaysian youth community. Furthermore, the current study examined the relationship between technological exposure and cyberbullying, as well as the relationship between aggressive behavior and cyberbullying. The findings show that cyberbullying is positively associated only with aggressive behavior. As studied, online settings have facilitated opportunities for cyber aggression and cyberbullying, with an added point on the power of anonymity of Internet users affording the opportunity of cyberbullying. Additionally, this study is particularly important and worthwhile as the phenomenon of cyberbullying is considered to be of global significance today. More awareness and attention from parents, peers, and teachers may be needed to reduce the probability of the association of cyberbullying with aggressive behavior. Particularly, visually impaired youths are categorized as a vulnerable community that is more likely at risk of exposure to cyberbullying. The distinctiveness of this study is its ability to shed light on the visually impaired Malaysian youth community in understanding the nature and extent of cyberbullying by acknowledging the risk factors and effects of cyberbullying.

Author Contributions: Conceptualization, W.L.K., P.B.O. and N.N.C.; methodology, W.L.K. and P.B.O.; software, W.L.K.; validation, P.B.O.; formal analysis, W.L.K.; investigation, W.L.K.; resources, P.B.O.; data curation, P.B.O.; writing—original draft preparation, W.L.K.; writing—review and editing, P.B.O. and N.N.C.; visualization, W.L.K.; supervision, P.B.O. and N.N.C.; project administration, W.L.K., P.B.O. and N.N.C.; funding acquisition, P.B.O. All authors have read and agreed to the published version of the manuscript.

Funding: The project is funded by Fundamental Research Grant Scheme (FRGS), grant number FRGS/1/2020/SS0/SYUC/02/, awarded by the Ministry of Higher Education (MOHE) in late 2019.

Institutional Review Board Statement: The study was conducted with the approval of the Institutional Review Board (or Ethics Committee) of SUNWAY UNIVERSITY (protocol code PGSUREC2021/028 with a date of approval of 2 September 2021) for studies involving humans.

Informed Consent Statement: Informed consent was obtained from all subjects involved in the study.

Data Availability Statement: Not applicable.

Acknowledgments: The authors would like to express their gratitude to all participants who took part in this study and to the Ministry of Higher Education (MOHE) Malaysia under the Fundamental Research Grant Scheme (FRGS/1/2020/SS0/SYUC/02/2) for the funding support.

Conflicts of Interest: The authors declare no conflict of interest.

References

1. Hamm, M.P.; Newton, A.S.; Chisholm, A.; Shulhan, J.; Milne, A.; Sundar, P.; Ennis, H.; Scott, S.D.; Hartling, L. Prevalence and Effect of Cyberbullying on Children and Young People: A Scoping Review of Social Media Studies. *JAMA Pediatr.* **2015**, *169*, 770–777. [CrossRef]
2. Wok, S.; Idid, S.; Misman, E. Social media use for information-sharing activities among youth in Malaysia. *J. Mass Commun.* **2012**, *2*, 1029–1047. [CrossRef]
3. Goodman-Deane, J.; Mieczakowski, A.; Johnson, D.; Goldhaber, T.; Clarkson, P.J. The impact of communication technologies on life and relationship satisfaction. *Comput. Hum. Behav.* **2016**, *57*, 219–229. [CrossRef]
4. Abaido, G.M. Cyberbullying on social media platforms among university students in the United Arab Emirates. *Int. J. Adolesc. Youth* **2020**, *25*, 407–420. [CrossRef]
5. Chan, N.N.; Ahrumugam, P.; Scheithauer, H.; Schultze-Krumbholz, A.; Ooi, P.B. A hermeneutic phenomenological study of students' and school counsellors' "lived experiences" of cyberbullying and bullying. *Comput. Educ.* **2020**, *146*, 103755. [CrossRef]
6. Hruška, J.; Maresova, P. Use of Social Media Platforms among Adults in the United States—Behavior on Social Media. *Societies* **2020**, *10*, 27. [CrossRef]
7. Ybarra, M.L.; Mitchell, K.J. Youth engaging in online harassment: Associations with caregiver–child relationships, Internet use, and personal characteristics. *J. Adolesc.* **2004**, *27*, 319–336. [CrossRef]
8. Malaysia has 28 million social media users as of January 2021, says Comms Ministry sec-gen. *The Star*, 22 September 2021.
9. Wong, A.; Ho, S.; Olusanya, O.; Antonini, M.V.; Lyness, D. The use of social media and online communications in times of pandemic COVID-19. *J. Intensiv. Care Soc.* **2021**, *22*, 255–260. [CrossRef]

10. Martiniello, N.; Lehane, C.; Eisenbarth, W.; Johnson, A.; Wittich, W. Exploring the use of smartphones and tablets among people with visual impairments: Are mainstream devices replacing the use of traditional visual aids? *Assist. Technol. Off. J. RESNA* **2019**, *34*, 34–45. [CrossRef]
11. Wrzesińska, M.; Tabała, K.; Stecz, P. The online behavior of pupils with visual impairment: A preliminary report. *Disabil. Health J.* **2016**, *9*, 724–729. [CrossRef]
12. Wu, S.; Adamic, L. Visually impaired users on an online social network. In Proceedings of the CHI'14: CHI Conference on Human Factors in Computing Systems, Toronto, ON, Canada, 26 April–1 May 2014. [CrossRef]
13. Whittaker, E.; Kowalski, R.M. Cyberbullying Via Social Media. *J. Sch. Violence* **2015**, *14*, 11–29. [CrossRef]
14. Heiman, T.; Olenik-Shemesh, D. Cyberbullying Involvement of Adolescents with Low Vision Compared to Typical Adolescents, as Related to Perceived Social Support. *J. Aggress. Maltreatment Trauma* **2017**, *26*, 105–115. [CrossRef]
15. Smith, P.K.; Mahdavi, J.; Carvalho, M.; Fisher, S.; Russell, S.; Tippett, N. Cyberbullying: Its nature and impact in secondary school pupils. *J. Child Psychol. Psychiatry* **2008**, *49*, 376–385. [CrossRef]
16. Slonje, R.; Smith, P.K. Cyberbullying: Another main type of bullying? *Scand. J. Psychol.* **2008**, *49*, 147–154. [CrossRef]
17. Wang, C.-W.; Musumari, P.M.; Techasrivichien, T.; Suguimoto, S.P.; Tateyama, Y.; Chan, C.-C.; Ono-Kihara, M.; Kihara, M.; Nakayama, T. Overlap of traditional bullying and cyberbullying and correlates of bullying among Taiwanese adolescents: A cross-sectional study. *BMC Public Health* **2019**, *19*, 1756. [CrossRef]
18. Lai, C.S.; Mohaffyza, M.; Lee, M.F.; Mohd Salleh, K.; Sulaiman, N.; Rosli, D.O.; Chang, W.V.S. Prevalence of Cyberbullying among Students in Malaysian Higher Learning Institutions. *Adv. Sci. Lett.* **2017**, *23*, 781–784. [CrossRef]
19. Saharrudin, N.; Akmar, G.; Abu Samah, A.; Ahmad, A.; Abdullah, H. Cyberbullying among Malaysian Youth: The Case of Selangor. *Int. J. Acad. Res. Bus. Soc. Sci.* **2019**, *9*, 1060–1070. [CrossRef]
20. Yusuf, S.; Al-Majdhoub, F.; Mubin, N.; Chaniago, R.; Khan, R. Cyber Aggression-Victimization Among Malaysians Youth. *Asian J. Univ. Educ.* **2021**, *17*, 240–260. [CrossRef]
21. Hanaffi, F. Malaysia Is Ranked Second Most Aggressive Country for Cyber-Bullying in Asia. *World of Buzz*, 26 July 2020.
22. Ng, C.G.; Kanagasundram, S.; Ann, Y.H.; Tan, L.H.; Teoh, K.M. Cyber Bullying—A New Social Menace. *Asean J. Psychiatry* **2016**, *17*, 104–115.
23. Hinduja, S.; Patchin, J.W. Bullying, cyberbullying, and suicide. *Arch. Suicide Res.* **2010**, *14*, 206–221. [CrossRef]
24. Nixon, C.L. Current perspectives: The impact of cyberbullying on adolescent health. *Adolesc. Health Med. Ther.* **2014**, *5*, 143–158. [CrossRef]
25. Brinkley, J.; Tabrizi, N. A Pilot Study Examining the Online Behavior of Web Users with Visual Impairments. *Int. J. Adv. Comput. Sci. Appl.* **2013**, *4*, 150–155. [CrossRef]
26. Hafiar, H.; Subekti, P.; Nugraha, A. Internet Utilization by the Students with Visual Impairment Disabilities. *Int. J. Emerg. Technol. Learn. (iJET)* **2019**, *14*, 200–207. [CrossRef]
27. Rosner, Y.; Perlman, A. The Effect of the Usage of Computer-Based Assistive Devices on the Functioning and Quality of Life of Individuals who are Blind or have low Vision. *J. Vis. Impair. Blind.* **2018**, *112*, 87–99. [CrossRef]
28. Soni, D.; Singh, V. See No Evil, Hear No Evil: Audio-Visual-Textual Cyberbullying Detection. *Proc. ACM Hum.-Comput. Interact.* **2018**, *2*, 1–26. [CrossRef]
29. Bashir, L.; Gupta, S. Social Networking Usage Questionnaire: Development and Validation in an Indian Higher Education Context. *Turk. Online J. Distance Educ.* **2018**, *19*, 214–227. [CrossRef]
30. Buss, A.H.; Perry, M. The aggression questionnaire. *J. Pers. Soc. Psychol.* **1992**, *63*, 452–459. [CrossRef]
31. Holfeld, B.; Leadbeater, B.J. The Nature and Frequency of Cyber Bullying Behaviors and Victimization Experiences in Young Canadian Children. *Can. J. Sch. Psychol.* **2014**, *30*, 116–135. [CrossRef]
32. Cohen, J. Set Correlation and Contingency Tables. *Appl. Psychol. Meas.* **1988**, *12*, 425–434. [CrossRef]
33. Ngonso, B.F. Effect of Social Media on Teenagers and Youths: A Study of Rural Nigerian Teenagers and Youths in Secondary Schools. *Glob. Media J.* **2019**, *17*, 1–6.
34. Hu, B.Y.; Johnson, G.K.; Wu, H. Screen time relationship of Chinese parents and their children. *Child. Youth Serv. Rev.* **2018**, *94*, 659–669. [CrossRef]
35. den Hamer, A.; Konijn, E.A.; Keijer, M.G. Cyberbullying behavior and adolescents' use of media with antisocial content: A cyclic process model. *Cyberpsychol. Behav. Soc. Netw.* **2014**, *17*, 74–81. [CrossRef]
36. Runions, K.C.; Bak, M. Online Moral Disengagement, Cyberbullying, and Cyber-Aggression. *Cyberpsychol. Behav. Soc. Netw.* **2015**, *18*, 400–405. [CrossRef] [PubMed]
37. Law, D.M.; Shapka, J.D.; Domene, J.F.; Gagné, M.H. Are Cyberbullies really bullies? An investigation of reactive and proactive online aggression. *Comput. Hum. Behav.* **2012**, *28*, 664–672. [CrossRef]
38. Peker, A.; Nebioglu, M. Mediating Role of Self-Control in the Relationship between Aggressiveness and Cyber Bullying. *Psychiatry Behav. Sci.* **2021**, *11*, 40. [CrossRef]
39. Calvete, E.; Orue, I.; Estévez, A.; Villardón, L.; Padilla, P. Cyberbullying in adolescents: Modalities and aggressors' profile. *Comput. Hum. Behav.* **2010**, *26*, 1128–1135. [CrossRef]
40. Wrzesinska, M.A.; Knol-Michałowska, K.; Stecz, P.; Kopytowska, M.; Binder-Olibrowska, K. Internet risky behaviours among youth with visual impairment. *PeerJ* **2021**, *9*, e12376. [CrossRef]

proceedings

Proceeding Paper

Validity and Reliability Testing of Geographical Information System (GIS) Quality and User Satisfaction towards Individual Work Performance †

Mohd Saiful Rahman, Shamila Mohamed Shuhidan *, Mohamad Noorman Masrek and Mohammad Fazli Baharuddin

Faculty of Information Management, Universiti Teknologi MARA, Shah Alam 40150, Malaysia
* Correspondence: shamila@uitm.edu.my
† Presented at the International Academic Symposium of Social Science 2022, Kota Bharu, Malaysia, 3 July 2022.

Abstract: This article aims to validate the instrument's reliability and validity in measuring the influence of geographical information system (GIS) quality and user satisfaction on its impact on individual work performance. Based on an extensive literature review and input from experts, 68 items were constructed focusing on GIS quality, user satisfaction, and individual work performance to ascertain the opinions of GIS users were established to be the findings of the study. The instrument was composed and then reviewed by panelists from subject matter experts. Later, a pilot study was conducted and participated by 30 respondents. Concerning the analysis of the data, statistical analysis was performed, the analysis output then demonstrated that the instrument was dependable and that no abnormalities had occurred in the data.

Keywords: geographical information system; GIS quality; user satisfaction; individual work performance; software quality

Citation: Rahman, M.S.; Shuhidan, S.M.; Masrek, M.N.; Baharuddin, M.F. Validity and Reliability Testing of Geographical Information System (GIS) Quality and User Satisfaction towards Individual Work Performance. *Proceedings* **2022**, *82*, 68. https://doi.org/10.3390/proceedings2022082068

Academic Editor: Mohamad Rahimi Mohamad Rosman

Published: 20 September 2022

1. Introduction

Geographical information system (GIS) is a solution that contributes to the understanding of the ability to collect, visualize, and produce analysis [1]. GIS is software that works in the spatial area focused on geographic information. In addition, the ability of GIS covers the capturing, analyzing, and presenting of all types of information [2]. Furthermore, GIS can also translate information to give a more comprehensive understanding through a combination of spatial and business data [3]. However, the quality of the GIS is still in a grey area and the evidence with regards contributing factors in GIS quality is still minimal. Therefore, a standard produced by the International Organization for Standardization (ISO), namely the ISO-25010 Software Quality Model, was used in measuring GIS quality for this study.

The main objective of this study is to validate the instrument's reliability and validity for the researcher to proceed with the actual study in identifying the determinant of GIS quality that are anticipated to influence user satisfaction, and how user satisfaction may affect individual work performance. Literature analysis revealed eight factors that have been identified, namely functional suitability, performance efficiency, compatibility, usability, reliability, security, maintainability, and portability. Hence, this paper aims to validate the instrument's reliability and validity in measuring the influence of geographical information system (GIS) quality and user satisfaction on its impact on individual work performance.

2. Literature Review

GIS is a powerful solution in giving ideas to users via the ability in providing information based on location-centric called geospatial information [4]. A simple definition of GIS is a system that focuses on data capture, storing, managing, analyzing, and presenting all forms of geographically referenced information identified according to locations [5]. GIS operates differently from other software. The uniqueness of GIS provides functionalities in terms of maps and comprehensive analysis; in some cases, it was able to give solutions regarding business directions. The solutions that GIS bring come with different ways of doing things, especially when the element of "Where" is added as a variable in making decisions. It is able to boost users to undergo innovation in daily operation through spatial information [6]. Spatial information refers to attributes that provide context and meaning when combined with business data [5]. Some spatial information comes in points of locations or shapes of area. Another interesting area offered in GIS is the ability to perform a query and produce analysis faster through a visualization dashboard for users to have an earlier understanding of when data were collected and combined with spatial information. It gives location intelligent analysis that provides proper insight, especially in planning and understanding the nature of the specific areas of businesses [7].

Many software in the market were produced and the measurement of quality were still in grey area. GIS is also a software that needs to be validated in terms of the quality provided to the industry. Quality in GIS is very subjective and depends on the nature of use. Different organizations or users have different objectives towards the use. Software quality is elaborated as the elements that contribute to pleasing and meeting the needs of users [8]. Many models have been developed to measure software quality [9–11]. The evolutions of software quality were started by McCall (1977), Boehm (1978), Dromey (1995), with the FURPS model, the ISO-9126 (2001) Software Quality Model and the most updated standard followed by the industry is the ISO-25010 Software Quality Model. For the context of this study, ISO-25010 was chosen as it was the revised version and used vastly in measuring the quality of software.

User satisfaction is one of the major factors contributing to the acceptance of software. The acceptance of GIS has been a subject of debate in the community [12]. According to [13], user satisfaction refers to the perception of GIS's usefulness, trust, pleasure, and level of comfort when using it. From a usefulness context, GIS was determined by its ability, features, and functionality in producing output [14]. The perspective of trust focuses on whether the GIS is able to produce trusted information and the ability to improve performance as per the user's expectation. Any situation where a lack of trust occurred will be a big issue hampering any implementation of GIS [15,16]. Hence, it is important to ensure that trust is in place so that users have strong beliefs that GIS is able to perform as per users' expectations. Meanwhile, pleasure in GIS are situations where users evaluate the software through actual experience in using it. The experience during use can be affected by the internal or external characteristics of GIS.

One of the critical aspects of measuring software implementation's success was measuring individual work performance [17,18]. Individual work performance is a concept in measuring the quality produced by individuals. Thus, any organization needs an individual who has the capabilities to exhibit a high work performance in achieving goals to increase organization productivity and competitiveness. Individual work performance is defined as actions or behaviors of individuals made in achieving goals [19,20]. In addition, performance is defined as a high result of work efficiency, effectiveness, and quality of work [20–22]. The higher an individual's performance will result in satisfaction and mastery regarding personal tasks or jobs.

3. Instrument and Data Collection

For this study, 68 items were constructed based on a three-domain questionnaire derived from a comprehensive literature review. It was organized to obtain users' viewpoints on GIS quality accommodated by the organization for use. In addition, for pre-testing, the instrument was reviewed and given feedback by six subject matter experts in software development, and academicians, including the GIS users, were successfully acquired. Several suggestions were raised, including the use of relevant words and verbs to strengthen question clarity and the inclusion of examples to help responders comprehend the questions.

Through feedback and suggestions, amendments were implemented in response to the constructive feedback. Subsequently, this pilot study aimed to ensure that the respondents understood the questionnaire items and that no questions were misleading. This pilot study, which included 30 respondents, was examined using Statistical Package for Social Sciences (SPSS) version 26 by International Business Machines Corporation (IBM, Armonk, NY, USA). Finally, Cronbach's alpha was utilized to analyze the questionnaire's scale reliability. Table 1 lists the items for each dimension.

Table 1. Items for each Dimension.

Variables	Dimensions	Item
GIS Quality	Functional Suitability	The functions provided in GIS are suitable for me. The functions provided in GIS meet with my work requirements. The information provided in GIS is accurate. The results of the analysis provided from GIS are precise. Functions in GIS facilitate me in the business process. Functions provided in GIS assisting me in meeting organizational goals.
	Performance Efficiency	GIS is able to provide good response time. GIS is fast at doing data processing. GIS requires minimal close guidance from system administrators. Users are not limited to access GIS via personal computer (PC) only. The capacity of data storage provided in GIS is scalable. GIS can perform even in high numbers of user access.
	Compatibility	GIS is able to work with existing available systems. GIS is able to function without giving any negative impact to other systems. GIS enables me to interact with other systems. The output from GIS can be used by other systems.
	Usability	GIS is software that is appropriate for my use. GIS does not require a lot of time to learn how to use it. The training provided is enough for me. I can easily operate GIS without any guidance. GIS requires minimal technical capabilities. GIS is able to notify users required actions in avoiding errors. The GIS interface is well organized. The interface of GIS is interesting. GIS can be accessed by every member of my organization.
	Reliability	GIS has a low rate of failure. GIS is always able to return information needed by me. GIS is accessible when it is needed. GIS maintains its performance even if there are data faults. I am able to recover information in GIS even after failure occurs.
	Security	GIS is able to prevent access from unauthorized users. GIS provides proper data encryption. GIS is able to prevent any modifications of data. GIS is able to trace sources of access so it can control manipulation of actions. All activities in GIS were uniquely traceable. GIS is able to identify the genuine identity of users.
	Maintainability	GIS can be customized according to my needs. Any changes in GIS will have no impact on other systems. The concept in GIS can be reusable to other systems. I am able to identify errors when they happen. GIS is able to notify the cause of error. GIS is stable even if any modifications are made. I can easily test any new changes made in GIS.
	Portability	GIS can be adapted for the use of other departments in my organization. GIS is able to evolve according to needs of the organization. Process installing GIS on a user's devices is easy. GIS can be easily removed from other user devices. GIS has the potential to replace other systems in my organization. GIS has the ability to be the main visualization platform in my organization.

Table 1. *Cont.*

Variables	Dimensions	Item
User Satisfaction		GIS empowers me to accomplish tasks well. GIS improves my work performance. GIS increases my work productivity. All information produced from GIS is trusted. GIS has its anticipated effects. The experience of using GIS is enjoyable. GIS provides a pleasant experience. GIS causes no stress when using it.
Individual Work Performance	Task Performance	I am able to plan and organize all my work tasks. I have an excellent number of work completions. I fulfill all work quality criteria for my tasks. I complete my work tasks within the timeframe. I solve all work-related problems efficiently. I always update my knowledge for my work.
	Adaptive Performance	I always come up with creative ideas that are able to provide a new perspective about my tasks. I am able to rearrange my tasks goals according to the current situation. I am able to learn new tasks by using a combination of existing software and other technologies. I am always flexible and open minded in determining the way of delivering my tasks. I am able to quickly analyze potential solutions that could be used in completing my tasks. I always come up with creative ideas that are able to provide a new perspective about my tasks.

4. Pre-Testing, Validity, and Reliability of Research Instrument

Upon the instrument's establishment, a content validation process must be executed to evaluate the instrument. Hence, the questionnaire was distributed to six experts consisting of software development company representatives, GIS users, and academicians to examine and identify any concerns with the questions. Eventually, they all responded and suggested suitable feedback. The experts comprise two experienced GIS users, two professionals from GIS companies, and two academicians with Ph.D. Table 2 presents the information regarding the experts who contributed and were involved in this study. Several concerns were identified, including the need for more clear and precise alternatives when providing respondents with options, the use of more appropriate words and verbs to increase question clarity, and the inclusion of examples to assist respondents' understanding of the questions. Their response was valuable as it incorporated all insightful comments and recommendations into the current questionnaire, with suitable revisions and additions.

Table 2. Panel of Experts.

Panel of Experts	Organization
IT Manager	Organization A
GIS User	Organization A
Academician	University A
Academician	University B
GIS Project Manager	Organization A
Application Analyst	Organization B

The pilot study validated the questionnaire, allowing the investigational questions to be answered and appropriately verified [23,24]. The pilot study was conducted to validate that the respondents understood the questionnaire items and that no unclear questions were present. The data from such an initial survey were examined using SPSS version 26 and featured 30 individuals. Finally, Cronbach's alpha was used to define the scale reliability of the questionnaire, or how closely linked a group of questions is.

The questionnaires were designed into four sections. The first section has five items related to demographic information. The second section of the questionnaire consists of 49 items that focus on GIS quality, which includes functional suitability, performance efficiency, compatibility, usability, reliability, security, maintainability, and portability. The third section focuses on user satisfaction with eight items. The fourth section contains 11 items that focus on individual work performance. As a result, 73 items were subsequently included in the questionnaire before the execution of the actual study. Lastly, the questionnaire comes with two open-ended questions to allow respondents to respond subjectively about the challenges of GIS and suggestions to improve the system.

5. Reliability Analysis Result

Referring to Table 3, the survey collected 30 responses from a total of 45 respondents. Based on the responses captured in the survey, answers regarding demographic background indicated that 77% of respondents were male and 23% were female. In the context of age, the majority of the users were between 21 to 30 years old, the second largest group was between 41 to 50 years, and the third group was 31 to 40 years old. For education level, 63% of respondents had a bachelor's degree, 23% had a diploma, 10% had a master's degree, and only 4% had SPM. In terms of working experience, 40% of respondents were in the range of 1–5 years' experience, 23% had 11 to 15 years, 20% had 6–10 years' experience, 13% had 16 to 20 years' experience, and 4% had above 20 years. The findings also indicated the highest frequency of use is a few times a week, with once a week as the second-highest, and once a month as the lowest.

Table 3. Descriptive Statistic Demographic Result.

Item	Response	Percentage
Gender	Male Female	77% 23%
Age	21–30 years old 41–50 years old 31–40 years old	50% 30% 20%
Education Level	Bachelor's Degree Diploma Master's Degree Malaysia Education Certificate (SPM)	63% 23% 10% 4%
Working Experience	1–5 years 11–15 years 6–10 years 16–20 years Above 20 years	40% 23% 20% 13% 4%
Frequency of GIS Use	Few Times a Week Once a Week Once a Month	47% 40% 13%

Further analysis then proceeded by the identification of the Cronbach's alpha. Cronbach's alpha coefficients were evaluated in the investigation to establish the reliability and internal consistency of the scales used in the study, as shown in Table 4. There are different arguments on the acceptable values of the Cronbach's alpha value [23]. Any Cronbach's alpha values beyond 0.6 to 0.95 for all factors indicated that the degree of dependability was acceptable and normal [24]. Based on the output of the SPSS analysis, the Cronbach's alpha scores of all 68 items in the instrument for each dimension were between 0.942 and 0.975, according to the SPSS analysis findings. This outcome indicates that, at this point in the assessment, the overall index of the scale's internal consistency within the instrument is reliable, with no unexpected irregularities discovered in the data.

Table 4. Cronbach's Alpha Analysis Result.

Variables	Dimensions	Number of Items	Cronbach's Alpha Value
GIS Quality	Functional Suitability	6	0.861
	Performance Efficiency	6	0.806
	Compatibility	4	0.855
	Usability	9	0.892
	Reliability	5	0.896
	Security	6	0.952
	Maintainability	7	0.886
	Portability	6	0.899
User Satisfaction		8	0.942
Individual Work Performance	Task Performance	6	0.955
	Adaptive Performance	5	0.915

6. Discussion and Conclusions

Ultimately, all factors have been analyzed, and the findings compiled from questionnaire surveys on the quality of GIS used by the organization, user satisfaction, and individual work performance have elaborate the readiness to proceed with the next steps of this study. The main objective of this paper was to have clarity on the reliability of the instrument. Based on Table 3, any values above 0.6 to 0.95 indicated that the items were reliable and could proceed to actual data collection [23–25]. For GIS quality, the lowest Cronbach's alpha value is 0.806, and the highest value in the GIS quality items is 0.952. Meanwhile, for user satisfaction, the value is 0.942. Individual work performance results also indicated that the Cronbach's alpha value for task performance was 0.955, while adaptive performance was 0.915. Based on Cronbach's alpha value from the analysis, all items are above 0.6, and it indicates that there are no items that need to be deleted for this study [25,26]. Furthermore, the overall Cronbach's alpha value of the study is 0.984, which is reliable and can proceed to the actual study.

Therefore, this paper aims to summarize current research progress on the influence of GIS quality. The research findings are intended to provide substantial evidence and insights into the critical factors in GIS quality, namely functional suitability, performance efficiency, compatibility, usability, reliability, security, maintainability, and portability. This study also will provide an evidence on the influence of user satisfaction towards individual work performance and identifies the qualities needed in the GIS solution. Before the actual study, the instrument was pre-tested to identify potential errors and establish the degree of understandability of its items. In addition, six subject matter experts evaluated the instruments and gave constructive feedback. A pilot study was also undertaken, with 30 users participating. The data analysis results indicate that the overall index of the scale's internal consistency within the instrument is reliable, with no unexpected irregularities identified in the data.

Author Contributions: Conceptualization, M.S.R., S.M.S. and M.N.M.; methodology, M.S.R., S.M.S. and M.N.M.; validation, M.S.R. and M.F.B.; formal analysis, M.S.R., S.M.S., M.N.M. and M.F.B.; investigation, M.S.R.; resources, M.S.R.; data curation, M.S.R.; writing—original draft preparation, M.S.R.; writing—review and editing, M.S.R. and S.M.S.; visualization, M.S.R.; supervision, S.M.S. and M.N.M.; project administration, M.S.R.; funding acquisition, M.S.R. All authors have read and agreed to the published version of the manuscript.

Funding: This research received no external funding.

Institutional Review Board Statement: Not applicable.

Informed Consent Statement: Not applicable.

Data Availability Statement: Not applicable.

Acknowledgments: The researchers would like to extend their gratitude to Universiti Teknologi MARA (UiTM)—Faculty of Information Management Puncak Perdana for providing the necessary funding for this publication.

Conflicts of Interest: The authors declare no conflict of interest.

References

1. Gobakis, K.; Mavrigiannaki, A.; Kalaitzakis, K.; Kolokotsa, D.D. Design and Development of a Web-Based GIS Platform for Zero Energy Settlements Monitoring. *Energy Procedia* **2017**, *134*, 48–60. [CrossRef]
2. Kanickaraj, L. Geographic Information Systems (GIS) Defined. Available online: https://www.arcweb.com/blog/geographic-information-systems-gis (accessed on 14 May 2022).
3. Blišťan, P.; Kovanič, Ľ.; Kovaničová, M. The Importance of Geographic Information Systems Education at Universities in the Process of Building a European Knowledge-Based Society. *Procedia-Soc. Behav. Sci.* **2015**, *191*, 2458–2462. [CrossRef]
4. Martindale, J. Research Guides: Mapping and Geographic Information Systems (GIS): What Is Gis? Available online: https://researchguides.library.wisc.edu/GIS (accessed on 15 June 2022).
5. Arshad, Y.; Mohd Sani, M.S.; Syed Ibrahim, S.N. Geographical Information System (GIS) as an Innovative Adoption to Ease Customers in Locating SME Business Premises in Malacca. *MATEC Web Conf.* **2018**, *150*, 05045. [CrossRef]
6. Viana, C.M.; Abrantes, P.; Rocha, J. Introductory Chapter: Geographic Information Systems and Science. In *Geographic Information Systems and Science*; IntechOpen: London, UK, 2019; pp. 3–17.
7. Skopenko, N.; Yevsieieva-Severyna, I. Modern Information Systems and Technologies Implementation in Management for Increasing Competitiveness of Enterprises. *Sci. Works Natl. Univ. Food Technol.* **2020**, *26*, 58–70. [CrossRef]
8. Wahyuningrum, T.; Mustofa, K. A Systematic Mapping Review of Software Quality Measurement: Research Trends, Model, and Method. *Int. J. Electr. Comput. Eng.* **2017**, *7*, 2847–2854. [CrossRef]
9. Miguel, J.; Mauricio, D.; Rodríguez, G. A Review of Software Quality Models for the Evaluation of Software Products. *Int. J. Softw. Eng. Appl.* **2014**, *5*, 31–53. [CrossRef]
10. Pratiwi, W. Mujadilah, N. Effect of Accounting Information Systems, System Quality and Service Quality on User Satisfaction of Mobile Banking-Based Applications. *J. Econ. Financ. Manag. Stud.* **2021**, *4*, 383–396. [CrossRef]
11. Nistala, P.; Nori, K.V.; Reddy, R. Software Quality Models: A Systematic Mapping Study. In Proceedings of the 2019 IEEE/ACM International Conference on Software and System Processes (ICSSP), Montreal, QC, Canada, 25 May 2019.
12. Qin, L.; Qu, Q.; Zhang, L.; Wu, H. Platform Trust in C2C e-Commerce Platform: The Sellers' Cultural Perspective. *Inf. Technol. Manag.* **2021**, 1–11. [CrossRef]
13. Gao, M.; Kortum, P.; Oswald, F. Psychometric Evaluation of the Use (Usefulness, Satisfaction, and Ease of Use) Questionnaire for Reliability and Validity. *Proc. Hum. Factors Ergon. Soc. Annu. Meet.* **2018**, *62*, 1414–1418. [CrossRef]
14. Rahrouh, M.; Taleb, N.; Mohamed, E.A. Evaluating the Usefulness of e-Learning Management System Delivery in Higher Education. *Int. J. Econ. Bus. Res.* **2018**, *16*, 162–181. [CrossRef]
15. Truong, N.B.; Lee, H.; Askwith, B.; Lee, G.M. Toward a Trust Evaluation Mechanism in the Social Internet of Things. *Sensors* **2017**, *17*, 1346. [CrossRef] [PubMed]
16. Bano, M.; Zowghi, D.; da Rimini, F. User Satisfaction and System Success: An Empirical Exploration of User Involvement in Software Development. *Empir. Softw. Eng.* **2016**, *22*, 2339–2372. [CrossRef]
17. Widyastuti, T.; Hidayat, R. Adaptation of Individual Work Performance Questionnaire (IWPQ) into Bahasa Indonesia. *Int. J. Res. Stud. Psychol.* **2018**, *7*, 101–112. [CrossRef]
18. Koopmans, L.; Bernaards, C.M.; Hildebrandt, V.H.; Schaufeli, W.B.; de Vet Henrica, C.W.; van der Beek, A.J. Conceptual Frameworks of Individual Work Performance. *J. Occup. Environ. Med.* **2011**, *53*, 856–866. [CrossRef]
19. Koopmans, L.; Coffeng, J.K.; Bernaards, C.M.; Boot, C.R.L.; Hildebrandt, V.H.; de Vet, H.C.W.; van der Beek, A.J. Responsiveness of the Individual Work Performance Questionnaire. *BMC Public Health* **2014**, *14*, 513. [CrossRef]
20. Kalia, N.; Bhardwaj, B. Contextual and Task Performance: Do Demographic and Organizational Variables Matter? *Rajagiri Manag. J.* **2019**, *13*, 30–42. [CrossRef]
21. Park, Y.; Lim, D.H.; Kim, W.; Kang, H. Organizational Support and Adaptive Performance: The Revolving Structural Relationships between Job Crafting, Work Engagement, and Adaptive Performance. *Sustainability* **2020**, *12*, 4872. [CrossRef]
22. Loughlin, E.M.; Priyadarshini, A. Adaptability in the Workplace: Investigating the Adaptive Performance Job Requirements for a Project Manager. *Proj. Leadersh. Soc.* **2021**, *2*, 100012. [CrossRef]
23. Jain, S.; Angural, V. Use of Cronbach's Alpha in Dental Research. *Med. Res. Chron.* **2017**, *4*, 285–291.
24. Taber, K.S. The Use of Cronbach's Alpha When Developing and Reporting Research Instruments in Science Education. *Res. Sci. Educ.* **2017**, *48*, 1273–1296. [CrossRef]
25. Yanyun, Y.; Green, S.B. Coefficient Alpha: A Reliability Coefficient for the 21st Century? *J. Psychoeduc. Assess.* **2011**, *29*, 377–392. [CrossRef]
26. Saunders, M.; Lewis, P.; Thornhill, A. *Research Methods for Business Students*; Pearson: Harlow, UK, 2016.

proceedings

Proceeding Paper

Web-Based Parametric Effort Estimation for Mobile Application Development †

Nur Ida Aniza Rusli [1,*], Nur Atiqah Sia Abdullah [2], Fatin Nabila Abd Razak [3] and Nemo Menton Mufriz [2]

1 Faculty of Computer and Mathematical Sciences, Universiti Teknologi MARA, Cawangan Negeri Sembilan, Kampus Kuala Pilah, Kuala Pilah 72000, Negeri Sembilan, Malaysia
2 Faculty of Computer and Mathematical Sciences, Universiti Teknologi MARA, Shah Alam 40450, Selangor, Malaysia
3 Academy of Language Studies, Universiti Teknologi MARA, Cawangan Negeri Sembilan, Kampus Kuala Pilah, Kuala Pilah 72000, Negeri Sembilan, Malaysia
* Correspondence: idaaniza@uitm.edu.my
† Presented at the International Academic Symposium of Social Science 2022, Kota Bharu, Malaysia, 3 July 2022.

Abstract: Estimation methods are continuously being adapted to obtain better and clearer estimations needed to achieve development goals. Some estimation methods were invented before the modern mobile application technology that is currently available. Thus, these methods are unable to cater to the requirements for estimating modern mobile application features. The objective of this paper is to propose a web-based system as a method to estimate the effort and cost of developing a mobile application. The key idea behind this study is to identify cost drivers that can be applied in mobile application development through literature review. From the analysis, 19 cost drivers are found to fit the vision of this study. In addition, this study also seeks to investigate the price range of cost drivers acquired from existing similar systems. The total price range is accumulated, and the mean value of each cost driver is obtained, which is then inserted further into the new estimation metric. The proposed system is then evaluated by comparing the obtained results with six similar systems according to basic user needs requirements in an application. The results demonstrate that the proposed system is a more enhanced cost estimation software that contains more cost driver options, which users can utilize to estimate mobile application development costs.

Keywords: software effort estimation; parametric estimation; mobile application; estimation system

Citation: Rusli, N.I.A.; Abdullah, N.A.S.; Abd Razak, F.N.; Mufriz, N.M. Web-Based Parametric Effort Estimation for Mobile Application Development. *Proceedings* **2022**, *82*, 69. https://doi.org/10.3390/proceedings2022082069

Academic Editor: Mohamad Rahimi Mohamad Rosman

Published: 20 September 2022

1. Introduction

Effort estimation is the procedure carried out to anticipate the most sensible measure of effort required to create or maintain software. Effort estimation is a key project management activity needed for project planning, staff resources estimation, cost estimation, quality control, and benchmarking [1]. Enhancing the estimation techniques available to project managers would encourage more successful control of time and spending plans in software development [2].

The ever-growing need for better functionality and hope for a better way of life has brought forth a whole new mobile application development industry. Despite the accessibility of many versatile applications, software developers create many new applications to fulfill the interest of mobile device users worldwide [1]. This results in developers seeking the most efficient techniques for effort estimates for project plans, cycle designs, spending plans, investment analysis, pricing processes, and bidding rounds. Inappropriate software development effort estimation can result in project failures due to budget overruns and slips in scheduling [1].

Furthermore, existing methods such as Function Point, Object Point, and COSMIC Full Function Point (COSMIC FFP) have limitations, as they are prone to these inclinations: individual experience, political points, resources, time weight, and memory recall [3–5]. In addition, these estimation methods were invented before the modern mobile application technology available now, and they are most likely unable to cater to current features.

Therefore, new estimation methods are needed to estimate mobile application development efforts to overcome the estimation problems. Therefore, this study proposes a web-based parametric effort estimation system as an option for software developers or other users to estimate the cost of mobile application development. This paper is organized as follows: Section 2 describes the related works, Section 3 illustrates the proposed system, Section 4 summarizes the results, while Section 5 draws the conclusion.

2. Related Works

The following section reviews existing effort estimation models, consisting of mobile application estimation models and effort estimation systems.

2.1. Mobile Application Estimation Models and Systems

The effort estimation of mobile applications is a complex issue, and no specific model or process exists. It has been demonstrated from natural considerations that mobile application development suffers from effort estimation syndrome. Therefore, there have been several attempts in the last few years to address characteristics and techniques in the field of effort estimation across mobile applications.

Shahwaiz et al. [1] proposed a parametric model for assessing the effort necessary to create mobile applications. The regression-based model is measured using information from 161 mobile application characteristics and validated using the k-fold cross-validation method. In addition, the expected precision of this mobile application's particular model is contrasted with the standard precision of the general-purpose COCOMO II model. The correlation result demonstrates that this model is more precise than the COCOMO II model. Initially, there were 16 cost drivers, which resulted in seven categories of effort predictors upon calibration.

Altaleb and Gravell [6] depicted the results of a Systematic Literature Review (SLR) with respect to size estimation and effort models in mobile application development. This is followed by an outline of estimation techniques utilized crosswise over mobile applications gathered from 64 papers and presents the suggested 25 cost drivers.

Accuracy and efficiency are critical factors in ensuring a successful effort estimation model. Thus, many organizations have developed effort estimation systems to support the calculation process. This study reviews seven effort estimation systems specifically for mobile application development, including Estimate My App [7], How Much to Make an App [8], VenturePact [9], BuildFire [10], Otreva [11], Cleveroad [12], and Andreas Ley Calculator [13].

2.2. Comparison of Cost Drivers

This study has identified 40 cost drivers from the literature review in [1,6] and used further to identify the most significant cost drivers for our proposed method.

From the comparison in Table 1, this study includes all the cost drivers, with the majority scoring 3/7 or higher. However, the number of screens characteristic is discarded as a potential factor in the proposed method. This is because the number of screens factor is considered a young research discipline.

Software metric researchers, for example, are still trying to find a range for the number of screens required to classify a mobile app's complexity (either in small, medium, or large) [14,15]. Thus, the general effort costing cannot be derived, as no proper evaluation has been conducted for this factor.

Table 1. Comparison of cost drivers towards seven estimation systems.

Characteristics	Estimate My App	How Much to Make an App	Venture Pact	BuildFire	Otreva	Cleveroad	Andreas Ley Cost Calculator	Total
Function point size	/	-	-	-	-	-	-	1/7
UML diagram	-	-	-	-	-	-	-	0/7
Supported platform type	/	/	/	/	/	/	/	7/7
Supported device	-	-	-	-	-	-	/	1/7
Supported device	-	-	-	-	-	-	/	1/7
Back-end system availability and server config. flexibility	-	-	-	/	-	-	-	1/7
Development team skills	-	-	-	-	-	-	-	0//7
App development flexibility and complexity	-	-	-	-	-	-	-	0//7
Team communication, process, complexity and experience	-	-	-	-	-	-	-	0/7
Push notification	/	-	-	/	/	/	/	5/7
Landscape and portrait mode	-	-	-	-	-	-	-	0/7
Data storage and memory opt. complexity	-	-	/	/	-	-	-	2/7
Number of screens	-	-	/	-	-	/	/	3/7
Number of API parties	/	/	-	/	/	-	-	4/7
Support code reusability	-	-	-	-	-	/	/	2/7
Technology maturity	-	-	-	-	-	-	-	0/7
Battery and optimisation	-	-	-	-	-	-	-	0/7
Connection	-	/	-	/	-	-	/	3/7
Booking and reservation	/	-	-	-	/	-	-	2/7
Calendar and time	/	-	-	/	/	/	-	4/7
Map and localisation	/	-	/	/	/	/	/	6/7
Social sharing	/	/	-	/	/	/	-	5/7
Searching contents	/	-	/	-	/	-	-	3/7
Messaging	/	-	-	/	/	/	-	4/7
Deadline date	/	-	-	-	-	/	-	2/7
Number of functionalities	/	-	-	-	-	-	-	1/7
Registration and login	/	/	/	/	/	/	/	7/7
Chronological list	-	-	-	-	-	-	-	0/7
Number of files, classes, methods, statements, and LOC	-	-	-	-	-	-	-	0/7
Chronological list	-	-	-	-	-	-	-	0/7
File upload	/	-	-	-	-	/	-	2/7
Comment feature	/	/	-	-	-	-	-	2/7
Navigation	-	-	-	-	-	-	/	1/7
Interrupt handling	-	-	-	-	-	-	-	0/7
Security analysis support	/	-	/	/	-	-	/	4/7
Budget for the project	-	-	-	-	-	-	-	0/7
Compatibility with previous version	-	-	-	-	-	-	/	1/7
Multi language support	/	-	-	-	-	-	/	2/7
Media support	/	-	/	/	/	-	/	6/7
Pay process user feedback	/	/	/	/	/	/	/	7/7

Next, a total of two systems applied factors of data storage and memory opt. complexity, support code reusability, booking and reservation, deadline date, file upload, comment feature, and multi-language support. Among these factors, booking and reservation, file upload, comment feature, and multi-language support are included in this study because these factors are considered necessary for mobile business modelling [16–19].

Only one system applied factors such as function point size, supported device, back-end system availability, and server config. flexibility, number of functionalities, navigation, and compatibility with the previous version. From these factors, function point size is considered as an important factor, although only the Estimate My App system applied this factor for mobile app effort estimation. This study also includes this factor because function point size is considered a stable procedure to derive effort estimation [20].

Lastly, none of the selected systems applied cost factors such as UML diagram, development team skills, app development flexibility and complexity, team communication, process, complexity and experience, landscape and portrait mode, technology maturity, battery and power optimization, number of files, classes, methods, statements and LOC, chronological list, hardware access, interrupt handling, and budget for the project. Thus, these criteria are not considered essential and are discarded in the proposed system.

3. Proposed System

This section overviews of the proposed web-based parametric effort estimation for mobile application development.

3.1. Cost Drivers

From the comparison (as shown in Section 2.2), 19 cost drivers are obtained for further use in the proposed system. The estimated costing range for each factor is obtained from the reviewed system. The factors are included in this study to suggest the mobile application category and its respective cost. The estimated cost of each cost driver is shown in Table 2.

Table 2. Cost driver in proposed system.

Cost Driver (Proposed System)	Description	Estimated Cost (USD)
Function Point Size	Small (has around 2–3 key features); Medium (has around 4–7 key features); Large (has around 8–12 key features).	Small (USD 4500); Medium (USD 13,500); Large (USD 22,500)
Supported Platform	iOS (iPhone/iPad app); Android (Android phone/Tablet app)	USD 1200–USD 9600
User Interface Quality and Complexity	MVP (Minimum Viable Product. Very raw but functional); Basic (Still quite basic but pleasing to the eye); Polished (Professional bespoke UI design. May also have some animations and transitions etc.)	MVP (USD 400–USD 5850); Basic (USD 1200–USD 6750); Polished (USD 2400–USD 9600)
Push Notification	Real-time notifications between users, e.g., unread message counts, notifications of editing, etc.	Yes (USD 1350–USD 5000); No (USD 0)
Number of API Parties	Connect to one or more third party services (An information feed that you must incorporate with or an accomplice application); SMS Messaging; (Allow your app to send SMS messages); Phone Number Masking (Calls conducted by your app have masked phone numbers).	Connect to one or more third party services (USD 1350–USD 5000); SMS Message (USD 1800); Phone Number Masking (USD 1800)
Connection	Bluetooth (Use Bluetooth to communicate and transfer data between devices); Wireless (App does not need to connect via internet)	Bluetooth (USD 3840–USD 10,000); Wireless (USD 0);
Booking and Reservation	Managing capacity, choosing the start and end dates etc.	Yes (USD 2250–USD 5250); No (USD 0)
Calendar and Time	Display data in a calendar format	USD 2700–USD 3000;
Map and Localisation	Showing a map with data point e.g., driver locations, venue locations, etc.	Yes (USD 1350–USD 5000); No (USD 0)
Social Sharing	Ability to share pieces of information in a controlled way on social media account to drive engagement	Yes (USD 450–USD 3000); No (USD 0);
Searching and Contents	Users would be able to search content	Yes (USD 1350–USD 4500); No (USD 0);
Messaging	Allowing users within the app to send message to other Account users or group of users	Yes (USD 2250–USD 3750); No (USD 0);
File Upload	Users can upload video, photo content or audio	Yes (USD 1800); No (USD 0);
Registration and Login	Email/Password; Facebook; Twitter; Google	Email/Password (USD 320–USD 9600); Facebook (USD 800–USD 4000) Twitter (USD 800–USD 4000) Google (USD 800–USD 4000)
Comment Feature	Classic forum functionality for account users or simple commenting on information	Yes (USD 2250–USD 3000); No (USD 0);
Security Analysis Support	Security not important (Use Bluetooth to communicate and transfer); Basic Security Measures; Complete Protection (Protection against XSS & SQL Injection)	Security not important (USD 0); Basic Security Measures (USD 400–USD 1300); Complete Protection (USD 2400–USD 7800);
Multi-language Support	Provide support for multiple languages for your app	Yes (USD 1800); No (USD 0)
Media Support	Users are able to modify video, photo content or audio on their gadget (e.g., Filters).	Yes (USD 1800–USD 3770); No (USD 0)
Paying Process User Feedback	You will process ad-hoc or regular payments from users and manage refunds, etc.	Yes (USD 1440–USD 6000); No (USD 0);

3.2. System Interface

Figure 1 shows the proposed user interface for the system. It consists of 19 cost drivers to allow the user to choose based on their requirement specification. Figure 1 shows a part of the main interface for the system.

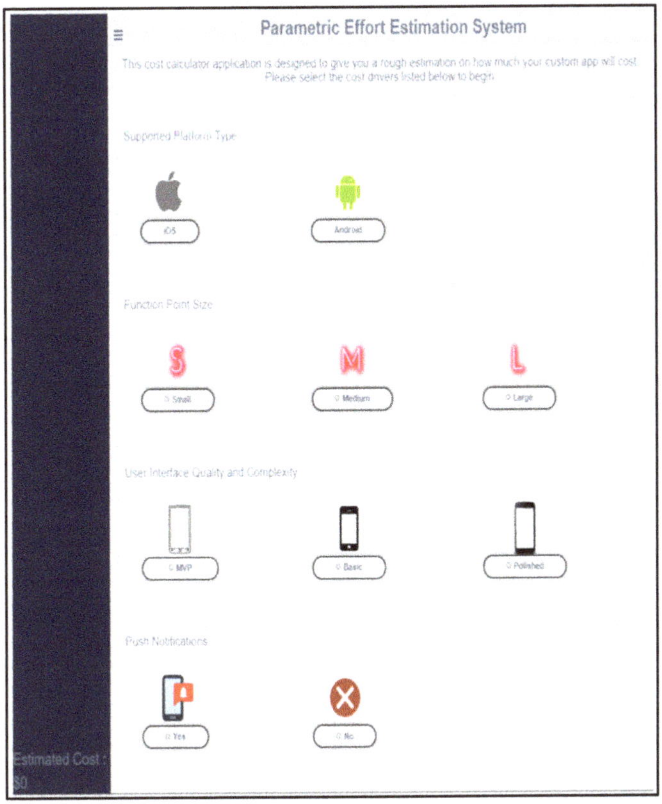

Figure 1. Main interface of proposed system.

There are 19 form groups in the main interface representing 19 cost drivers, as stated in the previous section. Each form group provides two to four options for the user to choose from. These options are types of radio buttons and checkboxes, depending on the cost driver.

Meanwhile, Figure 2 shows an example of a radio button used for the Function Point Size cost driver. This button allows the user to only make one selection for this type of cost driver. The selected option is changed as the user clicks a different option. Every change will deduct the previous value of the cost driver and update the new value of the selected cost driver in the bottom left corner. The total estimated cost will be calculated throughout the 19 cost-driver selections. The estimated cost is displayed at the bottom left of the interface.

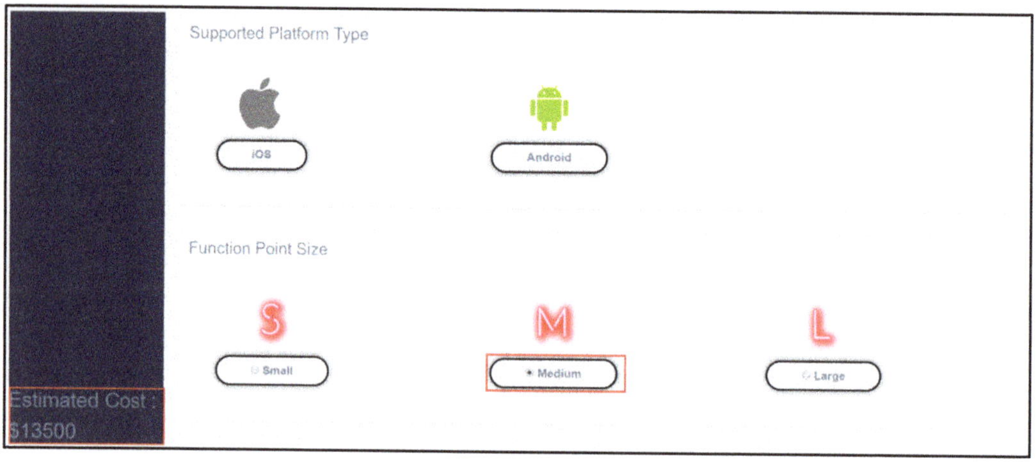

Figure 2. User interface for estimated cost.

Figure 3 shows an example of the checkbox button used for the number of API Parties cost driver. This button allows the user to make more than one selection or remove the selection if the user wishes to do so. For every choice the user makes, the system prompts the total estimated cost on the bottom left corner of the screen. The system will update this value according to the user's selections. The proposed system also provides the tooltips function (refer to Figure 4) to help the user further understand what each cost driver refers to. The tooltip will appear whenever a user hovers over the icons.

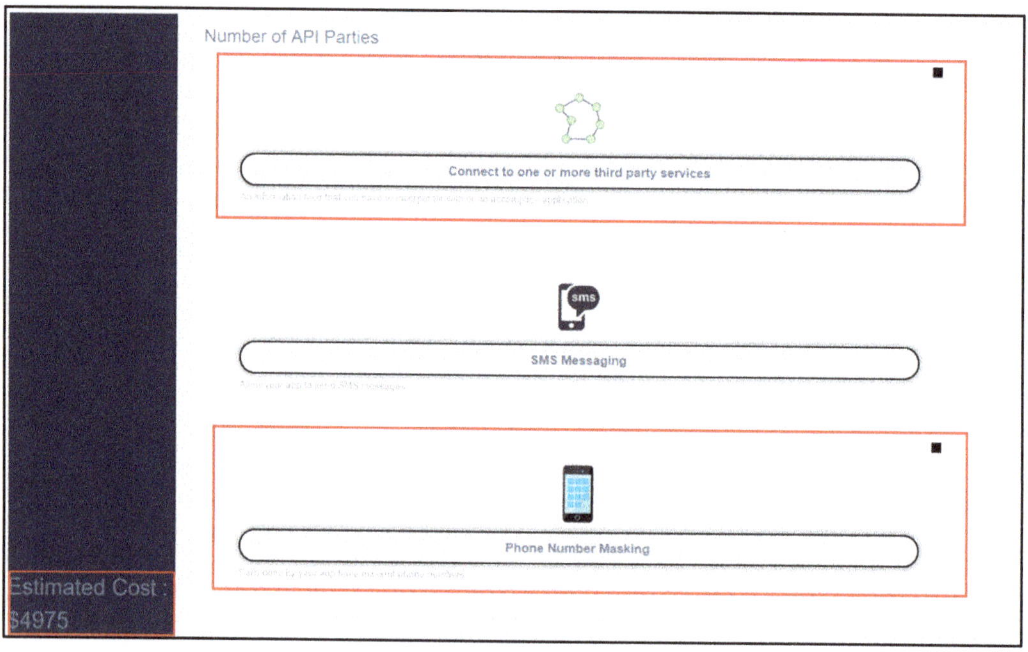

Figure 3. User interface for checkbox function.

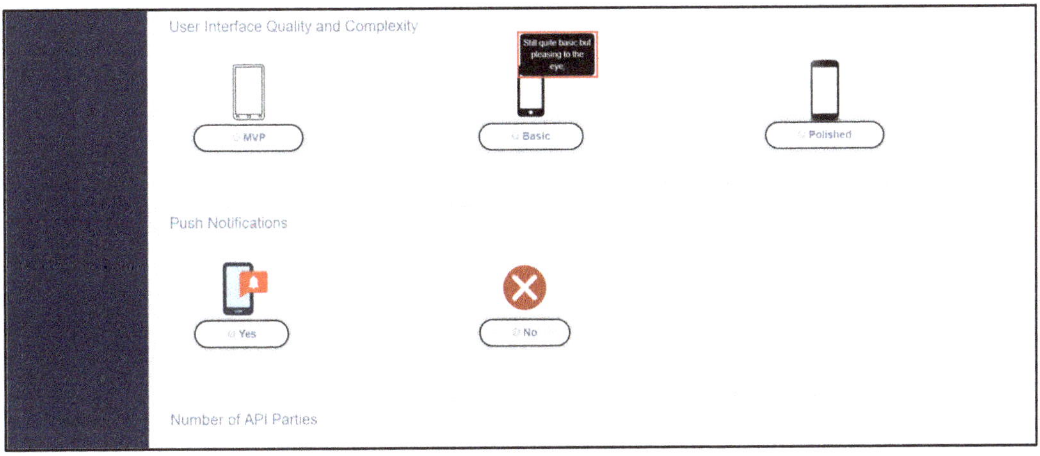

Figure 4. User interface for tooltip function.

4. Result and Discussion

This study evaluated the proposed system's credibility by comparing the proposed system's total value estimate against six other similar systems. The Cleveroad system is excluded from the evaluation since it does not provide any price ranges for the listed cost drivers. As a result, this study omits the Cleveroad system while formulating the cost range of the cost drivers.

Table 3 shows the estimated cost range of similar systems using the standard evaluation criteria. From the result, the percentage of difference between the proposed system and Estimate My App, How Much to Make an App, and Andreas Ley cost calculator are within a range of +10%–+20%.

Table 3. Percentage and difference in terms of cost range.

Title 1	Estimate MyApp	How Much to Make an App	VenturePact	BuildFire	Otreva	Andreas Ley Cost Calculator
	USD 49,500	USD 30,600	USD 24,180	USD 164,868	USD 66,003	USD 31,000
Proposed System	USD 55,300	USD 35,340	USD 18,025	USD 48,550	USD 44,080	USD 36,350
Difference in USD	USD 5800	USD 5340	USD −6155	USD −116,318	USD −21,923	USD 5350
Percentage of Difference	+11.72%	+17.80%	−25.45%	−70.55%	−33.22%	+17.26%

Estimate My App matches almost all the cost drivers (18 out of 19). The system in this study is 11.72% more costly than Estimate My App. The significant difference in percentage collected in the Table 3 is caused by the BuildFire's system having very high charges compared with the other six systems that considered the proposed system's price range. This is justified by seeing that the proposed system is −70.55% lower in cost in comparison with the BuildFire system.

After conducting research into the systematic literature review by Altaleb and Gravel [6], 40 cost drivers were mentioned as important and needed to be accounted for when performing the cost estimation of a mobile application. Moreover, these cost drivers were deemed relevant based on the current needs of e-commerce processes. In this study, only 19 cost drivers were selected for inclusion in this system due to the comparison table that was constructed between similar systems and the systematic literature review.

For future work, the estimated cost range collected in this study may have caused the results to be less appealing. This is because the BuildFire system was included as part in formulating the cost range despite having a much higher rate for their cost driver prices. The BuildFire system, however, is still included in this analysis due to the study's goal of identifying the most significant cost drivers for current mobile applications. This issue serves as a caution to avoid the future proposed system having extremely high-cost ranges in their cost drivers.

5. Conclusions

This paper has presented a web-based system to estimate mobile application development efforts and costs. The main objective of this research is to identify cost drivers relevant to modern mobile application development. Forty cost drivers were identified from the literature review. However, after analyzing seven systems, only 19 were considered potential factors. In addition, the costing factors were determined using the values offered by the reviewed systems. Furthermore, this study conducted a simple evaluation process to test the functionality of the developed system. This phase was conducted by comparing the total estimated cost of the proposed system with six other similar systems. A comprehensive table that contained the cost range difference in absolute values and percentages was constructed to analyze the results further. In conclusion, the system functions accordingly with a 20–30% significant difference between similar systems.

Author Contributions: Conceptualization, N.I.A.R. and N.A.S.A.; methodology, N.A.S.A. and N.M.M.; software, N.I.A.R., N.A.S.A. and N.M.M.; validation, N.I.A.R., N.A.S.A. and N.M.M.; formal analysis, N.A.S.A. and F.N.A.R.; investigation, N.I.A.R. and N.A.S.A.; resources, N.I.A.R. and N.A.S.A.; data curation, N.I.A.R., F.N.A.R. and N.M.M.; writing—original draft preparation, N.I.A.R. and N.M.M., writing—review and editing, F.N.A.R.; visualization, N.I.A.R. and F.N.A.R.; supervision, N.I.A.R. and N.A.S.A.; project administration, N.I.A.R. and N.A.S.A.; funding acquisition, N.I.A.R. and N.A.S.A. All authors have read and agreed to the published version of the manuscript.

Funding: This research was funded by MyRA Research Grant Scheme File No.: 600-RMC/GPM LPHD 5/3 (180/2021), under Universiti Teknologi MARA internal grant scheme.

Institutional Review Board Statement: Not applicable.

Informed Consent Statement: Not applicable.

Data Availability Statement: Not applicable.

Acknowledgments: The authors would like to thank the Ministry of Higher Education of Malaysia (MOHE) for funding the research project under MyRa Grant Scheme. The authors gratefully acknowledge the Faculty of Computer and Mathematical Sciences, Universiti Teknologi MARA for supporting the publication of this paper.

Conflicts of Interest: The authors declare no conflict of interest.

References

1. Shahwaiz, S.; Malik, A.; Sabahat, N. A parametric effort estimation model for mobile apps. In Proceedings of the 19th International Multi-Topic Conference (INMIC), Islamad, Pakistan, 5–6 December 2016; pp. 1–6.
2. Finnie, G.; Wittig, G.; Desharnais, J.M. A comparison of software effort estimation techniques: Using function points with neural networks, case-based reasoning and regression models. *J. Syst. Softw.* **1997**, *39*, 281–289. [CrossRef]
3. Albrecht, A.J. Measuring application development productivity. In Proceedings of the Joint Share/Guide/IBM Application Development Symposium, Monterey, CA, USA, 14–17 October 1979; pp. 83–92.
4. Banker, R.D.; Kauffman, R.J.; Wright, C.; Zweig, D. Automating output size and reuse metrics in a Repository-Based Computer-Aided Software Engineering (CASE) environment. *IEEE Trans. Softw. Eng.* **1994**, *20*, 169–187. [CrossRef]
5. Meli, R.; Abran, A.; Ho, V.T.; Oligny, S. On the applicability of COSMIC-FFP for measuring software throughout its life cycle. In Proceedings of the 11th European Software Control and Metrics Conference, Munich, Germany, 18–20 April 2000.
6. Altaleb, A.R.; Gravell, A.M. Effort estimation across Mobile app platforms using Agile processes: A systematic literature review. *J. Softw.* **2018**, *13*, 242–259. [CrossRef]
7. Oozou. Estimate My App. Available online: https://estimatemyapp.com/ (accessed on 28 October 2020).
8. Crew. How Much to Make an App? Available online: http://howmuchtomakeanapp.com/ (accessed on 28 October 2020).
9. VenturePact. Looking to Build an App? Available online: https://venturepact.com/mobile_app_price_calculator (accessed on 28 October 2020).
10. BuildFire. How Much Will Your Mobile App Project Cost You? Available online: https://buildfire.com/how-much-to-make-a-mobile-app-calculator/# (accessed on 28 October 2020).

11. Otreva. How Much Does It Cost to Build an App? Available online: https://www.otreva.com/calculator/# (accessed on 28 October 2020).
12. Cleveroad. How Much Does It Cost to Make an App? Available online: https://www.cleveroad.com/mobile-app-development-cost-calculator (accessed on 28 October 2020).
13. Ley, A. Cost Calculator for Mobile App Development. Available online: https://www.andreasley.ch/en/costcalculator/ (accessed on 28 October 2020).
14. Iida, N. New Technique for estimating development scale: Implementing the function scale method. *Fujitsu Sci. Tech. J.* **2010**, *46*, 149–157.
15. Rijwani, P.; Jain, S.; Santani, D. Software effort estimation: A comparison-based perspective. *Int. J. Appl. Innov. Eng. Manag.* **2014**, *3*, 18–29.
16. Hebig, R.; Derehag, J.; Chaudron, M.R. Identifying metrics' biases when measuring or approximating size in heterogeneous languages. In Proceedings of the ACM/IEEE International Symposium on Empirical Software Engineering and Measurement, Beijing, China, 22–23 October 2015; pp. 1–4.
17. Gomes, J.V.; Inácio, P.R.; Freire, M.M.; Pereira, M.; Monteiro, P.P. Analysis of peer-to-peer traffic using a behavioural method based on entropy. In Proceedings of the IEEE International Performance, Computing and Communications Conference, Austin, TX, USA, 7–9 December 2008; pp. 201–208.
18. Mendes, E.; Mosley, N.; Counsell, S. Investigating web size metrics for early web cost estimation. *J. Syst. Softw.* **2005**, *77*, 157–172. [CrossRef]
19. Van Den Berg, K.; Dekkers, T.; Oudshoorn, R. Functional size measurement applied to UML-based user requirements. In Proceedings of the 2nd Software Measurement European Forum (SMEF), Rome, Italy, 16–18 March 2005; pp. 69–80.
20. Lavazza, L.; Del Bianco, V. A case study in COSMIC functional size measurement: The rice cooker revisited. In *International Workshop on Software Measurement*; Springer: Berlin/Heidelberg, Germany, 2000; pp. 101–121.

proceedings

Proceeding Paper

Web Accessibility Testing for Visually Impaired People in Indonesia †

Maria Cahyahadi and Nina Setiyawati *

Department of Informatics Engineering, Universitas Kristen Satya Wacana, Salatiga 50715, Central Java, Indonesia
* Correspondence: nina.setiyawati@uksw.edu

† Presented at the International Academic Symposium of Social Science 2022, Kota Bharu, Malaysia, 3 July 2022.

Abstract: An equal society requires respect for differences and attention to minority groups. Web inclusivity for disabled people is closely related to how they can understand, use and utilize websites like non-disabled people. The objective of this research was to analyze whether web accessibility has been implemented in Indonesia based on the availability and functionality of the requirements for visually impaired people. Tests were performed manually and automatically on 30 samples. Automated tests were performed by validating the HTML DOM elements of the web page. The results showed that 222 out of 240 tests failed because specified requirements were not found. Based on this research, web accessibility in Indonesia is deficient. Websites should provide accessibility features to assist visually impaired people access information and services.

Keywords: accessibility; automation testing; website; visually impaired; HTML DOM

check for updates

Citation: Cahyahadi, M.; Setiyawati, N. Web Accessibility Testing for Visually Impaired People in Indonesia. *Proceedings* **2022**, *82*, 72. https://doi.org/10.3390/proceedings2022082072

Academic Editor: Mohamad Rahimi Mohamad Rosman

Published: 21 September 2022

1. Introduction

People with disabilities have higher vulnerability to experiencing exclusion in society. The leading cause is their elaborate particular needs that are more intricate to apply to public services and facilities than the standard design used for non-disabled people. National development plans often do not meet their particular needs. There is a low possibility that disabled people can enjoy the results of national development, and it is difficult for them to play an active role in national development [1–3]. The government and private businesses need to actualize disabled people's needs so they can be at ease without discrimination [4,5].

Disabled people's awareness of their rights and their commitment to embody it must come from society. The government shows support for people with disabilities by enacting and issuing laws and regulations that set equal rights for them [1,6–8]. However, actualizing equal rights for disabled people has not been maximally implemented [1,8–10]. The biggest challenge to creating inclusivity for disabled people is the lack of their participation and their chances to socialize to make their needs acknowledged [11–13].

Information technology has been a significant support in people's daily lives during the COVID-19 pandemic, where most activities are carried out online [14–16]. Therefore, information and public services have also shifted to digital platforms through websites [17,18]. Web applications are often the only available way for people to access some services, such as health, public administration, banking, education and entertainment.

Web inclusivity for disabled people is closely related to how they can understand, use and utilize web applications like non-disabled people. Web visual and functional design should not reduce their ability to access information and services. However, Indonesia is ranked 66th out of 120 other countries on the Inclusive Internet Index [19], with an accessibility score of 69.75, which does not meet the Web Content Accessibility Guideline [20] standards. This poor score is due to faint web accessibility implementation in Indonesia for people with disabilities on government and university websites, of which 95% do not support accessibility for disabled people [21].

This research focuses on the availability of web modalities for disabled people with limited vision. One of the hindrances of typical web design for visually impaired people is that web information is presented in small text/objects without audio information assistance. For instance, nearsighted users find it difficult to read information in small text sizes that non-disabled users can generally read.

Requirements that can be implemented on the web to help visually impaired users maximize the use of the web are object magnification, contrast settings, information in the form of sound and adjustable text sizes [20,22,23]. The purpose of this research is to analyze whether web accessibility requirements for users with limited vision have been implemented in a sample of websites that Indonesians widely use to obtain information and services. Hopefully, this research can provide information on web accessibility in Indonesia as a part of technology service improvements for visually impaired people.

2. Method

Based on the research flow in Figure 1, the first stage of the research is Problem Identification. At this stage, secondary research was conducted on the condition of disabled people's access to information. Based on the 2018 National Socio-Economic Survey (SUSENAS), access to information for disabled people using mobile phones or laptops was only 34.89%, and internet access for people with disabilities was only 8.5% [24]. Considering Undang-Undang No. 8 of 2016 on Persons with Disabilities, the Ministry of Social Affairs seeks to provide disabled people with equal opportunities, one of which is equal access to information [25]. However, most government and university websites in Indonesia are not disabled friendly, and the accessibility of information technology has not been widely applied [21]. Therefore, this research analyzes the accessibility of several websites in Indonesia for users with limited vision.

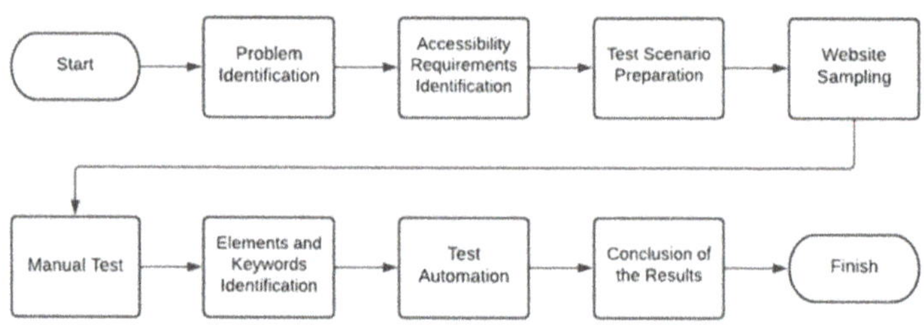

Figure 1. Research workflow.

The second stage is Accessibility Requirements Identification. Web accessibility requirements for visually impaired users were collected and selected through a literature review. Requirements related to visual impairment were obtained through sources in guidelines, laws and previous research. Then, these requirements were outlined and grouped based on similarities. The requirements listed in two or more sources were selected to be tested.

The third stage is Test Scenario Preparation. A test scenario was designed to verify the existence of web elements that implement the requirements' functionality. The fourth stage is Website Sampling. This stage was carried out randomly on the websites that Indonesians use to obtain information and services in education, health, civil, employment, banking, law and entertainment.

The fifth stage is a Manual Test in the form of exploratory tests. The Manual Test was carried out to find unknown web elements in each sample. Once discovered, several accessibility feature usage scenarios were tested for functionality. The sixth stage is Elements and Keywords Identification. For each accessibility feature that passed the Manual Test, its HTML element was identified to find keywords that point to that element. Keywords were used as coordinates in the test automation script. The seventh stage is Test Automation. The automation was performed by accessing the HTML DOM of a web page, and it verified the presence of elements based on coordinates.

The last stage is the Conclusion of the Results. The results obtained from manual and automated testing were compared. If there were differences in the results between manual and automated tests, then the anomalies that affected the differences in results were rechecked. Then, a descriptive analysis was conducted on what requirements were applied and how they affected web accessibility in Indonesia.

3. Results and Discussion

Relevant data regarding web accessibility requirements for visually impaired users were collected from three sources. Table 1 lists information on the requirements identified in the sources reviewed. We gathered five requirements, each was given a representative alphabetical code, and visual impairment types included blindness, low vision and color-blindness. Table 2 shows the association between the requirements and the party issuing the guidelines. An O indicates that the requirement was found at the source.

Table 1. Web accessibility requirements for visually impaired users.

Requirement	Type of Visual Impairment	Code
Auditory information	Blindness; Low vision	A
Adjustable text size	Low vision	B
Object magnification	Low vision	C
Contrast settings	Low vision	D
Written information on a color-based action	Color blindness	E

Table 2. Accessibility standards and guidelines.

Resource	A	B	C	D	E
Americans with Disability Act (ADA)	O	O	O		
The European Parliament and the Council of the European Union	O	O	O	O	O
Web Content Accessibility Guidelines (WCAG) 2.1	O	O		O	

Based on the requirements appearing in two or more sources, four were selected from Table 2 to be used in this research. The requirements are as follows:

- R1: Auditory information;
- R2: Adjustable text size;
- R3: Object magnification;
- R4: Contrast settings.

The test scenario template was designed as below.

1. Open browser;
2. Open sample web page;
3. Visually verify the availability of the requirements on the homepage;
    - 3.a. If one or more requirements are found;
        - 3.a.1. Check if the features can be utilized;
            - R1 Check if the audio can be played and produce a clear sound;
            - R2 Try all the available text size options and see if they work;
            - R3 Try playing with the magnifying feature and see if the object looks bigger than before;
            - R4 Try all the available contrast options and see if the visuals change;
        - 3.a.2. Record the results and proceed to step 4;
    - 3.b. If no requirements are found;
        - 3.b.1. Perform an exploratory test on some sample web pages to look for the existence of the requirements;
            - 3.b.1.1If the requirements are found, proceed to step 3.a.1;

3.b.1.2If none are found, proceed to step 4;

4. Close the browser.

The samples that were used were 30 HTML elements of all samples that passed the manual tests, and they were examined to acquire keywords. The collected keywords are listed in Table 3. The keywords were found in Indonesian and English. Each keyword was used with lowercase, uppercase and capitalization to increase the probability of finding the requirement, because the coordinates in the automation script were case sensitive. The implementation of the R1 requirement was not found. Instead, we used the <audio> tag; this was referred from [26] (Figure 2).

Table 3. Keywords of the requirements.

Keyword	Requirement
resizer; font-smaller; font-larger; font-normal; font-readable; compact mode; text magnifier; readable font; adjust font sizing	Adjustable text size (R2)
content scaling	Object magnification (R3)
contrast; contrast = default; contrast = night; contrast = high contrast; tema; theme; terang; bright; gelap; dark; default perangkat; device default; id = "menu-darkmode"; class = "icon-svg icon-darkmode"; data-dark = "system"; data-dark = "dark"; data-dark = "light"; dark mode; highlight titles; highlight links; color adjustments; dark contrast light contrast; high contrast; monochrome; high saturation; low saturation; appearance; device theme; dark theme; light theme	Contrast settings (R4)

```
1   *** Settings ***
2   Suite Setup         Open Browser        browser=[insert_browser_name]
3   Suite Teardown      Close Browser
4   Test Teardown       Capture Page If Test Failed
5   Resource            ../../resources.robot

6   *** Variables ***
7   ${sampleweb}        https://[sample_web_address]/
8   ${path}             [insert_path_with_keyword]
9   ${timeout}          [insert_time_in_seconds]

10  *** Test Cases ***
11  Verify Accessibility Element
12  Go To      ${sampleweb}
13  SeleniumLibrary.Wait For Condition      return document.readyState == 'complete'
14        timeout=${timeout}
15  # Some steps might be required before verifying the element, depending on user
    interface of each sample
16  Wait Until Element Is Visible      xpath=${path}      timeout=${timeout}
```

Figure 2. Automation script.

The automation script was written using the Robot Framework (Robot Framework Foundation, Helsinki, Finland). Lines 1–5 define the general settings of the test, such as a set of actions before–after running a suite, after running a test and the resource/library used. Lines 6–9 define variables to store constantly changed values for the sample's URL, a path with each keyword in Table 3, and the timeout based on the responsibility of the web sample. Lines 10–16 define the test case developed based on the manual test scenario. First, it goes to a specified web address and then waits for the page to load completely. Next, it can directly verify the element, or it needs to go through a few steps before the element becomes visible. For example, the element was nested in a burger menu, so it clicked on the burger menu element before verifying the keyword element.

The results of the manual and automated tests are attached in Table 4. Code X states that the requirements were not found, NW means that the requirements were found but did not work, and O means that the requirements were found and could be used. There were differences in the manual and automated testing results in the Google, Kompas and Facebook Indonesia samples. Details of the anomaly in the results are described in Table 5.

Table 4. Manual and automated test results.

Num.	Sample	Manual	Automated
1	The Indonesian Government's Official Website Regarding COVID-19	X	X
2	Ministry of Communication and Information of the Republic of Indonesia	X	X
3	Tax Official Website of Indonesia	X	X
4	BPJS Kesehatan (Health Insurance)	X	X
5	Bank Rakyat Indonesia	X	X
6	Rumah belajar of the Ministry of Education, Culture, Research and Technology of the Republic of Indonesia	X	X
7	PT Kereta Api Indonesia (Indonesian Railways)	X	X
8	Kampus Merdeka	X	X
9	Indonesian National Police	X	X
10	Pedulilindungi	X	X
11	Indonesian Immigration Official Website	X	X
12	Population and Civil Registration Office of Salatiga City	X	X
13	Salatiga City District Court	X	X
14	Indonesian Agency for Meteorological, Climatological and Geophysics	X	X
15	Central Java Job Market	X	X
16	Alodokter	X	X
17	Pinterest	X	X
18	Tokopedia	X	X
19	Blog Ruangguru	X	X
20	Google Indonesia	R1 X R2 X R3 X R4 O	X
21	Persatuan Tunanetra Indonesia	R1 X R2 O R3 X R4 O	R1 X R2 O R3 X R4 O
22	Yayasan Mitra Netra	R1 X R2 O R3 X R4 O	R1 X R2 O R3 X R4 O
23	Youtube	R1 X R2 X R3 X R4 O	R1 X R2 X R3 X R4 O
24	CNN Indonesia	X	X
25	Kompas	R1 X R2 X R3 X R4 O	X
26	LinkedIn Indonesia	X	X
27	Kapanlagi	X	X
28	Halodoc	X	X
29	Facebook Indonesia	R1 X R2 O R3 X R4 O	X
30	Bank Central Asia	X	X

Table 5. Anomalies of differences in test results.

Sample	Requirement	Anomaly
Google	Contrast settings (R4)	Failed to detect the element of contrast settings because Google blocked the robot used in the automated test.
Kompas	Contrast settings (R4)	The keyword id = "menu-darkmode" did not lead to one specific object, so the robot failed to detect it. This problem was resolved by changing the keyword to class = "icon-svg icon-darkmode" through XPath and had to be written the same for it to be detectable.
Facebook Indonesia	Adjustable text size (R2) and Contrast settings (R4)	The element's path could not be read or encrypted. Example: class = "d2edcug0 hpfvmrgz qv66sw1b c1et5uql lr9zc1uh a8c37 × 1j fe6kdd0r mau55g9w c8b282yb keod5gw0 nxhoafnm aigsh9s9 d3f4 × 2em mdeji52x a5q79mjw g1cxx5fr lrazzd5p oo9gr5id hzawbc8m"

The results show that 80% of the samples were not accessible to visually impaired people. Samples number 1–15, government-owned websites in education, health, civil, employment, banking and law, did not implement any of the specified requirements. Samples number 16–19, 24, 26–28 and 30 were also absent of these requirements.

Based on the results, various parties pay little attention to visual impairment accessibility when developing websites. Furthermore, websites owned by organizations focusing on visually impaired people only implement two of the requirements, which are adjustable text sizes (R2) and contrast settings (R4).

The key to optimizing the proposed automation testing is building the precise path from the keyword used for locating the accessibility element. Finding the keyword that leads to a specific element can avoid the anomaly of the machine not being able to verify the accessibility element despite it being present on the web page.

4. Conclusions

The implementation of web accessibility requirements for visually impaired users is deficient. Based on the results, only 14 tests passed out of 240 tests, and only 4 failed due to technical anomalies of automated testing. Therefore, the government and various parties are encouraged to implement accessible features to help visually impaired users access information and services optimally. In the future, it is recommended to research other requirements and other types of disabilities to support information dissemination and the awareness of poor web accessibility in Indonesia. To achieve website inclusivity, we must care about accessibility for minority groups.

Author Contributions: Conceptualization, M.C. and N.S.; methodology, M.C. and N.S.; software, M.C.; validation, M.C. and N.S.; formal analysis, M.C. and N.S.; investigation, M.C. and N.S.; resources, M.C. and N.S.; data curation, M.C.; writing—original draft preparation, M.C.; writing—review and editing, M.C.; visualization, M.C.; supervision, N.S.; project administration, M.C. and N.S.; funding acquisition, N.S. All authors have read and agreed to the published version of the manuscript.

Funding: This research received no external funding.

Institutional Review Board Statement: Not applicable.

Informed Consent Statement: Not applicable.

Data Availability Statement: Not applicable.

Acknowledgments: This paper was presented during a panel at the 2022 International Academic Symposium of Social Science (IASSC 2022).

Conflicts of Interest: The authors declare no conflict of interest.

References

1. Ndaumanu, F. Hak Penyandang Disabilitas: Antara Tanggung Jawab dan Pelaksanaan oleh Pemerintah Daerah. *J. HAM* **2020**, *11*, 131–150. [CrossRef]
2. Pugu, M.R.; Lumentut, D.T.; Pelamonia, Y.G. Perempuan Pedesaan dan Disabilitas di Provinsi Papua dan Papua Barat (Suatu Analisa Hubungan Internasional dalam Pelaksanaan Otsus Papua). *Syntax Lit. J. Ilm. Indones.* **2022**, *7*, 3464–3482. [CrossRef]
3. Siregar, N.A.; Purbantara, A. Melawan Stigma Diskriminatif: Strategi Pemberdayaan Penyandang Disabilitas di Desa Panggungharjo. *J. Pemberdaya Masy. Media Pemikir. Dakwah Pembang.* **2020**, *4*, 27–50. [CrossRef]
4. Istifarroh; Nurgoho, W.C. Perlindungan Hak Disabilitas Mendapatkan Pekerjaan di Perusahaan Swasta dan Perusahaan Milik Negara. *Mimb. Keadilan* **2019**, *12*, 21–34. [CrossRef]
5. Wulansari, Y.; Prabawati, I. Implementasi Kebijakan Perlindungan dan Pemberdayaan Penyandang Disabilitas (Studi Aksesibilitas Bangunan Umum Di Stasium Malang Kota Baru). *Publika* **2021**, *9*, 307–320. [CrossRef]
6. Marsitadewi, K.E.; Widnyani, I.A.P.S. Peran Pemerintah Kota Denpasar dalam Mengakomodasi Hak Penyandang Disabilitas dalam Memperoleh Pekerjaan. *Politicos J. Polit. Dan Pemerintah.* **2021**, *1*, 98–112. [CrossRef]
7. Septiana, F.I.; Effendi, Z.R. Mewujudkan Akses Pendidikan Tinggi Bagi Penyandang Disabilitas. *Incl. J. Spec. Educ.* **2019**, *5*, 11–18.
8. Shaleh, I. Implementasi Pemenuhan Hak Bagi Penyandang Disabilitas Ketenagakerjaan di Semarang. *Kanun J. Ilmu Huk.* **2018**, *20*, 63–84. [CrossRef]
9. Kristiyanti, E. Model Penyelenggaraan Pendidikan Inklusif Bagi Penyandang Disabilitas Intelektual: Studi Kasus di DKI Jakarta. *Indones. J. Relig. Soc.* **2019**, *1*, 67–79. [CrossRef]
10. Widinarsih, D. Penyandang Disabilitas di Indonesia: Perkembangan Istilah dan Definisi. *J. Ilmu Kesehat. Sos.* **2019**, *20*, 127–142.
11. Anisa, L.N. Pemenuhan Kebutuhan Dasar Bagi Penyandang Disabilitas Tunadaksa di Paguyuban Difabel Sehati Sukoharjo. Bachelor's Thesis, Sekolah Tinggi Pembangunan Masyarakat Desa (APMD), Yogyakarta, Indonesia, 2022.
12. Arisal, A.A.; Syukur, M. Diskriminasi Sosial Anak Penderita Kusta di Kota Makassar. *Phinisi Integr. Rev.* **2020**, *3*, 297–304. [CrossRef]
13. Erissa, D.; Widinarsih, D. Akses Penyandang Disabilitas Terhadap Pekerjaan: Kajian Literatur. *J. Pembang. Mns.* **2022**, *3*, 22. [CrossRef]
14. Fauziah, A.; Nuraini, A.; Ningsih, F.N.; Kurniawan, T.W.A. Tetap Produktif Juga Kreatif Pada Masa Pandemi COVID. Available online: http://openjournal.unpam.ac.id/index.php/PD/article/view/15828/8486 (accessed on 4 May 2022).
15. Indrawan, I.G.A.; Hidayat, R.A.L.; Juitania. Survei Kualitas Pembelajaran dan Kepuasan Belajar Mahasiswa Terhadap Pembelajaran Daring Di Masa Pandemi COVID-19. *J. Pendidik. Dan Konseling* **2022**, *4*, 355–364.
16. Salsabila, U.H.; Sari, L.I.; Lathif, K.H.; Lestari, A.P.; Ayuning, A. Peran Teknologi dalam Pembelajaran di Masa Pandemi Covid-19. *Al-Mutharahah J. Penelit. Dan Kaji. Sos. Keagamaan* **2020**, *17*, 188–198. [CrossRef]
17. Hikmatyar, M.A.; Sudiarjo, H.L. Business Model Legal Aspects of E-Government Document and Personal Data Security. *J. Tek. Inform. Dan Sist. Inf. (JATISI)* **2022**, *9*, 633–647.
18. Lumbanraja, A.D. Ugensi Transformasi Pelayanan Publik melalui E-Government Pada New Normal dan Reformasi Regulasi Birokrasi. *Adm. Law Gov. J.* **2020**, *3*, 220–231.
19. The Economist. The Inclusive Internet Index. Available online: https://theinclusiveinternet.eiu.com/explore/countries/performance (accessed on 5 July 2022).
20. Web Content Accessibility Guidelines (WCAG) 2.1. World Wide Web Consortium. 2018. Available online: https://www.w3.org/TR/WCAG21/ (accessed on 4 May 2022).
21. Arini, F.D. How Accessible the University Websites in Indonesia for People with Disabilities? *Indones. J. Disabil. Stud.* **2020**, *7*, 164–169. [CrossRef]
22. Effective Communication, American with Disabilities Act. 2010. Available online: https://www.ada.gov/effective-comm.htm (accessed on 21 April 2022).
23. Accessibility Requirements for Products and Services, European Parliament. 2019. Available online: https://eur-lex.europa.eu/EN/legal-content/summary/accessibility-of-products-and-services.html (accessed on 6 May 2022).
24. Yuherni, Y. Realitas Self Disclosure Penyandang Tuli di Media Sosial (Studi Fenomenologi Pengguna Instagram Pada Anggota Gerakan Untuk Kesejahteraan Tunarungu Indonesia (GERKATIN) Kota Palembang). Bachelor's Thesis, Faculty of Social and Politic Science, Sriwijaya University, Palembang, Indonesia, 2021.
25. Qonita, I. Kemensos Dorong Aksesibilitas Informasi Ramah Penyandang Disabilitas. Kementerian Sosial Republik Indonesia. Available online: https://kemensos.go.id/kemensos-dorong-aksesibilitas-informasi-ramah-penyandang-disabilitas (accessed on 6 May 2022).
26. Sousa, C.C.S.; Oliveira, L.M.; Rodrigues, C.L.; Bulcão-Neto, R.F.; Ferreira, D.J. Web Accessibility Testing for Deaf: Requirements and Approaches for Automation. In Proceedings of the IEEE International Conference on Systems, Man, and Cybernetics (SMC), Toronto, ON, Canada, 10–14 October 2020; pp. 2734–2739.

proceedings

MDPI

Proceeding Paper

Performance in Virtual Teams: Towards an Integrative Model †

Ali El Idrissi * and Mohamed Fourka

Innovation and Management of Industrial System Research Team, Faculty of Sciences and Technologies Tangier, Abdelmalek Essaadi University, Tetouan 416, Morocco
* Correspondence: ali.elidrissi1@etu.uae.ac.ma
† Presented at the International Academic Symposium of Social Science 2022, Kota Bharu, Malaysia, 3 July 2022.

Abstract: Virtual teams (VTs) are groups of people who work interdependently with shared purpose across space, time, and organization boundaries, using technology to communicate and collaborate. This literature review examined the status of the published research on VTs functioning to identify the main factors impacting their performance. Our main findings are the conceptualization of a multi-level model integrating factors classified into six categories: (1) individual factors; (2) group dynamics or team members' interactions; (3) context factors; (4) technology-mediated communication (TMC); (5) trust; and (6) leadership. The framework elaborated from this literature review needs to be tested in different environments.

Keywords: virtual teams; technology-mediated communication; virtual collaboration; trust; leadership; team performance; literature review

Citation: El Idrissi, A.; Fourka, M. Performance in Virtual Teams: Towards an Integrative Model. *Proceedings* **2022**, *82*, 73. https://doi.org/10.3390/proceedings2022082073

Academic Editor: Mohamad Rahimi Mohamad Rosman

Published: 21 September 2022

1. Introduction

Globalization, business competition, and rapid advancement in information and communication technology have propelled the growing prevalence of VTs in the past decade [1,2]. The use of VTs allows organizations to remotely engage specialists regardless of their physical location and enables them to respond faster to market change [1,3]. The Coronavirus (COVID-19) crisis forced many employees around the world to work from home, which has suddenly increased the number of VTs [4].

Many researchers have shown the advantages of VTs, such as bringing diverse expertise and perspectives [5], reducing costs and stress, and saving time [3]. Moreover, geographically and temporally dispersed teams enable continuous 24/7 productivity [6]. However, the lack of non-verbal cues and social interaction when communicating through technological tools hinders team consensus and makes conflict resolution more difficult [7,8]. In VTs, individuals can also experience stress and isolation due to the characteristics of virtual collaboration such as technical problems, geographic and temporal distribution, and cultural differences [2]. Leaders also face many challenges in monitoring and managing VTs, and difficulties in building trust among and between team members [6].

Alaiad and his colleagues have conducted a systematic literature review of research papers published between 2007 and 2018. They stated that there is no specific standardized model able to describe virtual collaboration. Even the few existing models of VTs are either incomplete or not tested [1].

Our research question is: what are the main factors impacting VT performance?

2. Method

This literature review is based on a search of articles across scientific databases such as Google Scholar, Scopus, Web of Science, and Science Direct, etc., using keywords such as "virtual teams", "dispersed teams", "distributed teams", "remote teams", "virtual global teams", and "team performance". We used different combinations of these keywords. After analyzing 30 recent articles published between 2017 and 2021, which allowed us to reach saturation in terms of concept identification, we then drafted an initial model with different factors that impact team performance in a virtual environment and their relationships. In the next step, we deepened our understanding of each concept using the "snowballing" approach. We retained more than 160 papers as relevant to our research. After analyzing these papers, we completed the model. The more recent papers are prioritized for the best consistency with technological advancement and practice evolution over time. Finally, we created notifications on the scientific sites mentioned above, to receive by email new publications related to our topic.

3. Literature Review

Based on the literature review of more than 160 articles, we categorized factors impacting VT performance into five categories: (1) *individual factors* which are related to the VT members as individuals; (2) *group dynamics* that consider the team interactions; (3) *context factors* are related to the organizational, cultural, and technological context of the work; (4) *technology-mediated communication* (TMC); (5) *trust*; and (6) *leadership*.

3.1. Individual Factors

3.1.1. Team Members' Competencies

Belova and Mezhevov [9] has identified the core competencies of VT members: (a) *cognitive competence*, which is the knowledge and skills needed in the professional field; (b) *functional competence*, which focuses mainly on knowledge of digital technologies for professional use [10]; (c) *social competence*, which refers to the ability to create and maintain effective social interactions, close relationships, and respond in adaptive ways within a team [11], and (d) *meta-competence*, which is the higher-order skills and abilities upon which competencies are based. It includes initiative, responsibility, a positive perception of criticism, a high level of self-control, and conscientiousness. Conscientiousness is one of the most consistent predictors of individual performance [9,12].

3.1.2. Motivation

Vida Davidavičienė et al. [13] argued that motivation has a significant impact on VTs' knowledge sharing, which helps to achieve high VT performance. This has implicitly been confirmed by [14] who considered that enjoyment (i.e., intrinsic motivation) is one of the main predictors of knowledge sharing. Zhang et al. [15] have qualified two types of motivation: intrinsic motivation pertains to activities done for their inherent interest and enjoyment; in contrast, extrinsic motivation is related to external reasons including economic and social rewards, or punishment [13,16]. Davidavičienė and colleagues [13] concluded that organizations must implement strategies that promote motivation with rewards, collaboration, and trust between team members.

3.2. Group Dynamics

3.2.1. Shared Mental Models and Norms

Shared mental models are a common understanding or a form of team knowledge that members have regarding their tasks and how they need to interact in order to accomplish such tasks [17]. DeChurch and Mesmer-Magnus [18] have demonstrated the positive effect of shared mental models on team performance. In a high virtual context, shared mental models play an important role in enabling team members to work together effectively in the absence of direction from a singular leader [17]. On the other hand, shared norms are legitimate, socially shared standards that orient individual behavior. Shared norms enable self-managing VTs to more quickly develop agreements concerning the process of the team's work [17]. Moreover, setting clear team norms and expectations, and creating opportunities for members to share experiences through repeated interactions, help build trust among members [19]. Finally, when interpersonal norms are established, cultural differences are neglected and, in return, conflicts are reduced [20].

3.2.2. Team Awareness

Awareness in the VTs' context is an understanding of the activities of others, which provides a context for one's own activity [21]. Three types of team awareness are suggested in the literature: presence awareness, task knowledge awareness, and social awareness [22–24]. *Presence awareness* is a subjective feeling that individuals are physically available and accessible to each other as if they were co-located [23]. In VTs, ICT reinforces the feeling of being "in touch" or being connected to the team. This enables team members and leaders to monitor the work and render assistance to others who need it [5,23,25]. *Task knowledge awareness* facilitates coordination between team members through the knowledge gained about each other regarding who is doing what [22,23,25]. It also enhances team performance. *Social awareness* concerns knowledge about the team members' personal information, their social situation, and interactions [5,26]. Lim [5] listed many benefits of disclosure awareness (named here as "social awareness") in VTs, including: enhancing familiarity, liking, and interpersonal relationships, which are important for trust development and knowledge sharing.

3.2.3. Process Losses

The inability of VT members to observe each other's actual efforts tends to lead to a greater reliance on perceptions and assumptions that could be both biased and erroneously negative [3]. That can lead to significant motivation losses (such as social loafing) and coordination losses [12,25]. TMC reinforces the effect of team awareness on team performance in two ways: (1) being aware of their colleagues' effort, team members increase their own effort towards their common goals [12,25]; (2) using technology, team members can be able to better coordinate their activities in a virtual environment [25].

3.2.4. Team Experience

- Team resilience can be defined as the collective capacity to deal with adverse events and rebound as strengthened and more resourceful [27,28]. Open communication and the quality of relationships are important factors for team-resilience development [2,29,30]. In return, it contributes to reducing the level of relational conflict [2,31,32].
- Team familiarity can lower the barriers and communication concerns created by geographic, nationality, structural, and demographic differences [33]. Moreover, professional familiarity, rather than a personal one, is salient in shaping VT's information elaboration (i.e., exchanging, discussing, and integrating information), which has a positive effect on performance [34,35].

3.2.5. Knowledge Sharing

Knowledge sharing is impacted by several factors categorized into three areas: (1) *individual dimension*, including motivation, interpersonal relationships, and trust among team members; (2) *organizational dimension*, related to the organizational structure, leadership, and reward structure; (3) *technological dimension*, including tools, infrastructure quality (e.g., internet availability, hardware capacity), and system availability [13,36]. ICT plays a critical role in reducing information inconsistencies and misunderstandings by sharing needed information with team members. Consequently, this may improve team performance [1].

3.2.6. Conflict

Different perceptions, values, norms, and communication styles in VTs may cause conflict to occur [13]. Moreover, conflicts in VTs are more difficult to manage than in co-located teams [37]. Task conflict and process conflict are not necessarily "bad" and may encourage effective team functioning through an open discussion of different alternatives [38,39]. However, if not appropriately managed, it can degenerate into relationship conflict, which is harmful to team performance [37,40].

3.3. Context Factors

3.3.1. Team Virtuality and Configuration

Teams within organizations represent various levels of virtual communication and collaboration, as opposed to daily basis face-to-face communication and collaboration. Virtuality can be approached either from a team design perspective (i.e., geographic or temporal distance, and configurational dispersion) or a technology-use perspective (i.e., the extent to which technological tools are used among team members, media richness, and media synchronicity) [41,42]. For example, high site dispersion is associated with faultlines formation and conflict, low level of cohesion, and low presence and task knowledge awareness [3,43–46].

3.3.2. Task Complexity and Interdependence

High task complexity in a high level of virtuality can lead to misunderstandings, mistakes, and coordination losses. Shared mental models and norms are a key element to enhance performance in such a situation [47,48]. Meanwhile, more interdependent work requires frequent and complex communication among the group members, with short feedback loops and multiple streams of information [49].

3.3.3. Team Diversity

VTs are characterized by significantly different backgrounds and experiences, such as cultural differences, professional and organizational background, and demographic characteristics [50]. This can promote creativity and innovation. However, diverse teams are more likely to diverge in their preferences and interests [50,51]. In sum, diversity may negatively influence team performance when subgroup formation is rated as high, but not if subgroups are perceived to be absent or low [52].

3.4. Technology-Mediated Communication

Team performance is positively impacted by knowledge exchange in climates that enable an oscillation between cooperative and assertive communication [53]. To optimize the performance of the virtual team, the leader should find a balance in the frequency of communication with each member of the team and between the available communication tools [54]. Communication mediated by technological tools, associated with geographically distributed teams, often lacks support for nonverbal cues, body language, inflection, gestures, and social interaction. This may make interactions and coordination more difficult and could increase misunderstandings and conflict among team members, which are detrimental to team performance [1,3,5,55]. Communication technology helps VT members to feel "in touch" or connected with their teammates [25]. Rivera [56] has demonstrated that team psychological safety and perceived peer support increase when richer media technology is used. In addition, team visualization tools can process complex and extensive information regarding team members' contributions to common goals. Such collaborative platforms are accessible to all team members in visual format without specific wording that can carry any negative emotions [12]. This visibility of effort may reduce social loafing and the withholding of task-oriented effort. However, a high level of effort does not necessarily reflect a high level of performance. Therefore, leadership is important to monitor team members' contributions [12].

Another advantage that face-to-face communication does not support is traceability and duplication. ICT enables VT members to access, store, retrieve, and exchange information across time and space boundaries [57,58]. Furthermore, communication through technological tools can be registered, duplicated, and repeated at any time by any team member around the world [51,59,60]. Finally, the ability to electronically capture, store, and retrieve the team's process and outcomes contributes to the organizational knowledge and learning development, and ultimately may increase VT effectiveness [58,61].

3.5. Trust

One of the most important challenges faced by VTs is building trust among and between members due to the absence of personal interaction and emotional cues in communicating through ICT. Trust is a crucial element for all aspects of collaboration and affects team performance [3,62–64]. Essentially, trust within any organization and in everyday life is a mix of feeling and rational thinking [65]. McALLISTER [66] distinguished between two main forms of trust: cognitive and affective trust. While cognitive trust is beneficial for VT performance, affective trust may increase social loafing due to the reliance on one's colleagues to provide support when needed, or even the comfort in asking them to accomplish their own work [67]. Team monitoring can reduce the relationship between affective trust and social loafing.

Considering the time windows of virtual communication, exchange between team members is often explicit, formal, and work issues-oriented. This lack of spontaneous and informal communication induces low team cohesion and mistrust [3]. However, [68] found that trust can be built by alternating the use of rich ICT media (e.g., discussion forums and instant messaging) with lean ICT media (e.g., document sharing and presentation display).

3.6. Leadership

Leadership in a virtual environment is a process whereby a person influences others through TMC to achieve a common goal towards team performance [17,51]. Leaders may play two main roles: first, task-oriented (or directive) leadership, in which the leader's behavior is directed at managing and monitoring the task performance; second, a relationship-oriented (or supportive) leader's behavior that aims to improve team cohesion and facilitate good relationships between team members. The social role is critical in virtual settings to overcome the negative effects of the lack of physical proximity and direct supervision [5,17,58,69]. In their study, Carte and colleagues [70] found that high-performing VTs are very task-focused, but the monitoring behavior is shared among team members. However, the increased difficulties of geographical dispersion and cultural diversity hinder team members' abilities to coordinate activities, influence, and motivate others. Self-management may increase the team's flexibility in setting goals and keeping track of teamwork to achieve a higher level of performance. These teams are characterized by distributed power, authority and ownership [17,70]. Shared leadership can also be considered as a result of successful self-management in VTs [17]. It is generally more convenient for task-related leadership, while the responsibility of formal leaders may consist of building trust among VTs [19].

Leaders need to understand how to take advantage of the use of technology by selecting the appropriate tools and features adapted to the organizational context and objectives, and continually driving members toward a shared vision. In addition, digital tools, such as social media platforms, help to build relationships and communication channels with external actors (e.g., partners, customers, etc.). Digital tools allow access to a large number of individuals and the possibility to interact with them through immediate communication [51].

Finally, leaders play a critical role in managing conflict before it negatively impacts team performance [39]. Leaders should help team members adopt the appropriate technology and adapt it to reduce conflict and increase their productivity [38].

4. Research Model

To answer our research question, we retained, through this literature review, the most important factors impacting VT performance and then conceptualized a multi-level model. Our conceptual model in Figure 1 represents the twelve (12) factors and nineteen (19) propositions that describe the relationships between factors influencing team performance in virtual settings.

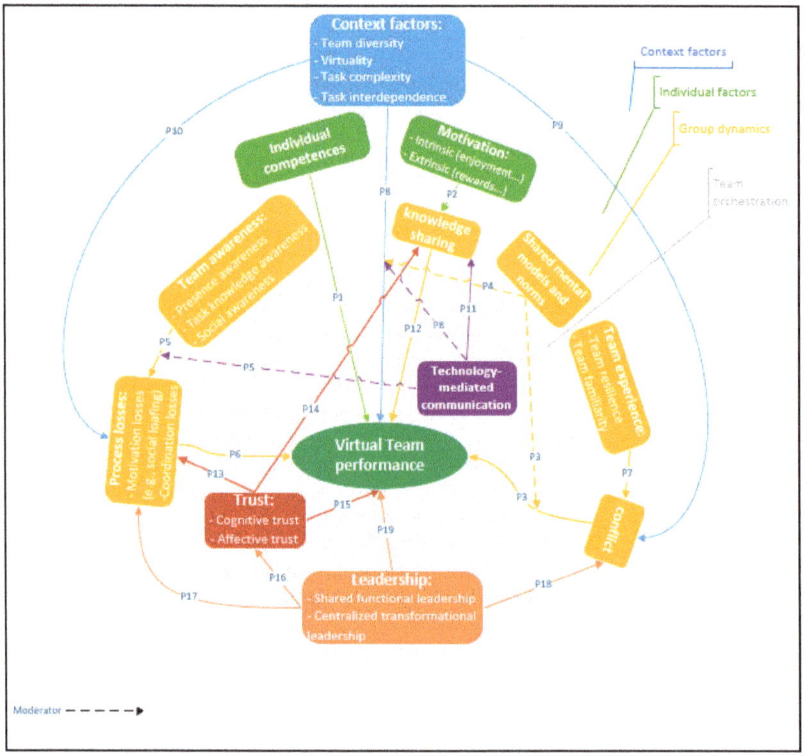

Figure 1. Multi-level model of team performance.

P1. Team members' competencies positively impact performance in VTs.

P2. Motivation positively impacts knowledge sharing in a virtual environment.

P3. Shared mental models and norms reduce the negative impact of conflict on VT performance.

P4. Shared mental models and norms reduce (moderate) the negative impact of context factors (cultural diversity, virtuality, and task complexity and interdependence) on team performance.

P5. TMC reduces (moderates) the negative relationship between team awareness and process losses in VTs.

P6. Process losses negatively impact team performance in virtual environments.

P7. Team experience (team resilience and familiarity) reduces the level of conflict between VT members.

P8. TMC reduces (moderates) the negative relationship between context factors and team performance.

P9. Context factors increase team conflict.

P10. Context factors increase process losses.

P11. TMC positively impacts knowledge sharing by storing, duplicating, and retrieving information.

P12. Knowledge sharing positively impacts team performance in VTs.

P13. Affective trust and process losses are positively related in VTs.

P14. Trust enhances knowledge sharing in VTs.

P15. Trust positively impacts team performance in VTs.

P16. Leadership increases trust level through leader–member exchange in a virtual setting.

P17. Leadership reduces process losses through objectives setting, planning, coordination, and monitoring in VTs.

P18. Leadership reduces team conflict through training, communication media selection, monitoring, establishing a safe communication climate, and conflict resolution in VTs.

P19. Leadership (shared functional leadership and centralized transformational leadership) positively impacts team performance by enabling continuous improvement of shared mental models and norms among VT members.

5. Conclusions

This literature review provides an insight into the factors influencing team performance in virtual settings. The main finding is that factors influencing team performance can be conceptualized as a multi-level model in which each level either adds complexity or helps to overcome challenges, or a mix of both. Context factors as inherent characteristics of VTs negatively impact team performance by hindering communication and knowledge sharing and increasing conflicts between VT members. As VT members start to work together, many challenges could emerge, such as conflict and process losses. These challenges can be mitigated by other factors such as team resilience, familiarity, knowledge sharing, referring to the same mental models and norms, and team awareness. Finally, TMC, trust, and leadership are the three important factors that orchestrate and regulate team behavior towards the best performance.

This study presents opportunities to VT managers, human resources managers, and top management in terms of team configuration, communication media selection, and important behaviors to reinforce in each stage of the project. However, this model needs to be tested in different environments, including cross-cultural contexts. Our future research work will focus on this proposed model test, and then we will strive to shed more light on the TMC and its role in improving VT performance.

Author Contributions: Conceptualization is delivered by A.E.I. and discussed with M.F.; writing and original draft preparation are made by A.E.I.; reviewing and editing are given by M.F., and methodology is constructed by A.E.I. and M.F. All authors have read and agreed to the published version of the manuscript.

Funding: This research received no external funding.

Institutional Review Board Statement: Not applicable.

Informed Consent Statement: Not applicable.

Data Availability Statement: Not applicable.

Conflicts of Interest: The authors declare no conflict of interest.

References

1. Alaiad, A.; Alnsour, Y.; Alsharo, M. Virtual Teams: Thematic Taxonomy, Constructs Model, and Future Research Directions. *IEEE Trans. Dependable Secur. Comput.* **2019**, *62*, 211–238. [CrossRef]
2. Peñarroja, V.; González-Anta, B.; Orengo, V.; Zornoza, A.; Gamero, N. Reducing Relationship Conflict in Virtual Teams With Diversity Faultlines: The Effect of an Online Affect Management Intervention on the Rate of Growth of Team Resilience. *Soc. Sci. Comput. Rev.* **2020**, *40*, 388–404. [CrossRef]
3. Morrison-Smith, S.; Ruiz, J. Challenges and barriers in virtual teams: A literature review. *SN Appl. Sci.* **2020**, *2*, 1096. [CrossRef]
4. Mysirlaki, S.; Paraskeva, F. Emotional intelligence and transformational leadership in virtual teams: Lessons from MMOGs, Leadersh. *Organ. Dev. J.* **2020**, *41*, 551–566. [CrossRef]
5. Lim, J.Y.-K. IT-enabled awareness and self-directed leadership behaviors in virtual teams. *Inf. Organ.* **2018**, *28*, 71–88. [CrossRef]
6. Dulebohn, J.H.; Hoch, J.E. Virtual teams in organizations. *Hum. Resour. Manag. Rev.* **2017**, *27*, 569–574. [CrossRef]
7. Glikson, E.; Erez, M. The emergence of a communication climate in global virtual teams. *J. World Bus.* **2020**, *55*, 101001. [CrossRef]
8. Peng, C.-H.; Lurie, N.H.; Slaughter, S.A. Using Technology to Persuade: Visual Representation Technologies and Consensus Seeking in Virtual Teams. *Inf. Syst. Res.* **2019**, *30*, 948–962. [CrossRef]

9. Belova, O.L.; Mezhevov, A.D. *Virtual Teams in Russian Organizations*; Springer International Publishing: Cham, Switzerland, 2020; pp. 1553–1562.
10. Enrique, G.G.; Joel, M.G. Best practices and opportunity areas for the intelligent management of virtual teams. *Manag. Sci. Lett.* **2020**, *10*, 3507–3514. [CrossRef]
11. Orpinas, P. Social Competence. In *The Corsini Encyclopedia of Psychology*; Weiner et, I.B., Craighead, W.E., John, É., Eds.; Wiley & Sons, Inc.: Hoboken, NJ, USA, 2010.
12. Glikson, E.; Woolley, A.W.; Gupta, P.; Kim, Y.J. Visualized Automatic Feedback in Virtual Teams. *Front. Psychol.* **2019**, *10*, 814. [CrossRef] [PubMed]
13. Davidavičienė, V.; Al Majzoub, K.; Meidute-Kavaliauskiene, I. Factors Affecting Knowledge Sharing in Virtual Teams. *Sustainability* **2020**, *12*, 6917. [CrossRef]
14. da Silva, F.P.; Mosquera, P.; Soares, M.E. Factors influencing knowledge sharing among IT geographically dispersed teams. *Technol. Forecast. Soc. Chang.* **2022**, *174*, 121299. [CrossRef]
15. Zhang, X.; de Pablos, P.O.; Xu, Q. Culture effects on the knowledge sharing in multi-national virtual classes: A mixed method. *Comput. Hum. Behav.* **2014**, *31*, 491–498. [CrossRef]
16. Ryan, R.M.; Deci, E.L. Intrinsic and extrinsic motivation from a self-determination theory perspective: Definitions, theory, practices, and future directions. *Contemp. Educ. Psychol.* **2020**, *61*, 101860. [CrossRef]
17. Eseryel, U.Y.; Crowston, K.; Heckman, R. Functional and Visionary Leadership in Self-Managing Virtual Teams. *Group Organ. Manag.* **2020**, *46*, 424–460. [CrossRef]
18. DeChurch, L.A.; Mesmer-Magnus, J.R. Measuring shared team mental models: A meta-analysis. *Group Dyn. Theory Res. Pract.* **2010**, *14*, 1–14. [CrossRef]
19. Liao, C. Leadership in virtual teams: A multilevel perspective. *Hum. Resour. Manag. Rev.* **2017**, *27*, 648–659. [CrossRef]
20. Adamovic, M. An employee-focused human resource management perspective for the management of global virtual teams. *Int. J. Hum. Resour. Manag.* **2018**, *29*, 2159–2187. [CrossRef]
21. Dourish, P.; Bellotti, V. Awareness and coordination in shared workspaces. In Proceedings of the 1992 ACM conference on Computer-supported cooperative work-CSCW '92, Toronto, ON, Canada, 1–4 November 1992; pp. 107–114. [CrossRef]
22. Espinosa, J.A.; Slaughter, S.A.; Kraut, R.E.; Herbsleb, J.D. Team Knowledge and Coordination in Geographically Distributed Software Development. *J. Manag. Inf. Syst.* **2007**, *24*, 135–169. [CrossRef]
23. Malhotra, A.; Majchrzak, A. Enhancing performance of geographically distributed teams through targeted use of information and communication technologies. *Hum. Relat.* **2014**, *67*, 389–411. [CrossRef]
24. Hunsaker, P.L.; Hunsaker, J.S. Virtual teams: A leader's guide. *Team Perform. Manag. Int. J.* **2008**, *14*, 86–101. [CrossRef]
25. Haines, R.; Vehring, N.; Kramer, M. Social Motivation Consequences of Activity Awareness Practices in Virtual Teams: A Case Study and Experimental Confirmation. In *Collaboration in the Digital Age*; Riemer, K., Schellhammer, S., Meinert, M., Eds.; Springer International Publishing: Cham, Switzerland, 2019; pp. 89–119. [CrossRef]
26. Daassi, M.; Jawadi, N.; Favier, M.; Kalika, M. Building Collective Awareness in Virtual Teams: The Effect of Leadership Behavioral Style. In *Leadership in the Digital Enterprise: Issues and Challenges*; IGI Global: Hershey, PA, USA, 2010; p. 23.
27. Walsh, F. *Strengthening Family Resilience*; Guilford Press: New York, NY, USA, 1998.
28. Gucciardi, D.F.; Crane, M.; Ntoumanis, N.; Parker, S.K.; Thogersen-Ntoumani, C.; Ducker, K.J.; Peeling, P.; Chapman, M.T.; Quested, E.; Temby, P. The emergence of team resilience: A multilevel conceptual model of facilitating factors. *J. Occup. Organ. Psychol.* **2018**, *91*, 729–768. [CrossRef]
29. Meneghel, I.; Salanova, M.; Martínez, I.M. Feeling Good Makes Us Stronger: How Team Resilience Mediates the Effect of Positive Emotions on Team Performance. *J. Happiness Stud.* **2014**, *17*, 239–255. [CrossRef]
30. Stephens, J.P.; Heaphy, E.D.; Carmeli, A.; Spreitzer, G.M.; Dutton, J.E. Relationship Quality and Virtuousness. *J. Appl. Behav. Sci.* **2013**, *49*, 13–41. [CrossRef]
31. Bowers, C.; Kreutzer, C.; Cannon-Bowers, J.; Lamb, J. Team Resilience as a Second-Order Emergent State: A Theoretical Model and Research Directions. *Front. Psychol.* **2017**, *8*, 1360. [CrossRef] [PubMed]
32. West, B.J.; Patera, J.L.; Carsten, M.K. Team level positivity: Investigating positive psychological capacities and team level outcomes. *J. Organ. Behav.* **2009**, *30*, 249–267. [CrossRef]
33. Haas, M.R.; Cummings, J.N. Barriers to knowledge seeking within MNC teams: Which differences matter most? *J. Int. Bus. Stud.* **2014**, *46*, 36–62. [CrossRef]
34. Huckman, R.S.; Staats, B.R.; Upton, D.M. Team Familiarity, Role Experience, and Performance: Evidence from Indian Software Services. *Manag. Sci.* **2009**, *55*, 85–100. [CrossRef]
35. Maynard, M.T.; Mathieu, J.E.; Gilson, L.L.; Sanchez, D.R.; Dean, M.D. Do I Really Know You and Does It Matter? Unpacking the Relationship Between Familiarity and Information Elaboration in Global Virtual Teams. *Group Organ. Manag.* **2018**, *44*, 3–37. [CrossRef]
36. Kaplan, A.M.; Haenlein, M. Users of the world, unite! The challenges and opportunities of Social Media. *Bus. Horizons* **2010**, *53*, 59–68. [CrossRef]
37. Furumo, K. The Impact of Conflict and Conflict Management Style on Deadbeats and Deserters in Virtual Teams. In Proceedings of the 41st Annual Hawaii International Conference on System Sciences (HICSS 2008), Waikoloa, HI, USA, 7–10 June 2008; p. 445. [CrossRef]

38. Chiravuri, A.; Nazareth, D.; Ramamurthy, K. Cognitive Conflict and Consensus Generation in Virtual Teams During Knowledge Capture: Comparative Effectiveness of Techniques. *J. Manag. Inf. Syst.* **2011**, *28*, 311–350. [CrossRef]
39. Wakefield, R.L.; Leidner, D.E.; Garrison, G. Research Note—A Model of Conflict, Leadership, and Performance in Virtual Teams. *Inf. Syst. Res.* **2008**, *19*, 434–455. [CrossRef]
40. Hinds, P.J.; Mortensen, M. Understanding Conflict in Geographically Distributed Teams: The Moderating Effects of Shared Identity, Shared Context, and Spontaneous Communication. *Organ. Sci.* **2005**, *16*, 290–307. [CrossRef]
41. Kirkman, B.L.; Mathieu, J.E. The Dimensions and Antecedents of Team Virtuality. *J. Manag.* **2005**, *31*, 700–718. [CrossRef]
42. Kramer, W.S.; Shuf, M.L. The world is not flat: Examining the interactive multidimensionality of culture and virtuality in teams. *Hum. Resour. Manag. Rev.* **2017**, *27*, 604–620. [CrossRef]
43. Alves, M.P.; Dimas, I.D.; Lourenço, P.R.; Rebelo, T.; Peñarroja, V.; Gamero, N. Can virtuality be protective of team trust? Conflict and effectiveness in hybrid teams. *Behav. Inf. Technol.* **2022**, 1–18. [CrossRef]
44. Chudoba, K.M.; Wynn, E.; Lu, M.; Watson-Manheim, M.B. How virtual are we? Measuring virtuality and understanding its impact in a global organization. *Inf. Syst. J.* **2005**, *15*, 279–306. [CrossRef]
45. Lü, M.; Watson-Manheim, M.B.; Chudoba, K.M.; Wynn, E. Virtuality and Team Performance: Understanding the Impact of Variety of Practices. *J. Glob. Inf. Technol. Manag.* **2006**, *9*, 4–23. [CrossRef]
46. O'Leary, M.B.; Cummings, J.N. The Spatial, Temporal, and Configurational Characteristics of Geographic Dispersion in Teams. *MIS Q.* **2007**, *31*, 433–452. [CrossRef]
47. Espevik, R.; Johnsen, B.H.; Eid, J.; Thayer, J.F. Shared Mental Models and Operational Effectiveness: Effects on Performance and Team Processes in Submarine Attack Teams. *Mil. Psychol.* **2006**, *18*, S23–S36. [CrossRef]
48. Marlow, S.; Lacerenza, C.N.; Salas, E. Communication in virtual teams: A conceptual framework and research agenda. *Hum. Resour. Manag. Rev.* **2017**, *27*, 575–589. [CrossRef]
49. Olson, G.M.; Olson, J.S. Distance Matters. *Hum. Comput. Interact.* **2000**, *15*, 139–178. [CrossRef]
50. Mello, A.S.; Ruckes, M.E. Team Composition*. *J. Bus.* **2006**, *79*, 1019–1039. [CrossRef]
51. Cortellazzo, L.; Bruni, E.; Zampieri, R. The Role of Leadership in a Digitalized World: A Review. *Front. Psychol.* **2019**, *10*, 1938. [CrossRef] [PubMed]
52. Op 't Roodt, H.; Krug, H.; Otto, K. Subgroup Formation in Diverse Virtual Teams: The Moderating Role of Identity Leadership. *Front. Psychol.* **2021**, *12*, 722650. [CrossRef] [PubMed]
53. Hajro, A.; Gibson, C.B.; Pudelko, M. Knowledge Exchange Processes in Multicultural Teams: Linking Organizational Diversity Climates to Teams' Effectiveness. *Acad. Manag. J.* **2017**, *60*, 345–372. [CrossRef]
54. Beirouty, Z.A.; Demirel, A.G. Enrichment of Virtual Teams Management through Communication. *Asian J. Soc. Sci. Manag. Technol.* **2022**, *4*, 15.
55. Lippert, H.; Dulewicz, V. A profile of high-performing global virtual teams. *Team Perform. Manag. Int. J.* **2018**, *24*, 169–185. [CrossRef]
56. Rivera, M. Team Virtuality and Psychological Safety: An Experiment. Master's Thesis, University of Central Florida, Orlando, FL, USA, 2022.
57. Hacker, J.V.; Johnson, M.; Saunders, C.; Thayer, A.L. Trust in Virtual Teams: A Multidisciplinary Review and Integration. *Australas. J. Inf. Syst.* **2019**, *23*. [CrossRef]
58. Opdenakker, R.; Cuypers, C. Introduction and Field Problem Concerning Virtual Project Teams. In *Effective Virtual Project Teams*; Springer International Publishing: Cham, Switzerland, 2019; pp. 1–15. [CrossRef]
59. Figl, K.; Saunders, C. Team Climate and Media Choice in Virtual Teams. *AIS Trans. Human-Computer Interact.* **2011**, *3*, 189–213. [CrossRef]
60. Stephens, K.K.; Rains, S.A. Information and Communication Technology Sequences and Message Repetition in Interpersonal Interaction. *Commun. Res.* **2011**, *38*, 101–122. [CrossRef]
61. Furst-Holloway, S.; Blackburn, R.; Rosen, B. Virtual team effectiveness: A proposed research agenda. *Inf. Syst. J.* **1999**, *9*, 249–269. [CrossRef]
62. De Jong, B.A.; Dirks, K.T.; Gillespie, N. Trust and team performance: A meta-analysis of main effects, moderators, and covariates. *J. Appl. Psychol.* **2016**, *101*, 1134–1150. [CrossRef] [PubMed]
63. Hung, Y.-T.; Dennis, A.; Robert, L. Robert, Trust in virtual teams: Towards an integrative model of trust formation. In Proceedings of the 37th Annual Hawaii International Conference on System Sciences, Big Island, HI, USA, 5–8 January 2004; p. 11. [CrossRef]
64. Kanawattanachai, P.; Yoo, Y. The Impact of Knowledge Coordination on Virtual Team Performance over Time. *MIS Q.* **2007**, *31*, 783–808. [CrossRef]
65. Lewis, J.D.; Weigert, A. Trust as a Social Reality. *Soc. Forces* **1985**, *63*, 967. [CrossRef]
66. McALLISTER, D.J. Affect- And Cognition-Based Trust As Foundations For Interpersonal Cooperation In Organizations. *Acad. Manage. J.* **1995**, *38*, 36. [CrossRef]
67. Robert, L.P. Behavior-Output Control Theory, Trust and Social Loafing in Virtual Teams. *Multimodal Technol. Interact.* **2020**, *4*, 39. [CrossRef]
68. Wang, X.; Wei, X.; Van Wart, M.; McCarthy, A.; Liu, C.; Kim, S.; Ready, D.H. The role of E-leadership in ICT utilization: A project management perspective. *Inf. Technol. Manag.* **2022**, 1–15. [CrossRef]

69. Kohntopp, T.; McCann, J. Leadership in Virtual Organizations: Influence on Workplace Engagement. In *The Palgrave Handbook of Workplace Well-Being, S. Dhiman*; Springer International Publishing: Cham, Switzerland, 2020; pp. 1–26. [CrossRef]
70. Carte, T.A.; Chidambaram, L.; Becker, A. Emergent Leadership in Self-Managed Virtual Teams. *Group Decis. Negot.* **2006**, *15*, 323–343. [CrossRef]

proceedings

MDPI

Proceeding Paper

Digitalization of Accounting Profession: An Opportunity or a Risk for Future Accountants? †

Yunita Awang [1,*], Shazalina Mohamed Shuhidan [2], Azuraidah Taib [1], Norfadzilah Rashid [3] and Mohd Sidki Hasan [4]

1 Faculty of Accountancy, Universiti Teknologi MARA, Dungun Campus, Dungun 23000, Terengganu, Malaysia
2 Faculty of Accountancy, Universiti Teknologi MARA, Puncak Alam Campus, Puncak Ala 42300, Selangor, Malaysia
3 Faculty of Business & Management, Universiti Sultan Zainal Abidin, Gong Badak Campus, Kampung Gong Badak 21300, Terengganu, Malaysia
4 Special Project Q3 Management Solution Sdn Bhd, Sungai Buloh 40160, Selangor, Malaysia
* Correspondence: yunita@uitm.edu.my
† Presented at the International Academic Symposium of Social Science 2022, Kota Bharu, Malaysia, 3 July 2022.

Abstract: IR4.0 has transformed the global industrial landscape including the accounting profession. Thus, accountants have significant opportunities to enhance their digitalized skills and knowledge but are at risk if they do not comprehend how the workplace is transforming. Different genders may have divergent perspectives concerning opportunities and risks posed by the digitalization process. This study examined both the perception of opportunity or risk about the digitalization of the accounting profession and the effect of gender on such perception. Online questionnaires were distributed to 546 accounting interns from the Malaysian top six accounting and finance public universities. Findings revealed that the digitalization of the accounting profession presents future accountants with both significant opportunities and risks, but no significant difference in gender regarding such perceptions.

Keywords: opportunity; risk; digitalization; accounting profession; gender

Citation: Awang, Y.; Shuhidan, S.M.; Taib, A.; Rashid, N.; Hasan, M.S. Digitalization of Accounting Profession: An Opportunity or a Risk for Future Accountants? *Proceedings* **2022**, *82*, 93. https://doi.org/10.3390/proceedings2022082093

Academic Editor: Mohamad Rahimi Mohamad Rosman

Published: 1 October 2022

1. Introduction

In previous decades, technological advancements occurred at a rapid rate. Given the exponential rate of technological advancement, it is not surprising that many aspects of human life are influenced by technology. Specifically, IR4.0 has transformed the global industrial landscape to rely heavily on digital software and the automation of robotic functions to replace human tasks [1]. The accounting profession is not an exception in this scenario. There are five stages of accounting evolution based on the technical and technological components: traditional manual, mechanized, automated, robotic, and artificial intelligence-assisted accounting [2]. Professionals appear to be receptive to the idea of automating relatively menial and repetitive tasks because the IT infrastructure enables them to assume more important responsibilities [3].

Due to digitalization, aspiring and experienced accountants must equip themselves to meet the changing demands of the accounting profession. There is a significant opportunity for accountants to increase their level of expertise, particularly in terms of digitalization skills and knowledge, which may give the impression that future accounting tasks will be more difficult [4]. Accountants, on the other hand, put themselves at risk as financial professionals at the core of organizations if they do not comprehend how the workplace is changing due to technology and digital transformation [5]. Given that technological innovation is driving the digitalization of the accounting profession, the gender of future accountants may influence their perception of the opportunity or risk posed by the digitalization process. Consequently, the objectives of this study are twofold: to evaluate the perception of opportunity or risk associated with the digitalization of the accounting profession and to investigate the effect of gender on this perception among future accountants. Accounting interns are considered future accountants because they are the accounting students who will soon enter the field [6]. Internships provide students with educational experiences related to specific jobs, positions, occupations, and professions [7]. Subsequently, they can increase their abilities through real-world experience and the knowledge gained from their employers.

This research is guided by the Technology Acceptance Model (TAM), a model developed by [8] based on a psychological theory known as the Theory of Reasoned Action (TRA). In explaining people's actions, the TRA recognized a causal relationship between various factors including beliefs, attitudes, intentions, and behaviors [9]. TAM is the most prevalent model for describing technology adoption at all organizational and individual levels [10]. According to TAM, the more willing users are to use a new system, the more time they will invest in learning and adopting it over the old system [11]. Perceived ease of use and perceived utility are the two proposed primary factors of IT adoption in this hypothesis. In other words, according to TAM, a person's intention to utilize technology is explained and predicted by his perception of IT's utility and simplicity.

The following section examines prior research on the digitalization of the accounting profession and the associated opportunities and risks. The sections that follow elaborate on the study's methodology, analysis, and discussion of findings, as well as its limitations and conclusion.

2. Literature Review

Digitalization is the use of digital technologies to modify a business model and generate new revenue and value-generating opportunities; it is the transition to a digital business [12]. It implies that more responsibilities will be delegated to Internet-connected software applications. With the push toward digitalization of the accounting profession, the accounting industry is anticipated to undergo parallel transformations [13]. To meet the requirements of a variety of business models, several digital technology advancements are available, such as its adoption in accounting, FinTech industries, Big data and data analytics (BDA), artificial intelligence (AI), and cloud adoption.

2.1. Digitalization of the Accounting Profession

The accounting industry is expanding and improving. Due to technological advancements and shifting consumer expectations, the scope of accountants' work is expanding. Thus, the scope of an accountant's job should expand and become more adaptable to technological advancements and digitalization, as the accounting profession now employs several automated systems that did not exist ten years ago [14]. With the implementation of digitalization in the accounting industry, accountants' work has radically shifted to become increasingly dependent on the development of modern equipment and technology.

As the effect of digitalization, people's thinking and accounting practices have also changed [15]. Together with the adoption of technology in the accounting industry, it is anticipated that the number of jobs offered for accountants with programming and analysis skills will decline. Therefore, it is necessary and desirable for businesses to provide current employees with sufficient retraining [10]. Appropriate training not only boosts employees' confidence but also improves the comprehension of their responsibilities and the information and skills required to perform their duties.

2.2. Opportunities and Risks in Digitalization

Due to technological advancements, globalization, and increased competition, careers are constantly changing. According to [16], 702 job titles are at risk of being automated, with accounting having the highest probability of being automated and digitized in the near future. In the coming years, many accounting transactions will be handled by artificial intelligence and automation systems [14]. Digitalization could be viewed as an opportunity or a threat within the accounting profession. When accountants acquire new skills, particularly in engineering, that will aid the emergence of new types of accounting professionals, there is an opportunity [17]. If accountants do not comprehend how technology and digital transformation are transforming the workplace, they place themselves at risk as the financial experts at the core of organizations [5]. Table 1 summarizes the most prevalent digitalization opportunities and risks.

Table 1. Opportunities and risks in digitalization.

Opportunities	Risk
Creation of new jobs	Destruction of jobs
Working time reduction and increased work autonomy	Working-time extension—an increase of 'anytime, anywhere' work
New forms of collaboration and cooperation between workers and machines	Weakening of workers' representation and bargaining coverage
Better ergonomics due to support in performing heavy and complex work	Increased competition between workers to reduce costs
Smart factories—jobs that have been offshored to low-pay countries will come back	Work intensification, dependence on 'data masters' and surveillance
More gender equality	Increased inequality between workers
Sharing economy	Erosion of tax base and social insurance financing

(Source: [18]).

According to a World Bank study, Malaysian companies are less likely to invest in upskilling and innovation than companies in countries that have successfully achieved high-income status [19]. Only 18.5% of businesses offer staff training, compared to an average of 40% in countries that Malaysia should use as a benchmark. Consequently, digital capabilities will increasingly determine which businesses create or lose value [20]. Soon, technological progress and digitization will have a significant impact on the accounting profession [17], and therefore, the digital journey is significant, as it is a crucial element of the company's success and the new accounting system. Thus, the purpose of this paper is to evaluate, from the perspective of future accountants, the perception of opportunity or risk associated with the digitalization of the accounting profession.

2.3. Gender and Technology

Gender is one of the most frequently studied personal variables in the literature on business ethics [21–26]. Both the gender socialization and structural approaches explain gender differences in values and work interests [27]. The gender socialization method emphasized how early socialization influences the ethical perspectives of women and men [28]. In contrast, the structural approach asserted that occupational duties would mitigate any gender disparities resulting from early socialization, with males and females in specific occupations making comparable work-related decisions [29,30].

In a review of research on technology usage and intention to use technology from a gender perspective, gender was identified as a significant factor in understanding human acceptance of technology [31]. In a variety of areas, including email, information seeking, online learning, communication technology, and online shopping behavior, the differences between men and women have been studied, with the majority of research focusing on men [32]. Females are more concerned than males with the use of information technology (IT), and this trait influences females' self-efficacy, resulting in a greater awareness of the effort required to use computers. Concerning the use of information technology, females are more anxious than their male counterparts, and this anxiety affects their self-efficacy, resulting in a greater estimation of the effort required to use IT [8].

Females are also more prone to computer anxiety, ineffective computer use, and negative attitudes toward computer use ([33] as cited in [31]). Moreover, reference [34] (as cited in [31]) suggests that female entrepreneurs face greater obstacles in terms of sociocultural issues, education, and technology than their male counterparts. Males have a more positive outlook on the use of technology than females, according to [35], who summarized a meta-analysis of 17 years of prior research. Teachers can persuade male students to embrace technology in the classroom by highlighting its significance to their future [36]. In the correlation matrix, gender was linked to various variables of access, use, and literacy, just as age was. In the route analysis, however, the only direct gender influence is on risky encounters: boys are more likely than girls to encounter online dangers [37].

Statistically, there is no difference between male and female students' attitudes toward technology [38]. An online survey reveals that females and males are equally optimistic about embracing technology. According to [39], the notion that women are less willing to use technology is false. Gender has no direct influence on the number of online experiences pursued [37]. Understanding the reasons for gender differences in the acceptance of new technologies will aid in the overall development of the technology ([34] as cited in [31]). Consequently, this is an opportune time to investigate the effect of gender on the perception of opportunity or risk associated with the digitalization of the accounting profession.

Thus, there are two hypotheses developed for this study, which are:

H1. *There is a significant positive relationship between gender and perception of opportunity in the digitalized accounting profession.*

H2. *There is a significant positive relationship between gender and perception of risk in the digitalized accounting profession.*

3. Research Methodology

3.1. Sample and Data

Future accountants are used as the unit of analysis based on a non-probability purposive sampling method. This study viewed accounting interns as future accountants because they are accounting students who will soon enter the profession [6].

Participating in the internship program were senior accounting students from the top six public universities for accounting and finance in Malaysia. According to information gathered from Internship Coordinators, the total number of interns was 546. Thus, the study's sample size was 226 [40]. The online questionnaires were distributed through direct communication with the Internship Coordinators at each of the six public universities. The data collection period lasted approximately two months, with 187 of 546 questionnaires (or 34.25 percent) being returned.

3.2. Measures

The questionnaire assessed responses based on demographic data and respondents' perspectives on the digitization of the accounting profession. Five measurement items were used to evaluate the respondents' opinions from the perspectives of "opportunities for future accountants" and "risk for future accountants" Adapted from [18], all ten items were scored on a five-point Likert scale ranging from "1: Strongly disagree" to "5: Strongly agree". The items for the variables are listed in Table 2.

Table 2. List of items.

Opportunities for Future Accountants	1. Digitalization creates more new jobs. 2. Digitalization reduces working time. 3. Digitalization raises more work autonomy. 4. Digitalization provides better agronomics. 5. Digitalization enhances gender equality for women in work.

Table 2. *Cont.*

Risks for Future Accountants	1. Digitalization introduces challenges to the work. 2. Digitalization raises job instability. 3. Digitalization expands working time 4. Digitalization would increase the 5. Digitalization increases competition among employees.

(Source: adapted from [18]).

4. Analysis and Findings

SPSS was used to conduct analyses based on 187 (or 34.25 percent) of the returned questionnaires. These included an analysis of demographic data, a mean score to determine the level of opportunity and risk associated with the digitalization of the accounting profession, and an independent *t*-test to determine whether there is a statistically significant difference in opportunity and risk associated with digitalization between male and female respondents.

4.1. Respondents' Profile

A total of 149 (79.68 percent) of the respondents were females, whereas only 38 (20.32 percent) were males. The majority of respondents (157 or 83.96 percent) were between the ages of 20 and 24 and 182 of them (97.3 percent) had internships lasting between 3 and 6 months. The majority of respondents were Malay (148, or 79.1 percent), followed by Chinese (18, or 9.6 percent). All respondents were Bachelor of Accountancy students at their respective universities.

The majority of interns are attached to Audit firms (116 or 62.03 percent), followed by Private Firms (36 or 19.25 percent). Other employers include Federal Government Agencies, State Government Agencies, Financial Institutions, and Others, which include Tax firms and Essential services firms. Selangor (48 or 25.7 percent) and Kuala Lumpur were where the vast majority of employers are based (43 or 23 percent). Only 21 (11.2 percent) of employers were located in Johor, while less than 10 percent were located in other states. The demographic information demonstrates the credibility of the respondents as study participants.

4.2. Reliability of the Instrument

The measurement of reliability was Cronbach's Alpha, with values above 0.7 considered acceptable and values above 0.8 being preferred [41]. The values in Table 3 indicate that all questionnaire items were reliable for further analysis.

Table 3. Reliability statistics.

	Cronbach's Alpha	Cronbach's Alpha Based on Standardized Items	N of Items
The opportunity for the digitalized accounting profession	0.734	0.748	5
Risk of the digitalized accounting profession	0.848	0.848	5

4.3. Mean Score for Opportunity and Risk of the Digitalized Accounting Profession

The respondents' perspectives on the digitization of the accounting profession were evaluated based on the opportunities and risks for future accountants, as shown in Table 4. The interpretation of the mean score level adheres to the level used by ([42], as cited in [43,44]), with low (i.e., 1.00–2.33), medium (i.e., 2.34–3.67), and high (i.e., 3.68–5.00), respectively. Table 4 displays the mean scores for each item, as well as the overall mean scores for the opportunities and risks of digitalization of the accounting profession.

Table 4. Mean score.

	N	Min	Max	Mean	S.D.
1. Digitalization creates more new jobs.	187	1	5	3.67	0.925
2. Digitalization reduces working time.	187	1	5	3.84	0.896
3. Digitalization raises more work autonomy.	187	2	5	3.91	0.731
4. Digitalization provides better agronomics (support in performing heavy, dangerous, or complex work).	187	2	5	3.94	0.752
5. Digitalization enhances gender equality for women in work.	187	1	5	3.76	0.804
The opportunity for the digitalized accounting profession	187	2.40	5	3.82	0.574
1. Digitalization introduces challenges to the work.	187	1	5	3.87	0.747
2. Digitalization raises job instability.	187	1	5	3.63	0.860
3. Digitalization expands working time (work from anywhere and at any time).	187	1	5	4.00	0.823
4. Digitalization would increase the workload (work intensification, dependence on 'data masters' and surveillance).	187	1	5	3.81	0.877
5. Digitalization increases competition among employees.	187	1	5	3.87	0.833
Risk of the digitalized accounting profession	187	1.00	5	3.84	0.653

Note: N = population size; Min = Minimum; Max = Maximum; Mean = Average of a dataset; S.D = Standard Deviation. (The first highlighted row represents the opportunity and the second highlighted row represents the risk).

The results indicate the highest mean for opportunity (i.e., 3.94) is for item 4 (Digitalization provides better agronomics). The respondents perceived digitalization as an opportunity that provides support to complex work. As for risk, the highest mean (i.e., 4.00) is for item 3 (Digitalization expands working time). The respondents perceived digitalization as a risk that expands the working time because the digitalized environment enables work to be performed from anywhere and at any time.

Comparing the overall mean, the score of 3.82 for the opportunity and 3.84 for risk reflect the respondents' strong belief that the digitalization of the accounting profession presents future accountants with both high opportunities and risks. Consistent with the perception that accountants' jobs will become more difficult in the future, accountants have a significant opportunity to increase their level of expertise, particularly in digitalization skills and knowledge [4]. Digitalization will result in the automation of more mundane and repetitive tasks, allowing accountants to focus on value-adding opportunities for their organizations and their users [45].

As digitalization is unavoidable, future accountants should investigate all avenues for enhancing their technology knowledge and be prepared to use new technology in their personal and professional lives. Even though accountants will never become extinct, their numbers will decrease as their tasks become more automated and efficient [13]. Consequently, they will have to grow and change in tandem with technological advances. There are numerous opportunities for accountants to increase their level of expertise, particularly in terms of digitalization skills and knowledge, which could make their jobs appear more challenging in the future [4].

4.4. Gender Differences with Opportunity and Risk of the Digitalized Accounting Profession

An association between gender and opportunity of the digitalized accounting profession as well as between gender and risk of digitalized accounting profession among future accountants were examined and summarized in Table 5.

Male future accountants have a slightly higher perception of opportunity (M = 3.93; SD = 0.69) than female future accountants (M = 3.79; SD = 0.54), as shown in Table 5. Nevertheless, the difference is insignificant (Sig. 2-tailed value, $t = 1.129$; df = 49.24; $p = 0.265$). The magnitude of mean differences (mean difference = 0.14, 95 percent CI: −0.11 to 0.38) is very small (eta squared = 0.01).

In addition, male future accountants have a slightly higher risk perception (M = 3.98; SD = 0.59) than female future accountants (M = 3.79; SD = 0.59). However, it is not statistically significant (Sig. 2-tailed value, $t = 1.524$; df = 185; $p = 0.129$). The magnitude of the mean difference (mean difference = 0.18, 95 percent CI: −0.05 to 0.41) is very small (eta squared = 0.01).

Table 5. Independent *t*-test results.

Group Statistics					
	Gender	**N**	**Mean**	**Std. Deviation**	**Std. Error Mean**
The opportunity for digitalized accounting profession (D1OFA)	Male	38	3.9316	0.68856	0.11170
	Female	149	3.7960	0.54051	0.04428
Risk of digitalized accounting profession (D2RFA)	Male	38	3.9789	0.58962	0.09565
	Female	149	3.7987	0.66556	0.05452

Independent Samples Test										
		Levene's Test for Equality of Variances		***t*-test for Equality of Means**						
		F	Sig.	t	df	Sig. (2-tailed)	Mean Difference	Std. Error Difference	95% Confidence Interval of the Difference	
									Lower	Upper
D1OFA	Equal variances not assumed	3.927	0.049	1.129	49.239	0.265	0.13561	0.12016	−0.10583	0.37704
D2RFA	Equal variances assumed	2.559	0.111	1.524	185	0.129	0.18029	0.11832	−0.05315	0.41373

Consequently, the findings indicate that there is no significant difference between male and female future accountants in their perceptions of opportunity and risk associated with the digitalization of the accounting profession. As the majority of respondents in this study (157 or 84 percent) were between the ages of 20 and 24 years old, this result may be reflective of the existing group of future accountants who are members of Generation Z. This generation is generally eager to explore new opportunities due to digitalization and keen to participate in the environment's demanding nature.

Generation Z, born after 1995, is the post-millennial generation, who is more accustomed to using technology to communicate in social settings [46]. Generation Z, or Generation C [47] (connected, communicating, content-centric, computerized, community-oriented, clicking), is the first generation to be born between 1995 and 2009 [48] or 1995 and 2010 [49]. They are the first generation to grow up in a globally connected environment [50] where technology was readily available ([51] as cited in [52]). Given this generation's extensive exposure to technology, gender indifference regarding the perceived opportunity and risk of a digitalized accounting profession is not unexpected. They are familiar with technology, flexible, adaptable, and able to quickly transition to new platforms [46].

In contrast, the findings contradict previous research that identified gender differences in technology. In the context of information technology, including computers, email services, electronic data management systems, and other related technologies, gender is a factor in technology adoption, with men being more technologically adept than women [31]. Moreover, gender differences serve as a moderator in the adoption of multimedia for learning [36]. The effect of task-technology fit on perceived usefulness is stronger in men than in women, leading to the conclusion that when the task and technology are a good match, men are more likely to perceive the technology as advantageous. Consequently, the findings of this study contribute to the body of evidence indicating that males and females compete for the best career success and that females no longer view technological progress negatively.

As they are members of Generation Z and were exposed to technology at a young age, the digitalization of the accounting profession is viewed as both an opportunity and a risk, depending on how they adapt to the new environment. Even women were just as eager to embrace technology in the digital age [39]. This may then provide companies with information about gender indifference that facilitates the selection and hiring of new employees to meet industry demands. Employers in both commercial and public accounting firms may therefore focus on developing programs and training newcomers' abilities to use emerging ICT applications in line with digitalization.

5. Conclusions

This study discovered that the digitalization of the accounting profession presents both high opportunity and significant risk for future accountants, with the respective highest mean score items for the opportunity and risk being "Digitalization provides better agronomics—support in performing heavy, dangerous, or complex work" and "Digitalization expands working time—work from anywhere and at any time". However, gender differences did not affect the perceived opportunity or risk associated with the digitalization of the accounting profession.

The findings give beneficial insights into the educational systems and professional organizations. As future accountants viewed digitalization of the accounting profession as both high opportunity and high risk, the educational system could include more information on the opportunities and risks of digitalization in the curriculum to provide accounting graduates with better career planning guidelines. In addition, professional organizations may provide future accountants with greater exposure to the opportunities and risks of a digitalized environment. Thus, future accountants can have clearer perspectives and be better prepared for the actual digitalized workplace.

However, the current study is limited to students from some public universities, which may limit the applicability of the findings beyond the university setting. Future research could be extended to accounting interns at private universities, whose curricula may place a different emphasis than those at public institutions. Similarly, research may be conducted on current accounting practitioners, as they are potential mentors for newcomers. In addition, qualitative research methods, such as in-depth interviews, may be used to learn more about the anticipated opportunity or risk and changes in the roles of accountants in a digitalized environment. Finally, the findings may aid in the recruitment of new personnel, indicating that gender should not be the primary criterion for evaluating job applications from digital workplace newcomers. In other words, gender discrimination should not be an issue in the digitalization of the accounting industry.

Author Contributions: Conceptualization, Y.A., S.M.S. and A.T.; methodology, Y.A. and A.T.; software, Y.A. and S.M.S.; validation, S.M.S. and N.R.; formal analysis, Y.A.; investigation, Y.A., S.M.S., A.T., N.R. and M.S.H.; resources, Y.A., S.M.S., A.T., N.R. and M.S.H.; data curation, Y.A. and A.T.; writing—original draft preparation, Y.A.; writing—review and editing, Y.A., S.M.S. and A.T.; visualization, Y.A., A.T. and N.R.; supervision, Y.A. and N.R.; project administration, Y.A., S.M.S., A.T., N.R. and M.S.H.; funding acquisition, Y.A., S.M.S., A.T., N.R. and M.S.H. All authors have read and agreed to the published version of the manuscript.

Funding: This research was funded by the Research Collaboration Fund 2020 (600-UiTMCTKD (PJI/RMU 5/2/1) Jld. 4).

Institutional Review Board Statement: The research participants were provided with sufficient information to assist in an informed decision as to whether to take part in research (informed consent) through the cover letter of the questionnaire.

Informed Consent Statement: Informed consent was obtained from all subjects involved in the study.

Data Availability Statement: Not applicable.

Conflicts of Interest: The authors declare no conflict of interest.

References

1. Sima, V.; Gheorghe, I.G.; Subić, J.; Nancu, D. Influences of the industry 4.0 revolution on the human capital development and consumer behavior: A systematic review. *Sustainability* **2020**, *12*, 4035. [CrossRef]
2. Bakulina, G.; Kalinina, G.; Luchkova, I.; Pikushina, M.; Gracheva, A. Transformation of the accountancy profession during digitalization of agriculture. *BIO Web Conf.* **2020**, *17*, 00188. [CrossRef]
3. Andreea, C.D.; Mihaela, N.A.; Elena, C.N.; Mugurel, Ț.M.; Nadia, P.A. An Investigation of the Perceived Impact of IT on the Accounting Profession. Available online: https://cig.ase.ro/wp-content/uploads/2021/09/Codreanu.pdf (accessed on 14 March 2022).
4. Moll, J.; Yigitbasioglu, O. The role of internet-related technologies in shaping the work of accountants: New directions for accounting research. *Br. Account. Rev.* **2019**, *51*, 100833. [CrossRef]

5. ACCA. *The Digital Accountant: Digital Skills in a Transformed World*; ACCA Global: London, UK, 2020.
6. Suhaiza, I. Effect of Ethical Ideologies on Ethical Judgment of Future Accountants: Malaysian Evidence. *Asian Rev. Account.* **2014**, *22*, 145–158.
7. Adeosun, O.T.; Shittu, A.I.; Owolabi, T.J. University internship systems and preparation of young people for world of work in the 4th industrial revolution. *Rajagiri Manag. J.* **2021**, *16*, 164–179. [CrossRef]
8. Venkatesh, V.; Morris, M.G.; Davis, G.B.; Davis, F.D. User acceptance of information technology: Toward a unified view. *MIS Q.* **2003**, *27*, 425–478. [CrossRef]
9. Lai, Y.L.; Lee, J. Integration of Technology Readiness Index (TRI) Into the Technology Acceptance Model (TAM) for Explaining Behavior in Adoption of BIM. *Asian Educ. Stud.* **2020**, *5*, 10. [CrossRef]
10. Zhang, Y.; Xiong, F.; Xie, Y.; Fan, X.; Gu, H. The Impact of Artificial Intelligence and Blockchain on the Accounting Profession. *IEEE Access* **2020**, *8*, 110461–110477. [CrossRef]
11. Mian, S.H.; Salah, B.; Ameen, W.; Moiduddin, K.; Alkhalefah, H. Adapting Universities for Sustainability Education in Industry 4.0: Channel of Challenges and Opportunities. *Sustainability* **2020**, *12*, 6100. [CrossRef]
12. Gartner Glossary. Available online: https://www.gartner.com/en/information-technology/glossary/digitalization (accessed on 24 May 2022).
13. Duong, D.C.T.; Fledsberg, K. *Digitalization of the Accounting Industry: The Influence Of Digitalization on the Accountants' Role and Their Self-Understanding. An Exploratory Study Based on 13 Norwegian Accounting Firms*; University of Agder: Kristiansand, Norway, 2019.
14. Tekbas, I.; Nonwoven, K. The Profession of the digital age: Accounting Engineering. In *IFAC Proceedings Volumes*; Project: The Theory of Accounting, Engineering; International Federation of Accountants: New York, NY, USA, 2018. Available online: https://www.ifac.org/knowledge-gateway/preparing-future-ready-professionals/discussion/profession-digital-age-accounting-engineering (accessed on 14 March 2022).
15. Fettry, S.; Anindita, T.; Wikansari, R.; Sunaryo, K. The future of accountancy profession in the digital era. In *Global Competitiveness: Business Transformation in the Digital Era, Proceedings of the First Economics and Business Competitiveness International Conference (EBCICON 2018), Bali, Indonesia, 21–22 September 2018*; Routledge-Taylor & Francis Group: London, UK, 2019.
16. Frey, L.; Botan, C.; Kreps, G. *Investigating Communication: An Introduction to Research Methods*, 2nd ed.; Allyn & Bacon: Boston, MA, USA, 1999.
17. Gulin, D.; Hladika, M.; Valenta, I. Digitalization and the Challenges for the Accounting Profession. *Entren.-Enterp. Res. Innov.* **2019**, *5*, 428–437. [CrossRef]
18. Voss, E.; Riede, H. *Digitalisation and Workers Participation: What Trade Unions, Company Level Workers and Online Platform Workers in Europe Think*; ETUC–European Trade Union Confederation: Brussels, Belgium, 2018.
19. Yeap, C. Malaysia's digital economy transformation needs more than great headlines. *Edge Wkly* **2021**. Available online: https://www.theedgemarkets.com/article/malaysias-digital-economy-transformation-needs-more-great-headlines. (accessed on 14 March 2022).
20. Hirt, M.; Willmott, P. Strategic principles for competing in the digital age. *McKinsey Q.* **2014**, *5*, 1–13.
21. Craft, J.L. A review of the empirical ethical decision-making literature: 2004–2011. *J. Bus. Ethics* **2011**, *117*, 221–259. [CrossRef]
22. Ford, R.C.; Richardson, W.D. Ethical decision making: A review of the empirical literature. *J. Bus. Ethics* **1994**, *13*, 205–221. [CrossRef]
23. Loe, T.W.; Ferrell, L.; Mansfield, P. A review of empirical studies assessing ethical decision making in business. *J. Bus. Ethics* **2000**, *25*, 185–204. [CrossRef]
24. Lopez-Nicolas, C.; Nikou, S.; Molina-Castillo, F.J.; Bouwman, H. Gender differences and business model experimentation in European SMEs. *J. Bus. Ind. Mark.* **2020**, *35*, 1205–1219. [CrossRef]
25. Lu, J.; Ren, L.; Zhang, C.; Wang, C.; Petkeviciute, N.; Streimikis, J. Gender difference in corporate social responsibility implementation in Lithuanian SMEs. *Oeconomia Copernic.* **2020**, *11*, 549–569. [CrossRef]
26. O'Fallon, M.J.; Butterfield, K.D. A review of the empirical ethical decision-making literature: 1996–2003. *J. Bus. Ethics* **2005**, *5*, 375–413. [CrossRef]
27. Betz, M.; O'Connell, L.; Shepard, J.M. Gender differences in proclivity for unethical behavior. *J. Bus. Ethics* **1989**, *8*, 321–324. [CrossRef]
28. Peterson, D.; Rhoads, A.; Vaught, B.C. Ethical beliefs of business professionals: A study of gender, age and external factors. *J. Bus. Ethics* **2001**, *31*, 225–232. [CrossRef]
29. Feldberg, R.L.; Glenn, E.N. Male and female: Job versus gender models in the sociology of work. *Soc. Probl.* **1979**, *26*, 524–538. [CrossRef]
30. Lacy, W.B.; Bokemeier, J.L.; Shepard, J.M. Job attribute preferences and work commitment of men and women in the United States. *Pers. Psychol.* **1983**, *36*, 315–329. [CrossRef]
31. Goswami, A.; Dutta, S. Gender differences in technology usage—A literature review. *Open J. Bus. Manag.* **2015**, *4*, 51–59. [CrossRef]
32. Orji, R.O. Impact of gender and nationality on acceptance of a digital library: An empirical validation of nationality based UTAUT using SEM. *J. Emerg. Trends Comput. Inf. Sci.* **2010**, *1*, 68–79.

33. Jackson, L.A.; Ervin, K.S.; Gardner, P.D.; Schmitt, N. Gender and the Internet: Women communicating and men searching. *Sex Roles* **2001**, *44*, 363–379. [CrossRef]
34. Mayoux, L. *Jobs, Gender and Small Enterprises: Getting the Policy Environment Right*; International Labour Organization (ILO) Working Papers; International Labour Office: Geneva, Switzerland, 2001.
35. Cai, Z.; Fan, X.; Du, J. Gender and attitudes toward technology use: A meta-analysis. *Comput. Educ.* **2017**, *105*, 1–13. [CrossRef]
36. Park, C.; Kim, D.G.; Cho, S.; Han, H.J. Adoption of multimedia technology for learning and gender difference. *Comput. Hum. Behav.* **2019**, *92*, 288–296. [CrossRef]
37. Livingstone, S.; Helsper, E. Balancing opportunities and risks in teenagers' use of the internet: The role of online skills and internet self-efficacy. *New Media Soc.* **2010**, *12*, 309–329. [CrossRef]
38. Verma, C.; Dahiya, S. Gender difference towards information and communication technology awareness in Indian universities. *SpringerPlus* **2016**, *5*, 1–7. [CrossRef]
39. Blasko, D.G.; Lum, H.C.; Campbell, J. Gender differences in perceptions of technology, technology readiness, and spatial cognition. *Proc. Hum. Factors Ergon. Soc. Annu. Meet.* **2020**, *64*, 1395–1399. [CrossRef]
40. Krejcie, R.V.; Morgan, D.W. Determining Sample Size for Research Activities. *Educ. Psychol. Meas.* **1970**, *30*, 607–610. [CrossRef]
41. Pallant, J. *SPSS Survival Manual*, 4th ed.; McGraw Hill: Berkshire, UK, 2010.
42. Landell, K. *Management By Menu*, 1st ed.; Wiley and Sons Inc.: London, UK, 1997.
43. Abdul, H.A.; Rahman, S.N.S.A.; Hamzah, M.H. Metacognitive skills of Malaysian students in non-routine mathematical problem solving. *Bolema Bol. Educ. Mat.* **2017**, *31*, 310–322. [CrossRef]
44. Hairuzila, I.; Muhammad, R.; Tony, L.A. Implementation of PBL to enhance the soft skills of engineering students. *SHS Web Conf.* **2018**, *53*, 03008.
45. Richins, G.; Stapleton, A.; Stratopoulos, T.C.; Wong, C. Big data analytics: Opportunity or threat for the accounting profession? *J. Inf. Syst.* **2017**, *31*, 63–79. [CrossRef]
46. Gabrielova, K.; Buchko, A.A. Here comes Generation Z: Millennials as managers. *Bus. Horiz.* **2021**, *64*, 489–499. [CrossRef]
47. Friedrich, R.; Le Merle, M.; Peterson, M.; Koster, A. *The Rise of Generation C. Implications for the World of 2020: PwC*; Booz Co.: New York, NY, USA, 2010; p. 24.
48. McCrindle, M.; Wolfinger, E. *The ABC of XYZ: Understanding the Global Generations*; McCrindle Publication: Bella Vista, Australia, 2009.
49. Seemiller, C.; Grace, M. *Generation Z Goes to College*; John Wiley & Sons: San Francisco, CA, USA, 2016.
50. Cilliers, E.J. The challenge of teaching generation Z. *PEOPLE: Int. J. Soc. Sci.* **2017**, *3*, 188–198. [CrossRef]
51. Turner, A. Generation Z: Technology and social interest. *J. Individ. Psychol.* **2015**, *71*, 103–113. [CrossRef]
52. Kirchmayer, Z.; Fratričová, J. What motivates generation Z at work? Insights into motivation drivers of business students in Slovakia. In In Proceedings of the Innovation Management and Education Excellence through Vision, Milan, Italy, 25–26 April 2018; pp. 6019–6030.

proceedings

MDPI

Proceeding Paper

Digitalization among Refugees in Malaysia †

Mohd Ramlan Mohd Arshad [1,2,*] and Mohd Na'eim Ajis [2,3]

[1] Faculty of Administrative Science and Policy Studies, UiTM Negeri Sembilan, Seremban 70300, Malaysia
[2] Research Institute for Indonesia, Thailand, and Singapore (UUM-ITS), Sintok 06010, Malaysia
[3] School of Government, Universiti Utara Malaysia (UUM), Sintok 06010, Malaysia
* Correspondence: mramlan2957@uitm.edu.my

† Presented at the International Academic Symposium of Social Science 2022, Kota Bharu, Malaysia, 3 July 2022.

Abstract: There are issues with connectivity and networking for refugees during their stay in Malaysia. Thus, this study aims to examine digitalization activities among refugees. This qualitative study with purposive sampling involved three informants representing the government, private sector, and NGOs based in Putrajaya, Malaysia. The conducted analysis found that a digitization process has taken place, but related parties must ensure digital systems function evenly when addressing the refugee problem in Malaysia. The results could benefit the government and other parties directly involved in refugee issues such as the UNHCR, local NGOs, the private sector, and civil societies.

Keywords: digitalization; internet; networking; migrant; refugees

1. Introduction

Malaysia is a non-signatory to the 1951 Refugee Convention and the 1967 Protocol. Therefore, it is not the official responsibility of the government to address the problem of refugees entering the country either legally or illegally through any entrance at borders such as air, land, or water. However, the government signed the Universal Declaration on Human Rights (UDHR) in 1948 to ensure that human rights in Malaysia are protected, regardless of the background of the individuals. The refugee community is included in the group that has received compassion from the government through the concept of universal peace practiced by the government. Due to the government's lenient attitude towards refugees, most refugees from both the Southeast Asian region and outside the region have chosen to migrate to and seek refuge in Malaysia. This has led to an increase in the number of refugees in Malaysia. The situation is best described through statistics which show that there are a total of 179,830 refugees and asylum seekers who have registered with the UNHCR [1]. The breakdown of refugees is 155,030 people from Myanmar, with 103,090 Rohingya people, 22,450 that are Chin ethnic, and 29,330 representing other oppressed ethnicities in Myanmar. The remaining 24,800 people are refugees from various countries: 6690 people from Pakistan, 3700 people from Yemen, 3280 people from Syria, 3200 Somalis, 2730 from Afghanistan, 1680 Sri Lankans, 1200 Iraqis, and 770 Palestinians. The figures show a significant increase in previous years, with only 111,298 refugees in 2020 and 129,111 in 2019 [1]. Thus, the government's burden of managing refugees who enter the country is becoming heavier and needs to be resolved, as the issue of refugees has a direct impact on the sovereignty of the country. The efforts taken include ensuring human safety and the registration process. Therefore, the government has worked to ensure that every refugee who crosses the country's borders is officially registered. To officially register refugees, the government launched the Tracking Refugees Information System (TRIS), operated by a company under the Ministry of Home Affairs (KDN) called Barisan Mahamega Sdn. Bhd, on 1 April 2017.

As there is a large influx of refugees into Malaysia, it is difficult to maintain a manual registration process. This is because the process involves various elements that need to be

Citation: Arshad, M.R.M.; Ajis, M.N. Digitalization among Refugees in Malaysia. *Proceedings* **2022**, *82*, 106. https://doi.org/10.3390/proceedings2022082106

Academic Editor: Mohamad Rahimi Mohamad Rosman

Published: 20 October 2022

recorded by ministries, government departments, and NGOs, including the National Security Council (MKN), Ministry of Home Affairs (KDN), Ministry of Foreign Affairs (KLN), Ministry of Health Malaysia (MOH), and the Ministry of Defense Malaysia (MINDEF); non-governmental actors and parties such as the United Nations High Commission for Refugees (UNHCR); international non-governmental organizations (INGOs) such as Human Rights Watch (HRW) and Amnesty International (Amnesty); local NGOs such as Islamic Welfare Organizations Malaysia (PERKIM), Islamic Relief Malaysia (IRM) and Humanitarian Aid Selangor Society (Human Aid); and organizations run by refugees themselves, such as the Rohingya Society in Malaysia (RSM) and the Chin Refugee Committee (CRC). The parties directly involved are responsible for ensuring that every refugee who seeks protection goes through a complete documentation process to avoid the problem of dumping refugees and foreign migrants who do not have the status to continue to occupy the country. The existence of refugees has an impact on various socio-economic aspects, including the harmony of life between local communities and refugees. In this regard, the process of digitalization among refugees is an effort that coincides with the government's mission to ensure that every person who enters Malaysia is officially recorded. Therefore, this study focuses on the digitization process to ensure the smooth management and administration of refugees in Malaysia.

2. Digitalization and Refugees

Over the years, interventions from the authorities to ensure technological connectivity have been developed to provide internet access to refugees, and are funded by both public and private sector actors. In several European countries, these initiatives have proliferated and involved coordination at the central level in response to the refugee crisis using broadband, mobile connectivity, Wi-Fi, GSM, mesh networks, satellites, drones, space, and the use of technology [2]. However, the lack of government funding and a comprehensive framework has led to resources, support, and funding being provided by the private sector, despite having very different values and interests in supporting digitalization services for refugee groups. For example, internet connectivity programs have been conducted by non-profit organizations such as UNHCR's Connectivity for Refugees and private projects such as Net Hope [2]. However, there is still an ongoing gap between the ideal vision of what digital connections can achieve (globally) and evidence of how new connections can be created [3]. Looking back at the lessons learned over the past few years, many human rights-related practitioners and research teams have challenged the assumptions that support the mainstream connectivity discourse and questioned whether connectivity provides the immediate benefits promised. This is because if all parties involved want access to connectivity to benefit from it, the impact on social dynamics in the refugee community also needs to be addressed. Currently, refugee communities face many restrictions as they move through various modus operandi to escape persecution, violence, and cruelty. However, space for them is often limited and isolated, and the existence of safe digitalized networking is uncertain once they arrive in a host country. This is because refugee groups consist of women, youth, the elderly, and groups formed from various types of sexual orientations as well as diverse abilities and skills, and varying medical needs. This situation limits their access to the coverage or signals of the internet; all the resources and benefits that come with an internet connection, such as the power of digitization, contribute to a greater asymmetry of power in the community.

Restrictions on the internet and the digital realm contribute to the social backwardness of refugees [2]. This is because the realm of digitalization is often framed as a borderless and egalitarian space. Therefore, digitalization solutions need to be prepared and presented as an attractive and viable method to enable refugees to face various challenges, either internally while migrating from their country of origin, or while residing in transit countries or host countries. Thus, from a conceptual point of view, access to digital space is always limited by the constraints of mobility and access to certain physical spaces. As long as connectivity depends on the varying quality and reliability of internet signal strength, access to charging points, network administrators, and mobile agents, it is unlikely to offer benefits to users equally, especially for refugees. Some refugees continue to experience limited digital mobility because of lower digital literacy rates (especially among women, girls, and the elderly) and the lack of online content in their native language [3].

When we imagine and describe how internet connections work, as was the case with telephones and other communication tools before the high-tech era, several spatial metaphors are used [4]. Spatial metaphors such as "cyberspace", "information highway", "electronic boundaries" and "global villages" not only serve as useful communication tools for explaining digital work through common terms, but they also allow us to imagine interactions between people either face-to-face or online [5]. The biased ways and cultural views of practitioners have influenced the design of these interventions; their position is particularly important in the context of continuity interventions because "metaphors can guide our imaginations about new inventions as they influence what can happen even before they exist" [4]. As internet access for refugees is often carried out through a top–down approach, there is a need to question the concept and approach of connectivity used and who manages the impact on services designed for refugees.

Next, a major concern in the digitization process involving refugee groups is the ability of an internet connection to provide immediate benefits to the refugees. This is due to the challenge of justifying benefits to refugees. Scholars say that "if internet connectivity is universally beneficial, then who can argue to determine it and whether the instruments created are appropriate?" [3]. From a theoretical view, digital connectivity and access to technology may provide undeniable benefits to refugees, as this can improve health and provide access to education and information, as well as opportunities to acquire new skills or employment [6]. However, from a practical view of real life, the distractions and conflicting narratives produced by the interference of various parties in this digitalization effort need to be minimized or eliminated. Although internet connectivity is a main objective in both the Sustainable Development Goals and the Global Compact on Refugees, it is often described by humanitarian practitioners and the mass media as a one-size-fits-all solution to some of the problems faced by refugees [7]. The different narratives between the parties providing these actual services and applications have hampered digitization efforts among the refugee community.

Unraveling the issue of digitization from this perspective offers an opportunity to explore how the spatial imagery of practitioners and researchers in the field is changing "not just how we imagine connectivity but how we formulate it" [8]. Critically examining the spatial quality of relevant discourse and stakeholder interventions can be performed using frameworks from human geography, urban planning, and design and information science that allow critical analysis of work through the creation of "imaginary spaces" or simulations [6]. Therefore, understanding this narrative dimensional space of what technology can achieve is an important research task. This spatial dimension is just as important for drones and radars, as is the use of technology for the emerging "Internet of Things", for high-powered broadband, and for virtual reality spaces that add to the online experience, as each reconfigures the connection between the actors involved in the technology's operation [3]. Based on the conceptual framework related to digitization, this study will look at the issues and elements related to the connection between technology and the refugee community in Malaysia.

3. Method and Materials

This study uses qualitative methods through interviews and content analysis. An interpretive approach is used to understand the phenomena that occur by performing an analysis of the responses of the informants. Therefore, the interpretive method is recognized as the best method to explore this area. Administrators of refugee-related organizations are the social actors involved in this study, and were selected based on their experience in handling refugee communities in Malaysia. Bryman explains more about social actors as people who can significantly change social structures [9]. Several scholars propose interpretive methods, such Burrel and Morgan, who state that the traditional philosophy of knowing the world by dealing directly with issues and phenomena is considered a phenomenological technique [10]. Moreover, the method of interpretivism through a phenomenological approach in qualitative research is effective for studying social phenomena [11,12]. Thus, to meet these objectives, an in-depth study of social phenomena and phenomenology among the administrators of the organizations that manage refugees in Malaysia is needed. This technique is suitable for studying the process of digitization among refugee communities. In this study, qualitative design is used because the researcher must study, in-depth, the phenomena related to the effects of digitization that occur. Therefore, in-depth interviews as a tool provide more information related to the phenomena. Many scholars recommend interviews if researchers want to obtain valuable reasons for the phenomena that occur, such as Creswell and Poth, Flick, and Ryan [10,13,14]. Creswell and Poth suggest that high-value study designs can be created through qualitative information because the information from informants explains the causes of the social phenomena [13]. Moreover, Flick explains that it is best to use a qualitative design where assumptions from subjects are valued for the phenomenon [14]. Furthermore, Ryan also suggested a qualitative design in the interpretive approach, which is the assumption of the informants [10].

Therefore, this study used semi-structured in-depth interviews between January and November 2021 to obtain primary data from informants. Semi-structured interviews have allowed flexibility in answering questions and allowed informants to obtain primary data from informants. Semi-structured interviews have allowed flexibility in answering questions and allowed informants' "perspectives" to be explored [15]. Thus, the researchers chose semi-structured interviews to ensure that sufficient information was collected about the digitization process among refugees in Malaysia. Online interview sessions of approximately 45 min to 1 h were conducted with three representatives from organizations managing refugees from the government, UNHCR, and non-governmental organizations. An interview protocol outlining nodes such as using a smartphone and internet connection was used as a reference for the researchers. All the informants were fluent in English, so there was no language barrier in this study. To protect the confidentiality of the informants, fictional names were used in this study, namely, Informants A, B, and C. A purposive sampling technique was used because special considerations were made to only select informants who involved in managing refugee affairs in Malaysia. Researchers questioned the informants on how the digitization process in Malaysia impacts refugee groups in the short and long term. Informants were given full authority to explain and illustrate their views on the questions posed by the researchers. In addition, during the data analysis, the researchers used thematic analysis with the help of NVIVO 12 (QSR International, Burlington, MA, USA), which is software for qualitative research. In this method of data analysis, researchers shape the received information, summarize and analyze it into themes according to nodes, and report the findings in terms of a relevant thematic discussion and analysis [13]. This analysis meets the needs of qualitative scientific research, and the findings can be trusted in explaining the process of digitization among the refugee community in Malaysia.

4. Findings

4.1. Global Virtual World

The global virtual world is an interpretation of how internet connectivity has created a shared, egalitarian, and digital space that benefits refugees just like other individuals. The study by Poll and Informant A stated that the term "global village" has preceded internet-related issues and it is a general interconnected world phenomenon resulting from the spread of media and communication technology. It is also the dominant term to express global coexistence altered by transnational trade, migration, and culture [16,17]. This was also stated by Informant A: "This global village is easy to understand as the way internet connectivity works and what is offered is described as a two-way relationship between refugee individuals wherever they are" [17]. According to this delusion, simply bringing refugees into the digital or online space has the potential to re-create interpersonal relationships that have been weakened by long physical distances. However, the existence of virtual villages has bridged the interaction gap among refugees. This can be illustrated by the following explanation by UNHCR: "When refugees are forced to leave their homes, it leads to families being divided, community relationships being broken, and people being divided". The benefits of digitalization to refugees are clear: bridging communities online is the only way to communicate with families who left or went to another country. Informant B clarifies that it is the best way for them to access trusted sources of information about the asylum process and its procedural changes [18].

Informant C stated that the connectivity created through the global virtual world is not merely a prerequisite for reuniting separated refugee families; here, it acts more generally as a variable, and is capable of providing refugees with other extraordinary benefits exclusively in this digitally shared space [19]. For example, the private sector often describes how their services work throughout the global world. For example, referring to Mark Zuckerberg's statement that "promote access to the Internet as a variable of human rights and security because companies will help the UN bring internet connectivity to refugee camps". The inclusiveness in the connection involving the refugees will benefit all parties involved in this effort [20]. Thus, broadband services for refugees are a reflection of digitalization involving a growing population of refugees around the world, whether they live inside refugee camps or outside camps. Moreover, the process of digitization requires financial and non-financial investments in long-term solutions to address the issue of internet connectivity and broadband access for refugees [21]. These efforts are part of the idealistic digitalization of refugees. Reflections of idealistic delusion came in 2015 when the refugee crisis in Europe received a lot of media coverage. From 2015 to 2019, there is shift in digitalization as more organizations are working to improve refugee digital access, and it seems that every person or organization is developing apps, producing more than 1000 different apps to help refugees [22]. This was also emphasized by Informant B, who stated that "narratives related to the virtual world have become part of the conversation among the public, such as the use of digitization in conveying news related to refugees, which often issues related to this community will be blocked by the government". However, in the virtual world, information will almost certainly spread widely [18]. Informant C stressed that "the virtual world is not an isolated idea for the refugee community because they often use digital platforms to connect with family in their home country or connect with friends in the country or other host countries such as the USA, Canada, and Australia" [19]. A study by Paul proved that, in Japan, there are "internet cafes for refugees" used by the refugee community and homeless people to rent to obtain an internet connection [23]. Thus, this study found that the global virtual world has contributed to the process of digitization among refugee communities. Informant C said that, despite some physical barriers and broadband connectivity, refugees still benefit from sustainable development strategies that emphasize internet connectivity fairly and equitably [19]. However, the global virtual world alone is not enough to accelerate the process of digitization among refugees, so digital augmentation (enhancement) is needed to complete this effort.

4.2. Digital Augmentation (Enhancement)

Apart from the global virtual world, the rise of digitalization can be seen from a critical ICT perspective; there is no perfect 'cure' to the digitalization problem among refugees. Graham stated that digital augmentation looks at aspects of how to integrate existing technologies and advances into structures, networks, and locations to address the problems of refugee communities in the world [5]. Informant A clarifies that internet connectivity is now not only focused on connectivity alone, but has shifted to digital escalation, which involves the structure and network of virtual imaginations, new thinking, technology, and design, as well as encouragement from service providers and recipients [17]. This is because "no magic device can solve all communication problems and every need requires information from various angles as well as more advanced technology" [24]. It is agreed by Informant A stated that "I also think there is no solution to the problem related to digitalization involving these refugees because there are some ethical issues when third parties (service providers) network design that affect the dynamics of social refugees due to not seeking consent from them." [17]. From the spatial perspective, as explained by Informant B, "it can be understood that ICT needs to be developed through existing social networks among refugees as attempts to build a new network have less impact due to ethical problems and difficulty finding solutions to all problems" [18]. This is also stated by Smart et al., who stated that improvements in the use of digital and technology need to be made through existing relationships because building new structures is unprofitable [3]. This is because social connectivity among refugees needs to use existing beliefs, as it further facilitates the augmented digital process. Informant B stated that "the use of new systems in the flow of information will affect digitization and the physical ecosystem among refugees as a result of the things they replace and as a result of questionable third-party interventions on ethics and security levels" [18]. In addition, to ensure that digital augmentation among refugees can become a reality, community centers are key for meeting the physical space requirements that ensure success. Creating physical space and meeting software requirements can ensure better connections. Informant C explained that "the increase in digitization allows for greater professional development and employment opportunities for the refugee community" [19]. This is also explained by Balestra, who stated that access to digital learning requires more effort in line with the growing number of refugees, as life in refugee camps often has barriers to the internet [25]. Based on informants and scholars, it can be understood that digital augmentation is an important process of digitization involving the refugee community. This not only involves the appropriate physical space to meet the needs of digitization, but also other needs such as internet networks and software in the area of refugee camps. This needs to be completed to ensure that the digital divide between the refugee community and the population in a country can be bridged and achieve the goal of a sustainable development strategy that emphasizes digitalization at various levels of society. In addition, service providers must also meet ethic requirements in the provision of broadband procedures to ensure that personal and security issues are protected.

4.3. Scattered Information Landscape

Currently, there is no official platform that explains virtual images to the refugee community, so those who are users of this service need to identify a method that is most suitable for them to use. Therefore, it must be ensured that the disruption to the flow of information can be overcome to achieve digitalization among refugees. Hannides et al. said the information landscape can help in understanding disturbing information [26]. This is because the information landscape identifies the process of information dissemination that occurs through various communication media. However, the problem is that the scattered and decentralized information landscape is causing problems for refugees. This was explained by Informant A: "in addition to technical problems such as connection to the internet, refugees also have problems with fragmented information. No information center explains all the facilities and services to the refugees" [17]. This situation, according to Lloyd, has caused many refugees to be left out of the digitization process because they cannot obtain enough information from those responsible for managing refugee affairs in an area [18,27]. This matter is also explained by Informant B, who stated that "we have different ways of understanding information, either individually or in groups, through a normative cultural, political, and economic background". For example, references such as newspapers, community message boards, or certain posts that contain various types of information. We also personally go to certain places to find information, such as libraries, museums, schools, or traffic stations. We often rely on fellow refugees to explain certain things to each other [18]. Based on these findings, the information landscape for refugees is disorganized and makes it difficult to digitize information on virtual platforms such as TikTok, Facebook, Instagram, and others. This is supported by the UNHCR, which states that refugee settlements in Malaysia are scattered and information cannot be channeled in an organized manner [28]. The UNHCR also clarified that the majority of refugees do not have access to accurate information because of weaknesses in the digitization of information that occurs. This statement links with Informant C's response, which is, "There is a problem with the flow of information, especially among refugees who mostly do not get a formal education in their country of origin. The refugee community is also unfamiliar with the pattern of information disseminated in Malaysia. Refugees must re-learn how to understand local customs to obtain information when confronted with an unfamiliar and unreliable information environment [19]. As a result, this scattered information landscape has made it difficult for the refugee community to understand the digitization that is taking places; for example, the UNHCR and other government departments have used electronic and online approaches. In addition, scattered information makes this situation more difficult. Refugees require a new-centered information landscape to be created. Smart et al. support the creation-centered information landscape as a mechanism for communication between refugees and the parties directly involved [3]. Lloyd also explained that a study on consolidating all information on one platform is a good effort towards full digitization, i.e., refugee-related information needs to be re-adapted to their new environment through links and broadband connectivity to overcome information gaps, adjust and modify the way information is obtained, and to understand and reconstruct new information that is more complete and structured [27]. Therefore, it can be concluded that the scattered information landscape is one of the factors that can contribute to the success or failure of the digitization process among refugees in Malaysia. If the flow of information can be controlled and consolidated on a virtual platform, then the process of digitization for refugees will be easier, because refugees are largely illiterate in relation to technology and high-tech tools.

5. Discussion and Conclusions

Digitization is a catalyst for strong social connections and networks that benefited the refugee community in Malaysia. However, the limitations of broadband and internet networks provided by service providers have the potential to create a digital imbalance between the refugee community and local communities, especially in terms of limited information and different levels of technological literacy. This is aligned with the Informants' statements, which clarify that these digital gap leads to the virtual world being neglected among the refugee communities in Malaysia [17–19]. Moreover, the UNCHR states that the inability of refugee communities to gain access to information, education, and employment further enhances the digital divide that exists between them and local communities [28]. In addition, the informants' statements supported the problem of access to the internet and equipment, which causes refugees to lag behind in the ongoing digitization process [17–19]. According to Hannides et al., most studies on refugees show that low economic status, disability, gender, and age are the main factors influencing internet access and the ownership of technology-friendly devices, resulting in difficulties in the virtual world [26]. These factors have resulted in uneven access to connections and devices, making the digitization process difficult.

Furthermore, the informants stated that "the limited use of the internet and access to scattered information has facilitated the efforts of criminal groups such as smugglers and human traffickers to pass on false information to refugees trying to enter Malaysia" [17–19]. Due to limitations in information and internet access, refugee groups are easily deceived and trust the information provided by criminal groups, which causes them to be involved in human trafficking syndicates. In line with this, the parties involved in the management of refugees in Malaysia (both government and non-government actors) have to play their roles to ensure the refugee community is involved in the ongoing digitalization of Malaysia. Scattered information leads to a lack of privacy because refugee communities often share devices, which has compromised the quality of independent access to information [29]. According to Poole, barriers to individual freedom regarding access to mobile phones and the internet are key variables of human rights violations. For example, applications in the virtual world are often designed to provide sensitive information, such as health and safety information [30]. The informants agreed that scattered information makes it difficult to access true information [17–19].

In addition, the main question of who is responsible for providing a connection to the internet and technology is a matter of policy that needs to be resolved. The Informants argued that they need an augmenter (enhancer) to ensure they can cope with the digitization process in Malaysia [17–19]. Thus, the parties directly involved, such as the government and the UNCHR, as well as the indirect actors such as the private sector and NGOs, need to work together to ensure the goal of digitization among refugees can be achieved. If the government, private sector, and NGOs work together, digitalization among refugees will be easier because a holistic approach is the main catalyst in complex digitization efforts [31]. Sengupta stressed that if only the private sector, which is a provider of internet and broadband services, is involved, then the issue of privacy and information leakage has the potential to occur [20]. Therefore, the cooperation of public and private organizations and NGOs can reduce the burden on one party only. The government may be involved in making public policy, which is the main reference for digitization for the refugee community, while the UNHCR can be a facilitator in implementing it, and the private sector and NGOs can contribute in terms of financial support and workforce strength. This can reduce the imbalance of carrying out responsibilities in the digitization process in terms of refugee registration, temporary placement in Malaysia, and repatriation to third countries. When these related parties are seriously involved in the process of providing facilities, technology-friendly equipment, and innovation, the goal of digitization among the refugee community can be facilitated.

Lastly, a barrier to digitization exists because the refugee community is forced to share devices. Apart from that, there are digital literacy issues among refugees related to privacy and freedom issues. The refugees do not have the ability to speak English, have visual or physical disabilities, and are not familiar with information technology, causing them to have to rely on others to operate devices and surf the internet. Caswell suggests that humanitarian organizations should use their resources to help refugees operate their electronic devices and supply technological hardware to address privacy issues [29]. This is important because if the problem of imbalance in browsing the virtual world occurs, then the Sustainable Development Goal (SDG) of reducing the digitalization gap among refugees may not be achieved. There is a need to assess the opinions, experiences, and perspectives of the refugee community themselves in building a model of internet connectivity and information technology, as they are a target group with special needs. This allows for meaningful and critical dialogue spaces related to interventions designed to ensure internet connectivity is configured to the current state of refugee communities, who face problems such as not being able to interact with electronics, and low technological literacy.

Author Contributions: Conceptualization, M.R.M.A. and M.N.A.; methodology, M.R.M.A.; software, M.R.M.A.; validation, M.R.M.A. and M.N.A.; formal analysis, M.R.M.A. and M.N.A.; investigation, M.R.M.A. and M.N.A.; resources, M.R.M.A. and M.N.A.; data curation, M.R.M.A. and M.N.A.; writing—original draft preparation, M.R.M.A. and M.N.A.; writing—review and editing, M.R.M.A. and M.N.A.; visualization, M.R.M.A. and M.N.A.; supervision, M.R.M.A. and M.N.A.; project administration, M.R.M.A. and M.N.A. All authors have read and agreed to the published version of the manuscript.

Funding: This research received no external funding.

Institutional Review Board Statement: Not applicable.

Informed Consent Statement: Informed consent was obtained from all informants.

Data Availability Statement: Not applicable.

Acknowledgments: Gratitude to our institutions, Faculty of Administrative Science and Policy Studies, UiTM, School of Government, UUM Sintok, Research Institute for Indonesia, Thailand, and Singapore, and all related parties who were involved directly and indirectly in completing this research.

Conflicts of Interest: The authors declare no conflict of interest.

References

1. UNHCR Malaysia. Figure at Glance. Available online: https://www.unhcr.org/en-my/figures-at-a-glance-in-malaysia.html (accessed on 25 May 2021).
2. UNHCR. Space and Imagination: Rethinking Refugees' Digital Access. UNHCR Innovation Service; Luxembourg Aid and Development. Available online: https://www.unhcr.org/innovation/wp-content/uploads/2020/04/Space-and-imagination-rethinking-refugees%E2%80%99-digital-access_WEB042020.pdf (accessed on 10 November 2021).
3. Smart, C.; Donner, J.; Graham, M. Connecting the world from the sky: Spatial discourses around Internet access in the developing world. In Proceedings of the ICTD '16: Eighth International Conference on Information and Communication Technologies and Development, Ann Arbor, MI, USA, 3–6 June 2016; pp. 1–11.
4. Stefik, M. *Internet Dreams: Archetypes, Myths, and Metaphors*; MIT Press: Cambridge, MA, USA, 2021.
5. Graham, M. Contradictory connectivity: Spatial imaginaries and techno mediated positional ties in Kenya's outsourcing sector. *Environ. Plan A Econ. Space* **2015**, *47*, 867–883. [CrossRef]
6. UNICEF. Syria Crisis Mid-Year 2018 Humanitarian Results. Available online: https://www.unicef.org/appeals/files/UNICEF_Syria_Crisis_Situation_Report_Mid_Year_2018.pdf (accessed on 30 March 2022).
7. Maza, C. Refugees Want Wi-Fi More Than Food or Water, Says Turkish CEO. Available online: https://www.newsweek.com/refugees-want-wifi-more-food-or-water-says-turkish-ceo-790547 (accessed on 25 January 2022).
8. Graham, M.; Andersen, C.C.; Mann, L. Geographical imagination and technological connectivity in East Africa. *Trans. Inst. Br. Geogr.* **2014**, *40*, 334–349. [CrossRef]
9. Bryman, A. *Social Research Methods*; Oxford University Press: Oxford, UK, 2016.
10. Burrell, G.; Morgan, G. *Sociological Paradigms and Organizational Analysis: Elements of the Sociology of Corporate Life*; Routledge: Abingdon, UK, 2017.
11. Myers, M.D. *Qualitative Research in Business and Management*; Sage Publication: Washington, DC, USA, 2013.
12. Ryan, G. Introduction to positivism, interpretivism and critical theory. *Nurse Res.* **2019**, *25*, 14–20. [CrossRef] [PubMed]
13. Creswelland, J.W.; Poth, C.N. *Qualitative Inquiry and Research Design: Choosing among Five Approaches*; Sage Publications: Washington, DC, USA, 2017.
14. Flick, U. *An Introduction to Qualitative Research*; Sage Publications Limited: London, UK, 2018.
15. Daae, J.; Boks, C. A classification of user research methods for design for sustainable behaviour. *J. Clean. Prod.* **2015**, *106*, 680–689. [CrossRef]
16. Poll, R. *Afterward: The Global Village*; Rutgers University Press: New Brunswick, NJ, USA, 2021.
17. Informant A (In-Depth Interview, 20 June 2021). Unpublished Ph.D. Thesis, Universiti Utara Malaysia, Kedah, Malaysia, 2022.
18. Informant B (In-Depth Interview, 25 November 2021). Unpublished Ph.D. Thesis, Universiti Utara Malaysia, Kedah, Malaysia, 2022.
19. Informant C (In-Depth Interview, 25 November 2021). Unpublished Ph.D. Thesis, Universiti Utara Malaysia, Kedah, Malaysia, 2022.
20. Sengupta, S. Mark Zuckerberg Announces Project to Connect Refugee Camps to the Internet. *The New York Times*, 26 September 2015.
21. Levin, B. Global Broadband Plan for Refuge Inclusion: Winter 2019, Broadband 4 Refugees. Available online: https://static1.squarespace.com/static/5a0f82f67131a5ac3ca77f03/t/5c9cd941ee6eb02afe82469a/1553783109805/GBP4RI+March+FINAL_For+Posting.pdf (accessed on 20 February 2022).
22. Madianou, M. Technocolonialism: Digital Innovation and Data Practices in the Humanitarian Response to Refugee Crises. *Soc. Media Soc.* **2019**, *5*, 1–13. [CrossRef]
23. Paul, K.; The Japanese Workers Who Live in Internet Cafes. Motherboard. Available online: https://www.vice.com/en/article/mgbkdy/the-japanese-workers-who-live-in-internet-cafes2015 (accessed on 1 May 2022).
24. UNHCR. Trust Is Still One of the Biggest Barriers for Communicating with Refugees. Available online: https://medium.com/unhcr-innovation-service/trust-is-still-one-of-the-biggest-barriers-for-communicating-with-refugees56f35 (accessed on 3 March 2022).
25. Balestra, G.; When Innovation Is Yet Another Connected Community Centre: Connectivity at the Margins of Innovation. UNHCR Innovation Service Medium. 2019. Available online: https://medium.com/unhcr-innovation-service/when-innovation-is-yet-another-connected-community-centre-connectivity-at-the-margins-6bcb4227fc54 (accessed on 28 December 2021).
26. Hannides, T.; Bailey, N.; Kaoukji, D. Voices of refugees: Information and communication needs of refugees in Greece and Germany. *BBC Media Action*, 15 September 2016.
27. Lloyd, A. Researching fractured (information) landscapes Implications for library and information science researchers undertaking research with refugees and forced migration studies. *J. Doc.* **2017**, *73*, 35–47. [CrossRef]
28. UNHCR. Connectivity for Everyone. Available online: https://www.unhcr.org/innovation-year-in-review/essay-connectivity-for-everyone/ (accessed on 1 January 2022).

29. Casswell, J. The Digital Lives of Refugees: How Displaced Populations Use Mobile Phones and What Gets in the Way. Available online: https://www.gsma.com/mobilefordevelopment/wp-content/uploads/2019/07/The-Digital-Lives-of-Refugees.pdf (accessed on 3 November 2021).
30. Poole, D. *Refugee Connectivity: A Survey of Mobile Phones, Mental Health, and Privacy at a Syrian Refugee Camp in Greece*; Data & Society Research Institute: Chicago, IL, USA; Harvard Humanitarian Initiative: Cambridge, MA, USA, 2018.
31. Kaurin, D. *Data Protection and Digital Agency for Refugees*; World Refugee Council Research Paper No. 12; Centre for International Governance Innovation: Waterloo, ON, Canada, 2019.

proceedings

Proceeding Paper

The Perspective of Work Ergonomics on Employee Task Performance in Hotel and Tourism Industry, Malaysia †

Siti Nur Nadhirah Abdul Latip [1], Muhammad Safuan Abdul Latip [2,*], Masliana Tamrin [3] and Mohd Zulfakar Mohd Nawi [4,5]

1 Graduate of School Business, Universiti Tun Abdul Razak, Wilayah Persekutuan 50400, Kuala Lumpur, Malaysia
2 Faculty of Hotel and Tourism Management, Universiti Teknologi MARA Cawangan Terengganu Kampus Dungun, Sura Hujung, Dungun 23000, Terengganu, Malaysia
3 Faculty of Business and Management, Universiti Teknologi MARA Cawangan Melaka Kampus Alor Gajah, Alor Gajah 78000, Melaka, Malaysia
4 Faculty of Technical and Vocational Education, Universiti Tun Hussein Onn Malaysia, Parit Raja, Batu Pahat 86400, Johor, Malaysia
5 Kolej Komuniti Rompin, Kuala Rompin 26800, Pahang, Malaysia
* Correspondence: safuanlatip@uitm.edu.my

† Presented at the International Academic Symposium of Social Science 2022, Kota Bharu, Malaysia, 3 July 2022.

Abstract: Work ergonomics is considered one of the important issues in every organisation because it is related to employees' safety and health. The statistics recorded by the Department of Occupational Safety and Health (DOSH), Malaysia, showed an increasing number of accidents in the workplace related to the hotel and tourism industry. Thus, this study aims to examine the relationship between work ergonomics, including physical ergonomics, organisational ergonomics, and cognitive ergonomics, with employee performance. To conduct this study, the researchers will collect data from primary sources. The researchers will randomly distribute the questionnaire to those working in the hotel and tourism industry in Malaysia. All data will be analysed using Smart PLS. The study findings will be clarified using frequency analysis, descriptive statistics, correlation, and multiple regression analysis.

Keywords: work ergonomics; physical ergonomics; organisational ergonomics; cognitive ergonomics

Citation: Latip, S.N.N.A.; Latip, M.S.A.; Tamrin, M.; Nawi, M.Z.M. The Perspective of Work Ergonomics on Employee Task Performance in Hotel and Tourism Industry, Malaysia. *Proceedings* **2022**, *82*, 7. https://doi.org/10.3390/proceedings2022082007

Academic Editor: Mohamad Rahimi Mohamad Rosman

Published: 7 September 2022

1. Introduction

The workforce is the most valued asset of any organisation, and therefore, work environment conditions are critical to optimising productivity and performance. In addition, employee performance and productivity are now two of the most significant issues that an organisation's human resource department must handle. Employee performance is critical for an organisation since it ensures that the organisation's goals are achieved. Performance may be defined as the behaviour or actions that employees engage in, in accordance with their duties and roles, for the organisation's goals to be met [1]. The factors that influence productivity and performance among employees can be classified into two main categories: management-driven factors, which include developing an organisational plan such as division of responsibilities at all levels, having excellent working hours with enough time for rest, absence coverage, safety and health policies, advancing working system, and providing training [2]. In addition, the second category is related to workplace elements including workplace design, machinery and tools, temperature, lighting, humidity, and noise.

Ergonomics is the scientific study of the interactions between humans and other elements of a system to improve human well-being and overall system performance. The term ergonomics is derived from the Greek terms *ergon*, which means work, and *nomos*, which means rules; thus, ergonomics may be defined as the law of work [3]. Ergonomics is the study of equipment and jobs that are designed to be suitable for human capabilities and limitations. Ergonomics, therefore, matches the job to the individual, whether in the workplace or with a consumer product, and it provides opportunities for business by increasing human well-being, lowering costs, improving quality, and increasing productivity [4].

Dr Paula Cenni, a European Ergonomist, defined Ergonomics as a science of work which relates to a culture and a method beneficial for the development of a man-centred labour system aimed at psychophysical welfare and at safety and production efficiency [5].

Based on occupational accident statistics reported by the Department of Occupational Safety and Health, the number of occupational accidents in Malaysia rose year-by-year in all industries, including the manufacturing sector, hotel and tourism sector, utility sector, public service, and statutory authorities. Meanwhile, Social Security Organisation (SOCSO) statistics in 2013 showed that the number of cases related to ergonomics reached 694 from 2630 reported cases [6].

In addition, the Chairman of the National Institute of Occupational Safety and Health (NIOSH) stated that the number of occupational musculoskeletal disorder (MSD) cases in Malaysia is increasing year-by-year. For example, in 2005, there were only 10 MSD cases reported; however, in 2014, the number of cases reported were 675. In just 9 years, the number of cases has risen dramatically. The Social Security Organisation (SOCSO) showed that claims related to ergonomics were also higher than other occupational diseases.

As a developing country, Malaysia views human resources as a national asset, and national growth occurs when each employee's contribution to the economy is significant and beneficial. Like other developing countries, safety and health issues are important to maintaining a safe working environment and employees' safety while performing their tasks. Thus, the Malaysian government enacted a law in 1994 called the Occupational Safety and Health Act (OSHA 1994), which underlines the employer and employees' role and responsibilities to create a positive, safe, and healthy environment in the workplace. OSHA 1994 is designed to ensure employees are not abused, which can lead to an accident, disease, or illness.

As the specific Occupational Safety and Health Act 1994 has mentioned, the work environment should be secure and conducive to fulfilling the physiological and physical needs of the employees. Furthermore, due to employers' lack of concern and serious intervention on ergonomic issues, workplace illnesses and personal accidents are consistently high. Ergonomics have a lower priority and are less likely to be followed by many people because they are time-consuming, costly, and considered difficult for management to handle with the current economic situation [7]. However, employees also have the right to work in a secure and healthy workplace. Meanwhile, better working circumstances help to meet the demands of employees and increase their performance. Furthermore, it can reduce the number of absent employees due to health problems caused by a lack of workplace safety, and the business activities can run smoothly. Organisations must have training on the benefits of providing ergonomically appropriate facilities in their workplace. The more knowledge that the employer receives, the more protection they can have for their employees and assist them in remaining healthy, and they can possibly help achieve the best results from employees [8].

Human factors are highly complicated, especially in terms of technology. Ergonomics has received little debate and study among researchers, leading to a lack of awareness among the general community. Furthermore, surveys on occupational safety and health issues in the hotel and tourism industry are limited in Malaysia. Today, a growing number of cases of ergonomic illness is being recorded by the Department of Occupational Safety and Health (DOSH), and the study regarding it is crucial. Indirectly, it has a negative impact on employee morale, and employers will face an increasing cost-of-compensation problem.

Therefore, it is critical to understand what types of work ergonomics could potentially affect employees' performance. This includes: 1. identifying the relationship between physical ergonomics and employees' performance; 2. determining the relationship between organisational ergonomics and employees' performance; and 3. examining the relationship between cognitive ergonomics and employees' performance. Thus, the study focuses on the review of the work ergonomic perspective on employee task performance in the hotel and tourism industry as a conceptual analysis.

This model was developed by [9] and used mainly by previous researchers to study work environment and performance. The proposed developed the habitability pyramid to demonstrate the relationship between comfortable environmental work and physical comfort, functional comfort, and, lastly, psychological comfort (Figure 1). Physical comfort covers basic human needs such as security, cleanliness, and accessibility, all of which are required for a high standard of life. Furthermore, functional comfort is frequently defined as ergonomic assistance for people when doing work-related responsibilities and activities, such as lighting and meeting space. Meanwhile, psychological comfort stems from feelings of identification, ownership, and control over one's employment.

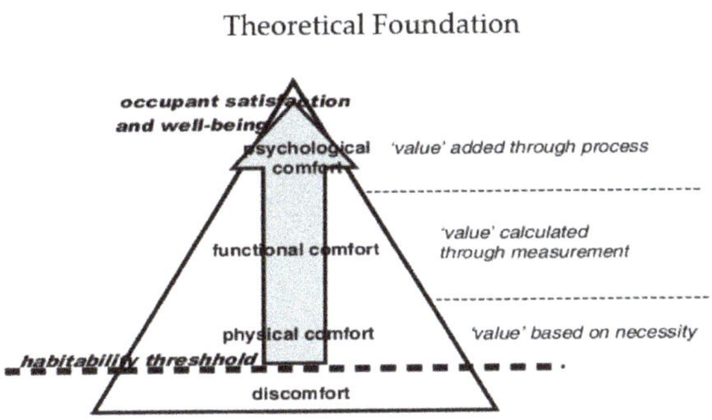

Figure 1. The habitability pyramid.

2. Literature Review

2.1. Work Ergonomics

Ergonomics is a combination of two Greek words, *Ergon* (work) and *Nomos* (natural laws). Ergonomics is defined as a holistic strategy where physical, cognitive, social, organisational, environmental, and other associated variables are considered when designing and evaluating tasks, products, settings, and systems to fully match employees' needs, skills, and limits [10]. From the Occupational Safety and Health Act (OSHA) perspective, ergonomics is a process where jobs are designed based on workplace layout, work setting, and workflow that is suitable for employees rather than being enforced on the employees. To align and meet the needs, capacities, and limitations in the workplace, ergonomics can be described as a holistic approach that considers physical, cognitive, social, organisational, environmental, and other associated factors [10].

A proper workplace design through ergonomics can help protect employees from any illness and injuries such as muscle pain or cumulative trauma disorder (CTD). Employee safety is inevitably crucial for every organisation because every country has its Occupational Safety and Health Act that employers must comply with to ensure the safety of employees in the workplace. According to [11], if employers do not adequately care for their employees, several safety and health issues can arise, which may cause unnecessary long-term stress among employees. For example, an office design that forces workers to sit in uncomfortable positions for an extended time may decrease commitment and increase lethargy and dissatisfaction among employees. Consequently, this may affect employees' bodies and nervous systems that ensures the plausibility of a musculoskeletal disorder (MSD) diagnosis.

The philosophy of ergonomics is concerned with recognising the relationship between people and other elements within a system, and developing theories, concepts, and strategies to accommodate workers, which would allow companies to fully optimise employee productivity and overall system efficiency [7]. One of the main objectives of ergonomics is to increase the level of employee task performance in the workplace. According to Taylor's theory, an organisation's management is responsible for designing job assignment and maintaining workers' safety and comfort to work efficiently in the workplace while also reaching a higher degree of productivity [12]. Taylor opines that one of the fastest ways to accomplish results is by investing in the right individuals and equipment.

According to [7], ergonomics concerns about interactions between people and other elements of a system and the formulation of theory, principles, and data for workers to increase human capital and system performance. This concept involves broader and all-encompassing use of structures that consider the possible attainable domain of specialisation within the area of ergonomics. The three domains of specialisation are physical, organisational, and cognitive ergonomics, which are the main focus of this study.

2.2. Physical Ergonomics

Physiological ergonomics incorporates human anatomical, anthropometric, neurological, and biomechanical factors linked to physical activity [7]. Physical ergonomic factors such as lighting, noise, temperature, and humidity can affect employees' productivity and performance. This statement aligns with a previous study by [13], stating that physical ergonomics, especially layout and design, influence employees' actions in the workplace and have a positive relationship with employees' performance. An excellent physical work environment will help to boost employees' performance and improve their productivity [14]. Physical ergonomics is important not just for employee performance, but also for their well-being, social relationships, and health.

Besides that, a good and safe work environment can help make a business more competitive in their industry. In the working environment concept, task performance and physical ergonomics are inextricably linked with the working environment [15]. Problems related to physical ergonomics will arise almost every day for quite a long time.

Additionally, it is crucial to promote and develop the ergonomics concept in various industries in Malaysia to help increase the performance and safety of workers in the workplace [16]. Most researchers refer to Herzberg's two-factor theory on "hygiene factors or job context", which emphasises the importance of 'work environment' and 'organisation policy' in helping people understand the relationship between physical space, ergonomics, and employee performance [17].

Previous research by [18] showed that physical work ergonomics has a positive and significant effect on Ibis Style Bali Denpasar Hotel employees. In addition, physical ergonomics (office layout, lighting, furniture, and equipment) also showed a positive and significant relationship with employees in a government agency in Kedah, Malaysia [19].

According to [20], who conducted a study on workplace ergonomics and academic staff performance in the College of Education at Umm Al-Qura University in Mecca, physical ergonomics has a positive relationship with employees' performance. This proved that if an organisation creates and provides an excellent ergonomic workplace, people will perform better while executing their tasks. As a response, the following hypothesis is developed to be tested in future study:

H1. *There is a significant relationship between physical ergonomics and employee task performance.*

2.3. Organisational Ergonomics

Macroergonomics is an organisational study that focuses on harmonising work systems at both the macro- and microlevels [10]. In ergonomics, the term "compatibility" has been described as "a fit" in an unstructured manner. Organisational ergonomics focuses on organisational structures, policies, and procedures to figure out how to improve the sociotechnical system [10], indicated that communication, job design, virtual organisation, teamwork, organisational culture, and quality management are all related to organisational ergonomics. The fundamental goal of organisational ergonomics is to adjust the organisation's whole work practice and working system to accommodate existing activities and available capabilities, talents, and individual constraints. It aims to assure the efficiency, convenience, and safety of the workforce.

Organisational ergonomics uses participatory ergonomics by including individuals in ensuring that all key aspects of their work environment are controlled, with adequate competence and authority to drive all procedures to achieve desired results. Based on [21], only 18 case studies were chosen out of 166 projects that went through a huge multinational company's vetting process. It was discovered that adopting an organisational ergonomics technique lowered the probability of work-related musculoskeletal disorders while increasing activity levels. This implies that there is a strong link between organisational ergonomics and employee performance.

According to a previous study [22], communication (one of the variables in organisational ergonomics) had a significant positive effect on employees' performance in Hotel Puri Saron, Seminyak, Kuta, Bali. In addition, another study [23] also showed that communication had a significant and positive relationship with workers' performance in North Sumatra Province. It can therefore be concluded that communication is an important variable that must be a priority for organisations to develop good organisational ergonomics in their workplace.

In addition, past research by [24] indicated that 73.5% of employees' performance is influenced by motivation and leadership (variables of organisational ergonomics) in Wahana Resources Ltd. North Seram District, Central Maluku Regency, Indonesia. This means that organisational ergonomics is crucial for every organisation included in the hotel and tourism industry to ensure the performance of employees will increase and directly affect business success. Thus, the following hypothesis is developed to be tested in future studies:

H2. *There is a significant relationship between organisational ergonomics and employees' performance.*

2.4. Cognitive Ergonomics

Cognitive ergonomics is the study of workers' mental processes and how they interact with other people and their electronic environments, emphasising the examination of individual cognitive processes and the limits of those processes [25]. According to [26], cognitive ergonomics is the study of cognition work and organisational settings to enhance human well-being and machine efficiency.

Moreover, components of cognitive ergonomics include logical reasoning, perception, motor reaction, and the relationship and interaction of other human beings in the workplace [10,27]. Additionally, training, workload, workplace tension, decision-making, and the human–system relationship are also variables related to cognitive ergonomics. However, there is still a lack of study that is directly focused on the relationship between cognitive ergonomics and workers' performance among researchers.

Previous studies revealed a significant relationship between training and employees' performance in the hotel industry [28,29]. In addition, training was also found to have a positive relationship with employees' performance in the oil and gas industry [30,31]. This means that one of the variables related to cognitive ergonomics is training, which considered an important factor that helps to boost employees' performance and must be properly planned by human resources management in every organisation regardless of the size of business and industry types. When employees have proper training to conduct their tasks, this helps to reduce their workload that would otherwise be large due to a lack of idea or skill to make it.

Another work of research, [32], showed that one of the main factors that influences and has a significant effect on employees' performance in the hotel industry is workload. The workload is one of the main variables related to cognitive ergonomics in an organisation. Task complexity, complementation of task in a particular period, and effort needed to complete the task all contribute to the workload of a person. Thus, an organisation needs to ensure that workload is matched with the individual capacity to ensure they can perform well and increase their performance. However, this study was terminated by [32] since it revealed that workload has a negative impact on performance among academics in Turkey. Thus, the following hypothesis is developed to be tested in future study:

H3. *There is a significant relationship between cognitive ergonomics and employee performance.*

3. Conclusions and Discussion

Based on findings made during the literature review, many studies have been conducted to observe the relationship between physical ergonomics and employees' performance from an employee's perspective. However, there is a lack of study regarding two other main variables related to work ergonomics, which are organisational ergonomics and cognitive ergonomics. The main problem of injuries and accidents that happened in the workplace is related to work ergonomics that have a negative effect on employees' performance. In addition, work ergonomics can also lead to mental health, bone disease, and muscle issues for employees, which can have a negative effect on employees' performance. The researchers considered it necessary to examine and assess the knowledge of what sorts of work ergonomics are affecting employees' performance, due to numerous concerns that have arisen in the workplace relating to employees' safety and health, including in the hotel and tourism industry. Furthermore, future research should consider a comparative study with other countries to identify the extent to which ergonomics affects the performance of workers in this industry and includes the mediator variables, such as employee job satisfaction, to better understand the employees' perspective regarding work ergonomics.

The Occupational Safety and Health Act 1994 is a piece of Malaysian legislation that was gazetted in 1994 to ensure safety and health for workers in Malaysia. As a basis, Malaysian businesses and industries must obey the act's regulations. Thus, employers must ensure they provide a working environment free of risks that might threaten their employees' safety and health. Some examples and conditions that employers and management in the organisation must implement in their workplace include dividing tasks according to employees' ability, ensuring the workplace environment is in good condition, and providing training and supervision to guarantee workplace safety and health. This is very crucial because it can affect employees' performance and, in the end, can lead to a negative effect on business activities and efficiency. In addition, employees must also obey all regulations regarding safety and health in their workplace to avoid facing any unwanted issues while conducting their tasks.

Author Contributions: S.N.N.A.L. conceptualisation, data curation, formal analysis, funding acquisition, and writing original drive. M.S.A.L. funding acquisition, methodology, supervision, software, validation, and writing original drive. M.T. funding acquisition, writing (review and editing), and resources. M.Z.M.N. funding acquisition, writing (review and editing), and resources. All authors have read and agreed to the published version of the manuscript.

Funding: The registration fees are funded by Pembiayaan Yuran Prosiding Berindeks (PYPB), Tabung Dana Kecemerlangan Pendidikan (DKP), Universiti Teknologi MARA (UiTM), Malaysia.

Institutional Review Board Statement: Not applicable.

Informed Consent Statement: Not applicable.

Data Availability Statement: Not applicable.

Conflicts of Interest: The authors declare no conflict of interest.

References

1. Priansa, D. *Perencanaan & Pengembangan SDM*; Alfabeta: Bandung, Indonesia, 2014.
2. Gerald, C. *Personnel and Human Resource Management*, 5th ed.; Continuum Publisher: London, UK, 2002.
3. Schelussel, A.; Maykel, J. Ergonomics and musculoskeletal health of surgeon. *Clin. Colon Rectal. Sur.* **2019**, *32*, 424–434. [CrossRef] [PubMed]
4. MacLeod, D. *The Ergonomics Edge: Improving Safety, Quality, and Productivity*, 1st ed.; Van Nostrand Reinhold: New York, NY, USA, 1995.

5. Mohamed, N.; Tuah, S.; Othman, A. The dark sides of workstation: Ergonomics aspects and occupational diseases towards job performance at General Hospital in Malaysia. *J. Eng. Health Scin.* **2020**, *4*, 149–160.
6. Boon, P. Ergonomics-Related Injury Cases on the Rise'. Available online: https://www.theborneopost.com/2017/11/20/ergonomics-related-injury-cases-on-the-rise/ (accessed on 20 November 2017).
7. Ismaila, S. A study on ergonomics awareness in Nigeria. *Aust. J. Basic Appl. Sci.* **2010**, *4*, 731–734.
8. Limerick, R. Participatory ergonomics: Evidence and implementation lesson. *Appl. Ergon. J.* **2018**, *68*, 289–293. [CrossRef]
9. Vischer, J. The effects of the physical environment on job performance: Towards a theoretical model of workspace stress. *Stress Health* **2007**, *23*, 175–184. [CrossRef]
10. International Ergonomic Association. *Definition and Domains of Ergonomics*; Retrived; IEA: Santa Monica, CA, USA, 2016.
11. Cumulative Trauma Disorder. Available online: https://www.safetyandhealthmagazine.com/articles/cumulative-trauma-disorder (accessed on 1 June 2013).
12. Tietjen, M.; Myers, R. Motivation and job satisfaction. *Manag. Decis.* **1998**, *36*, 226–231. [CrossRef]
13. McGuire, D.; McLaren, L. The impact of physical environment on employee commitment in call center: The mediating role of employee well-being. *Team Perform. Manag.* **2009**, *15*, 35–48. [CrossRef]
14. Samson, G.; Waiganjo, M.; Koima, J. Effect of workplace environment on the performance of commercial banks employees in Nakuru Town. *Int. J. Manag. Stud. Res.* **2015**, *3*, 76–89.
15. Nanzushi, C. The Effect of Workplace Environment on Employee Performance in the Mobile Telecommunication Firms in Nairobi City County. Master's Thesis, University of Nairobi, Nairobi, Kenya, 2015.
16. Loo, H.; Richardson, S. Ergonomics issues in Malaysia. *J. Soc. Sci.* **2012**, *8*, 61–65. [CrossRef]
17. Herzber, F. *One More Time: How Do You Motivate Employees?* Harvard Business School: Boston, MA, USA, 1987.
18. Indah, N.; Riana, I. The influence of physical work environment and organizational culture on work motivation and employee performance at Ibis Styles Bali Denpasar Hotel. *Am. J. Humanit. Soc. Sci. Res.* **2020**, *4*, 43–51.
19. Hamidi, N.; Mansor, F.; Hashim, M.; Muhammad, N.; Azib, W. The relationship between physical workplace environment and employees' performance. *J. Contemp. Soc. Sci. Res.* **2020**, *4*, 56–67.
20. Alzahrani, N. Workplace ergonomics and academic staff performance in College of Education in Umm Al-Qura University in Makkah. *Am. J. Educ. Res.* **2019**, *7*, 604–617. [CrossRef]
21. Larson, N.; Wick, H.; Hallbeck, S.; Vink, P. Corporate ergonomics programs: Identifying value through a company award process. *IIE Trans. Occup. Ergon. Hum. Factors* **2015**, *3*, 9–23. [CrossRef]
22. Idayanti, E.; Ayu, D.; Piartrini, P. The effects of communication, competency, and workload on employee performance in Hotel Puri Saron, Seminyak, Kuta, Bali. *Am. J. Humanit. Soc. Sci. Res.* **2020**, *4*, 29–37.
23. Prayogi, M.; Taufik, M.; Siregar, L. The influence of communication and work discipline to employee performance. *Adv. Soc. Sci. Educ. Humanit. Res.* **2019**, *343*, 423–426.
24. Paais, M.; Pattiruhu, J. Effect of motivation, leadership, and organizational culture on satisfaction and employee performance. *J. Asian Financ. Econ. Busine* **2020**, *7*, 577–588. [CrossRef]
25. Carayon, P.; Karsh, B.; Gurses, A.; Holden, R.; Hoonakker, P.; Hundt, A.; Montague, E.; Rodriguez, J.; Wetterneck, T.B. Macroergonomics in healthcare quality and patient safety. *Sage J.* **2013**, *8*, 4–54.
26. Christy, V.; Duraisamy, S. Ergonomics and employee psychological wellbeing. *Int. J. Manag.* **2020**, *11*, 435–438.
27. Zolotova, M.; Giambattista, A. Designing cognitive ergonomics features of medical devices. Aspects of cognitive interaction. *Des. J.* **2019**, *22*, 463–474. [CrossRef]
28. Khalaf, S.; Morsy, M.; Ahmed, G.; Ali, N. Impact of effective training on employee performance in hotel establishments. *J. Fac. Tour Hotel Fayoum Univ.* **2016**, *10*, 92–109.
29. Lin, L.; Yang, L. A Study of the effect of perceived training and development on employee performance at selected hotels in Sarawak. *J. Penyelid Sains Sos.* **2020**, *12*, 12–24.
30. Karimi, S.; Nejad, A. The effect of organizational training on jab satisfaction and individual performance of Ahwaz Oil Company employees. *Rev. Publicando* **2018**, *5*, 385–401.
31. Naji, G.M.A.; Isha, A.S.N.; Alzoraiki, M.; Sharafuddin, O.; Al-Mekhlafi, A.A. Enhancement of employees performance via professional training and development: A study on oil and gas companies operating in Yemen. *Int. J. Sci. Technol. Res.* **2020**, *9*, 934–940.
32. Akca, M.; Kucukoglu, T. Relationships between mental workload, burnout, and job performance: A research among academicians. In *Valuating Mental Workload for Improved Workplace Performance*; IGI Global: Hershey, PA, USA, 2020; pp. 49–68.

proceedings

MDPI

Proceeding Paper

The Mediating Role of Perceived Constraints between Human–Pet Relationships and Willingness to Travel with Pets: A Theoretical Framework †

Wei Ying Chong [1,*], Xin Hui Choong [2] and Toong Hai Sam [3]

1 Taylor's Business School, Taylor's University, Subang Jaya 47500, Malaysia
2 Graduate School of Business, SEGi University, Petaling Jaya 47810, Malaysia
3 Faculty of Business and Communications, Inti International University, Nilai 71800, Malaysia
* Correspondence: weiying.chong@taylors.edu.my
† Presented at the International Academic Symposium of Social Science 2022, Kota Bharu, Malaysia, 3 July 2022.

Abstract: Many modern families involve pets in most of their daily social activities. Yet, many of them do not involve pets when undertaking tourism activities. Drawing upon theories of planned behaviour and leisure constraint models, this study intends to examine how perceptions of their relationship with pets and perceived constraints may influence the willingness of owners to travel with pets. Focusing on reviewing multidisciplinary literature on human–pet relationship-oriented leisure trip activities with a method of theoretical analysis, this paper proposed an integrated theoretical framework based on recognized and synthesized patterns and similarities within the literature.

Keywords: theory of planned behaviour; leisure constraint model; human–pet relationship; perceived constraints; willingness to travel with pets

Citation: Chong, W.Y.; Choong, X.H.; Sam, T.H. The Mediating Role of Perceived Constraints between Human–Pet Relationships and Willingness to Travel with Pets: A Theoretical Framework. *Proceedings* **2022**, *82*, 13. https://doi.org/10.3390/proceedings2022082013

Academic Editor: Mohamad Rahimi Mohamad Rosman

Published: 8 September 2022

1. Introduction

The role of pets in modern families is undergoing growing changes [1,2]. Pets are playing an ever more significant role in the lives of humans as close companions that breaks the traditional view of human–animal relations [1]. The phenomenon has undoubtedly been accentuated by the growth of the middle class, aging populations and shifting attitudes toward human–pet bonding. In modern families, pet owners involved pets in most of their social activities [3]. Indeed, many of them view their pets as "animal-companions with needs, wants, and rights comparable to those of other family members" [4] (p. 549). An owner's intention to have their furry family members as companions during holidays is high, although the actualization of the dream is rather low [5].

Many studies have been conducted to investigate pet owners' travel intentions with their pet and constraints they have encountered over the last decade [5]. Nevertheless, little has been achieved in terms of investigating human–pet relationships and the pet owner's travel intentions with their pet [6]. This study aims to address the research gap by providing an integrated framework that examines the mediating role of perceived constraints between human–pet relationships and willingness to travel with pets. A clear theoretical research framework can uncover potential research avenues to bring pet owners' perspectives and experiences to the core of human–pet relationship research.

2. Literature Review

This section discusses the relevant literature of the study. It starts with a discussion on the underlying literature on human–pet relationships, follow with an overview of previous studies on perceived pet-related tourism constraints. Then, it examines the willingness of owners to travel with pets. Finally, the section concludes with elaborations about how a theory of planned behaviour and a leisure constraint model support the proposed theoretical framework.

2.1. Human–Pet Relationship

Pet ownership is common in this modern living era, which is evidently seen with tremendous increments in pet ownership around the world. Since the 1990s, the term "pet" has gradually replaced the term "companion animal", acknowledging its value derived from the relationship instead of emphasizing their practical or economic benefits [2]. Pets do not merely provide companionship; the unconditional love and support given to their owners were perceived as sources of emotional support that fulfil both physical and mental health benefits [1].

The important function of pets [7] has explained why pet ownership is welcomed despite the costs. Previous studies have shown that pets are conceptualized as social facilitators that help people in establishing friendships, and they are well known for reducing the chronic stress of loneliness, anxiety and depression in both children and adults [7]. Furthermore, it is also observed that the intensity of pet owners' grief after their pets died is very similar to their grief for the death of a close human [8], suggesting the human–pet relationship is an authentic and meaningful emotional bond. In this study, the human–pet relationship refers to "accompanying and associating with one's pet and the relationship between the owner and the pet that results from such interaction" [9].

Human–pet relationships have been studied in various fields of knowledge since decades ago, where pets are recognized as family members [10]. Pets served as a secure base which led their owners to pursuing activities [7], risk taking [11] and exploring the world confidently [12]. In addition, pets are safe havens where they readily offer their owners with love, support, relief, and comfort whenever they are needed [12]. Indeed, pet owners feel close to their pets. They enjoy this closeness [1].

Many pet owners consider travelling with their pets, yet in reality not many of them involve pets when undertaking tourism activities [5]. Many studies have been conducted to understand the phenomena [6,13,14]. Ref. [13] have shown that the human–pet relationship will have impact on an owner's intention to travel with a pet. Those traveling with their pets are more likely to see pets as members of the family [5]. Thus, we proposed:

Hypothesis 1. *Human–pet relationships will have a positive impact on willingness to travel with a pet.*

2.2. Perceived Pet-Related Tourism Constraints

The idea of taking pets along on leisure activities was first studied by Greenebaum [15] after observing the increasing degree of attachment pet owners had towards their pets. Greenebaum [15] investigated pet owners' behaviours and the decision processes that they considered when taking their pets along to participate in leisure activities tailored for pets. The study suggested a few elements that pet owners should evaluate before making decisions, including a pet's preferences and its abilities, as well as the venue environment and other participants' characteristics [15]. Miller and Howell [16] further studied the reaction of other participants when they see pets participating in leisure activities where the participant's resentments are the primary focus of the study. These resentments could be due to a previous undesirable experience, such as being chased or bitten by pets, and irresponsible owners who failed in cleaning up their pet's waste [16]. However, the sample sizes of these two studies were small and the activities rather routine, despite lacking general applicability, and yet they provided useful insight for further research.

As per guidance, it is essential for owners to contemplate pet-related tourism constraints, such as their pet's physical abilities, the influences of their pets that might impact on other people as well as the extra time and expenses involved [17]. Apart from the mentioned aspects, both owners and pets should undertake some training activities prior to participating in tourism activities, to shape the owner's leisure participation behaviour [18]. Both studies had narrowed the gaps in tourism literature significantly.

In this study, perceived pet constraints refer to "factors that inhibit pet owners from including their pets in tourism activities and the obstacles that owners encounter when they decide to include their pets" [19]. Chen et al. [19] recognized pet constraints as reasons that deter a pet owner's decisions on whether to include their pets in tourism activities; these pet constraints are then categorized into three categories: a pet's specific constraints, a pet's interpersonal constraints, and a pet's structural constraints, by altering Crawford and Godbey's [20] typology. This study also revealed that pet owners are likely to be demotivated to include pets in tourism activities if they perceive that they cannot overcome the constraints. Chen et al. [19] also proposed in their study that perceived constraints have discouraged and negatively influenced owners' intentions to travel with a pet. Prior studies have revealed that the roles of perceived constraints have an influence on pet owners' travel intentions and have been well documented in the literature on human–pet interaction [5,21]. In conclusion, pet constraints will have an impact on an owner's intentions to travel with a pet. Thus, we proposed:

Hypothesis 2. *Human–pet relationships will have a positive impact on perceived pet-related tourism constraints.*

Hypothesis 3. *Perceived pet-related tourism constraints will have a negative impact on an owner's intention to travel with a pet.*

2.3. Theory of Planned Behaviour

Most human behaviour is performed under one's own control, with rational and intention [22]. The propensity to behave sensibly is described as behavioural intention. Ajzen [22] has explained this phenomenon via the theory of planned behaviour (TPB) model. He further pointed out that behavioural intention is determined by attitude, subjective norms, and perceived behavioural control [22].

Attitude refers to the psychological emotion and the positive or negative assessment that an individual conducted when engaging in certain behaviours [23], as illustrated in the TPB model. In other words, his/her behavioural intention will be more positive when he/she has a more positive attitude, and vice versa. On the other hand, subjective norms refer to the degree of social pressure felt about their behaviour, and the perceived opinions of those who are deemed important to an individual, capable of influencing his/her decision making [22]. There is a positive relationship between subjective norms and behavioural intention [24], indicating that the more positive subjective norms other people perceived, their stronger their behavioural intention. Meanwhile, Ajzen (1991) defined perceived behavioural control as the perception of the possible difficulties that arise from various factors beyond one's control, such as time, money, chance, etc., that an individual has while performing a specific behaviour. Therefore, it could be understood as the likelihood of an individual in performing a specific behaviour is higher when he/she has more control over the opportunities and resources. In summary, all these three factors are illustrated in the TPB model in predicting an individual's behavioural intentions [22].

Peng et al. (2014) [24] applied the theory of planned behaviour and acknowledged that a pet owner's attitudes, subjective norms, and perceived behavioural control impacted their intentions in including pets in leisure travel. The applicable theory of planned behaviour in this model is further supported by Moghimehfar et al. [25]. Moghimehfar et al. [25] have included perceived constraints in the theory of planned behaviour, and revealed that perceived constraints have an effect on intention

2.4. Leisure Constraint Model

Leisure constraints act as barriers to participation in leisure activities, which leads to non-participation whenever constraints are encountered [26]. Therefore, a holistic view of constraints in participating in leisure activities is required within the growing travel and tourism industries and the increasing number of research studies upon this topic of interest is evidently seen. It began with the leisure constraints model developed by Crawford and Godbey [20] and further elaborated by Crawford et al. [27], which explains the underlying constraints inhibiting an individual's desire for participation by categorising three dimensions: intrapersonal constraints, interpersonal constraints, and structural constraints.

Crawford and Godbey [20] described how intrapersonal constraints arise when a participating decision is purely affected by individual psychological states that influenced preference, leading to non-participation; this could be due to anxiety, religiosity, stress, and perhaps lack of interest. The fact that intrapersonal constraints are derived from psychological state, they are considered relatively unstable and may alter within a short time frame [28].

In contrast to intrapersonal constraints, interpersonal constraints interact with both preferences and participation, and they occur due to the unavailability of other people (friend, family member, or partner), preventing an individual from participating when a strong preference for a at least one co-participant is required in the activities of interest [20]. Interpersonal constraints are probably due to change across different life stages, which are highly dependent on marital status, family size, and types of activities [28].

Lastly, structural constraints are the prevailing factors between leisure preference and participation [20]. Lack of budget, time, and opportunity, or even bad weather could be contributors to structural constraints [28]. Apart from describing the dimensions embedded in leisure constraints, the leisure constraints model also suggested that they are hierarchical in nature, therefore intrapersonal constraints must be negotiated first, followed by interpersonal constraints and structural constraints [27].

As guided by the hierarchical model of leisure constraints established by Crawford, Jackson and Godbey [27], detailing intrapersonal, interpersonal, and structural constraints, a pet-related tourism constraints model was proposed by Chen et al. [19], where the constraints were modified for the following three categories: a pet's specific constraints, a pet's interpersonal constraints and a pet's structural constraints. These modifications were necessary because the role of pets is to be included in examining the possible constraints, rather than focusing solely on pet owners. Therefore, a pet's specific constraints refer to a pet's abilities; a pet's interpersonal constraints describe other participants' perceptions and reactions towards pets, and lastly a pet's structural constraints indicate the extra resources required to travel with pets. Thus, we proposed:

Hypothesis 4. *Perceived constraints mediating the human–pet relationship and willingness to travel with pet.*

3. Theoretical Framework

The preceding discussion reveals a theoretical framework with hypotheses capturing the formation of pet owners' intentions to travel with pets (Figure 1). This framework is an extension of the theory of planned behaviour model and leisure constraint model, wherein the human–pet relationship and perceived pet-related tourism constraints are incorporated to capture how these two factors can impact pet owner willingness to travel with pets. Previous studies have shown that the theory of planned behaviour model [17] is a suitable model to examine the intention to travel with pets. Authors have used the theory of planned behaviour model and leisure constraint model in this study.

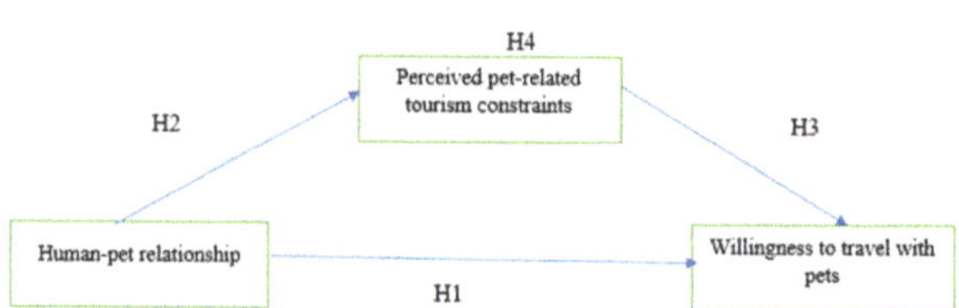

Figure 1. Theoretical framework with hypotheses.

4. Discussion

Figure 1 introduced a theoretical framework with hypotheses for modelling travelling with pets based on existing literature. It has emphasized the need to consider human–pet relationships when trying to understand pet–owner intentions to travel with pets. It is proposed that the relationship among these key variables can be understood in terms of the theory of planned behaviour model and leisure constraint model. In this study, the human–pet relationship is an emotional mechanism that facilitates pet-oriented leisure trip activities.

Scholars can empirically test the proposed framework by either using a quantitative research methodology approach, such as a questionnaire survey, or qualitative methods. They can also use mixed methods to compare the findings and provided useful insights for further research.

5. Conclusions

In conclusion, we have systematically reviewed and theoretically analysed the construct of the human–pet relationship, perceived pet-related tourism constraints, and willingness to travel with pets. We have drawn on different theoretical perspectives to develop an integrative conceptual framework to understand human–pet relationship-oriented leisure trip activities. A theoretical framework has been developed to describe the key components and relationships. We hope that our review provides a clear framework for future research on human–pet relationship-oriented leisure trip activities and will spur further scholarly investigation into these three constructs.

Author Contributions: Conceptualization, W.Y.C. and X.H.C.; methodology, T.H.S.; software, T.H.S.; validation, W.Y.C., X.H.C. and T.H.S.; formal analysis, X.H.C.; investigation, X.H.C.; resources, X.H.C.; data curation, X.H.C.; writing—original draft preparation, W.Y.C. and X.H.C.; writing—review and editing, W.Y.C.; visualization, X.H.C.; supervision, W.Y.C. and T.H.S.; project administration, X.H.C. All authors have read and agreed to the published version of the manuscript.

Funding: This research received no external funding.

Institutional Review Board Statement: Not applicable.

Informed Consent Statement: Not applicable.

Data Availability Statement: Not applicable.

Acknowledgments: Paper presented during the International Academic Symposium of Social Science 2022, organized, and hosted by the Faculty of Information Management, Universiti Teknologi MARA Kelantan Branch, Malaysia.

Conflicts of Interest: The authors declare no conflict of interest.

References

1. Allen, R.; McConnell, E.; Paige, L.; Humphrey, B.T. We are family: Viewing pets as family members improves wellbeing. *Anthrozoös* **2019**, *32*, 459–470.
2. Fudge, E. *Pets*; Routledge: Oxford, UK, 2016.
3. Stokes, L.; Wright, I. The Pet Travel Scheme (PETS) and parasite protection for the travelling pet. *Vet. Nurse* **2015**, *6*, 60–70. [CrossRef]
4. Holbrook, M.B. Pets and people: Companions in commerce? *J. Bus. Res.* **2008**, *61*, 546–552. [CrossRef]
5. Matijevic, J.; Kiatkawsin, K. Pet owners' travel intention and perceived constraints. In *Asia Pacific Tourism Association Annual Conference*; Asia Pacific Tourism Association: 2021; pp. 70–74. Available online: https://www.apta2021.org/_files/ugd/e5cd8f_a2a291e2ab904199954a0e73e6f1c834.pdf#page=75 (accessed on 14 May 2022).
6. Hung, K.P.; Chen, A.; Peng, N. Taking dogs to tourism activities: Incorporating attachment into a pet-related constraint-negotiation model. *J. Hosp. Tour. Res.* **2016**, *40*, 364–395. [CrossRef]
7. Bures, R.M. Integrating pets into the family life cycle. In *Well-Being Over the Life Course: Incorporating Human–Animal Interaction*; Springer: Berlin/Heidelberg, Germany, 2021; pp. 11–23.
8. Behler, A.M.C.; Green, J.D.; JoyGaba, J. We lost a member of the family. Predictors of the grief experience surrounding the loss of a pet. *Hum. Anim. Interact. Bull.* **2020**, *8*, 54–70.

9. Dotson, M.J.; Hyatt, E.M. Understanding dog–human companionship. *J. Bus. Res.* **2008**, *61*, 457–466. [CrossRef]
10. Rauktis, M.E.; Rose, L.; Chen, Q.; Martone, R.; Martello, A. Their pets are loved members of their family: Animal ownership, food insecurity, and the value of having pet food available in food banks. *Anthrozoös* **2017**, *30*, 581–593. [CrossRef]
11. Trigg, J.; Thompson, K.; Smith, B.; Bennett, P. Archetyping relationships with companion animals to understand disaster risk-taking propensity. *J. Risk Res.* **2019**, *22*, 475–496. [CrossRef]
12. Hayden-Evans, M.; Milbourn, B.; Netto, J. Pets provide meaning and purpose: A qualitative study of pet ownership from the perspectives of people diagnosed with borderline personality disorder. *Adv. Ment. Health* **2018**, *16*, 152–162. [CrossRef]
13. Chen, A.; Kuang-peng, H.; Peng, N. A cluster analysis examination of pet owners' consumption values and behavior–segmenting owners strategically. *J. Target. Meas. Anal. Mark.* **2012**, *20*, 117–132. [CrossRef]
14. Ying, T.; Tang, J.; Wen, J.; Ye, S.; Zhou, Y.; Li, F.S. Traveling with pets: Constraints, negotiation, and learned helplessness. *Tour. Manag.* **2021**, *82*, 104183. [CrossRef]
15. Greenebaum, J. It's a dog's life: Elevating status from pet to "fur baby" at yappy hour. *Soc. Anim.* **2004**, *12*, 117–135. [CrossRef]
16. Miller, R.; Howell, G.V.J. Regulating consumption with bite: Building a contemporary framework for urban dog management. *J. Bus. Res.* **2008**, *61*, 525–531. [CrossRef]
17. Chen, H.; Hung, K.P.; Peng, N. Planned leisure behaviour and pet attachment. *Ann. Tour. Res.* **2011**, *38*, 1657–1662. [CrossRef]
18. Hultsman, W.Z. Couple involvement in serious leisure: Examining participation in dog agility. *Leisure Studies* **2012**, *31*, 231–253. [CrossRef]
19. Chen, A.H.; Peng, N.; Peng, H.K. Developing a pet owners' tourism constraints scale—The constraints to take dogs to tourism activities. *Int. J. Tour. Res.* **2014**, *16*, 315–324. [CrossRef]
20. Crawford, D.W.; Godbey, G. Reconceptualizing barriers to family leisure. *Leis. Sci.* **1987**, *9*, 119–127. [CrossRef]
21. Carr, N.; Cohen, S. Holidaying with the family pet: No dogs allowed! *Tour. Hosp. Res.* **2009**, *9*, 290–304. [CrossRef]
22. Ajzen, I. The theory of planned behavior. *Organ. Behav. Hum. Decis. Processes* **1991**, *50*, 179–211. [CrossRef]
23. Eagly, A.H.; Chaiken, S. *The Psychology of Attitudes*; Harcourt Brace Jovanovich College Publishers: San Diego, CA, USA, 1993; pp. 1–794.
24. Peng, N.; Chen, A.; Hung, K.P. Including pets when undertaking tourism activities: Incorporating pet attachment into the tpb model. *Tour. Anal.* **2014**, *19*, 9–84. [CrossRef]
25. Moghimehfar, F.; Halpenny, E.A.; Walker, G.J. Front-country campers' constraints, negotiation, and pro-environment behavioral intention: An extension to the theory of planned behavior. *Leis. Sci.* **2018**, *40*, 174–193. [CrossRef]
26. Gu, Q.; Qiu, H.; King, B.E.; Huang, S. Understanding the wine tourism experience: The roles of facilitators, constraints, and involvement. *J. Vacat. Mark.* **2020**, *26*, 211–229. [CrossRef]
27. Crawford, D.W.; Jackson, E.L.; Godbey, G. A hierarchical model of leisure constraints. *Leis. Sci.* **1991**, *13*, 309–320. [CrossRef]
28. Nyaupane, G.P.; Andereck, K.L. Understanding travel constraint: Application and extension of a leisure constraints model. *J. Travel Res.* **2008**, *46*, 433–439. [CrossRef]

proceedings

MDPI

Proceeding Paper

Levels of Stress, Anxiety, and Depression in the Initial Stage of Movement Control Order in Malaysia: A Sociodemographic Analysis †

Mohd Khairi Ismail [1,*], Syamsulang Sarifuddin [2], Muhamad Zahid Muhamad [3,4] and Chamhuri Siwar [5]

1 Faculty of Business and Management, UiTM Cawangan Terengganu Kampus Dungun, Kuala Dungun 23000, Malaysia
2 Department of Business Studies, HELP University, Kuala Lumpur 50490, Malaysia
3 Faculty of Plantation and Agrotechnology, Universiti Teknologi MARA Cawangan Melaka Kampus Jasin, Merlimau 77300, Malaysia
4 Faculty of Agriculture, Universiti Putra Malaysia, Seri Kembangan 43000, Malaysia
5 Institute For Environment And Development (LESTARI), Universiti Kebangsaan Malaysia, Bangi 43600, Malaysia
* Correspondence: khairiismail@uitm.edu.my; Tel.: +60-1346-309-38
† Presented at the International Academic Symposium of Social Science 2022, Kota Bharu, Malaysia, 3 July 2022.

Abstract: Due to the COVID-19 epidemic, many have lost their source of income, causing them to be socially isolated and consequently limiting their participation in social events. This has led to the occurrence of mental health illnesses and has impacted the level of life satisfaction. This study assessed the effect of the introduction of movement control orders (MCOs) in Malaysia during the initial phase of the COVID-19 epidemic. Between 1 April 2020 and 30 May 2020, a Google form containing a questionnaire assessing socio-demographic information and the effects of MCO on mental health was made available to the public via email and a social media forum, and 762 responses were collected. Those who took the initiative to fill out the responses were therefore categorized as survey participants. City-dwelling Chinese-ethnic women between the ages of 18 and 25 who have mild to severe symptoms of sadness, anxiety, and stress were at the highest risk for developing mental health disorders, according to the study. The firms should be accorded appropriate consideration, acknowledgement, focus, and financial support. The goal of mental health programs for affected individuals should be to create a society free of the mental health difficulties, which can grasp the soul and mind and, eventually, improve the quality of life.

Keywords: mental health; movement control order; COVID-19

Citation: Ismail, M.K.; Sarifuddin, S.; Muhamad, M.Z.; Siwar, C. Levels of Stress, Anxiety, and Depression in the Initial Stage of Movement Control Order in Malaysia: A Sociodemographic Analysis. *Proceedings* **2022**, *82*, 14. https://doi.org/10.3390/proceedings2022082014

Academic Editor: Mohamad Rahimi Mohamad Rosman

Published: 8 September 2022

1. Introduction

Mental health is not merely health issue, but it can affect the well-being of living. This is incongruent with the definition by another piece of literature [1], which defines mental health as a state of well-being of individuals who are aware of the capabilities of themselves to withstand the pressures of life and, in turn, contribute to society and country. In fact, the global burden of diseases and disabilities is influenced by mental health, which can affect productivity and, in turn, hinder the economic development of a country [2].

Malaysia has carried out movement control orders to stop the COVID-19 virus from spreading. The impacts were substantial, as most of the people lost a source of income, causing social distancing and limiting social activities, which are among the major and apparent factors for the occurrence of mental health disorders and affect the well-being of life [3,4]. There is a scarcity of research in Malaysia on health and psychological health throughout the COVID-19 epidemic. Theoretically, the pandemic was projected to have both positive and negative effects on Malaysia's mental health [5]; therefore, it is substantial to be studied, as there is not yet a quantitative study on health and mental wellbeing in Malaysia's community [6].

Among the positive effects during the period of movement control expected by the author is that people at home are thus encouraged to spend time with family, which can increase the family's state of harmony and encourage a balanced life, which ultimately has a favorable effect on Malaysians' overall mental health. However, theoretically, the pandemic is expected to provide positive and negative effects on mental health in Malaysia [5]. Since there has not been a quantitative study that looks at the mental health of the Malaysian community during the COVID-19 pandemic, research on mental health in Malaysia is crucial [6]. This study obtained quantitative evidence of mental health problems using a Depression Anxiety Stress Scale 21 (DASS-21) analysis and further examined the relationship between mental problems and sociodemographic factors such as race, age, and gender. DASS-21, in both the Malay and English version, was used to assess mental health. The DASS has been used widely as an instrument that is recognized for determining symptoms of depression, anxiety, and stress [7]. In addition, the Depression Anxiety Stress Scale 21 (DASS-21) was employed in many cultures and other nations during COVID-19 as a screening tool to identify a person's level of depression, anxiety, and stress [8–10].

As a result, this study aimed to investigate the physical and mental health of the Malaysian society during the period of Movement Control Orders in order to combat the COVID-19 pandemic.

2. Literature Review

A pandemic known as the Novel Coronavirus 2019 (COVID-19) has had a tremendous impact on our societies, economies, health, and human behaviour [11]. The COVID-19 pandemic caused many countries to carry out sanctions and the closure of socio-economic activities [12]. This includes Malaysia, which implemented movement control directives to stop the COVID-19 virus from spreading. The impacts were substantial, as most of the people lost a source of income, causing social distancing and limiting social activities, which are among the major and apparent factors for the occurrence of mental health disorders and affecting the well-being of life [3,4].

Disease outbreaks are able to affect the state of mental health of the people, culture and environment. It is because this pandemic has been spreading quickly to the whole world that it has led to great fear, concern, and anxiety, especially to certain groups, e.g., the elderly people and people with comorbid disorders [13]. It has the potential to affect existing diseases and can lead to psychiatry-related symptoms, which may be related to impaired mental and interaction immunity [14].

3. Research Design

This study was quantitative in nature. This quantitative research only displays descriptive data. Using an internet questionnaire, the information was gathered through a survey. Before distributing questionnaires, it was required to establish the population to ensure that sampling could be undertaken. For this study, the population consisted of B40 and M40 household members earning less than RM9620 per year. This study used a purposive sampling strategy. A purposeful sample, also known as a judging or expert sample, is a nonprobability sample type. The fundamental objective of a purposive sample is to produce a statistically representative sample of the population. A purposive sample is one whose characteristics are specified for a study-relevant goal. Participants in the final sample represented 14 states in Malaysia (Table 1).

Table 1. Respondents' Profile.

Gender	%	Race	%
Male	53.9	Malay	68.2
Female	46.1	Chinese	14.7
State		Indian	12.3
Perlis	1.4	Others	4.7
Kedah	7.7	**Education level**	
Pulau Pinang	14.2	Doctor of Philosophy (PhD)	6.4
Perak	8.1	Master	13.8
Selangor	27.8	Bachelor's degree	29.7
Johor	9.2	Diploma	22.2
Negeri Sembilan	2.8	STPM/Certificate	4.9
Melaka	2.1	SPM/MCE	18.1
Pahang	3.1	PMR/SRP	3.0
Terengganu	3.1	UPSR/Completed Primary 6	0.8
Kelantan	5.5	No formal education	1.2
Sabah	3.7	**Employment sector**	
Sarawak	2.4	Government sector	32.0
W.P. Labuan	0.3	Private sector	46.3
W.P. Putrajaya	1.2	Self employed	21.7
W.P. Kuala Lumpur	7.3	**Category of income**	
Area		Based on hourly/daily/weekly	13.3
Urban	66.9	Based on monthly	69.7
Rural	33.1	Based on piece rate	17.1
Age		**Monthly salary**	
18–25 years old	15.7	RM580 and below	5.0
26–30 years old	14.2	RM580–RM980	6.3
31-40 years old	39.5	RM981–RM2614	27.6
41-60 years old	28.7	RM2615–RM4360	27.3
61 years old and above	1.8	RM4361–RM9619	33.9

From April 1 to May 30 of 2020, 762 replies in total were gathered throughout the data collection period. During the movement control order period, online surveys were used to perform the research for two months. The Google forms containing the study's questions were sent publicly via email and platforms for social media, like Facebook and WhatsApp. Those who took the initiative to fill out the responses were therefore categorized as survey participants. Thus, the final sample included respondents from 14 Malaysian states.

The online survey was distributed using Google Forms to contacts and contacts of contacts, in accordance with the snowball and simple sampling methods. Contacts were urged to widely distribute the survey to their networks. Studies were analyzed using a descriptive analysis and the Depression, Anxiety, and Stress Scale (DASS). The Depression, Anxiety, and Stress Scale (DASS) is a screening test for identifying a person's level of depression, anxiety, and stress. With this screening test, you can find out your mental health status and whether you are stressed, worried, or depressed. The DASS is an instrument that is often used to assess the level of an individual for the analysis of depression and anxiety. DASS has no implications for patients or individuals in the classification system, such as the discrete diagnostic manual and Mental Disorder Statistics (DSM) and any disease classification. The DASS only evaluates the symptoms that are associated with depression, anxiety, and stress [15]. In the early stages of using DASS, it contained 42 items but was modified to 21 items. The DASS was much used in psychology-related studies, in which its reliability and validity have been recognized in various fields of study. Thus, the DASS is an instrument that is recognized for determining symptoms of stress, anxiety, and depression [7]. During COVID-19, the DASS-21 was utilized in several cultures and countries as a screening tool to identify a person's level of depression, anxiety, and stress [8–10].

It is also worth noting that this study was employed in the scope of social science and not into psychology studies, by technical means. This research does not involve any patients of known mental health; instead, it was conducted generally on community basis. The data collected originated from the society; therefore, they were not adhered with ethical or confidentiality issues. The respondents' responses were only used for academic purposes alone.

4. Analysis and Discussion

4.1. Respondent's Profile

The profile of the study's respondents is shown in Table 1. The majority of respondents were males of Malay ethnicity who represented respondents from all states in Malaysia. Most respondents were working in the of private sector and represented all levels of education, from no formal education to having a doctorate degree. Furthermore, the majority of respondents were located in the city. In the aspects of age and salary, respondents represented the age of engaging in work actively and coming from a B40 and M40 category in Malaysia.

4.2. DASS-21 Score Analysis

The DASS-2 Score Analysis in Table 2 reveals that nearly a quarter of Malaysian respondents during the era of the mobility control order suffered from mental health issues. As many as 23.1% of Malaysians have at least a mild mental health problem to a very bad problem. If the components of the mental health problems of the Malaysian community are detailed, almost 10% of the respondents experienced severe and very severe symptoms for both mental problem components of depression and anxiety during the movement control period in Malaysia to combat the COVID-19 pandemic.

Table 2. DASS-21 score and Mental Health Problem Components.

DASS-21 Scoring	Percentage (%)
Normal	76.9
Mild	10.9
Moderate	8.7
Severe	2.9
Very severe	0.7
Depression Level	
Normal	71.1
Mild	8.9
Moderate	10.8
Severe	5.0
Very severe	4.2
Anxiety Level	
Normal	71.4
Mild	5.5
Moderate	13.3
Severe	4.5
Very severe	5.4
Stress Level	
Normal	75.9
Mild	8.1
Moderate	8.9
Severe	6.3
Very severe	0.8

4.3. Cross-Tabulation Analysis

Table 3 shows the rate and the percentage-level of depression and whether respondents showed symptoms of depression or not within the different socio-demographic variables. The results of the study found that respondents living in urban areas showed more symptoms of depression (30%) than those living in rural areas (26.6%). In addition to that, the findings of the study found that respondents that were aged 18–25 years showed symptoms of depression that were much higher, namely 45%, and this is consistent with the conclusions of [16] in Spain. The study's findings also revealed that, in Northern Spain, during the COVID-19 pandemic's emergency period, one-quarter of respondents involved in the study had mental health problems. Respondents experienced depression (27.5%), had symptoms of anxiety (26.9%), and experienced stress (26.5%), respectively. This group which is the most active in socializing may have been experiencing a higher probability of depression symptoms due to the closure of social engaging places, such as central shopping malls and entertainment centers, where the percentage is nearly twice as much compared to the other age groups.

Table 3. Rate and Percentage of Depression, Anxiety, and Stress Levels for demographic factors [1].

Variables	Category	Depression		Anxiety		Stress	
		Normal	Not Normal *	Normal	Not Normal *	Normal	Not Normal *
Gender	Male	292 71.0%	119 29.0%	300 73.00%	111 27.00%	319 77.60%	92 22.40%
	Female	250 71.2%	101 28.8%	244 69.50%	107 30.50%	259 73.80%	92 26.20%
	Total	542 71.1%	220 28.9%	544 71.40%	218 28.60%	578 75.90%	184 24.10%
Area of Living	Urban	357 70.0%	153 30.0%	365 71.60%	145 28.40%	386 75.70%	124 24.30%
	Rural	185 73.4%	67 26.6%	179 71.00%	73 29.00%	192 76.20%	60 23.80%
	Total	542 71.1%	220 28.9%	544 71.40%	218 28.60%	578 75.90%	184 24.10%
Age	18–25 years old	66 55.0%	54 45.0%	63 52.50%	57 47.50%	84 70.00%	36 30.00%
	26–30 years old	83 76.9%	25 23.1%	78 72.20%	30 27.80%	82 75.90%	26 24.10%
	31–40 years old	216 71.8%	85 28.2%	225 74.80%	76 25.20%	228 75.70%	73 24.30%
	41–60 years old	165 75.3%	54 24.7%	167 76.30%	52 23.70%	173 79.00%	46 21.00%
	61 years old and above	12 85.7%	2 14.3%	11 78.60%	3 21.40%	11 78.60%	3 21.40%
	Total	542 71.1%	220 28.9%	544 71.40%	218 28.60%	578 75.90%	184 24.10%
Race	Malay	389 74.8%	131 25.2%	396 76.20%	124 23.80%	404 77.70%	116 22.30%
	Chinese	61 54.5%	51 45.5%	56 50.00%	56 50.00%	75 67.00%	37 33.00%
	Indian	69 73.4%	25 26.6%	66 70.20%	28 29.80%	69 73.40%	25 26.60%
	Others	23 63.9%	13 36.1%	26 72.20%	10 27.80%	30 83.30%	6 16.70%
	Total	542 71.1%	220 28.9%	544 71.40%	218 28.60%	578 75.90%	184 24.10%
Employment Sector	Government	196 80.3%	48 19.7%	188 77.00%	56 23.00%	192 78.70%	52 21.30%
	Private	235 66.6%	118 33.4%	232 65.70%	121 34.30%	257 72.80%	96 27.20%
	Self-employed	111 67.3%	54 32.7%	124 75.20%	41 24.80%	129 78.20%	36 21.80%
	Total	542 71.1%	220 28.9%	544 71.40%	218 28.60%	578 75.90%	184 24.10%
Category of income	Hourly/daily/weekly basis	62 61.4%	39 38.6%	61 60.40%	40 39.60%	78 77.20%	23 22.80%
	Monthly basis	401 75.5%	130 24.5%	394 74.20%	137 25.80%	404 76.10%	127 23.90%
	Piece rated basis	79 60.8%	51 39.2%	89 68.50%	41 31.50%	96 73.80%	34 26.20%
	Total	542 71.1%	220 28.9%	544 71.40%	218 28.60%	578 75.90%	184 24.10%

Table 3. *Cont.*

Variables	Category	Depression		Anxiety		Stress	
		Normal	Not Normal *	Normal	Not Normal *	Normal	Not Normal *
Monthly income	RM580 and below	30 78.9%	8 21.1%	30 78.90%	8 21.10%	32 84.20%	6 15.80%
	RM580 to RM980	32 66.7%	16 33.3%	34 70.80%	14 29.20%	35 72.90%	13 27.10%
	RM981 to RM2614	148 70.5%	62 29.5%	155 73.80%	55 26.20%	161 76.70%	49 23.30%
	RM2615 to RM4360	147 70.7%	61 29.3%	145 69.70%	63 30.30%	158 76.00%	50 24.00%
	RM4361 to RM9619	185 71.7%	73 28.3%	180 69.80%	78 30.20%	192 74.40%	66 25.60%
	Total	542 71.1%	220 28.9%	544 71.40%	218 28.60%	578 75.90%	184 24.10%

[1] Number of respondents, n = 762; * Not Normal = Respondents at least being mildly to severely affected.

Table 3 also illustrates the level of anxiety related to the demographic variables. In terms of gender disparities, severe symptoms of depression were more prevalent in women, with statistics showing that 30.5% of women in Malaysia suffered from at least a few anxiety symptoms during the COVID-19 pandemic, compared to only 27% of male respondents. The study's findings are consistent with the results of the study by [17] in Taiwan and [18] in Italy, who found that the occurrence of the symptoms of anxiety is higher for women than men. Despite the fact that mental issues do not favor race or skin color and can happen to anyone, this study found that the respondents from a Chinese ethnicity suffer from anxiety symptoms the most (50%) compared to other ethnicities during the movement control order.

Table 3 also shows the rate and the percentage of the impact of if respondents experienced stress symptoms or not on different sociodemographic variables. Relatively, women who lived in the city and were aged 18–25 years were the demographic that were most prone to stress symptoms, compared to other demographic factors. From the aspect of the employment sector, available respondents who worked in the private sector who suffered from the risk of stress symptoms was high, contributing 27% compared to respondents who worked in the government or even worked alone. In addition, of respondents who received wages based on job (piece rated)/freelancer/working online, food delivery employees working through the phone applications and e-hailing drivers were those who demonstrated stress symptoms, contributing to 26.2% of the respondents at least experiencing symptoms of stress either mildly or severely affected compared to respondents who received monthly or even a weekly salary while the pandemic struck.

5. Conclusions

This study did not consider medical cases and therefore did not require ethical consent. This study is in line with studies conducted by many researchers in the same field [12,19–21]. Therefore, this study does not require any ethical concern. This was a cross-sectional study that relied mostly on self-reported questionnaires to quantify psychiatric symptoms; no clinical diagnosis was made. The gold standard for mental diagnosis consists of a structured clinical interview and functional neuroimaging [21,22].

The study's findings demonstrate that communities in Malaysia experience impaired mental health as a result of the COVID-19 epidemic, as evidenced by signs of stress, anxiety, and sadness. The results of this study found that almost a quarter of respondents need to be given attention, as there is a probability that this figure may increase while the world is making the best efforts to reduce the number of deaths brought on by the COVID-19 pandemic and COVID-19 positive cases.

With the deteriorating economic situation, of course, many companies went bankrupt, and many individuals will lose their source of income. Individuals should manage their personal financial situation and, at the same time, their mental health in a stable state. Economic uncertainty and loss of employment may not only cause a person to lose sanity but also cause indirect costs to immediate family members. This is because family members must bear the patient's cost financially and have to sacrifice their rest time to pay attention and care for the patient. Therefore, the crucial moral support from individuals to family and friends is much needed.

In addition to the disabled and the elderly who need to be given attention, women also need support during this pandemic. The findings of the study found that female respondents living in the city is a demographic group that shows the highest mental health problems, at least to a small extent to very severe, either through depression, anxiety, or stress. This is probably because the women who are married and living in the city are working women. With the state of closing economic activities and the need to work from home, these women need to be juggling to do their work activities and "work" at the home at the same time. This is exacerbated by children's online school activities, which need to be given attention for as long as they are indoors due to the closure of the school during the movement control order period in Malaysia.

Author Contributions: Conceptualization, M.K.I. and C.S.; methodology, M.Z.M. and S.S.; software, M.K.I.; validation, formal analysis, investigation, writing—original draft preparation, writing—review and editing, M.K.I., S.S., M.Z.M. and C.S.; project administration, funding acquisition, M.K.I. All authors have read and agreed to the published version of the manuscript.

Funding: This research was funded by the Research Collaboration Fund 2021 (600-UiTMCTKD) (PJI/RMU 5/2/1) and SIGni-Finomics (600-UiTMCTKD (PJI/RMU 5/1) SIG 1/2022 (20)).

Institutional Review Board Statement: Not applicable.

Informed Consent Statement: Informed consent was obtained from all subjects involved in the study.

Data Availability Statement: The datasets generated and analyzed for the study are available from the corresponding author on reasonable request.

Acknowledgments: The authors wish to thank the Universiti Teknologi MARA Cawangan Terengganu for funding this study under the Research Collaboration Fund 2021 (600-UiTMCTKD) (PJI/RMU 5/2/1) and SIGni-Finomics (600-UiTMCTKD (PJI/RMU 5/1) SIG 1/2022 (20)).

Conflicts of Interest: The authors declare no conflict of interest.

References

1. World Health Organization. *Promoting Mental Health: Concepts, Emerging Evidence, Practice: A Report of the World Health Organization, Department of Mental Health and Substance Abuse in Collaboration with the Victorian Health Promotion Foundation and the University of Melbourne*; World Health Organization: Geneva, Switzerland, 2005; ISBN 9791157467679.
2. World Health Organization. *Depression and Other Common Mental Disorders: Global Health Estimates*; World Health Organization: Geneva, Switzerland, 2017; pp. 1–24.
3. Leigh-Hunt, N.; Bagguley, D.; Bash, K.; Turner, V.; Turnbull, S.; Valtorta, N.; Caan, W. An overview of systematic reviews on the public health consequences of social isolation and loneliness. *Public Health* **2017**, *152*, 157–171. [CrossRef] [PubMed]
4. Verme, P. Happiness, freedom and control. *J. Econ. Behav. Organ.* **2009**, *71*, 146–161. [CrossRef]
5. Shanmugam, H.; Juhari, J.A.; Nair, P.; Chow, S.K.; Ng, C.G. Impacts of COVID-19 Pandemic on mental health in Malaysia: A Single Thread of Hope. *Malays. J. Psychiatry* **2020**, *29*, 78–84.
6. Rathakrishnan, B.; Kamaluddin, M.R.; Singh, S.S.B. Mental health issues during COVID-19 pandemic: Treatment and how to overcome. *Malays. J. Psychiatry* **2020**, 29. [CrossRef]
7. Antony, M.M.; Cox, B.J.; Enns, M.W.; Bieling, P.J.; Swinson, R.P. Psychometric properties of the 42-item and 21-item versions of the depression anxiety stress scales in clinical groups and a community sample. *Psychol. Assess.* **1998**, *10*, 176–181. [CrossRef]
8. Wang, C.; Chudzicka-Czupała, A.; Grabowski, D.; Pan, R.; Adamus, K.; Wan, X.; Hetnał, M.; Tan, Y.; Olszewska-Guizzo, A.; Xu, L.; et al. The Association between physical and mental health and face mask use during the COVID-19 pandemic: A comparison of two countries with different views and practices. *Front. Psychiatry* **2020**, *11*, 569981. [CrossRef] [PubMed]
9. Le, H.T.; Lai, A.J.X.; Sun, J.; Hoang, M.T.; Vu, L.G.; Pham, H.Q.; Nguyen, T.H.; Tran, B.X.; Latkin, C.A.; Le, X.T.T.; et al. Anxiety and depression among people under the nationwide partial lockdown in Vietnam. *Front. Public Health* **2020**, *8*, 589359. [CrossRef] [PubMed]
10. Tee, M.L.; Tee, C.A.; Anlacan, J.P.; Aligam, K.J.G.; Reyes, P.W.C.; Kuruchittham, V.; Ho, R.C. Psychological impact of COVID-19 pandemic in the Philippines. *J. Affect. Disord.* **2020**, *277*, 379–391. [CrossRef] [PubMed]
11. Tran, B.X.; Nguyen, H.T.; Le, H.T.; Latkin, C.A.; Pham, H.Q.; Vu, L.G.; Le, X.T.T.; Nguyen, T.T.; Pham, Q.T.; Ta, N.T.K.; et al. Impact of COVID-19 on economic well-being and quality of life of the Vietnamese during the national social distancing. *Front. Psychol.* **2020**, *11*, 565153. [CrossRef] [PubMed]

12. Xiong, J.; Lipsitz, O.; Nasri, F.; Lui, L.M.W.; Gill, H.; Phan, L.; Chen-Li, D.; Iacobucci, M.; Ho, R.; Majeed, A.; et al. Impact of COVID-19 pandemic on mental health in the general population: A Systematic review. *J. Affect. Disord.* **2020**, *277*, 55–64. [CrossRef] [PubMed]
13. Dong, L.; Bouey, J.; Bouey, J. Public mental health crisis during covid-19 pandemic, China. *Emerg. Infect. Dis.* 2020. [CrossRef]
14. COVID-19 Disrupting Mental Health Services in Most Countries, WHO Survey. Available online: https://www.who.int/news/item/05-10-2020-covid-19-disrupting-mental-health-services-in-most-countries-who-survey (accessed on 9 November 2020).
15. Lovibond, S.H.; Lovibond, P.F. *Manual for the Depression Anxiety Stress Scales; Behaviour Research and Therapy*; Elsevier: Amsterdam, The Netherlands, 1995; ISBN 7334-1423-0.
16. Ozamiz-Etxebarria, N.; Dosil-Santamaria, M.; Picaza-Gorrochategui, M.; Idoiaga-Mondragon, N. Stress, anxiety, and depression levels in the initial stage of the COVID-19 outbreak in a population sample in the Northern Spain. *Cad. Saude Publica* **2020**, *36*, e00054020. [CrossRef] [PubMed]
17. Wong, L.P.; Hung, C.C.; Alias, H.; Lee, T.S.H. Anxiety symptoms and preventive measures during the COVID-19 outbreak in Taiwan. *BMC Psychiatry* **2020**, *20*, 376. [CrossRef] [PubMed]
18. Gualano, M.R.; Lo Moro, G.; Voglino, G.; Bert, F.; Siliquini, R. Effects of COVID-19 lockdown on mental health and sleep disturbances in Italy. *Int. J. Environ. Res. Public Health* **2020**, *17*, 4779. [CrossRef] [PubMed]
19. Zhang, M.W.B.; Ho, R.C.M. Moodle: The cost effective solution for internet cognitive behavioral therapy (i-cbt) interventions. *Technol. Health Care* **2017**, *25*, 163–165. [CrossRef] [PubMed]
20. Soh, H.L.; Ho, R.C.; Ho, C.S.; Tam, W.W. Efficacy of digital cognitive behavioural therapy for insomnia: A meta-analysis of randomised controlled trials. *Sleep Med.* **2020**, *75*, 315–325. [CrossRef] [PubMed]
21. Ho, C.S.; Chee, C.Y.; Ho, R.C. Mental health strategies to combat the psychological impact of COVID-19 beyond paranoia and panic. *Ann. Acad. Med. Singap.* **2020**, *49*, 155–160. [CrossRef] [PubMed]
22. Husain, S.F.; Yu, R.; Tang, T.B.; Tam, W.W.; Tran, B.; Quek, T.T.; Hwang, S.H.; Chang, C.W.; Ho, C.S.; Ho, R.C. Validating a Functional Near-Infrared Spectroscopy Diagnostic Paradigm for Major Depressive Disorder. *Sci. Rep.* **2020**, *10*, 9740. [CrossRef] [PubMed]

proceedings

MDPI

Proceeding Paper

Dealing with the Locality at the Islamic Tourism Destination from the Islamic Perspective in Malaysia †

Zulaipa Ruzulan [1,2,*], Mohammad Aizat Jamaludin [2] and Khairusy Syakirin Has-Yun Hashim [3]

1 Academy of Contemporary Islamic Studies (ACIS), Universiti Teknologi MARA, Shah Alam 40450, Selangor, Malaysia
2 International Institute for Halal Research and Training, (INHART), International Islamic University Malaysia, Jalan Gombak 53100, Selangor, Malaysia
3 Kulliyyah of Architecture and Environmental Design (KAED), International Islamic University Malaysia, Jalan Gombak 53100, Selangor, Malaysia
* Correspondence: zulaipa3198@uitm.edu.my
† Presented at the International Academic Symposium of Social Science 2022, Kota Bharu, Malaysia, 3 July 2022.

Abstract: Government authority plays a significant role in ensuring the aim of being a famous Islamic tourism destination is achievable with proper planning and implementation. However, challenges to sustain Islamic tourism destination could be triggered if the government fails to manage localities where it will probably face participatory unwillingness among the society in terms of tourism development. Therefore, this paper carried out an exploratory study on the significance of considering the locality to promote an Islamic tourism destination from the Islamic perspective in Malaysia. Document analysis was chosen in this qualitative research that involves retrieving the data from the journal articles, government reports, and any reliable sources on websites. This study found that the spirit of governing tourism destinations aligns with Shariah principles as well as other sustainability motives, which leads to success where understanding the local context plays a significant role in realizing this mission.

Keywords: *al-'urf*; Islamic perspective; tourism destination; locality; tawhidic paradigm

Citation: Ruzulan, Z.; Jamaludin, M.A.; Hashim, K.S.H.-Y. Dealing with the Locality at the Islamic Tourism Destination from the Islamic Perspective in Malaysia. *Proceedings* **2022**, *82*, 39. https://doi.org/10.3390/proceedings2022082039

Academic Editor: Mohamad Rahimi Mohamad Rosman

Published: 14 September 2022

1. Introduction

Branding a destination as an Islamic tourism destination means that all the attributes associated with available tourism products should be insured as Shariah-compliant. This is owing to any associated terms related to Islam, such as "Islamic", "halal", "Shariah", and "Muslim-friendly tourism" under the purview of Islamic law. This implies that once there is a motive to promote the products and services in the Islamic tourism industry, the tourism operators need to have a sense of Islamic religious observance in managing the tourism activities at the destination.

In pertaining to these matters, Islamic tourism destination branding is a driving factor for local authorities due to the economic prospect that could lead to a good chance for development. This is because revenue from the tourism sector is expected to increase significantly by 2025. The growth rate is about 24.05% annually, resulting in a market value of USD 90884400, and online sales will reach 72% of the overall revenue of the tourism and travel sector [1]. As a result, the increase in travel by Muslim population could be a potential for many destinations to offer Shariah-compliant products in the tourism market segment.

Therefore, Islamic tourism planning at destinations is among the key indicators leading to a sustainable Islamic tourism destination. The planning involves all stakeholders to realize the mission, including locality considerations such as the public, tourism operators, political conditions, and customs. The Islamic religion recognizes society and reality at the highest position in Islamic law, where the five essential values (*maslahah*) must be observed to avoid harm to the people. However, there is an imbalance in managing Islamic tourism destinations where economic growth and its planning become the most significant concern rather than establishing the goal in accordance with Islamic principles. Therefore, this paper explores the significance of locality considerations at Islamic-tourism destinations from the Islamic perspective to ensure the development of Islamic-tourism destinations in line with Islamic teaching.

2. Literature Review

2.1. Tourism and Islam

Islam emphasizes the betterment of all humankind, and the holy texts of the Quran and Sunnah provide comprehensive guidance to everyone. This is to ensure and protect society's well-being. With this, tourism is part of the corpus of human activities that Islam recognizes and encourages humankind to be involve in [2]. One study emphasized that tourism provides many advantages, and among them is to increase the level of *taqwa* (piety) towards Allah SWT. This is shown by several verses that encourage travel. The verses are as follows.

> "Do they then not travel through the Earth so that their minds gain wisdom, and their ears thus learn to hear? For surely it is not the eyes that are blind, but blind are the minds which are in the foremost."
>
> (Al-Hajj, 22:46)

> "Have you not seen the ships speeding through the sea, carrying God's provisions so that He might show you some of His wonders? Herein, behold, there are messages indeed for all who are patient, grateful."
>
> (Luqman, 31:31)

> "And how many a Message in the Heavens and the Earth which they pass by, yet they turn away from it!"
>
> (Yusuf, 12:105)

Indeed, these verses encourage us to travel and ponder the bounties Allah SWT has provided to His creatures. Apart from that, tourism is not just a matter of a short escape and relaxation. It is part of increasing obedience to Allah SWT, since many advantages and wisdom can be obtained from traveling.

In this vein, tourism was practiced a long time ago, during which time Islamic tourism was more under the purview of religious tourism, where Muslims traveled to sacred places to carry out a pilgrimage because it was a mandatory requirement of the Islamic religion. Therefore, the scope and the discussion of Islamic tourism in the early stage were more focused on traveling to sacred places. The interpretation is based on religious tourism being associated with the religion itself, such as Jewish tourism, Christian tourism, Buddhist tourism, and any other faiths, including Islamic tourism, wherein travel by Muslims involves visiting the holy lands of Mecca and Medina [3]. However, this segmented Islamic tourism has gone beyond religious understanding as time has gone by. It includes a single pace of activities as long as they do not contradict Islamic law [4], wherein this industry focuses on economic growth and at the same time makes available the faith's basic needs at the tourism destination. These efforts ensure that Muslim travelers can efficiently perform their religious obligations when away from their home but still feel the same ambience as at home. This could ensure that many activities are generated in sequence to fulfil the right of the tourism stakeholder.

Therefore, activities of Islamic tourism need to be under Islamic governance. The *tawhidic* paradigm (Unity of God) of seeking the pleasure of Allah must be adopted to ensure all the activities are implemented following Shariah requirements [5]. When the tourism stakeholders understand that *tawhid* is a foundation of Islamic religious belief, it will influence the actions of dealing with Islamic tourism development. Since Islamic tourism draws in more Muslim travelers, this concept has been introduced as a niche tourism segment in the tourism industry. Research has revealed that tour operators use several terms associated with the Islamic religion to attract prospective travelers to visit the destination. Popular terms are "Islamic tourism", "halal tourism", "Shariah-compliant tourism", and "Muslim-friendly tourism." These terms play significant roles in tourism destinations in attracting potential travelers [6]. A group of researchers has emphasized that tourism in Islam adheres to Islamic principles and involves Muslims who desire to preserve their religious practices while traveling. Halal tourism is described as Muslims traveling for pleasure or business to tourist sites in non-Islamic and Islamic nations that are not explicitly prohibited by Islamic Shariah [7,8]. This definition certainly is from the perspective of travelers. The tourism planners' side describes accountability in providing Islamic tourism products and services according to Islamic principles, where the *tawhidic* paradigm takes precedence in strategizing the destination. For this reason, the Islamic spirit is the priority and catalyst for the sustainability of a destination.

In addition [9], the branding is not limited to the Muslim travelers only, but also includes non-Muslims as part of an escape to experience non-ordinary feelings. Other than that [6], there is no limitation to offering an Islamic tourism product, including in non-Muslim countries, as long as the Islamic tourism product remains under Islamic observance. Thus, Islamic attributes should be available at Islamic tourism destinations to increase the satisfaction of Muslim travelers, leading to loyalty and revisiting the Islamic-tourism destination. In this part, the Islamic tourism products should adhere to Shariah principles, wherein the drive is not only for economic development, but also observing Shariah requirements on those particular matters. The tourism stakeholders at the destinations need to ensure that the offered products or activities do not go against any Islamic teachings, such as mingling between genders without a *mahram*, providing halal food, and managing the tourism activities in such a way that they do not tarnish the image of the Islamic religion itself. Therefore, meticulous monitoring from the government and the scholars (*ulama'*) is imperative to ensure that the implementation of the Islamic tourism industry at the destination remains with the Islamic spirit within the locality.

2.2. *Al-'Urf as a Consideration to Sustain the Islamic Tourism Destination*

Consideration of locality is significant in Islamic law, where it is translated as *al-'urf* in the Arabic language. Jurists have recognized *al-'urf* as one of the principles of Islamic rule under secondary sources other than preference (*istihsan*), the presumption of continuity (*istishab*), extended analogy (*maslahah mursalah*), blocking the means (*sadd al-dharai'*), and any other secondary sources that can be used to deal with arising issues [10]. Although there is no uniformity in terms of this recognition, and some parts of the *mazahib* (school of thoughts) disagree with the position of *al-'urf* as the secondary source of Islamic law, the consideration of locality still plays a significant role as a tool to help jurists to deal with Shariah issues by recognizing the maxim of *al-'adah muhakkamah* (custom is the basis of judgement) as the establishment of customary practice principles that the entire *mazahib* has agreed on. From this, it has been denoted that Islam celebrates the life of the people, wherein what has been customary in one's life should be respected if it does not contradict Shariah principles.

Al-'urf refers to customs or matters that are customary in the life of a society, such as actions, habits, and beliefs. Linguistically, *al-'urf* means "a commonly known matter that refers to good deeds" [11]. Technically, from the Islamic law perspective, *al-'urf* refers to "something that has always been practiced continuously by the people, whereby the practice is customary and accepted in ordinary life and they can easily and reasonably understand the method" [12,13]. In this regard, *al-'urf* is a standard practice in a society where it occurs naturally and continuously. This has resulted in *al-'urf* becoming a societal value that is difficult to remove from the community's life. Therefore, understanding *al-'urf* is significant because it concerns societal needs [14].

In a nutshell, *al-'urf* covers something that can be seen and includes most people's paradigms, beliefs, and thinking styles at a particular place and time. Thus, to ensure that the objective of Shariah (*maqasid al-shariah*) is achievable, Islam recognizes the local customs since those practices are customary and are not easy to ward off. If no elements lead to non-Shariah compliance, Islam allows the action. Certain *maslahah* need to be protected, including religion, intellect, lineage, property, and life. If not, it will lead to difficulties in society in many aspects of life, and *fitnah* (slander) could occur in the religion.

This has been shown in many areas of *fiqh mualamat* (man-to-man matters), including halal industries, where scholars consider the reality and the circumstances in which there is a need to resolve Shariah-related issues. Indeed, this benefits all the stakeholders in halal areas, including Islamic tourism products and activities at a destination, since all essential needs are thoroughly protected in line with the objectives of Shariah.

2.3. *Sustainability from the Islamic Perspective*

The concept of sustainability is not new in Islam, where clear messages in the divine laws encourage prosperity on Earth and lead towards *al-falah* (success) here and hereafter. Islam stresses the attainment of the objectives of Shariah as the ultimate purpose of the lawgiver by fulfilling the five essential values of Shariah (*maslahah*) namely religion, life, mind, offspring and wealth [15]. This means that the commandment of Allah SWT is to provide easiness to the people (*raf' al-harj*) and to avoid harm. In conclusion, Islam guides humankind comprehensively for its betterment here and hereafter.

Similarly, this concept aligns with the sustainable development goals (SDGs) that purport to preserve and maintain well-being. With these key elements of the SDGs, Islam provides basic principles to achieve the goals. Indeed, sustainability is intimately tied to development, which involves a manner that drives people to balance their economic and social lives with the available ecological resources for human development [16]. This shows that applying sustainability could ensure resilience towards the evolution of times and circumstances at a tourist destination. The sustainability concept has become the main focus of the World Tourism Organization (UNWTO), where sustainable development in the corpus of the tourism takes complete account of its current and future economic, social, and environmental impacts, addressing the needs of visitors, the industry, the environment, and host communities [17]. The UNWTO has further emphasized that all relevant tourism stakeholders must be well informed on the mission to remain sustainable since sustainability development is a continuous process. Thus, as the sole authority, the government is accountable for overseeing the implementation to reach the goal. Strong leadership needs to be established at Islamic tourism destinations where it can be sustained by maintaining a high level of tourist satisfaction that can build a meaningful experience. With this, community engagement is needed to increase participation in sustainable tourism.

In Islam, the concept of *dar'u mafasid awla min jalbi al-masalih* (preventing evil takes precedence over any consideration of benefits) is appropriate for sustainable tourism. This has been proven by the approach of Islamic scholars in governance practices. Among the essential criteria is looking toward the public interest (*maslahah ammah*), which already covers the three elements of the tourism industry's sustainable development goals: economy, social, and environment. This falls under the basic principle of all Islamic law to ensure equilibrium of life in society. The issuance of guidelines at the federal level for determining Islamic legal ruling in 2017, clauses 42 and 43, emphasized the preservation criteria of the locality [18].

> 42: "The public interest is the goodness and benefits to a society that certified its validity by the Mufti or the Fatwa Committee to achieve *maqasid syariah*. *Maqasid syariah* is the goodness (*maslahah*) that is to be achieved by Shariah for the benefit of mankind in this world and the hereafter. It includes observance of the five main necessities (*al-daruriyyat al-khamsah*), i.e., safeguarding religion, life, lineage, intellect, and property.
>
> 43: The public interest criteria are:
>
> Coinciding with the *maqasid syariah*.

(i) Real matters (*haqiqi*) or strong conjectures (*ghalabat al-zann*), not imagined ones.

(ii) In accordance with local *'urf* and practices recognized by Shariah.

(iii) Comprehensive (*al-umum*) or mostly comprehensive (*al-ghalabah*) [19].

In addition, the ordinance of the enactment at the state level in Malaysia also has considered public interest. Examples are in Section 54. *Qaul muktamad* is to be followed by [20].

> "(1) In issuing any Fatwa under Section 48 or certifying an opinion under Section 53, the Fatwa Committee shall ordinarily follow *qaul muktamad* (the accepted views) of the Mazhab Syafie.
>
> (2) If the Fatwa Committee is of the opinion that following the *qaul muktamad* of the Mazhab Syafie will lead to a situation which is repugnant to the public interest, the Fatwa Committee may follow the *qaul muktamad* of the Mazhab Hanafi, Maliki or Hanbali.
>
> (3) If the Fatwa Committee is of the opinion that none of the *qaul muktamad* of the four Mazhabs may be followed without leading to a situation which is repugnant to the public interest, the Fatwa Committee may make the Fatwa according to ijtihad without being bound by the *qaul muktamad* of any of the four Mazhabs."

Moreover, the Malaysian government also introduced the Total Planning and Development Doctrine in 1997 to cater to the sensitivity and the needs for human development in the national development plan, where the development focus is on moderation between physical and human development [21]. In this guideline, the element of *taqwa* becomes the main spiritual value, which consists of the relationship between man and God, man and man, and man and other creatures. Based on the main principles of planning and development highlighted, it is denoted that the sustainable progress of the country can only be achieved if sensitivity and human needs are being considered accordingly.

Based on the highlighted clauses, considering the life of the society as long as it does not contravene Shariah requirements becomes the main preference in ensuring that the decision that has been planned or enforced in the society is a reasonable and practical solution for them. This is indeed part of making an Islamic tourism destination more resilient to the changing times and circumstances. This is supported by a study [4] that stressed Islamic culture and values among the significant considerations for developing Malaysia as an Islamic tourism destination where the tourism stakeholders need to ensure the originality and uniqueness of Malaysian culture. Furthermore, a review study [22] emphasized that the pillars of sustainability, namely, society, economy, and environment, determine the positive or negative impact on the Islamic tourism destination. If these pillars are well managed, they will positively impact the destination since the local people's willingness to participate in sustainable tourism development is an indicator. This is because the continuous process of Islamic tourism destination development is under the coverage of the spiritual first, where the blessing of God is the utmost goal to achieve [23].

3. Methods

This study adopted a qualitative research methodology wherein an exploratory approach was involved to comprehend the significance of considering a locality to promote the establishment of Islamic tourism destination branding from the Islamic perspective in Malaysia. For the data collection, document analysis was chosen, which involved retrieving the data from journal articles, government reports, online newspapers, and any reliable sources on websites, covering topics such as Islamic tourism destinations, localities from the perspective of Islamic law, travelers' satisfaction and loyalty, and the willingness of the local community to participate in Islamic tourism activities. Thereafter, the data obtained were concurrently analyzed by using content analysis techniques until the objective was addressed accordingly.

4. Results and Discussion

In Malaysia, the Muslim population is the leading cause of Muslim friendly tourism development, which has led to the government's recognition that this industry is an important segment contributing to economic growth. Many Islamic tourism products are available and influence foreign and domestic travelers to visit destinations with this advantage. In this vein, the population is creating higher demand in the realm of Muslim-friendly tourism at the destination and is assisting in establishing tangible and intangible Islamic tourism attributes such as mosques, halal food, Islamic entertainment, and any other Shariah-compliant activities.

Furthermore, government planning for tourism development also considers the culture and beliefs of the society. This could be seen in several states in Malaysia that has adopted the Islamic concept in the administration of the state, including tourism. Terengganu, for instance, has adopted *'Maju, Berkat, Sejahtera'* in governing the tourism industry, where any activities must comply with Islamic teachings. To be in line with the mission, amusement guidelines were introduced to ensure that the entertainment organized in the state would not tarnish the image of Islam. The event organizer and the hotel and resort host must observe the potential client who intends to run the event at the locations.

Similarly, the Kelantan state is well known as the *'Serambi Mekah'*, which has led the state government to choose to manage the state in line with Islamic teaching. There are municipalities like Bachok, which is branding itself as an Islamic tourism city to attract more potential travelers to visit Kelantan [24]. Indeed, these states have many similarities in their political approach, particularly in the commitment to implementing Islamic teaching following Shariah. This is because many of the residents in these states are Muslims, and the society is very particular and sensitive towards the implementation of Shariah.

With this common political condition in both states, outsiders also view both states as synonymous with Islam. Many concerns related to Shariah have been raised by society, or even by cybertroopers, especially regarding elements that go against Shariah. The topics of wearing a bikini at the beach [25–27], organizing a *'dangdut'* (a type of music for dancing) at a hotel event [28], and banning celebrities that tend to influence immoral behavior to provide entertainment [28] are among the controversial issues that have created urgency for the state government to provide clarification. This is due to the states' Islamic image already having been made known by society. It is significant for the government to preserve the community's status quo by ensuring that there is no immoral behavior that could taint the existing practices of society. If the issues cannot be resolved and managed accordingly, it will create a negative impact on the tourism development itself due to the dissatisfaction of the society with the failure of the government to manage the issue accordingly.

Moreover, the effort from the federal government based on the establishment of the Islamic Tourism Centre under the Ministry of Tourism, Arts and Culture (MOTAC) is another layer to dig into regarding the potential of the Muslim-friendly tourism attributes in a particular destination in Malaysia in terms of the potential to promote it as an Islamic tourism destination. Sabah, for instance, has become a recent potential destination promoted as an Islamic tourism destination because of its unique geographical location for ecotourism activities. It has been stressed by Tuan Haji Muhammad Zamri Muhamed Shapek, Director of IPDAS, that there is a need to correct the misunderstanding of Middle East travelers' intention to visit the destination, wherein the travelers tend to visit luxurious places only. Still, he has refuted the concept that Middle Eastern travelers are also interested in adventurous activities that involve nature, like what Sabah could offer [29]. This shows that ecotourism products provide a map with the three pillars of sustainable tourism. Sustainability is in line with a significant aspect of Islam, which offers advantages for tourism development.

The above discussion shows that the locality, or *al-'urf*, plays a significant role as an agent for the sustainability of an Islamic tourism destination. Indeed, the sustainability concept introduced recently is already in line with Islamic principles, whereby local consideration is among the main preference for consideration to protect society, economy, and environment. With this, the federal and state levels must be concerned with the reality at the destination to increase the participatory willingness among the society to sustain Islamic tourism destinations.

5. Conclusions

In conclusion, dealing with localities to brand and promote Islamic tourism destinations is crucial since Muslims are the majority population in Malaysia. With the beliefs, customs, and practices that exist in their life, the government is accountable for preserving the well-being of local needs. The available tourism activities that have the potential to be promoted as Islamic tourism products at a destination could be monitored accordingly to avoid repellence from public interest, which would damage the locality. As a result, the aims of sustainable Islamic tourism would not be achievable if non-Shariah compliant activities are not mitigated accordingly based on the *tawhidic* paradigm. Therefore, this study would like to propose that the government start looking thoroughly into the establishment of a Shariah governance approach in Islamic tourism management to preserve the rights of Muslims in general and, at the same time, help the industry develop according to Islamic teaching. With this clear governance framework, the tourism stakeholders will mutually understand their functions and consequently preserve and maintain the elements of the locality accordingly based on the Shariah perspective.

Author Contributions: Conceptualization, Z.R. and M.A.J.; methodology, Z.R.; software, Z.R.; validation, M.A.J. and K.S.H.-Y.H.; formal analysis, Z.R.; investigation, Z.R.; resources, M.A.J.; data curation, M.A.J.; writing—original draft preparation, Z.R.; writing—review and editing, Z.R.; visualization, Z.R.; supervision, M.A.J.; project administration, M.A.J.; funding acquisition, Z.R. All authors have read and agreed to the published version of the manuscript.

Funding: This research received no external funding.

Institutional Review Board Statement: Not applicable.

Informed Consent Statement: Not applicable.

Data Availability Statement: Not applicable.

Acknowledgments: The authors would like to thank Universiti Teknologi MARA because the registration fee was funded by Pembiayaan Yuran Prosiding Berindeks (PYPB), Tabung Dana Kecemerlangan Pendidikan (DKP), Universiti Teknologi MARA (UiTM), Malaysia. In addition, many thanks to the Academy of Contemporary Islamic Studies (ACIS), UiTM, Malaysia; and the International Institute for Halal Research and Training, (INHART), International Islamic University Malaysia, Malaysia, for the full support to complete this article.

Conflicts of Interest: The authors declare no conflict of interest.

References

1. Ghorbanzadeh, D.; Zakieva, R.R.; Kuznetsova, M.; Ismael, A.M.; Ahmed, A.A.A. Generating destination brand awareness and image through the firm's social media. *Kybernetes* 2022, *ahead-to-print*.
2. Ruzulan, Z. Pelancongan dan hospitaliti dalam Islam. In *Industri Halal Malaysia*, 1st ed.; Ishak, A.H., Ruzulan, Z., Eds.; UiTM Press: Shah Alam, Malaysia, 2022; pp. 125–146. Available online: https://mindappz.com/Store/ProductDetail?i=a%2FzHkp9xrug%3D (accessed on 27 April 2022).
3. Seyer, F.; Muller, D. Religious tourism. In *The Long Tail of Tourism*; Papanthanisis, A., Ed.; Gabler: Wiesbaden, Germany, 2011.
4. Abdullah, A.A.; Awang, M.D.; Abdullah, N. Islamic tourism: The characteristics, concept and principles. *KnE Soc. Sci.* **2020**, *4*, 196–215. [CrossRef]
5. Sofyan, A.S.; Abror, A.; Putra, T.W.; Muslihati, M.; Sofyan, S.; Sirajuddin, S.; Katman, M.N.; Darussalam, A.Z. Crisis and disaster management for halal tourism: A systematic review. *Tour. Rev.* **2021**, *77*, 129–145. [CrossRef]
6. Carboni, M.; Perelli, C.; Sistu, G. Is Islamic tourism a viable option for Tunisian tourism? Insights from Djerba. *Tour. Manag. Perspect.* **2014**, *11*, 1–9. [CrossRef]
7. Boğan, E.; Sarıışık, M. Halal tourism: Conceptual and practical challenges. *J. Islamic Mark.* **2019**, *10*, 87–96. [CrossRef]
8. Suban, S.A.; Madhan, K.; Shagirbasha, S. A bibliometric analysis of Halal and Islamic tourism. *Int. Hosp. Rev.* 2021, *ahead-of-print*.
9. Battour, M.; Hakimian, F.; Ismail, M.; Boğan, E. The perception of non-Muslim tourists towards halal tourism Evidence from Turkey and Malaysia. *J. Islamic Mark.* **2018**, *9*, 823–840. [CrossRef]

10. Nyazee, I.A.K. *Islamic Jurisprudence*; The Other Press: Petaling Jaya, Malaysia, 2003.
11. Manzur, M.M. *Lisān Al-ʿarab*; Dar Iḥyaʾal-Turath al-ʿArabī: Beirut, Lebanon, 1997.
12. Al-Suyūṭī, J.; ʿAbdul, R. *Al-Ashbāh Waal-Naẓaʾir*; Dār al-Salām: Cairo, Egypt, 2011.
13. Nujaym, I. Al-Ashbāh Wa Al-Naẓāʾir. Dar al-Kutubal-ʿIlmiyyah: Beirut, Lebanon, 1993.
14. Majid, M.Z.A. *Tranformasi Fikah Semasa*; PTS Islamika Sdn. Bhd.: Selangor, Malaysia, 2011.
15. Jalil, A. The significances of maslahah concept and doctrine of maqasid (objectives) Al-Shari'ah in project evaluation. *J. Muamalat Islam. Financ. Res.* **2006**, *3*, 171–202.
16. Mensah, J. Sustainable development: Meaning, history, principles, pillars, and implications for human action: Literature review. *Cogent Soc. Sci.* **2019**, *5*, 1653531. [CrossRef]
17. UNWTO. Sustainable Development. Available online: https://www.unwto.org/sustainable-development (accessed on 27 April 2022).
18. Jabatan Kemajuan Islam Malaysia. *Garis Panduan Pengeluaran Fatwa di Malaysia*; JAKIM: Putrajaya, Malaysia, 2017.
19. Jabatan Kemajuan Islam Malaysia. *The Guidelines of Determining Islamic Legal Ruling*; JAKIM: Putrajaya, Malaysia, 2017. Available online: http://e-smaf.islam.gov.my/e-smaf/assets/files/garis_panduan/a7e239efb5ffed79171b923ee948c010.pdf (accessed on 13 September 2022).
20. PNMB-LawNet. Administration of the Religion of Islam (State of Selangor) Enactment. 2003. Available online: http://www2.esyariah.gov.my/esyariah/mal/portalv1/enakmen2011/Eng_enactment_Upd.nsf/f831ccddd195843f48256fc600141e84/311f05623bfbe1a748256f34000d561d?OpenDocument (accessed on 13 September 2022).
21. Jabatan Perancangan Bandar dan Desa Semenanjung Malaysia. *Garis panduan perancangan dan pembangunan sejagat*; Kementerian Perumahan dan Kerajaan Tempatan: Putrajaya, Malaysia, 1997.
22. Hamid, M.A.; Ismail, N.; Iffah, Z.; Fuza, M.; Othman, N. Review of Islamic tourism and sustainability. *Int. J. Soc. Sci. Res.* **2022**, *4*, 447–456.
23. Yunus, A.; Noor, M.; Ainshah, N.; Shah, M. Islamic tourism in Langkawi Island: Enhancing sustainability of Islamic value. *Int. J. Appl. Bus. Econ. Res.* **2017**, *15*, 59–71. Available online: http:www.serialsjournal.com (accessed on 27 April 2022).
24. Habsah, I. Malaysia's 'Nami Island' a Popular Tourist Attraction. *The Malaysian Reserve*, 15 September 2020. Available online: https://themalaysianreserve.com/2020/10/08/malaysias-nami-island-a-popular-tourist-attraction/ (accessed on 3 June 2022).
25. Battour, M.; Ismail, M.N.; Battor, M. The impact of destination attributes on Muslim tourist's choice. *Int. J. Tour. Res.* **2011**, *13*, 527–540. [CrossRef]
26. Jo Timbuong and Fatimah Zainal. Terengganu Cancels "Immoral" Redang Music Festival, But Say Organisers Can Appeal—Nation | *The Star Online*. The Star [Internet]. 2018. Available online: https://www.thestar.com.my/news/nation/2018/10/03/terengganu-cancels-immoral-redang-music-festival-but-say-organisers-can-appeal/ (accessed on 30 July 2019).
27. Buletin Online. Bikini; Kerajaan Terengganu Ambil Pendekatan Cermat, Halus Dan Mendidik. *Buletin Online [Internet]*. 24 June 2018. Available online: https://buletinonline.net/v7/index.php/bikini-kerajaan-terengganu-ambil-pendekatan-cermat-halus-dan-mendidik/?fbclid=IwAR0MIKRFd0GoR_uHRKRHLN_LGK3Or2KwbDGqYsyEsjLFWVwd8Gg3Y20D1KQ (accessed on 29 July 2019).
28. Rahman, H.A. Tak lulus permit island music Festival. *Harian Metro [Internet]*. 2018. Available online: https://www.hmetro.com.my/mutakhir/2018/10/382943/tak-lulus-permit-island-music-festival (accessed on 29 April 2022).
29. Islamic Tourism Centre of Malaysia. Sabah Shows Potential in Muslim-Friendly Tourism and Hospitality Sector. 2022. Available online: https://itc.gov.my/sabah-shows-potential-in-muslim-friendly-tourism-and-hospitality-sector/?fbclid=IwAR1Ll5oKOimqDg5B8IrWcjPSnUP5mLI2dIajvuOwco7bnU4w2cIXyxK-H_g (accessed on 29 April 2022).

proceedings

MDPI

Proceeding Paper

Reassessing Malaysian Poverty Measurement after COVID-19: A Multidimensional Perspective †

Mohd Khairi Ismail [1,*], Vikniswari Vija Kumaran [2], Syamsulang Sarifuddin [3], Siti Nurul Munawwarah [4], Sharmila Thinagar [5], Nor Zuriati Amani Ab Rani [6] and Muhamad Zahid Muhamad [7,8]

1 Faculty of Business and Management, UiTM Cawangan Terengganu Kampus Dungun, Kuala Dungun 23000, Malaysia
2 Faculty of Business and Finance, Universiti Tunku Abdul Rahman, Kampar 31900, Malaysia
3 Faculty of Business, Economics and Accounting, HELP University, Kuala Lumpur 50490, Malaysia
4 School of Accounting and Finance, Faculty of Business & Law, Taylor's University, Subang Jaya 47500, Malaysia
5 Faculty of Business, Economics, and Accounting, HELP University, Shah Alam 40150, Malaysia
6 Faculty of Business & Management, UiTM Cawangan Kelantan, Machang 18500, Malaysia
7 Faculty of Agriculture, Universiti Putra Malaysia, Seri Kembangan 43000, Malaysia
8 Faculty of Plantation and Agrotechnology, Universiti Teknologi MARA Cawangan Melaka Kampus Jasin, Merlimau 77300, Malaysia
* Correspondence: khairiismail@uitm.edu.my

† Presented at the International Academic Symposium of Social Science 2022, Kota Bharu, Malaysia, 3 July 2022.

Abstract: Poverty is a multifaceted phenomenon that has always existed historically. In addition to discussions of poverty issues, aspects of poverty measurement are essential topics. A prominent poverty measurement system is the unidimensional measurement based on poverty line income (PLI). Therefore, the use of multidimensional poverty measurements is proposed, specifically the multidimensional poverty index (MPI), in assessing poverty, including non-monetary aspects. This study discusses the concept of unidimensional and multidimensional poverty measurement and the implementation of these concepts in Malaysia during the COVID-19 outbreak. Furthermore, it aims to contribute to the public debate on COVID-19 policy responses by quantifying the potential impact on global multidimensional poverty using the Global Multidimensional Poverty Index (MPI), which captures concurrent or overlapping deprivation at the household level. This study recommends policies by which members of the community and industry experts would be included in the poverty reduction program, in line with Maslow's requirements. Overall, this research focuses on planning and orienting policy responses from a multidimensional perspective that integrates health, social and economic goals.

Keywords: poverty; multidimensional poverty index; COVID-19

Citation: Ismail, M.K.; Kumaran, V.V.; Sarifuddin, S.; Munawwarah, S.N.; Thinagar, S.; Rani, N.Z.A.A.; Muhamad, M.Z. Reassessing Malaysian Poverty Measurement after COVID-19: A Multidimensional Perspective. *Proceedings* **2022**, *82*, 48. https://doi.org/10.3390/proceedings2022082048

Academic Editor: Mohamad Rahimi Mohamad Rosman

Published: 15 September 2022

1. Introduction

Poverty is a multidimensional phenomenon [1]. In discussing the issue, the very concept of poverty should be understood. There was no general definition of poverty at the national level until the World Summit for Social Development in 1995, which defined absolute poverty as a lack of basic needs, including food, clean water supply, sanitation facilities, health, clothing, education, and information. In addition, poverty includes deficiencies in assets, income, health, education, residence, environmental safety, social levels, and culture.

Citing [2], who used the World Bank definition, poverty is a state of shortage of food, clothing, access to health facilities, primary education, employment opportunities, clean water, and social facilities. The study of [3] explains that the poor and extremely poor are trapped in a state of shortage due to poor health levels, family problems, poor source distributions, infrastructure imperfections, and various forms of limited employment opportunities. Dimensional poverty measurement is a qualitative leap forward in poverty care, as it involves shifting from measures based simply on income to others based on several social, labor, and environmental variables, which are extremely useful for defining public policy [4].

In addition, COVID-19 has also harmed shared prosperity, which is defined as income growth for the lowest 40% of a country's population. Average worldwide shared prosperity is expected to stagnate or possibly decrease between 2019 and 2021 as average income growth slows. The pandemic-induced slowdown in economic activity is expected to disproportionately affect the poorest individuals, perhaps leading to even lower shared prosperity indices in future years [5].

Most of B40 and M40 has been involved in SME businesses, working as full-time employees in the industry. However, the COVID-19 pandemic in 2020 reduced the main sources of their income, especially paid employment, and self-employment, which was recorded as negative at 16.1% and negative at 9.7%, respectively. This reduction in income was caused by job loss and contributed to a decrease in working hours and an increase in skill-related underemployment. The median monthly household gross income registered a negative decline of 11.3% as compared to 2019. The value of the median monthly household gross income in 2020 was RM5209 (2019: RM5873) [6].

Globally, efforts to eradicate poverty have been paid attention to for a lengthy period. Currently, efforts to eradicate poverty continue through the implementation of the Sustainable Development Goals (SDGs). Their implementation has targeted 17 key goals in achieving sustainable development and poverty eradication is one of the main agendas by 2030. Poverty eradication in the context of SDG not only considered basic requirements but also considers multidimensional aspects. The diversity of the dimensions of poverty not only refers to income, but to basic facilities, education, social discrimination, and readiness in decision making.

In the context of Malaysia, the measurement of poverty uses a money matrix approach, i.e., the poverty lines income (PLI) system of measurement has been used for a long time. This measurement determines individual poverty according to the minimum income required for daily life. For example, if PLI is set at RM950 per month, individuals who earn less are categorized as poor. However, through the Eleventh Malaysia Plan (RMK-11), multidimensional poverty measurement (MPI) has begun to be introduced [7]. Here, the measurement of poverty is quite different from that discussed earlier and combines money and non-money elements.

Therefore, this paper's main aim is to discuss the concept of unidimensional and multidimensional poverty measurement and, subsequently, focus on the measurement of poverty in Malaysia. As the pandemic spreads, understanding the extent of the threat it poses in terms of reversing development gains is critical for developing and implementing public policy. Therefore, this study can contribute to the study of poverty, especially in the body of literature related to the issue, and to the measurement of poverty after the COVID-19 pandemic. Besides, this paper aims to contribute to the public debate on COVID-19 policy responses by quantifying the potential impact on global multidimensional poverty as measured by the global Multidimensional Poverty Index (MPI) developed by [8], which captures concurrent or overlapping deprivation at the household level.

This paper is divided into four parts; the first discusses unidimensional poverty measurement or the money metric approach. The second part focuses on multidimensional poverty measurements. The next focuses on the measurement of poverty in Malaysia. Finally, this paper also examines issues that have led to the use of multidimensional poverty measurements in Malaysia.

2. Multidimensional Poverty Measurement

The cost of measurement of unidimensional poverty in measuring poverty has led to efforts to measure poverty from various dimensions. This readiness is clearly seen when a unidimensional approach only measures poverty using earnings to determine individual poverty. Multidimensional Poverty Approaches can overcome the weaknesses of the unidimensional approach as they consider non-monetary aspects such as health, education, and living standards in measuring individual poverty. Based on the concept of the "Capability Approach ', specified by [9], poverty not only refers to a lack of money but shortcomings in other social commands, for example, lack of health, education, and living standards. In other words, individuals with enough income but still lacking in non-monetary aspects can still be categorized as poor. This idea has led to a broader determination of poverty.

There are three main dimensions in forming MPI: Education, Health, and well-being. The study by [10] found that MPI methods reflected the widespread shortcomings faced in basic needs and core human functioning for people in 104 countries. Despite the availability of data, MPI measurements show different results compared to unidimensional poverty measurements. MPI measurements classify individuals as poor if they earn more than 30% of the defined deficiency. The difference in poverty rates is because, although individuals obtain an income that exceeds the minimum value set, they may experience shortages of other aspects and thus are categorized as poor if measured using MPI. The difference in MPI's measurement compared to the earlier measurement of income is also evidenced by the studies of [3,4,8].

MPI methods are seen to be more flexible as they can measure poverty using primary and secondary data, although primary data usage is more often used [3]. The study by [11] has discussed the basic calculations of MPI using primary data in detail. In the basic calculations of MPI, two main computational components have been considered, poverty (H) and average poverty (A). The rate of poverty, (H), refers to the ratio of individuals with multidimensional poverty:

$$M_0 = H \times A \tag{1}$$

where Q is the number of individuals categorized as poor and n is the population. Average poverty, A, illustrates the rate of the component of the indicator, D, on average, indicating individuals with deficiencies. For poor households, shortage scores will be summed up and divided by the number of indicators and poor individuals:

$$H = \frac{\sum_1^q c}{QD} \tag{2}$$

where C is the amount of weight loss experienced by poor individuals and D is the number of components of the indicator taken. Table 1 shows MPI data for respondents to calculate MPI.

Table 1. Multidimensional Poverty Index (MPI) Calculation Data.

Indicators	Respondent				Weightage
	1	2	3	4	
Household size	4	7	5	4	
All 17-60-year-old household members have less than 11 years of schooling	0	1	1	0	1/4 = 0.25
Distance to health facilities exceeding 5 km and no mobile health clinic facilities available	0	0	0	0	1/4 = 0.25
Apart from the water supply in the house and the public water pipe/standing pipe	1	1	0	0	1/4 = 0.25
SCORE CI (sum of each deprivation multiplied by its weight)	0.25	0.5	0.5	0	
Household/poor respondent if less 2 indicators	No	Yes	Yes	No	
Censored Score CI (k)	0	0.5	0.5	0	

Score for respondents or second household:

$$ci = (1 \times 1/4) + (1 \times 1/4) = 0.5$$

$$\text{Poverty ratio (H)} = [(7 + 5)/(4 + 7 + 5 + 4)] = 0.6$$

The value of 0.60 shows 60% of poor households.

$$\text{Average poverty gap (a)} = \lfloor((0 \times 4) + (0.5 \times 7) + (0.5 \times 5) + (0 \times 4))/(4 + 7 + 5 + 4)\rfloor = 0.3$$

Average poverty (a) = 0.3 for individuals with a shortage is 30% of indicator weights.

MPI value is designated through the value of the coordinated poverty rate ratio, Mo = MPI × A.

$$\text{MPI} = \text{H} \times \text{A}$$

MPI, Mo = 0.6 × 0.3 = 0.18.

MPI value is 0.18. The MPI range is from value 0 to value 1. The higher the MPI value, the higher the value of dimensional poverty. Overall, MPI is a measurement that measures the multidimensional poverty faced by individuals, considering the average poverty faced by these individuals.

3. Poverty Measurement in Malaysia

The measurement of poverty in Malaysia began at the beginning of independence and has reflected a continuing pattern to this day. The implementation of the New Economic Policy (NEP) in 1970 was an aggressive measure of the government in eradicating poverty [4]. As a result, the poverty rate decreased from 49.3 percent in 1970 to 16.5 percent in 1990. The poverty rate continued to show a decrease to 3.6 percent in 2007 and its rate decreased to 0.6 percent in 2014 [12]. The decline in poverty rates involves urban and rural areas. Overall, the decline in poverty rates was due to the efforts of the government in implementing various poverty eradication policies and programs. Among the poverty eradication programs are land development, agricultural-based development, development of rural areas, and basic facilities and utilities [13].

In the context of poverty in Malaysia, poverty is described through financial perspectives or income [14]. This concept of poverty is divided into two, absolute poverty and relative poverty, as studied by [15]. Absolute poverty refers to the circumstances in which an individual or household cannot afford to meet the basic needs of enjoying minimal and reasonable welfare of life. These basic requirements are food, clothing, dwelling, education, and medicine. Absolute poverty is measured by comparing the comparison of a group of households with a level called the poverty line income (PLI). This PLI is determined based on the living standards of the country. PLI is a minimum income amount to allow households, on average, to accommodate food requirements, clothing, and basic expenses such as rent, fuel and energy, transportation and communication, health, and recreation, while poverty is determined based on the income level received with PLI, i.e., poor or not poor.

The official use of PLI in measuring poverty began in 1976, when the calculation of PLI is based on food requirements and basic needs [16]. Food items are based on Recommended Daily Allowances, calculated by a technical group of the Food Division of the Ministry of Health appointed by the Economic Planning Unit (EPU). Items for basic needs are based on demographic, gender, and age factors and include spending on education and recreation, transportation and communications, rentals, oil, and energy, clothing, and health care, as non-food items These basic needs are also guided by the World Bank's Living Standard Measurement Study introduced by [17]. Among the items used by the EPU are clothing, rent and energy, transportation, and communication. Determination of minimum value for these items is based on income investigation and household expenses (HIES). Therefore, the measurement of poverty used is based on the needs of households in meeting basic food and non-food needs to ensure that every household leads an active, healthy, and proactive life.

In detail, there are two types of PLI calculations in Malaysia, i.e., PLI food and non-food. For the method of determining the food PLI, the requirements for the basic components according to the composition and size of the household will be identified by gender, age, recommended dietary allowance (RDA), nutrient component, and Physical Activity Level (PAL) [1]. Subsequently, the RDA value will be determined by considering the group, age, and nutrient components. Food retail price information is obtained by taking into account seven types of foods: (1) grain products, (2) meat, fish, eggs and chicken, (3) milk, (4) legumes, (5) oil and fats, (6) sugar and (7) vegetables and fruits. For the determination of non-food PLI, the type of goods and services is determined by reference to items (1) clothing and footwear, (2) gross rent, fuel, utilities, (3) furniture, equipment, and home furnishings, (4) transportation and services and (5) non-food items.

Although PLI involves a single value generation to measure poverty, the value is always reviewed and changes according to the current situation. The Government of Malaysia updates PLI values based on the Consumer Price Index (CPI). However, due to the differentiation in living standards between the Peninsula, Sabah, and Sarawak, the Government uses different PLIs. This means that the value of PLI for the peninsula is different compared to that for Sabah and Sarawak. For example, the Economic Prevention Unit (UPE) sets the annual PLI for the peninsula as RM950, whereas the PLI for Sabah and Sarawak is RM1160.

However, the concept and measurement of poverty as described are considered too narrow by [1], who claim that this concept ignores other important aspects of poverty such as facilities and social isolation. Efforts need to consider various factors, processes, and obstacles to enhance the lives of the poor for the better, whether economically, ecologically, or socially. Hence, there is an effort to assess the measurement of poverty so that these problems can be overcome.

Multidimensional Poverty Index in Malaysia

In Malaysia, traditionally poverty is measured using PLI. This approach categorizes household wealth based on certain levels of income. In 2010, the United Nations Development Program (UNDP) used a multidimensional poverty index (MPI in Human Development Report (HDR)). MPI is a method of relative poverty measurement practiced by developed countries. In line with this, Malaysia has also developed measurements of MPI at the appropriate national level for socioeconomic well-being within the National Development Framework. The use of MPI to ensure the transition of policy considerations beyond the mere issue of poverty includes all households affected. MPI measurement will complement PLI to measure and monitor poverty from multiple dimensional perspectives. Hence, the multidimensional poverty index (MPI) was introduced in the Eleventh Malaysia Plan (RMK-11), completing PLI and helping the government address B40 issues, as this measurement provides more complete information for the formulation of policy documents [7].

Currently, there are about 2.7 million B40 households with an average monthly income of RM2537. In the event that Malaysia continues to develop, B40 households should not be marginalized in enjoying opportunities resulting from the country's prosperity. If B40 households continue to stay at their current socioeconomic position, this will result in social costs to the country through a reduction in the number of skilled workers required, affecting the growth of the country's output. In addition, this will also cause continuous urban imbalance and affect the potential of rural and suburban economic growth. Employment opportunities, access to healthcare and education services and SSN can be expected to ensure that B40 households obtain the opportunity to enjoy a better life.

In the Malaysian context, MPI can measure access to basic facilities and services that help social mobility, enable intensity-based analysis, and subsequently identify the needs of low-income households more clearly. MPI is a measurement at the appropriate national level of socio-economic well-being for the country's development framework. The use of MPI will ensure the transition of the consideration of policy beyond the issue of poverty, including all households affected. MPI will complement PLI to measure and monitor poverty from multidimensional perspectives. This measurement can reflect both multidimensional dissemination incidents (the number of inhabitants experiencing various aspects of life in certain populations) and intensity (how many limitations of the measurements are experienced on average at a time).

The measurement and study of multidimensional poverty provide insight into the susceptibility of individuals and families to various risk factors. This includes the COVID-19 pandemic, which constitutes health, humanitarian, and socioeconomic disaster. This measurement identifies a subset of the population that is more vulnerable to economic losses or, rather, a vulnerability in its various forms and manifestations. Individuals and businesses are unavoidably vulnerable to what economists refer to as "shocks" or unfavorable occurrences that have the potential to produce significant reductions in living standards. People in situations of vulnerability and marginalization are the main victims [18].

According to [19], the impoverished find it difficult to comply with restrictive measures such as social isolation or lockdown. Furthermore, the history of pandemics such as SARS and Ebola has demonstrated that economic poverty plays a significant role in disease transmission [20]. Thus, poverty may increase the chance of contracting COVID-19, just as infection increases the probability of sliding into poverty [21]. Aside from these monetary-based conclusions, multidimensional measurements can give more precise information on the poor's incapacity to monitor health measures or their degree of vulnerability to a contagious disease such as COVID-19. People who are impoverished on several levels are deprived of actual possibilities or capacities [9].

4. Policy Recommendation

From the point of view of researchers and economists, it is time for the Malaysian government to hand over the reduction of poverty to members of the community and industry experts in various fields. After 64 years of independence, in 2021 the community's direct poverty management method meets a conflict of interest, as politicians are also resource holders in distributing poverty aid. The poor are seen as the 'Subject of Content' and there is an opportunity for them to be prosecuted merely to meet the demands of social media and mainstream media. Assistance often fails to completely help the needy communities. Researchers suggest that the National Leader is sincere in managing the poverty of the community from the grassroots level, but they need to be involved indirectly and simply become moderators or whistleblowers in managing poverty. Leaders need to find a 'Key Person' or 'Key Organization' in every field that has the best Track Record and is open to all, with clear terms and goals in meeting the MPI dimensions mentioned above.

This idea is based on certain principles. The first is the principle of Maslow's requirement, where people who have succeeded in a field will need to serve and give back to the community. For example, many businesspeople become philanthropists and set up their own community bodies and Corporate Social Responsibility (CSR) campaigns. Therefore, leaders need to offer platforms, such as assigning local areas with high poverty rates to these businessmen who might offer employment opportunities (earnings dimension) or educational scholarship assistance (education dimension), for example, and offering tax relief, adopting the village on behalf of these philanthropists. This and other initiatives can improve the level of satisfaction and requirements, as in Maslow's theory of needs.

The second principle is that of effectiveness and specialization. 'Key Persons' or 'Key Organizations' appointed are those with expertise and specialization, which is outcome-based, and who understand the needs of the community in their field. When the poverty management campaign is implemented by its members, the effectiveness of managing poverty can be increased from the grassroots. The privatization policy of the 1980s has made significant differences in several institutions/organizations after they were privatized, making statutory bodies more efficient and functioning more effectively than before.

Funds that are supposed to be given directly to the poverty program can be saved and used for the purpose of development expenditure and other national management schemes. However, all the selection processes for the 'Key Person' or 'Key Organization' should be transparent and audited according to the objectives and evaluations set annually. Requirements and Feedback Community members should also be considered in ensuring that the management process can be improved in the future.

In the above MPI index, there are four dimensions: Education, Living Standards, Health, and Income. If we take education, for example, and the Key Organization is Telekom Malaysia, offering an internet plan for a village area in the interior of the Sabah state, such as during the COVID-19 Pandemic, not only improves the quality of online education for the students but also reduces future poverty rates. Not only that but, in the face of the pandemic challenges, there was a belief that online education can be implemented. Therefore, the government should implement Education for Everyone. University entry policy should be increased by 100%, because the online mode does not require a large physical infrastructure. Some education can be implemented online and can be carried out with the involvement of community members or expert organizations that are made into 'Key Persons' or 'Key Organizations', meeting the needs of the education dimension.

This proposal can also reduce the "Menu Costs" in the management of community poverty. This is because every time a leader changes, the poverty campaign will also change according to the manifesto or the campaign of the politician. The issue of poverty should be the most important issue and should be governed by a capable person with no conflict of interest and with continuously implemented objectives, processes, and dimensions that have been developed in eradicating poverty from the bottom, directly empowering the needs and interests of the 'Key Person' or 'Key Organization' in releasing their contribution to the community in the field in which they are involved. Community members should also enjoy necessary assistance without any suspension according to the influence of the party or politician.

5. Conclusions

This paper has discussed two types of poverty measurement, unidimensional and multidimensional, based on previous studies. The discussion has also considered the basic concept of measurement and the means of calculating both measurements. In addition, this paper has discussed in detail the measurement of poverty in Malaysia used since the 1970s in terms of the concept, formation, and calculation of the poverty line income (PLI). Currently, attention to the B40 issue that has become the focus in Malaysia has led to the introduction of the multidimensional poverty index (MPI) via RMK-11. The use of this MPI should ensure that all relevant elements and quality of life are accounted for as an additional measurement of income, which is relevant to the current COVID-19 pandemic.

Policies are recommended in this study, and it is suggested that members of the community and industry experts are involved in the poverty reduction program and that this is aligned with Maslow's requirements. This study also recommends that the key person selection policy must be transparent and audited annually in accordance with the objectives and evaluations set forth. Overall, this research can help to plan and orient policy solutions from a multidimensional approach, integrating health, social and economic priorities. Poverty reduction involves a variety of techniques, all of which are focused on improving not only income, but also non-monetary factors such as education, health and living standards.

Author Contributions: Conceptualization, M.K.I. and V.V.K.; methodology, S.S.; validation, V.V.K.; formal analysis, S.N.M.; investigation, S.S.; resources, M.K.I.; data curation, N.Z.A.A.R.; writing—original draft preparation, S.N.M.; writing—review and editing, S.T.; visualization, S.T.; supervision, N.Z.A.A.R.; project administration, writing, review, editing, M.Z.M.; funding acquisition, M.K.I. All authors have read and agreed to the published version of the manuscript.

Funding: The research was funded by Pembiayaan Yuran Prosiding Berindeks (PYPB), Tabung Dana Kecemerlangan Pendidikan (DKP), Universiti Teknologi MARA (UiTM), Malaysia.

Institutional Review Board Statement: Not applicable.

Informed Consent Statement: Not applicable.

Data Availability Statement: Not applicable.

Conflicts of Interest: The authors declare no conflict of interest.

References

1. Siwar, C.; Ahmed, F.; Idris, N.D.M.; Idrus, S.; Ghan, A.H.A.; Zahari, S.Z. Poverty Mapping and Assessing Socio-Demographic Characteristics of the Households: A Case Study in Pahang, Malaysia. *J. Appl. Sci.* **2014**, *14*, 2632–2640. [CrossRef]
2. Braithwaite, J.; Mont, D. Disability and poverty: A survey of World Bank Poverty Assessments and implications. *Alter* **2009**, *3*, 219–232. [CrossRef]
3. Alkire, S.; Santos, M.E. Acute Multidimensional Poverty: A New Index for Developing Countries. *SSRN Electron. J.* **2010**. [CrossRef]
4. Siwar, C.; Hasan, S.K.; Chamhuri, N. *Ekonomi Malaysia*; Pearson Malaysia: Kuala Lumpur, Malaysia, 2005.
5. McKibbin, W.; Fernando, R. *Economics in the Time of COVID-19*; Baldwin, R., Weder di Mauro, B., Eds.; Centre for Economic Policy Research Press: London, UK, 2020.
6. Department of Statistics Malaysia. *Press Release Malaysia Economic Performance Fourth Quarter 2020*; Department of Statistics Malaysia: Putrajaya, Malaysia, 2021.
7. Economic Planning Unit (EPU). Rancangan Malaysia Kesebelas (Eleventh Malaysia Plan): 2016–2020. 2015. Available online: http://rmk11.epu.gov.my/book/eng/Elevent-Malaysia-Plan/RMKe-11Book.pdf (accessed on 1 September 2022).
8. Alkire, S.; Santos, M.E. Measuring Acute Poverty in the Developing World: Robustness and Scope of the Multidimensional Poverty Index. *World Dev.* **2014**, *59*, 251–274. [CrossRef]
9. Kurien, C.T.; Sen, A. Poverty and Famines: An Essay on Entitlement and Deprivation. *Soc. Sci.* **1983**, *11*, 66. [CrossRef]
10. Calvo, C.; Fernandez, F. *Measurement Errors and Multidimensional Poverty*; OPHI Working Paper: Oxford, UK, 2012.
11. Ibrahin, N.N.; Nabilla, F.; Husain, M.; Rahman, R.A. *The Multidimensional Poverty Index (MPI) and Its Application In Malaysia: A Case Study UiTM Students, Shah Alam Campus. In Proceedings of the Prosiding PERKEM VI*; Universiti Kebangsaan Malaysia: Selangor, Malaysia, 2011.
12. Siwar, C.; Zahari, S.Z.; Ismail, M.K. Kemiskinan Dan Agihan Pendapatan Di Malaysia: Status Dan Kadar Kemiskinan Mengikut Etnik Dan Strata Serta Golongan Berpendapatan Isi Rumah 40 Peratus Terendah (B40). In *Proceedings of the Prosiding Persidangan Kebangsaan Kecemerlangan Melayu II (Ekonomi Melayu)*; Universiti Malaya: Kuala Lumpur, Malaysia, 2015.
13. Economic Planning Unit (EPU). Tenth Malaysia Plan (2011–2015). 2011. Available online: https://www.epu.gov.my/en/economic-developments/development-plans/rmk/tenth-malaysia-plan-10th-mp (accessed on 1 September 2022).
14. Jamil, N.; Mat, S.H.C. Realiti Kemiskinan Satu Kajian Teoritikal. *J. Ekon. Malays.* **2014**, *48*, 167–177. [CrossRef]
15. Economic Planning Unit (EPU). Fifth Malaysia Plan (1986–1990). 1986. Available online: https://www.epu.gov.my/en/economic-developments/development-plans/rmk/fifth-malaysia-plan-1986-1990 (accessed on 1 September 2022).
16. Abdul Rasool, M.S.; Mohd Harun, M.F.; Salleh, A.M.; Haji Idris, N.A. Poverty Measurement in Malaysian Zakat Institutions: A Theoretical Survey. *J. Ekon. Malaysia* **2011**, *45*, 123–129.
17. Ravallion, M. Expected Poverty Under Risk-Induced Welfare Variability. *Econ. J.* **1988**, *98*, 1171. [CrossRef]
18. United Nations Development Programme (UNDP). Africa Human Development Report 2016 Accelerating Gender Equality and Women's Empowerment in Africa. 2016. Available online: https://www.undp.org/publications/africa-human-development-report-2016 (accessed on 1 September 2022).
19. Buheji, M.; da Costa Cunha, K.; Beka, G.; Mavrić, B.; Leandro do Carmo de Souza, Y.; Souza da Costa Silva, S.; Hanafi, M.; Chetia Yein, T. The Extent of COVID-19 Pandemic Socio-Economic Impact on Global Poverty. A Global Integrative Multidisciplinary Review. *Am. J. Econ.* **2020**, *10*, 213–224. [CrossRef]
20. Abramowitz, S.; McLean, K.E.; McKune, S.L.; Bardosh, K.L.; Fallah, M.; Monger, J.; Tehoungue, K.; Omidian, P.A. Community-Centered Responses to Ebola in Urban Liberia: The View from Below. *PLoS Neglected Trop. Dis.* **2015**, *9*, e0003706. [CrossRef] [PubMed]
21. Diwakar, V.; Shepherd, A. Sustaining escapes from poverty. *World Dev.* **2021**, *151*, 105611. [CrossRef]

proceedings

MDPI

Proceeding Paper

Legal Reformulation of the Problematics of Doctor's Medical Licenses in Indonesia †

Rianda Dirkareshza *, **Nada Prima Dirkareshza** and **Mouva Putri Ramadhita**

Law Faculty, Pembangunan Nasional Veteran Jakarta University, D.K.I Jakarta 12450, Indonesia
* Correspondence: riandadirkareshza@upnvj.ac.id
† Presented at the International Academic Symposium of Social Science 2022, Kota Bharu, Malaysia, 3 July 2022.

Abstract: This paper intends to analyze the regulation of doctors' medical licenses in Indonesia, as well as find regulation loopholes in facing factual problems regarding these medical licenses. This paper uses socio-legal and juridical–normative research, as well as a case approach that scopes doctor medical license regulations and their problematics, especially in Indonesia. The findings of this research show that doctor profession ethics are the main key to receiving a medical license, and that collaboration is also needed to reformulate the regulation of medical licenses in Indonesia.

Keywords: medical license; doctor; profession ethics

check for updates

Citation: Dirkareshza, R.; Dirkareshza, N.P.; Ramadhita, M.P. Legal Reformulation of the Problematics of Doctor's Medical Licenses in Indonesia. *Proceedings* **2022**, *82*, 49. https://doi.org/10.3390/proceedings2022082049

Academic Editor: Mohamad Rahimi Mohamad Rosman

Published: 13 September 2022

1. Introduction

Doctors belong to the most essential profession that has a big role in preventing disease and providing direct health services to the community. In carrying out their profession, doctors are allowed to perform actions in the form of medical interventions on the physical human body. As a result of this, before carrying out medical work, it is necessary for a doctor to have a legal registration certificate and a medical license in order to provide medical services in Indonesia [1]. Providing medical services to the community requires a knowledge and understanding of Indonesian laws and regulations that are currently applicable in Indonesia [2]. In order to maintain doctor quality in Indonesia, a series of competency exams are arranged for doctors to take from a theoretical and practical point of view by the professional organization of the Indonesian Doctors Association (IDI).

In March 2022, a national-scale polemic regarding medical licenses occurred in Indonesia [3]. It started with the recommendation letter of the Honorary Court of Medical Ethics (MKEK) Number 0280/PB/MKEK/02/2022, which was issued and addressed to the chairman of the IDI general board, containing the results of an MKEK decision following the plenary meeting of the MKEK at the IDI headquarters on 8 February 2022. The results of an MKEK decision was written because a doctor with the initial 'T' (hereinafter referred as Dr. T/Doctor T), the former head of the Gatot Soebroto Army Hospital and the former Minister of Health for the term of 23 October 2019–23 December 2020, was dismissed with the reason of considered to have violated severe ethics codes and conducted several controversies throughout 2018-2022 [4]. As for the main facts regarding the dismissal of Dr. T, he has not submitted evidence that he carried out ethical sanctions following MKEK Decree No. 009320/PB/MKEK-Decree/02/2018, dated 12 February 2018. Second, Dr. T promoted the Nusantara vaccine to the wider community before his research was completed. Third, Dr. T acted as chairman of the Indonesian Association of Clinical Radiology Specialists (PDSRKI), which was formed without going through the necessary procedures in accordance with the IDI Procedures and Organizations (PRTALA), or the endorsement process at the IDI Congress. Fourth, issued circular letter number 163/AU/Sekr.PDSRKI/XII/2021, dated 11 December 2021, contained instructions to all branch leaders and PDSRKI members throughout Indonesia to not respond to or attend PB IDI events. Fifth, Dr. T applied for a membership transfer from the Central Jakarta branch of the IDI to the West Jakarta branch of the IDI [5].

The dismissal of Dr. T led to the revocation of Dr. T medical license for the doctor's profession. To this, the Minister of Law and Human Rights, Mr. Yasonna Laoly, responded, considering the IDI to have an overly broad power in affecting the ownership of a doctor's medical license. This issue has become a public controversy because Dr. T is a former government official. There are many opinions within the government community that argue subjectively, and a lot of civilians who also argue that the IDI is still needed to maintain the quality of doctors who will practice in the industrial world. Sociologically, the IDI is present to ensure the implementation of quality medical services [6]. Philosophically, the IDI maintains and enhances the dignity and honor of the medical profession [7]. Juridically, there is an expansion of the IDI's authority through Constitutional Court Decision No. 10/PUU-XV/2017, which causes disharmony within the regulations regarding medical licenses in Indonesia.

According to observer, there is legal uncertainty in Medical Practice Law that harms their constitutional rights due to article 14, paragraph (1), regarding the number of KKI members who come from the IDI, namely two people. The IDI board is also the commissioner of the KKI, which cause a potential conflict of interest. This issue is due to regulations that have been made, and applies to IDI administrators who are concurrently in office. In addition to this, there are also articles regarding the medical education law, whose contents should be included in the responsibilities of the medical collegium. This issue is because the medical education law enters the academic realm and the Indonesian academic body, which is the medical collegium of the IDI. However, in these articles, there is a statement of "professional organization". The articles are as follows: article 1, number 20; article 5, paragraph (2); article 7, paragraph (8); article 8, paragraph (4); article 11, paragraph (1); article 24, paragraph (1); article 36, paragraph (3); and article 39, paragraph (2).

Then, the applicant also discussed article 1, number 4 of the medical practice law which states: "Certificate of competence is a letter of recognition of the ability of a doctor or dentist to practice medicine throughout Indonesia after passing the competency test." This article forces new medical graduates to conduct second competency tests, which are carried out by the IDI for recertification. In fact, the applicant already achieve the competence of a doctor with education and training they experienced. Moreover, doctors formally obtain a professional certificate and a certificate of competence simultaneously upon completion of their medical education and the national competency test [8]. According to the applicant, proof of competence of a doctor is enough to obtain a professional certificate, which they receive after they pass the competency test of the doctor profession program. This issue is mentioned in article 29, paragraph (3), regarding the requirements to obtain a doctor's registration certificate, which include: a doctor's diploma as a specialist, dentist, or specialist dentist; a statement that they have conducted the Hippocratic Oath; a certificate of health; a certificate of competence; and a statement that they will comply and implement the provisions of professional ethics. This indicates that there will be a second competency test. In addition, the IDI the one that considered to have conducted monopoly practices in the issuance of professional certification, continuously facing a lawsuit, now even declared by the Constitutional Court as the only legal organization in which one can be a medical professional in Indonesia [9]. Furthermore, the IDI also does not have a supervisory board. According to Member of Commission IX of the House of Representatives, Irma Suryani Chaniago, the IDI requires its own supervisory board due to the dismissal of Doctor T. The IDI has failed to achieve its goals, of which there are at least three: prospering, developing, and protecting its members [10]. Therefore, it can be concluded that the power of the IDI is too broad and is considered a super-body by the government.

2. Legal Problems of Doctor's Medical License in Indonesia

On 25 March 2022, the results of the special session of the Honorary Council of Medical Ethics (MKEK) declared the permanent dismissal of doctor T. Chairman of the Presidium,

Abdul Azis, said that he decided, determined, and forwarded the results of the MKEK special meeting. As a result of the meeting, the MKEK decided on the permanent dismissal of colleague Dr. T as a member of the IDI, which was to be carried out within 28 working days. Consequently, the case became an internal organizational problem; thus, the Ministry of Health was only able to be a mediator between the two parties [11]. Reasons for the dismissal of Dr. T were shown, and one of those is regarding the *Terawan theory*, which concerns the medication of stroke by using a brainwashing method of the name Digital Subtraction Angiography (DSA).

Responding to this issue, various professions believe that the *Terawan theory* cannot be scientifically proven [12]. In 2017, there was a study conducted regarding doctors who advertise and the limitaion. According to the *Terawan theory*, doctors are not allowed to use their medical degrees and attributes to claim the well-being of medication and its products, and there should also be no element of self-praise [13]. This is also stipulated by the Indonesian Medical Code of Ethics (KODEKI), in which article 4 states "A doctor must avoid acts that are self-praise". Additionally, article 6 states "Every doctor shall be careful in announcing or applying any medical discovery of new techniques or treatments that have not been tested for truth and against things that can cause public unrest". Based on those articles, Doctor T violated two KODEKI articles. They cannot be subjected to arbitrary sanctions in the absence of community complaints of ethical transgressions or opposition from civilians to new medical discoveries made by IDI members.. However, right after Doctor T eventually completed a dissertation on his findings, the MKEK invited Doctor T regarding the IDI's subsequent objections. The MKEK served a summons to Doctor T five times. However, based on regulation, if a member has been summoned three times and does not attend, the MKEK sues in absentia [14].

In the organization and work procedures guidelines (ORTALA) of the MKEK, sanctions that are given to doctors who have committed ethical violations can be sanctioned via the lightest penalties, which take the form of advisory, oral, or written warnings, behavioral coaching, and re-education. Alternatively, they can receive the maximum penalty, which takes the form of an either temporary or permanent dismissal [15]. Doctor T himself was sanctioned based on MKEK Decree No. 009320/PB/MKEK-Decree/02/2018, which stated that the sanction imposed upon Doctor T was in the form of a temporary dismissal from the MKEK IDI for 12 months. This is one of the facts regarding the dismissal of Doctor T in SK MKEK 0280/PB/MKEK/02/2022, which was due to the fact that Doctor T did not submit evidence that he carried out the sanctions stated in the MKEK No. 009320/PB/MKEK-Decree/02/2018 decree. The temporary dismissal sanction is the third category of sanction in violation of medical ethics, and Doctor T may lose his rights and authority as a doctor and IDI member within the period of implementation of his sanctions. Doctor T can lose his rights and authority by having his license recommendation practice revoked. As a result, Dr. T will not able to conduct medical practices. In addition, his rights, authorities, and positions as IDI member and Organization under IDI position will also be removed [16].

During the period of the development of COVID-19 vaccines, Doctor T repeated his mistake when he initiated the *Terawan theory*. Doctor T is the initiator of the Nusantara vaccine, and in its progress, the Nusantara vaccine experienced problems, including clinical trials in which data were not coherent, as well as the resignation from the University of Gadjah Mada (UGM). The resignation was due to the non-involvement of UGM researchers in the clinical trial process. In addition, the Nusantara vaccine development process was also considered to not be in accordance with medical rules stipulated by the Food and Drug Supervisory Agency (BPOM). Moreover, if the vaccine progress remains in disagreement with research standards, the BPOM will not approve the existence of the next stage of clinical trials. Although the clinical trials have not been completed, Doctor T has injected the Nusantara vaccine into Indonesian officials [17]. Although Doctor T's research has not been completed, he conveyed to the public that the Nusantara vaccine has a high level of efficacy in dealing with the COVID-19 disease.

The third reason for Doctor T's dismissal regards the ethical sanctions that he made. Doctor T was appointed as chairman of the Central Radiology Specialist Association (PDSRI). However, this appointment was not in accordance with regulations; thus, it can be concluded that Doctor T was appointed as chairman unconstitutionally. Moreover, Doctor T, as chairman, also maneuvered by changing the name of the PDSRI to the Indonesian Association of Clinical Radiology Specialists (PDSRKI): this change did not go through the process of ratification at the IDI congress, and did not go through the necessary procedure in accordance with the IDI Procedures and Organizations (PRTALA). During his tenure as chairman of the PDSRKI, Doctor T also issued a leaflet that was memorably disobedient to the IDI, which was in the form of instructions that all branch chairmen and PDSRKI members throughout Indonesia should not respond to or attend events organized by the PB IDI. Promptly, the letter was discovered by the MKEK, which then became the fourth reason for the dismissal of Doctor T from IDI membership [18]. The last reason regards the transfer of membership that was requested by Doctor T to move from the Central Jakarta branch to the West Jakarta branch, which has conditions that should have been met in the membership mutation form, such as providing evidence in the form of a statement that has undergone sanctions [19]. After the announcement of the dismissal of Doctor T, the issue became the center of attention. Surely, this is due to the impact of the dismissal from IDI membership with Dr. T's medical license. Dr. T's medical license is still valid until 2025; however, his dismissal will make it difficult for him to extend his medical license. This is due to the need for a registration certificate to obtain a medical license which, according to article 29, paragraph (3) of medical practice law, requires the following conditions:

1. Ownership of a doctor's diploma as a specialist, dentist, or specialist dentist; ownership of an affidavit regarding the Hippocratic Oath; ownership of a health certificate;
2. Ownership of a certificate of competence;
3. Provision of a statement letter that states a willingness to comply with and implement the provisions of professional ethics.

The certificate of competence, or the SERKOM, must be renewed every 5 years by fulfilling 250 of the professional credit unit (SKP) within 5 years. In order to get the SKP, the doctor ought to become an IDI member. The reason for this is that, in order to receive and recommend the SKP for an extension of the SERKOM, which is a requirement for the extension of the STR and medical licenses, the IDI is necessary. By that, it can be concluded that a doctor is not capable of doing anything without being an IDI member [20]. This formed the assumption that the IDI is a *super-body* organization by various parties, including the government. In the general hearing meeting between Commission IX and the IDI, Irma Suryani Chaniago, who is a member of Commission IX of the House of Representatives, argued that the IDI needs to be dissolved, which due to IDI that made several doctors unemployed, after failing to attend the competency examination, and the dismissal of Doctor T, indicating that the IDI does not prosper its members [21].

As a result, the government has intervened in the form of a proposal for a reassessment by the Minister of Law on the IDI's authority, regarding the permit of doctor's practices so that medical practice law and medical education law can be improved/revised. According to Rahmad Handoyo as a member of Commission IX of the House of Representatives, in the regulations of the IDI, they are given great authority, even though the IDI is a professional organization that is beyond the executive realm. Moreover, the government is also not able to supervise the IDI [22]. Due to the government's helplessness in supervising and handling the IDI, the proposal regarding the establishment of the IDI supervisory board was made by Irma Suryani Chaniago, a member of Commission IX of the House of Representatives. According to Irma, "IDI should act more mature and all issues should not necessarily be determined by IDI. There should be an oversight over the IDI that corrects, gives advice, etc. to the organization of this profession. As a result, it does not become a 'superbody' and elitist" [23].

A section of the consideration point and of the medical practice law states "that health development is aimed at increasing awareness, willpower, and the healthy living ability for everyone in order to realize optimal health degrees as one of the elements of the general welfare as intended in the preamble of the Constitution of the Republic of Indonesia in 1945". Development in the field of health and providing general welfare is, in essence, the fulfillment of health rights. Society became one of the goals during the making of the law on the practice of medicine. Article 28H, paragraph (1) of the 1945 Constitution states that "Everyone has the right to live a prosperous life born and inner, live, and get a good and healthy living environment and entitled to health services." The article mentions the rights that must be fulfilled by each individual, one of which is the right to obtain health services. For the fulfillment of these rights, it is necessary for the convenience for each individual to find and reach health facilities. Indonesia itself still lacks human resource health (HRK) and, until 2025, Indonesia is estimated to remain lacking in health workers by as much as 40–50%. With a fairly large population, human resources are needed in health development in Indonesia because HRK is also a determining factor for the success of health development [24].

In 2021, the ratio of health workers in the form of specialists and general doctors to the Indonesian population was 0.67%: 1000 population. The ratio was very visible when patients were exposed to the virus that spread in every hospital during the COVID-19 pandemic. Even in health centers and other facilities in Maluku and Papua, 50% still do not have a doctor health worker. In a webinar held on 18 December 2021, entitled *Towards COVID-19 Endemic Readiness*, by the public health faculty of the University of Indonesia, Deputy Minister of Health Dante Saksono Harbuwono stated that there are 12 thousand doctor graduates per year, although there is a need to reach 140 thousand personnel; therefore, it can be declared that this is still not enough [25]. This is further aggravated by these graduate doctors who will be unemployed because of the difficulty in obtaining a license to practice. With the increasing number of infected patients and the lack of health workers, especially doctors, to cope with it, will worsen the psychological condition of health workers due to too large a workload [26].

3. Legal Comparison of Medical Licenses in Malaysia and the UK

Health is an important aspect of human life. Nevertheless, health workers, especially physicians, have an important role to play in maintaining, creating, and developing healthcare systems. People's perception of medical personnel is generally positive because people tend to often use the health system. This refers to the case of the COVID-19 pandemic that emerged at the end of 2019. The flow of public health services has increased due to the 153 million confirmed cases of COVID-19 by the WHO. Therefore, it is proven that doctors have a strong role in society, especially in the field of health [27,28].

A medical license is required for all health workers to be eligible to practice legally. In the implementation of the medical license, there are provisions that must be implemented and prepared. A doctor's license applies to all health workers, especially doctors, around the world. Of course, there are significant differences regarding the procedures and provisions of the medical license. These regulatory differences arise due to differences in legal systems in each country. In addition, a medical license is necessary in order to qualify for practices to be conducted to protect citizens in the fulfillment of qualified medical personnel by carrying out their duties in the health sector [29].

On the Asian continent, the ratio between doctors and the population is 1.6:1000. Based on this, the number of doctors who have a license to practice throughout the Asian continent is quite large. In Malaysia, all doctors who have a medical license are registered in the [30] *Malaysian Medical Council* system and can be accessed through the *Medical Register Information and Technical System* (MeRITS) website. In Malaysia, to be able to obtain a doctor's license legally, health practitioners, including doctors, must be registered with the Medical Council of Malaysia [31].

The Malaysian Medical Council was formed under *the Medical Act* in section 4E and section 3 of *the Medical Act 1971*, which was amended in 2012. This institution was created to conduct the evaluation and implementation of registration of medical practitioners, which is regulated in section 4A of the Medical Act 1971, which has been amended. Based on section 3 of *the Medical Act*, the legislation addresses regulations relating to the registration of medical practitioners, the regulation of medical licenses, the regulation of certain provisions on the working period of medical practitioners in performing public services, and contains provisions on matters related to them.

The Malaysian Medical Council requires prospective doctors to conduct an examination for Provisional Registration (EPR) in order to obtain a license to practice in Malaysia. This is stipulated in the Medical Act 1971. Before the prospective doctors participate in the EPR, the Malaysian Medical Council attaches some provisions before conducting the exam. In addition, for doctors who are going to do further studies, in order to support their medical practices, based on the Medical Law (Amendment) 2012, the *National Specialist Registration* (NSR) must be carried out by candidates with regard to existing provisions.

Based on section 2 *of the Medical Act*, the medical license refers to the certificate of practice, which is further based on section 20, article 1. This must be extended to no later than December 1st of each year by applying in the form of a predetermined form and making payments that are also determined for the certificate of practice as a doctor.

The Medical Act also strictly addresses the sanctions for doctors who commit violations of the law. Article 7 of section 20 explains that, for people who practice medicine but do not have a certificate of practice in connection with their applicable self, partner with a person who has been registered as a doctor but does not have a certificate of practice, hire a person who has been registered as a doctor but does not have a certificate of practice and is hired to run the business of a medical practitioner on behalf of a person who does not have a certificate of practice, are declared to have violated the law and are not entitled to their rights in the form of rewards or payments.

Unlike on the Asian continent, in Europe, the ratio between doctors and EU population in 2018 was 1:3700. In one of the countries in continental Europe, the United Kingdom, the implementation of a medical license is regulated by a special institution called *the General Medical Council (GMC)*. The GMC was formed under the *Medical Act*, which was established because, in 1841, about one-third of doctors in England were not competent, and the GMC was formed to regulate the standards of doctors in England. Not only does the GMC regulate the licensing of doctors' practices in the UK, but it also serves as an institution authorized to revoke doctors' licenses [32].

The GMC in issuing licenses to doctors in England is always guided by the *Medical Act* 1883, which is contained within 8 chapters and 57 articles concerning the practice of medicine in England. The registration and qualifications of doctors in the UK is stipulated in chapter 3 of the law, namely *registration by virtue of the primary United Kingdom or primary European qualifications*. Prospective physicians must take or complete a qualified medical program, and subsequently take a nationally held test in order to be eligible for registration as a fully registered medical practitioner.

In article 2 of section 10 of the *Medical Act* 1883, the regulatory system for prospective doctors to obtain a license to practice in England was reaffirmed. Prospective doctors are required to take resident practice within the specified period of time in the designated institutions. Furthermore, article 3 again explains that prospective doctors can apply for a certificate of practice if they have met the conditions specified by the GMC.

The Medical Act of 1883 also clearly explains the sanctions imposed on doctors if they commit violations. Under section 36 of article 1, if there is a violation committed by a doctor, then the state can remove the perpetrator's name from the EU medical database. Additionally, the perpetrator's license certificate can be suspended for no more than 12 months, or no more than 3 years if the violation committed endangers the community.

Asia and Europe are two continents that have different characteristics in terms of region and culture, which have their own uniqueness. Regarding the health sector, especially in the medical license section, the income of the permit is not very significant. For example, both Malaysia and the UK remain based on applicable legislation and require several stages to obtain a medical license, such as tests conducted nationally. Therefore, it is appropriate for every prospective doctor to have the same ease and right in obtaining their medical license.

4. Legal Reformulation to The Problematics of Doctor's Medical Licenses

The revocation of a medical license is not only be based on the doctor's error in treating the patient, but also on all work-related acts and omissions, i.e., actions and omissions that are closely related to actual medical activity and, depending on the severity of the violation, criminal offenses outside the scope of judicial activities. This explains that the ethics of the doctor's profession becomes one of the determinants in their right to hold a medical license in the profession. Not only ethics in behavior, but also in the ethics of a publication or theory found must be tested in accordance with the standardization of national professional organizations and international regulations that overshadow the profession of a doctor [33].

Looking at the results of the analysis in the previous discussion that compares existing regulations in Indonesia, Malaysia, and the United Kingdom related to the process of obtaining a medical license, it can be seen that only Indonesia still involves the executive. In Indonesia, the health office determines whether a doctor will get a medical license after being given a recommendation from professional organizations and a registration permit from the Indonesian Medical Council. The governments of Malaysia and the United Kingdom provide legal protection to professional organizations in selecting doctors before they provide medical services to the community. From this, it can be interpreted that Malaysia and the United Kingdom are fully confident in professional organizations in their ability to produce qualified doctors, without direct intervention from the government. However, the government is limited to only being a regulator to prevent occurrences that are detrimental to professional organizations and doctors, as well as patients, later.

Health is a key factor that can be used as one of the parameters of people's welfare. Therefore, all holders of power, both executive and legislative, should be able to collaborate to produce better legal protection in producing qualified doctors. A study assessed the impact of partnership alliances in improving health quality, concluding that cross-sectoral collaboration to promote health between the public, private, and non-governmental sectors has always been successful [34]. *The Penta-helix* theory is the most appropriate theory to dissect and produce a reformulation of the problem of doctors' license practices in Indonesia (Figure 1), in which this theory demands collaboration to achieve a goal which is, in this case, a *healthy state* and a *welfare state* [35].

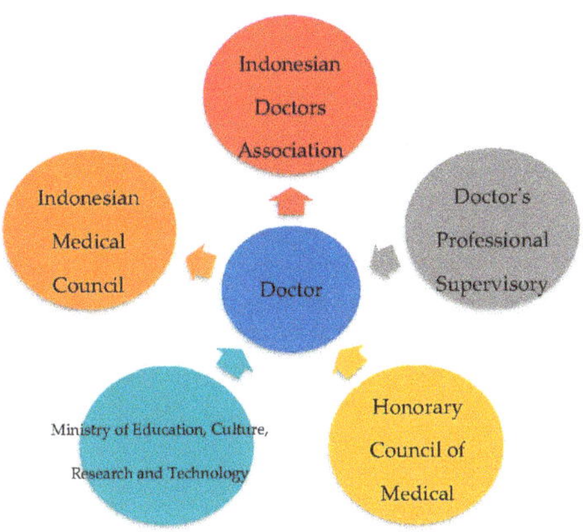

Figure 1. Reformulation of medical license with *Penta-helix* theory.

The collaboration above aims to produce a qualified doctor profession to serve the community to realize a *healthy state*, without direct intervention from the government. The author firstly hopes that the IDI, as a professional organization, maintains the authority to combine all potential doctors from all over Indonesia, maintain and improve the dignity and honor of the medical profession, develop medical science and technology, and improve the health of the Indonesian people in order to lead a healthy and prosperous society. Secondly, the KKI regulates, endorses, assigns, and fosters doctors who practice medicine, in order to improve the quality of medical services internally. Third, the Ministry of Culture, Research, and Technology (Kemendikbud) equalizes the curriculum at the university level in order to produce equal quality doctors, without any overlap. Scientific-level equality should not only be based on the comparison of subjects or study content, but should also correspond; in the context of scientific equality assessment, the effectiveness of teaching scientific content must also be taken into account for each prospective doctor. Fourth, the MKEK is an ethics institution that examines, prosecutes, and makes decisions on any ethical conflicts that have the potential for medical disputes between devices and the ranks of the IDI, and any medical disputes between its complainant doctors that have not been or are not handled by the Honorary Assembly of the Indonesian Medical License. Fifth, the doctor profession supervisory board is a new institution to supervise the IDI, KKI, and MKEK so as not to become a super-body [36] in order to guarantee the fundamental rights of a doctor. In addition, it is expected that the supervisory board of the doctor profession can also become an advanced mediator or decide the results of the decision from the advanced MKEK, or it can be called to an appeal to determine whether the dispute can be decided appropriately or not. As an institution resulting from collaboration between the executive and the legislature, it is hoped that the institution can provide legal certainty for all doctors from all tasks that have been given to the IDI, KKI, and MKEK. Therefore, a reconsideration and evaluation are needed in order to adjust Law No. 29 on Medical Practice of 2004 (UUPK) in Indonesia.

5. Conclusions

The case of Doctor T has become a very hotly discussed case in Indonesia. However, cases that arose due to the use of *digital subtraction angiography (DSA)* are the root cause of the revocation of Doctor T's practice license from. Regardless of the case, a license to practice is very important for a doctor to be able to medically practice, which applies to all doctors in the world. In the implementation of the medical license, there are provisions that must be implemented and prepared, including in Malaysia and the United Kingdom, with little visibility regarding the income of the permit, either regarding the governing law or the national institution that oversees the task. In addition, the professional code of ethics is closely related to the granting of the right to a license to practice to doctors. Specialized institutions are needed to obtain a license to practice from, without involving the executive, to produce qualified doctors without direct intervention from the government. In addition, collaborations based on the *Penta-helix* theory can be conducted to produce an increasingly qualified doctor profession to realize a *healthy state* within the community.

Author Contributions: Conceptualization, R.D. and N.P.D.; methodology, M.P.R.; software, N.P.D.; validation, R.D., N.P.D. and M.P.R.; formal analysis, M.P.R.; investigation, N.P.D.; resources, R.D.; data curation, R.D.; writing—original draft preparation, R.D.; writing—review and editing, M.P.R.; visualization, N.P.D.; supervision, R.D.; project administration, N.P.D.; funding acquisition, R.D. All authors have read and agreed to the published version of the manuscript.

Funding: This research received no external funding.

Informed Consent Statement: Not applicable.

Data Availability Statement: Not applicable.

Acknowledgments: Thanks to the Faculty of Law, Universitas Pembangunan Nasional Veterans Jakarta, for providing resources to the author to have the opportunity to attend international seminars.

Conflicts of Interest: The authors declare no conflict of interest.

References

1. Effendi, F. *Pertanggungjawaban Pidana Terhadap Dokter Yang Tidak Memiliki Surat Izin Praktik (Studi Perkara No. 1110 K/Pid.Sus/2012 PN Madiun)*; Universitas Bung Hatta: Padang, Indonesia, 2021.
2. Erawan, D.I. *Tinjauan Yuridis Tindak Pidana Dengan Sengaja Mempekerjakan Dokter Yang Tidak Memiliki Surat Izin Praktik Yang Dilakukan Oleh Pimpinan Klinik Kesehatan (Studi Kasus Putusan Nomor 451/Pid.Sus/2020/PN Jkt.Utr)*; Universitas Hasanuddin: Makassar, Indonesia, 2021.
3. DAL. IDI Balas Yasonna Soal Terawan: Surat Izin Dokter Dibuat Pemerintah. *CNN Indonesia*. 31 March 2022. Available online: https://www.cnnindonesia.com/nasional/20220331184623-20-778591/idi-balas-yasonna-soal-terawan-surat-izin-dokter-dibuat-pemerintah (accessed on 10 June 2022).
4. Fundrika, B.A. Beredar Surat Keputusan MKEK soal Dokter Terawan, Ada 5 Alasan yang Diduga Jadi Sebab Pemecatannya dari Anggota IDI. *Suara.com*. 28 March 2022. Available online: https://www.suara.com/health/2022/03/28/155000/beredar-surat-keputusan-mkek-soal-dokter-terawan-ada-5-alasan-yang-diduga-jadi-sebab-pemecatannya-dari-anggota-idi (accessed on 10 June 2022).
5. Nurhadi. 5 Fakta di Balik Alasan Pemecatan Terawan dari Keanggotaan IDI. *Tempo.co*. 12 April 2022. Available online: https://nasional.tempo.co/read/1581355/dewas-kpk-telisik-dugaan-lili-pintauli-terima-gratifikasi-saat-nonton-motogp (accessed on 10 June 2022).
6. Junadi, P.; Sedyawan, J.R.; Akbar, M.; Lestarini, D.; Purnama, D.A.; Kamal, K.; Bustamam, I.; Putri, Y.E.; Nasution, N.P.; Agustina, C.F.; et al. *Pedoman Pelaksanaan Program Pengembangan Pendidikan Keprofesian Berkelanjutan*; Armila, R., Herlambang, P., Harsa, B.T., Martadipura, B.M., Eds.; Pengurus Besar Ikatan Dokter Indonesia: Jakarta, Indonesia, 2021; Available online: http://idionline.org/wp-content/uploads/2022/02/BUKU-UNGU-3-mei-2021-Update.pdf (accessed on 10 June 2022).
7. Kurniawati Ramadhani, A. *Peran Ikatan Dokter Indonesia Dalam Penyelesaian Sengketa Medik*; Universitas Hasanuddin: Makassar, Indonesia, 2021.
8. Komalawati, V.; Kurniawan, D. Kompetensi dan Kewenangan Praktik Kedokteran: Perspektif Hukum Di Indonesia. *J. Ilm. Huk. De'Jure* **2018**, *3*, 152.
9. MK Sahkan IDI Sebagai Organisasi Tunggal Kedokteran. Kanal Pengetahuan Universitas Gadjah Mada: Yogyakarta, Indonesia. Available online: https://kanalpengetahuan.fk.ugm.ac.id/mk-sahkan-idi-sebagai-organisasi-tunggal-kedokteran/ (accessed on 11 June 2022).
10. Subarkah, T. Dinilai Superbody, IDI Disarankan Miliki Dewan Pengawas. *MediaIndonesia*. April 2022. Available online: https://mediaindonesia.com/humaniora/484621/dinilai-superbody-idi-disarankan-miliki-dewan-pengawas (accessed on 8 June 2022).
11. Aditya, N.R. Soal Kasus IDI dan Terawan, Menkes: Enggak Baik Waktu dan Energi Habis untuk Perdebatan. *Kompas*. 2022. Available online: https://nasional.kompas.com/read/2022/03/31/19345991/soal-kasus-idi-dan-terawan-menkes-enggak-baik-waktu-dan-energi-habis-untuk (accessed on 8 June 2022).
12. Nurhadi. Deretan Kontroversi Terawan yang Berujung Pemecatan dari IDI. *Tempo*. 2022. Available online: https://nasional.tempo.co/read/1576245/deretan-kontroversi-terawan-yang-berujung-pemecatan-dari-idi (accessed on 8 June 2022).
13. Santosa, F.; Permana, M.Y.; Baharuddin, M. Sponsorship Pendidikan Kedokteran: Batasan yang Sering Terabaikan. *J. Etika Kedokt. Indones.* **2018**, *2*, 16. [CrossRef]
14. Soemoele, S.A.; Sampurna, S.; Alwi, S. Implikasi Hukum Teknologi Brainwash (Studi Kasus Dr. dr. Terawan Agus Putranto, SpRad (K). *J. Ilmu Huk.* **2020**, *9*, 123.
15. Rozaliyani, A.; Meilia, P.D.I.; Libritany, N. Prinsip Penetapan Sanksi bagi Pelanggaran Etik Kedokteran. *J. Etika Kedokt. Indones.* **2018**, *2*, 21. [CrossRef]
16. Ramdhani, A.; Ramdhani, M.A. Konsep Umum Pelaksanaan Kebijakan Publik. *J. Publik J. Ilm. Bid. Ilmu Adm. Negara* **2017**, *11*, 1–12.
17. Farisa, F.C. Kontroversi Terawan dan Vaksin Nusantara yang Berujung Rekomendasi Pemecatan dari IDI. *Kompas*. 2022. Available online: https://nasional.kompas.com/read/2022/03/29/07144781/kontroversi-terawan-dan-vaksin-nusantara-yang-berujung-rekomendasi-pemecatan (accessed on 8 June 2022).
18. Pratama, R.A. Perhimpunan Dokter Radiologi Tolak Pemecatan Terawan. *Nasional Katadata*. March 2022. Available online: https://katadata.co.id/rezzaaji/berita/623eb3abe8d29/perhimpunan-dokter-radiologi-tolak-pemecatan-terawan (accessed on 8 June 2022).
19. Nugroho, S. Mazdjo Pray: Dokter Terawan yang Cerdas dan Berprestasi Dipecat, Sunardi yang Teroris Malah Dibela IDI. *Fajar Indonesia Network*. March 2022. Available online: https://fin.co.id/read/92271/mazdjo-pray-dokter-terawan-yang-cerdas-dan-berprestasi-dipecat-sunardi-yang-teroris-malah-dibela-idi/15 (accessed on 8 June 2022).
20. KKI Soal Terawan: Tanpa Rekomendasi IDI, Dokter Tak Bisa Praktik. *CNN Indonesia*. March 2022. Available online: https://www.cnnindonesia.com/nasional/20220329190132-20-777637/kki-soal-terawan-tanpa-rekomendasi-idi-dokter-tak-bisa-praktik (accessed on 8 June 2022).
21. Kasus Terawan, Amarah Meledak hingga Seruan Bubarkan IDI Muncul di DPR. *CNN Indonesia*. April 2022. Available online: https://www.cnnindonesia.com/nasional/20220405062058-20-780273/kasus-terawan-amarah-meledak-hingga-seruan-bubarkan-idi-muncul-di-dpr (accessed on 8 June 2022).

22. Oktaveri, J.A. IDI Pecat Dokter Terawan, DPR: Izin Praktik Kedokteran Seharusnya Wewenang Penuh Pemerintah. *Kabar24 Bisnis*. April 2022. Available online: https://kabar24.bisnis.com/read/20220404/15/1518889/idi-pecat-dokter-terawan-dpr-izin-praktik-kedokteran-seharusnya-wewenang-penuh-pemerintah (accessed on 8 June 2022).
23. Tsarina Maharani. DPR Serukan Pembentukan Dewan Pengawas IDI: Agar Tak Elitis Super Body. *CNN Indonesia*. April 2022. Available online: https://www.cnnindonesia.com/nasional/20220419185637-32-786827/dpr-serukan-pembentukan-dewan-pengawas-idi-agar-tak-elitis-super-body (accessed on 8 June 2022).
24. Lette, A.R. Jumlah Dan Kebutuhan Sumber Daya Manusia Kesehtan Di Fasilitas Kesehatan Kota Kupang. *J. Publ. Kesehat. Masy. Indones.* **2020**, *7*, 10.
25. M Iqbal Al Machmudi. Indonesia Masih Kekurangan SDM Kesehatan. *Medcom*. December 2021. Available online: https://www.medcom.id/nasional/politik/5b2Glxnk-indonesia-masih-kekurangan-sdm-kesehatan (accessed on 11 June 2022).
26. Titasari, N.A.; Fani, T. Dampak Psikologis Pandemi COVID-19 Pada Petugas Rekam Medis. *Pros. Diskusi Ilm.* **2021**, *1*, 75.
27. Hardie, E.A. Public perceptions of Australia's doctors, hospitals and health care systems. *Med. J. Aust.* **2008**, *189*, 210–214. [CrossRef] [PubMed]
28. World Health Organization. World Health Statistics 2021: Monitoring Health For The SDGs. 2021. Available online: https://apps.who.int/iris/bitstream/handle/10665/342703/9789240027053-eng.pdf?sequence=1&isAllowed=y (accessed on 8 June 2022).
29. Yeo, I.S.; Park, Y.; Lee, K.L.; Park, H.W. A history of Medical License in Korea. In *Ui Sahak*; 2002. Available online: https://pubmed.ncbi.nlm.nih.gov/12828143/ (accessed on 11 June 2022).
30. OECD and World Health Organization. Health at a Glance: Asia/Pacific 2020. November 2020 (Health at a Glance: Asia/Pacific). Available online: https://www.oecd-ilibrary.org/social-issues-migration-health/health-at-a-glance-asia-pacific-2020_26b007cd-en (accessed on 9 June 2022). [CrossRef]
31. Registration—Malaysian Medical Council (MMC). Malaysia. Available online: https://mmc.gov.my/registration/ (accessed on 8 June 2022).
32. Healthcare Personnel Statistics—Physicians—Statistics Explained. Eurostat Statistics Ecplainded. Available online: https://ec.europa.eu/eurostat/statistics-explained/index.php?title=Healthcare_personnel_statistics_-_physicians&oldid=460643 (accessed on 8 June 2022).
33. Häser, I. Higher requirements for doctors—Revocation of the license to practice medicine also in criminal offenses outside of the professional sphere of influence. *Klinikarzt* **2011**, *40*, 332–335. [CrossRef]
34. Caraka, R.E. Connecting climate and communicable disease to penta helix using hierarchical likelihood structural equation modelling. *Symmetry* **2021**, *13*, 657. [CrossRef]
35. Forss, K.S. Collaborating in a penta-helix structure within a community based participatory research programme: 'Wrestling with hierarchies and getting caught in isolated downpipes'. *Arch. Public Health* **2021**, *79*, 27. [CrossRef] [PubMed]
36. Schiller, M. The professional access of foreign doctors: A systematized presentation of the prerequisites for the license to practice medicine in cases with foreign reference under consideration of the current jurisprudence. *Medizinrecht* **2010**, *28*, 79–86. [CrossRef]

proceedings

MDPI

Proceeding Paper

Graduates' Competencies and Employability: A Conceptual Framework †

Siti Aishah Mokhtar [1,*], Noor Arina Md Arifin [1], Mohamad Rahimi Mohamad Rosman [1], Nurulannisa Abdullah [1], Faizal Haini Fadzil [1] and Azman Mat Isa [2]

1 Faculty of Information Management, Universiti Teknologi MARA, Kelantan Branch, Machang 18500, Malaysia
2 Faculty of Information Management, Universiti Teknologi MARA, Shah Alam 40450, Malaysia
* Correspondence: aishah835@uitm.edu.my; Tel.: +60-133869763

† Presented at the International Academic Symposium of Social Science 2022, Kota Bharu, Malaysia, 3 July 2022.

Abstract: Higher education is one of the institutions that have been widely recognized as major contributors to sustainable economic growth. In the global world today, changes occur frequently from time to time, and it has become more difficult to keep up with current trends. Through the process of educational studies from higher educational universities, people's knowledge is increasing, and they have to make the right decision in choosing the careers that are most suitable for them. Employability skills play an important role in order to ensure that they will be acceptable in the job market. Employability skills can be defined as those basic skills required for getting, keeping, and doing well on a job or task in which they responsible for. Generally, employability skills are needed by students to prepare themselves to meet the needs of job market requirements after their graduation. It is essential for any program of study at higher educational institutions to evaluate its curriculum regularly to ensure that its content remains relevant, is of high quality, and responds to the demand of the job market. Therefore, the purpose of this study is twofold: first, to determine the relationship between graduates' competencies and their employment and second, to propose a conceptual framework underlying the interrelationship between graduates' competencies and their employment. A structured literature review methodology was adopted, and a conceptual model was developed for further exploration of the topic

Keywords: graduates' employability; graduates' competencies; employability; job satisfaction; graduates

Citation: Mokhtar, S.A.; Arifin, N.A.M.; Rosman, M.R.M.; Abdullah, N.; Fadzil, F.H.; Isa, A.M. Graduates' Competencies and Employability: A Conceptual Framework. *Proceedings* **2022**, *82*, 54. https://doi.org/10.3390/proceedings2022082054

Academic Editor: Mohd Nasir Ismail

Published: 15 September 2022

1. Introduction

Nowadays, higher education is one of the institutions that have been widely recognized as major contributors to sustainable economic growth. The sector of higher educational institutions has produced graduates with certain skills that are needed by the industry and enable them to produce many research and development activities. This is important for the sustainable growth of companies and country development. In a global world today, changes occur frequently from time to time and it has become more difficult to keep up with current trends.

Graduates should have a set of achievements such as skills, knowledge, and others attributes in order to allow them to expand their employment and success in their work. It will give benefits to themselves, the organization in which they serve, and to the country's economy as a whole [1]. As stated by [2], it is important and compulsory for universities and graduates as they need to learn and adopt new skills to get their expected job. According to [3], employability has three major aspects, which include career identity, personal adaptability, and social and human capital. Career identity involves role identity, occupational identity, and organizational identity. Personal adaptability, which is willingness to change personal factors and social and human capital, is making investments in social and human capital in expectation of future returns in the workplace.

Through the process of educational studies from higher educational universities, people's knowledge is increasing and they have to make the right decision in choosing the careers that are most suitable for them. As we are in the global world, we are moving into the information world nowadays, or what we called the 'Information Age'. This change and transition is considered as one movement that will provide an emerging market for information professionals. Therefore, the purpose of this study is twofold: first, to determine the relationship between graduates' competencies and their employment and second, to propose a conceptual framework underlying the interrelationship between graduates' competencies and their employment.

2. Literature Review

2.1. Graduates' Competencies

Being new to the job market can be extremely tough, especially when employers and managers have plenty of choices of graduates applying for the job or position that they are offered. To ensure that employers will select them, the graduates must have good traits and attributes. As shown in the previous studies, it is important for the graduates to have good traits in order for them to be selected. As found by [4], responsibility traits, integrity traits, and being dependable and taking initiatives are the most important entry level of personality traits usually sought by employers. Building on that, [5] mentioned that the behaviors, attitudes, and work values of employees might be influenced by their historical, social, and economic experiences. Graduates' skills are more important than degree performance, as the recruitment process requires employees to apply their skills, and this concerns the employer as they want a worker who is equipped with some attributes such as interactive, personal [6], and other generic skills [7]. This finding has also been supported by [8], who have revealed that for some employers, a degree may now not represent anything more than a minimum requirement, in addition to other evidence of suitability. The concept of employability has been associated with workplace necessities and the higher education context also addressed the issues of graduate competencies or graduate employability [9].

2.2. Employability

The term employability is being used in correlation with other terms such as key skills, core skills, competencies, employability skills, generic skills, workplace know-how, critical enabling skills, transferable skills, key qualifications, trans-disciplinary goals, and process-independent qualifications [10]. According to [11], employability is a set of achievements, skills, and personal attributes that allow graduates to more easily gain employment, which benefits themselves, their workforce, the community, and the economy. In this present study, graduates are categorized according to the employability problem if they find it hard to gain employment after six months of their graduation. As stated by [12], employability is the situation in which the graduates are able to gain their initial employment, they can maintain their employment, and also can get new job if required. They are also capable to move and upgrade themself within the labor market in order for them to realize their potential through a sustainable environment. To be employed, an individual should have knowledge, skills, and attitudes and they should know how to use those assets to make themself well-presented in front of the employers.

3. Conceptual Model

Based on the literature search, five independent variables were identified, namely employment history/status, graduates' competencies, job satisfaction, challenges in the workplace, and labor market force. This research model adopts the theory of analytical dimensions of employability by [13]. This theory was chosen because it discusses how employability can be affected by the labor market institution, skills, and knowledge enhanced by education and training. In this study, five factors are considered as independent variables: graduates' competencies, job satisfaction, challenges in the workplace, employment history, and labor market force, and this research examined each of these variables as an element that should be focused on in conducting the study of the employability of undergraduate students. Figure 1 below shows the framework that was developed based on the literature review and research problem.

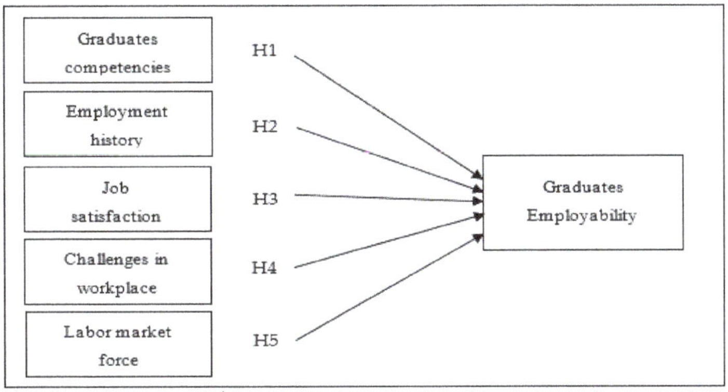

Figure 1. Conceptual Model.

Being new to the job market can be extremely tough, especially when employers and managers have plenty of choices of graduates applying for the job or position that they are offered. Employers also recognize that valuable employees are quickly becoming a rare product [14]. No matter the position, field, or profession, valuable employees will always be the choices of employers. It not only depends on the skills and talent that they have, but also from their good records, which are usually different and more established than their friends. Building on that, Ref. [5] mentioned that the behaviors, attitudes, and work values of employees might be influenced by their historical, social, and economic experiences. Attitude plays a vital role in every job we perform. Graduates who will engage in the job environment should also have teamwork skills. Teamwork is defined as the ability of people to work effectively individually as well as in a group, with the ability to be a team leader or manager [15]. Graduates should have a strong sense of self, confidence, high self-esteem, and a comprehensive knowledge of the company goals. It is supported by [16], as reported in their article. Therefore, it can be hypothesized that:

H1. *Graduates' competencies have a positive and significant relationship with graduate employability.*

After graduating, the student usually will find a job related to the course they took at university. However, not all of the graduates get a job relevant to the field of their study. To some extent, they have to grab any opportunities that come to them as long as they are employed. In today's labor market, the degree qualification they got from the university was not promising them the job that they want, even if there is a link between what is studied and the labor market. At the workplace, the graduates will learn new things other than what they have learnt in class, and it will give them opportunities to upgrade themselves for the highest position available. As mentioned by (14), from their survey on the employment experiences of graduates from the University of New South Wales LIS programs for the years 1997–2000, report data indicated that 25.8% of respondents occupied a librarian position in their first job experience, utilizing information management skills. Therefore, it can be hypothesized that:

H2. *Employment history has a positive and significant relationship with graduate employability.*

Job satisfaction has been of interest to organizational researchers, as it has a strong relationship with job performance and organizational commitment. As proven by [17], employees spend their time in doing their job and it affects their lives in general. It was supported by [18] in their study that promotion is the most significant factor affecting job satisfaction. According to [19], the personality style and job satisfaction of librarians are related in various types of libraries and other information professional fields. It was found that optimism, emotional strength, an ability to work in a team, visionary work style, and work drive are the personality factors that are positively related to job satisfaction. Therefore, it can be hypothesized that:

H3. *Job satisfaction has a positive and significant relationship with graduate employability.*

Employees in many sectors in which they are employed face challenges and demands mostly from the technological development in information services. Technology has changed the traditional practice and roles of the information professional, as it requires them to have basic computer skills in doing their job in order to provide good services to users. The rapid changes of technology allow professional workers to grab opportunities in order to grow their career development. The profession of library and information services is rapidly changing, with a lot of challenges, especially for fresh graduates, as they need to adapt to the new demands [20]. Building on [21], the application of information technology in the workplace seems to be a common phenomenon in the 21st century in some sectors such as banking, railways, insurance, and airways. Therefore, it can be hypothesized that:

H4. *Challenge in the workplace has a positive and significant relationship with graduate employability.*

The job market is defined as the total population of working people who are willing and able to work and have officially entered the labor market, whereas labor force is referring to the individuals who are actively seeking a job [22]. Higher education and the labor market have traditionally connected, although sometimes in loose and intangible ways [23]. The availability of employment opportunities is extremely influenced by the labor markets. Nowadays, many countries are facing the economy crisis and Malaysia is one of those countries. This crisis resulted in the issues of employability, as it has been discussed for a long time that job demands have been decreased and the employment rate has been reduced due to the current economic situations. Therefore, it can be hypothesized that:

H5. *Labor market force has a positive and significant relationship with graduate employability.*

4. Conclusions

This study was conducted with two purposes: first, to determine the relationship between graduates' competencies and their employment and second, to propose a conceptual framework underlying the interrelationship between graduates' competencies and their employment. A structured literature review methodology was adopted, and a conceptual model was developed for further exploration of the topic. Based on the literature search, five independent variables were identified, namely employment history/status, graduates' competencies, job satisfaction, challenges in the workplace, and labor market force. Next, the conceptual model of the study was proposed, and discussions were made on each hypothesis.

The next stage of this research is to validate the conceptual research model in a quantitative study. For this quantitative research, the format instrument will be used as the data can be easily generalized to the object being analyzed. Descriptive research design is integrated into the questionnaires, and this research will collect the data for the purpose of hypotheses testing in order to answer the questions of the current status of respondents. A tracer study will also be conducted as part of the data collection. SPSS will be used to analyze the data from the questionnaire. An appropriate instrument and research design will be tested through various statistical analysis tests, such as reliability analysis using descriptive and inferential methods, such as frequency, cross-tabulation, ANOVA, Cronbach's Alpha, Harmann Single-Factor Test, Pearson's Correlation, and Multiple Regression.

Author Contributions: Conceptualization, S.A.M. and M.R.M.R.; methodology, S.A.M. and N.A.M.A.; validation, M.R.M.R. and A.M.I.; formal analysis, N.A.; investigation, N.A., F.H.F. and S.A.M.; resources, N.A.; data curation, M.R.M.R.; writing—original draft preparation, S.A.M., M.R.M.R., N.A.M.A., N.A., F.H.F. and A.M.I.; writing—review and editing, S.A.M.; visualization, S.A.M.; supervision, A.M.I.; project administration, S.A.M. All authors have read and agreed to the published version of the manuscript.

Funding: This research received no external funding.

Institutional Review Board Statement: Not applicable.

Informed Consent Statement: Not applicable.

Data Availability Statement: Not applicable.

Acknowledgments: This research was partially supported by Universiti Teknologi MARA Kelantan Branch, Malaysia.

Conflicts of Interest: The authors declare no conflict of interest.

References

1. Yorke, M. *Employability in Higher Education: What It Is—What It Is Not. Learning and Employability Series One*; ESECT and HEA: York, UK, 2006. Available online: http://www.heacademy.ac.uk/assets/York/documents/ourwork/tla/employability/id116_employability_in_higher_education_336.pdf (accessed on 10 August 2012).
2. Hosain, M.S.; Mustafi, M.A.A.; Parvin, T. Factors Affecting the Employability of Private University Graduates: An Exploratory Study on Bangladeshi Employers PSU Research Review. 2021. Available online: https://doi-org.ezaccess.library.uitm.edu.my/10.1108/PRR-01-2021-0005 (accessed on 21 June 2012).
3. Hack-Polay, D.; Igwe, P.; Okolie, U.C. Room for improvement: A study of overconfidence in numerical skills among British graduates. *Ind. High. Educ.* **2020**, *34*, 50–61. [CrossRef]
4. Thompson, E. *Customer Relationship Management: Making the Vision A Reality*; Gartner. Inc.: Sofitel Rive Gauche, France, 2001.
5. Tay, A. Managing generational diversity at the workplace: Expectations and perceptions of different generations of employees. *Afr. J. Bus. Manag.* **2011**, *5*, 249–255.
6. Harvey, L. Defining and measuring employability. *Qual. High. Educ.* **2001**, *7*, 97–109. [CrossRef]
7. Lan, M.T.Q. Graduate generic competences from the perspective of VNU employers. *J. Teach. Learn. Grad. Employab.* **2020**, *11*, 131–145. [CrossRef]
8. Purcell, K.; Elias, P.; Davies, R.; Wilton, N. *The Class Of '99: A Study of The Early Labor Market Experience Of Recent Graduates*; DfES Research Report No. 691; DfES Publications: Nottingham, UK, 2004.
9. Rahman, S.A.; Lie, K.Y. Graduate Competencies: Issues and Solutions of Graduate Employability in Malaysia. In *Technology and Workplace Skills for the Twenty-First Century*; Neubauer, D.E., Ghazali, K., Eds.; International and Development Education; Palgrave Macmillan: New York, NY, USA, 2015. [CrossRef]
10. Shukran, A.R.; Munir, S. Graduate employability and graduateness: An overview of concepts, interpretations and emerging issue. *Int. Proc. Econ. Dev. Res.* **2011**, *23*, 183–188.
11. Lees, D. Graduates Employability-Literature Review. Available online: http://www.qualityresearchinternational.com/esecttools/esectpubs/leeslitreview.pdf (accessed on 21 April 2012).
12. Hillage, J.; Pollard, E. *Employability: Developing a Framework for Policy Analysis*; Research Report. No. RR85 (London, DfEE); Department for Education and Employment (DfEE): Suffolk, UK, 1998. Available online: https://www.voced.edu.au/content/ngv:40352 (accessed on 1 May 2012).
13. Psacharopoulos, G.; Schotter, M. Skills for Employability, Economic Growth and Innovation: Monitoring the Relevance of Education and Training Systems. Expert Network on Economics of Education (EENEE). 2004. Available online: http://www.eenee.de/docs/employability (accessed on 3 June 2012).
14. Willard, P.; Wilson, C.S.; Cole, F.T.H. Changing employment patterns: An Australian experience. *Educ. Inf.* **2003**, *21*, 209–228.
15. Zaharim, A.; Yusoff, Y.M.; Omar, Z.; Mohamed, A.B.; Muhamad, N.; Mustapha, R.; Ehsan, D. Employers Perception towards engineering employability skills in Asia. *WSEAS Trans. Adv. Eng. Educ.* **2009**, *9*, 306–315.
16. Raybould, J.; Sheedy, V. Are graduates equipped with the right skills in the employability stakes? *Ind. Commer. Train.* **2005**, *37*, 259–263. [CrossRef]
17. Rayton, B.A. Examining the interconnection of job satisfaction and organizational commitment: An application of the bivariate probit model. *Int. J. Hum. Resour. Manag.* **2006**, *17*, 139–154. [CrossRef]
18. Sharma, R.D.; Jyoti, J. Job satisfaction among school teachers. *IIMB Manag. Rev.* **2006**, *18*, 349–363.
19. Williamson, J.M.; Pemberton, A.E.; Lounsbury, J.W. An investigation of career and job satisfaction in relation to personality traits of information professionals. *Libr. Q.* **2005**, *75*, 122–141. [CrossRef]
20. Elisha, O.M. Contemporary issues in information management: A fresh look for information professionals. *Int. J. Libr. Inf. Sci.* **2009**, *1*, 082–091.

21. Thomas, V.K.; Satpathi, C.; Satpathi, J.N. Emerging challenges in academic librarianship and role of library associations in professional updating. *Libr. Manag.* **2010**, *31*, 594–609. [CrossRef]
22. Ingrassia, R. Labor Market Imbalances and Personnel Recruitment. *Am. J. Ind. Bus. Manag.* **2019**, *9*, 1376–1394. [CrossRef]
23. Johnston, B. The shape of research in the field of higher education and graduate employment: Some issues. *Stud. High. Educ.* **2003**, *28*, 413–426. [CrossRef]

proceedings

MDPI

Proceeding Paper

Arca Keris in Alor Gajah Town, Melaka: A SWOT Analysis of Design and Appeal Factor †

Azahar Harun [1,*], Khairunnisa Mohd Abdul Ghani [1], Sharmiza Abu Hassan [1], Mohd Hanif Mohd Omar [1], Nurhikma Mat Yusof [1] and Ohm Pattanachoti [2]

[1] College of Creative Arts, Universiti Teknologi MARA, Melaka 78000, Malaysia
[2] School of Social Sciences, Chiang Rai Rajabhat University of Thailand, Muang Chiang Rai 57100, Thailand
* Correspondence: azahar581@uitm.edu.my
† Presented at the International Academic Symposium of Social Science 2022, Kota Bharu, Malaysia, 3 July 2022.

Abstract: In 2020, the Alor Gajah Municipal Council (AGMC) embarked on a project to upgrade the town's public courtyard. One of its tasks was restoring and relocating a giant traditional Malay dagger replica known as the *Arca Keris* from the courtyard to an adjacent roundabout. Despite this initiative, most people are unaware of how such a change could improve the town's image and one's quality of life. In this sense, two questions arise: (1) what are the design specifications of the *Arca Keris*? and (2) how does the presence of the *Arca Keris* improve the town's image? In order to address this issue, the design and appeal factors of the *Arca Keris* were examined. A triangulation strategy was used, consisting of interviews and observation; this was followed by a SWOT analysis. The outcome demonstrates that the *Arca Keris* has both strengths and opportunities in terms of visibility and symbolization. However, it may have flaws, such as accessibility and anonymity. As a result, the general public was unable to appreciate the significance of the *Arca Keris* or comprehend the rationale for its relocation. This research should help us better understand how public art affects a town's image across the country.

Keywords: public art; municipal council; design and appeal

check for updates

Citation: Harun, A.; Ghani, K.M.A.; Hassan, S.A.; Omar, M.H.M.; Yusof, N.M.; Pattanachoti, O. Arca Keris in Alor Gajah Town, Melaka: A SWOT Analysis of Design and Appeal Factor. *Proceedings* **2022**, *82*, 55. https://doi.org/10.3390/proceedings2022082055

Academic Editor: Mohamad Rahimi Mohamad Rosman

Published: 16 September 2022

1. Introduction

Sculpture, environmental art, signs, murals, and ephemeral works are all examples of public art. In Malaysia, public art is frequently recognized, especially by local government authorities, as an amenity rather than expressive art [1]. This is in contrast to the avant-garde philosophy of "Art for art's sake", which maintains that the only thing that matters is the artist's self-expression [2]. In theory, public art, according to some scholars, has a lot to offer; this includes promoting society's core values, maintaining the ability of attractions to achieve desired goals, and adding value to urban environments and their inhabitants in terms of history, beauty, and purpose [3–6]. Therefore, careful planning and consideration should be taken to ensure that public art is not just made for utility purposes if it is to be placed in an urban area [7]. To put it another way, public art is the art that was created with the intention of igniting and honoring its viewers [8,9].

2. Problem Statements

Situated in the northern part of Melaka, Alor Gajah town is governed by the Alor Gajah Municipal Council (AGMC). As stated in the *Rancangan Tempatan Daerah Alor Gajah*, 2020 or local plan, AGMC is committed to implementing the sustainable development goals (SDGs) of ensuring the quality of life by providing good public service, boosting the town's image, and preserving the environment [10]. This pledge is believed to be fulfilled through altering and modernizing public facilities in and around Alor Gajah. For instance, a 2017 news report stated how AGMC and the faculty of Art and Design at the Universiti

Teknologi MARA Cawangan Melaka collaborated to create a mural art project to enhance the town's image [11]. Several old shop houses' walls are decorated with drawings that feature cultural themes to create a new image and unique experience, for the local and foreign visitors. AGMC began work on a development plan to upgrade a public courtyard in the heart of Alor Gajah town in 2020. In order to maximize the use of the public courtyard, AGMC decided to relocate a huge traditional Malay dagger replica known as *Arca Keris* to an adjacent roundabout (see Figure 1). Additional accessories, such as a new concrete base design, lighting, and landscaping were added to the final upgrade work in August 2020. Despite AGMC's efforts, no empirical study has yet been conducted on the *Arca Keris* in Alor Gajah, especially in terms of design and appeal. Scholars argue that very few public artworks are examined after they are displayed, which raises concerns about whether their design and appeal can reach the highest standards [2]. As a matter of fact, most public artworks in various Malaysian states are typically misinformed; hence, they fail to improve the quality of life as intended [6]. Taking into account all of the reasons for and against public art installations, this study intends to examine *Arca Keris* in Alor Gajah and answer the following questions: (1) what are the design specifications of the *Arca Keris*? and (2) how does the presence of the *Arca Keris* improve the town's image?

Figure 1. The Keris replica in Alor Gajah town.

3. Literature Review

Keris is a traditional Malay dagger known for its beauty and legend. In some circles, the *Keris* is even regarded as a sacred object endowed with magical powers. For example, it is known as '*Azimat*' in the Malay community and '*Tosan aji*' in the Javanese community [12]. One scholar noted that *Keris* is a form of traditional weapon that has unique features, such as being long or short, having two sharp eyes, having straight or winding blades, pointed at the end section and widening on the side base [13,14]. Technically, it was created with the intention of murdering an enemy in a stealthy, swift, and unexpected manner [15]. Despite this, the *Keris* is now considered one of the local cultural items and is no longer used for killing or self-defence [16]. As a result, it can be seen in Malay ceremonial occasions, particularly traditional weddings and the Malay martial art of *Silat* [1].

Keris can also be found in the design of the Jata Negara, or the Malaysian national coats of arms; which represent non-federated nations and serve as symbols of Malay rulers' sovereignty. It has consistently served as inspiration for local artists and is widely acknowledged as a significant work of art in public art initiatives [16,17]. The National Archive property, for example, houses *Pekaka*, a 1979 *Keris* sculpture by Arifin Ismail. *Pekaka* is made of copper with a decorative motif that is thought to have originated in Thailand [18]. Another example is Latif Mohiddin's *Keris*, which he built in 1998. This artwork can be found in front of Menara Maybank's main entrance in Kuala Lumpur. According to a public art scholar, the artist has successfully depicted *Keris*' concealed philosophy and method through the use of shape, lines, and style; thus, this provides the public with a distinct and novel experience [18]. Rosli Zakaria, a local sculptor and public art researcher, attempted to promote Malay culture by creating and installing a *Keris* sculpture at Changchun Sculpture

Park in China in 2006 [17]. Several local governments and town councils have also taken the lead in incorporating and promoting *Keris* sculpture as part of their urban beautification initiatives. *Tugu Taming Sari* in Kuala Kangsar, Perak, *Tugu Keris* in Klang, Selangor, and *Arca Keris* in Alor Gajah, Melaka are some of them.

Nonetheless, the *Keris* sculpture in front of Kuala Lumpur's Bukit Jalil National Stadium is undoubtedly the most popular. Ironically, this public art installation was not created by local artisans. In 1996, the Malaysian government commissioned Neil Lawson Baker, a well-known British sculptor, to create a giant *Keris* replica for the 16th Commonwealth Games in Kuala Lumpur [19]. Unlike the previous sculptures mentioned earlier, the *Keris* replica at the Kuala Lumpur National Stadium, which stands around seven metres tall and weighs more than ten tonnes, is an exact reproduction. Baker struggled to cast the *Keris*, at first; particularly the *Luk* or wavy-shaped blade, and the *Hulu* or the hilt. This is most likely due to a lack of knowledge and experience with *Keris'* ideology and approach. As a result, Ramlan Abdullah, a well-known local sculptor, was enlisted to assist in the alteration of the *Keris* [19].

Keris' significance as a symbol of Malay identity and Islam has always been acknowledged and appreciated, even by non-Malays. Thus, any attempt to minimize its significance, whether by a group or an individual, may be seen as disrespectful of Malay society. Take a look at the following scenarios. According to the International Islamic University of Malaysia (IIUM) university news, the new president, Datuk Dr. Mohd Abu Bakar had suggested replacing an existing *Keris* replica at the university entrance with a Qalam (book) replica during his speech at IIUM Takrim Day 2019 held at the IIUM Cultural Center (ICC) [20]. The present artwork, according to the president, is unfit to symbolize the university as a center of intellectual debate and enlightenment. He went on to say that the Qalam is more appropriate because it reflects the university's logo as well as the concept of Hidayah, or Allah's bounty. Former Minister of Culture, Arts, and Heritage Datuk Rais Yatim, on the other hand, expressed his dissatisfaction with the situation [21]. He was actually hinting that the current IIUM president's action could be misinterpreted as an insult to the Malay population. In the second scenario, during the 2006 UMNO general assembly, Datuk Hishamuddin Hussein, the party's youth leader, was photographed kissing a *Keris* while severely telling the Inter-faith Commission (IFC) not to play with Malay religious sentiment [22]. Hishamuddin Hussein's gesture was reportedly in response to two problematic events. The first was in response to an allegation made by the Islamic Consumer Association that a food product's packaging had a cross emblem. The second was a rumor that national yachtsman Azhar Mansor had left Islam, and was converting Malay Muslim children to Christianity at an Ipoh, Perak church [22]. Despite the changes in setting, these two scenarios clearly demonstrate *Keris'* close connection to Malay interests and faith.

4. Methodology

This study used a triangulation technique, which involves drawing conclusions by using data from multiple sources [23]. In this case, the study's initial phase included interviews and observation. Authorization from the UiTM Research Ethic Committee REC/06/2022 (ST/MR/108) was obtained before conducting the task. To execute this, aerial photographs were captured using a drone and a Digital Single Lens Reflex, or DSLR camera. For the interviews, we were able to speak with Tuan Haji Zainal Abidin Bin Abu Bakar, an AGMC technical assistant to the architect; who is involved in the early stage of the design and development of *Arca Keris*. Data from the observation and interview were coded, tabulated, and analyzed using a SWOT analysis; this stands for strengths, weaknesses, opportunities, and threats. This research tool can be effectively used for assessing the capabilities and shortcomings of the subject under study [24].

4.1. Analysis

The *Arca Keris* of Alor Gajah is located in a 40-feet wide × 60-feet long roundabout adjacent to a public courtyard, old shophouses, and a parking lot (see Figure 2). The roundabout acts as a transition that allows traffic from three main roads—Jalan Datuk Naning, Jalan Datuk Md Zain, and Jalan Datuk Dol Said—to move smoothly. Motorists and pedestrians can see the artwork from all sides. A row of old shophouses adjacent to it has been decorated with mural design. However, because it is close to traffic, access to this public art is limited due to safety concerns.

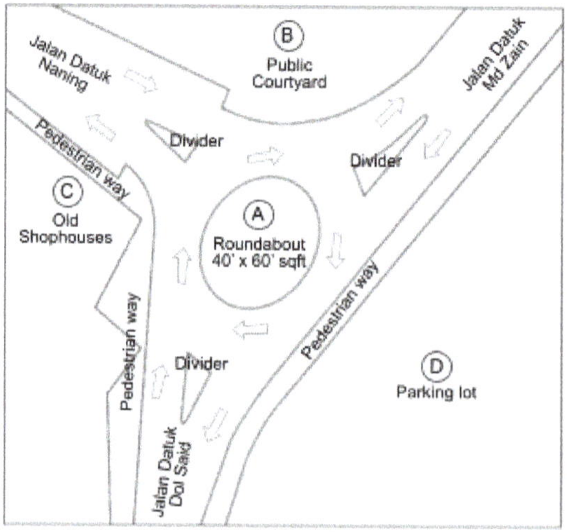

Figure 2. Alor Gajah town roundabout.

The *Arca Keris* is about 18-feet tall and 30-feet long (see Figure 3); in addition, it is made of composite material, namely ferro-cement and wire mesh, and coated in bright yellow and black paint. It is set on a new concrete foundation designed by Hyundai Engineering Malaysia Sdn Bhd as a gift to the AGMC. There is a large amount of clearance of 15 feet between the concrete base and the edge of the curb. Landscaping and lighting fixtures were added to the roundabout.

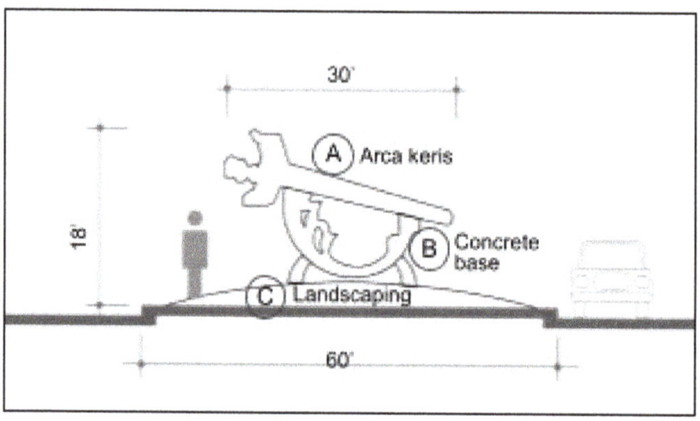

Figure 3. Detailed illustration of Arca Keris.

SWOT Analysis

- **Strength:** The *Arca Keris* of Alor Gajah town is a 15-feet tall structure that stands out in the crowd. Its design and construction began in 2002, according to AGMC's director of the Town Planning Department, Puan Nurul Adawaty; and was overseen by Haji Zainal Abidin Bin Abu Bakar, a technical assistant architect. The design motif was chosen as a symbol of the Alor Gajah region and the Melaka state. The state's symbol, which features five dagger blades representing the five renowned Malay warriors *Hang Tuah*, *Hang Jebat*, *Hang Kasturi*, *Hang Lekir*, and *Hang Lekiu*, is also representative of Melaka's valor and historical image. Furthermore, the placement of the *Arca Keris* in the Alor Gajah roundabout can also serve to raise awareness of the local history associated with the 15th-century *Naning War* against colonialists; this enables a stronger sense of place and identity for the municipality of Alor Gajah.
- **Weaknesses**: The *Arca Keris* is located on a 40-feet wide and 60-feet long roundabout where three streets meet: Jalan Datuk Naning, Jalan Datuk Dol Said, and Jalan Datuk Md Zain. The public is not permitted to enter the roundabout area due to heavy traffic and safety concerns; however, the public can observe the artwork from a safe distance. Because of these constraints, effective involvement and appreciation were unable to be achieved in a meaningful way. Public art in public outdoor places, according to public art experts Ramlan Abdullah and Rosli Zakaria (personal communication, 2022), makes the space appear more engaging and vibrant; this has a beneficial impact on urban identity. In this sense, public art must be made available to the general public in order to fulfil its duty and function. In terms of craftsmanship, its design has no apparent significance that is worth noticing. According to the crafter Haji Zainal Abidin Bin Abu Bakar (personal communication, 2022), its design was literally copied from a typical dagger; this is despite the fact that the original intention was to resemble the legendary Taming Sari. This is due to the fact that there was no visual reference of the legendary Taming Sari available at the time. As a result, the *Arca Keris* is regarded as anonymous because the general public is unable to determine its identity. In an interview, a sculpture expert remarked that when creating a public sculpture and placing it in a public space, it is necessary to have a larger perspective; particularly when creating public art based on a race's identity. This clearly demonstrates the significance of the sculptor's instructions. The aesthetic value of public art will be ignored and forgotten if it is not applied.
- **Opportunities:** The *Arca Keris* can aid in the construction of cultural identity by differentiating a society and attracting visitors to its uniqueness, as well as promoting local culture. This opportunity will definitely raise awareness, foster goodwill among stakeholders and artists, and eventually lead to the development of new public art projects.
- **Threats:** Regarding *Arca Keris's* location, there are two potential threats. First, the *Arca Keris* is situated near a roundabout with heavy traffic, particularly during rush hour. In order to prevent incidents, visitor safety must be addressed. Second, because it is situated in an open space, there are concerns that the weather will affect the 20-year-old work of public art's substance and color.

5. Conclusions

Without a doubt, the *Arca Keris* in Alor Gajah town is a popular three-dimensional landmark that reflects Malay identity and local history; therefore, it should be preserved as such. In this study, we observed that it has the ability not only to beautify the town, but also to improve its image; thus, creating a welcoming sense for both local and foreign visitors. Furthermore, the study discovered that the only flaws that detract from its appeal are aesthetic proportion, craftsmanship, and accessibility. In this regard, it is suggested that local governments, especially AGMC, pay greater attention to the design and appeal factor in order to ensure that the *Arca Keris* is better appreciated by the general public.

Author Contributions: Conceptualization, A.H., S.A.H., K.M.A.G., M.H.M.O., N.M.Y. and O.P.; methodology, A.H.; software, A.H.; validation, A.H., S.A.H. and K.M.A.G.; formal analysis, A.H., S.A.H. and K.M.A.G.; investigation, A.H., S.A.H. and K.M.A.G.; resources, A.H. and S.A.H.; data curation, A.H. and S.A.H.; writing—original draft preparation, A.H.; writing—review and editing, A.H., S.A.H. and K.M.A.G.; visualization, M.H.M.O. and N.M.Y.; supervision, A.H. and S.A.H.; project administration, A.H. and S.A.H.; funding acquisition, A.H. and S.A.H. All authors have read and agreed to the published version of the manuscript.

Funding: This research was funded by the MALAYSIAN MINISTRY OF HIGHER EDUCATION (MOHE), grant number 600-RMC/FRGS 5/3 (005/2021).

Institutional Review Board Statement: Not applicable.

Informed Consent Statement: Not applicable.

Data Availability Statement: Not applicable.

Acknowledgments: The authors would like to thank the research interest group, Public Art Creative Aesthetic and Quality (PACAQ) of UiTM Cawangan Melaka; and the Alor Gajah Municipal Councils (AGMC), particularly Puan Nurul Adawaty Hj Ismail, Director of Town Planning Department, and Tuan Haji Zainal Abidin Abu Bakar, Technical Assistant Architect, for assisting in the completion of this study.

Conflicts of Interest: The authors declare no conflict of interest.

References

1. Mustafa, M. Public Art in the Federal Territory of Putrajaya: Questions of Value and Role. *Wacana Seni J. Art Discourse* **2009**, *8*, 71–96.
2. Stevens, Q.; Lossau, J. *The Uses of Art in Public Space*; Routledge, Taylor & Francis Group: New York, NY, USA, 2015.
3. Chenhui, W. Public Art Has More Than Aesthetic Value. *Chinese Social Sciences Today*, 16 June 2020. Available online: http://www.csstoday.com/Item/7966.aspx (accessed on 18 April 2022).
4. Hall, T.; Robertson, I. Public Art and Urban Regeneration: Advocacy, claims and critical debates. *Landsc. Res.* **2001**, *26*, 5–26. [CrossRef]
5. Landi, P.J. Public Art—Purpose and Benefits: Exploring Strategy in the New England City of Pittsfield, Ma. In *Master Thesis of Landscape Architecture*; University of Massachusetts Amherst: Amherst, MA, USA, 2012.
6. Jasmi, M.F.; Mohamad, N.H.N. Roles of public art in Malaysian urban landscape towards improving quality of life: Between aesthetic and functional value. *Procedia Soc. Behav. Sci.* **2016**, *222*, 872–880. [CrossRef]
7. Bingham-Hall, J. Public art as a function of Urbanism. *Everyday Pract. Public Art* **2015**, 161–176. [CrossRef]
8. Kroiz, L. *Cultivating Citizen the Regional Work of Art in the New Deal Era*; University of California Press: Oakland, CA, USA, 2018.
9. Sharp, J.; Pollock, V.; Paddison, R. Just art for a just city: Public art and social inclusion in urban regeneration. In *Urban Studies*; Routledge: New York, NY, USA, 2020; Volume 42, p. 1001.
10. *Rancangan Tempatan Daerah Alor Gajah*; MPAG: Melaka, Malaysia, 2020.
11. Lukisan Mural Penambah Seri Alor Gajah. *Berita Harian*, 19 November 2020. Available online: http://api.bharian.com.my/hujung-minggu/kembara/2020/11/755801/lukisan-mural-penambah-seri-alor-gajah (accessed on 1 March 2022).
12. Mohamad, K.; Rahman, N.H.S.N.A.; Samian, A.L.; Ramli, Z. Keris: Artifak kejuruteraan spiritual Melayu. *Int. J. Malay World Civ. (J. Antarabangsa Alam Dan Tamadun Melayu)* **2013**, *1*, 39–49.
13. Hashim, H.J.; Mohd, N. *Keris: Peranannya Dalam Masyarakat Brunei*; Dewan Bahasa dan Pustaka Brunei: Bandar Seri Begawan, Brunei, 2004.
14. Daud, M.; Muhammad, A.; Ibrahim, A.N. *Proses Metalurgi Dalam Pembuatan Keris Tradisional. Seminar Senjata Tradisional: Sejarah dan Falsafah*; Jabatan Muzium Negara: Kuala Lumpur, Malaysia, 2009.
15. Mohamad, K.; Rahman, N.H.S.N.A.; Samian, A.L. Falsafah perkerisan dalam masyarakat Melayu. In *SARI: Jurnal Alam dan Tamadun Melayu*; Universiti Kebangsaan Malaysia: Selangor, Malaysia, 2012; Volume 30, pp. 105–119. ISSN 0127-2721.
16. Halim Husain, A. Regional identity of Malaysian sculptors based on Social Cultural System. In Proceedings of the 2nd Social Sciences, Humanities and Education Conference: Establishing Identities through Language, Culture, and Education (SOSHEC 2018), Surabaya, Indonesia, 21 July 2018. [CrossRef]
17. Hashim, A.; Lee, H.Y.; Mohd Nor, N. Rekabentuk 3-Dimensi Dalam Kesenian Melayu Tradisional. *Kupas Seni* **2022**, *10*, 1–11.
18. Awang, Z. Malaysian Art: A Search for Local and Islamic Identity through Sculpture. *SPAFA Dig.* **1990**, *11*. Available online: https://www.spafajournal.org/index.php/spafadigest/issue/view/2 (accessed on 22 April 2022).
19. Zakaria, R. Stail abstrak arca awam Kuala Lumpur: Refeleksi terhadap tegahan Agama Islam membina arca berbentuk figura bernyawa. *ISME Colloq.* **2016**, *3*, 144–158.
20. Nizam, S. President Suggests to Replace 'Keris' with 'Qalam'. *IIUM Today*, 26 July 2019. Available online: https://news.iium.edu.my/?p=130939 (accessed on 22 April 2022).

21. Asyraf, F. Rais Yatim Was Disappointed with IIUM's Relocation of Keris Monument. *Malaysiakini*, 9 November 2019. Available online: https://www.malaysiakini.com/news/499229 (accessed on 22 April 2022).
22. Fauzi Abdul Hamid, A.; Ismail, M.T. Islamist conservatism and the demise of Islam Hadhari in Malaysia. *Islam Christ. Muslim Relat.* **2014**, *25*, 159–180. [CrossRef]
23. Tashakkori, A.; Teddlie, C. *Mixed Methodology: Combining Qualitative and Quantitative Approaches*; Sage: Thousand Oaks, CA, USA, 1998.
24. Thompson, A.A.; Strickland, A.J.; Gamble, J.E. *Crafting and Executing Strategy-Concepts and Cases*, 15th ed.; McGraw-Hill/Irwin: Irvine, CA, USA, 2007.

proceedings

MDPI

Proceeding Paper

Tips of the Iceberg: Domestic Violence during COVID-19, Movement Control Order (MCO) in Malaysia †

Tengku Adida Tengku Zainal Mulok [1], Hamizah Abdul Rahman [1,*], Ravindran Nadarajan [2] and Nor Razinah Mohd. Zain [3]

1 Faculty of Law, UiTM Cawangan Kelantan, Bukit Ilmu, Machang 18500, Malaysia
2 Faculty of Business and Finance, Universiti Tunku Abdul Rahman, Jalan Universiti, Bandar Barat, Kampar 31900, Malaysia
3 Institute of Islamic Banking and Finance, International Islamic University Malaysia, Kuala Lumpur 53100, Malaysia
* Correspondence: hamizahabdrahman@uitm.edu.my

† Presented at the International Academic Symposium of Social Science 2022, Kota Bharu, Malaysia, 3 July 2022.

Abstract: Domestic violence can be defined as physical or emotional abuse of family members for many reasons, such as finances, drug abuse, jealousy, or other reasons that can result in injury or even the death of the victim. This paper explores several "tips of the iceberg" situations, which reflect cases of domestic violence during the COVID-19 Movement Control Order (MCO). This study was conducted primarily through a doctrinal study in elucidating relevant documents that comprised statutory laws and articles in legal journals. This paper also addresses the sufficiency of legal rules and regulations for protecting domestic violence victims. Lastly, the result of this study identified future initiatives to assist domestic violence victims.

Keywords: domestic violence; COVID-19; Movement Control Order; legal framework

Citation: Mulok, T.A.T.Z.; Rahman, H.A.; Nadarajan, R.; Zain, N.R.M. Tips of the Iceberg: Domestic Violence during COVID-19, Movement Control Order (MCO) in Malaysia. *Proceedings* **2022**, *82*, 60. https://doi.org/10.3390/proceedings2022082060

Academic Editor: Mohamad Rahimi Mohamad Rosman

Published: 16 September 2022

1. Introduction

Domestic violence is a global issue that occurs throughout the world. It is not isolated to Malaysia and is a widespread phenomenon among countries such as Pakistan, China, India, Brazil, and Japan. It was reported that in Dhaka, 60% of the housewives have been battered by their husbands [1]. Grossman and Lundy (2007) stated that domestic violence occurs across all ethnicities and racial groups, profoundly affecting women, who are most frequently the victims. Women were abused in various forms, either physically, socially, financially, sexually, psychologically, or emotionally, and contends that female victims of violence should be given special attention and comprehensive assistance. It is pointed out in [2] that in America, approximately, 4 million women are battered each year, with injuries caused by their husbands or partners. Up to 50 percent of homeless women and children are fleeing domestic violence. Malaysian survivors' need for shelter demands urgent attention, as domestic violence shelters are sorely lacking in Malaysia. International best practices recommend a minimum of one family place in a women's shelter per 10,000 people, but Malaysia only has an estimated one family place per 72,538 people. It is reported in [3] that shelter availability in the Klang Valley found that, even while the numbers of government domestic violence shelters are inadequate, they are simultaneously not being fully utilised by women in need. In contrast, NGO shelters were found to be consistently at or over capacity [4].

Women are the primary focus of domestic violence studies. It is stated in [5] that domestic violence is prevalent not only in all races but across all occupational and age groups as well, and that across all these ethnographic and demographic groupings, domestic violence profoundly affects women, who are the most frequent victims. Women are abused in various forms, either physically, socially, financially, sexually, psychologically, or emotionally. Ref. [6] argues that women who have been subjected to violence should be able to access assistance, and their cases should be given special attention. Following this, the government allocated RM21 million towards domestic violence support centres in Budget 2021. Other recommendations were also included in Budget 2021, including increased government social workers through the MySTEP program and financial support for childcare.

2. How Bad Is Domestic Violence in Malaysia?

In Ref. [2], domestic violence is defined as when a partner is treated violently, including physical, sexual, psychological, and social abuse. Domestic violence behaviour includes: (i) deliberately subjecting the victim to fear; (ii) subjecting them to physical injury; (iii) coercing their behaviour; (iv) threatening and/or committing sexual acts; (v) confining or detaining the partner involuntarily; (vi) treachery; or (vii) damaging property with the intent of causing grief or annoyance. (Domestic Violence Act 1994, s. 2;. In many cases, domestic violence is only revealed to others after the victim becomes severely injured or is even murdered by their abuser. Abuser identities are not restricted to partners, they can also be a grandparent(s), parent(s), child(ren), or a disabled family member. The diversity of victims is equally inclusive. The UN Trust Fund to End Violence against Women (UNTF EVAW) surveyed 122 of its grantees from 69 countries worldwide. They found that key factors exacerbating violence included food shortage, lockdowns, unemployment, and economic insecurity [7].

In Malaysia, ref. [8] highlights that before the enforcement of the Domestic Violence Act (1994), the number of domestic violence cases between 1991 to 1994 was less than 300 cases per year. However, according to the Royal Malaysian Police Force, it was reported that based on the statistics of violence against women in Malaysia from the years 2000–2006, the number of reported cases was more than 2500 per year [9]. This demonstrates that the number of domestic violence incidents remained very high, even after the enforcement of the Act in 1996. The former minister for Women, Family and Community Development, Datuk Seri Shahrizat Abdul Jalil, expressed her view that violence against women has increased alarmingly, despite the measures taken to tackle it [10]. She noted that despite the enforcement of the Act, there were still 3264 cases reported in 2006. She further stated that the statistics showed a worrying trend of rising cases of domestic violence. Considering that this issue has been extensively debated, and continues to be through electronic and print media, and contentious differences of opinion remain between certain quarters, researchers are prompted to propose research questioning how far the Domestic Violence Act (1994) has gone in providing protection for women.

Previously, compilations of research on social problems have raised the issue of domestic violence in Malaysia, dating back to the 1970s, including a book produced by university academicians titled "Transformasi Masyarakat: Cabaran Keluarga, Gender dan Sosiobudaya", which began to get scholarly attention by the mid-2000s. Later, ref. [11] found that domestic violence—either to children, spouses, elderly relatives, or disabled people, had become a new social problem that had shown an increased number of cases and had otherwise worsened in severity, and thus this phenomenon needed to be tackled throughout Malaysian society [12]. Statistics show an alarming trend of yearly increments in domestic violence cases in Malaysia—as reported by the Ministry of Women, Family and Society Development and the Malaysian Royal Police (PDRM) (see Figure 1 and Table 1 below).

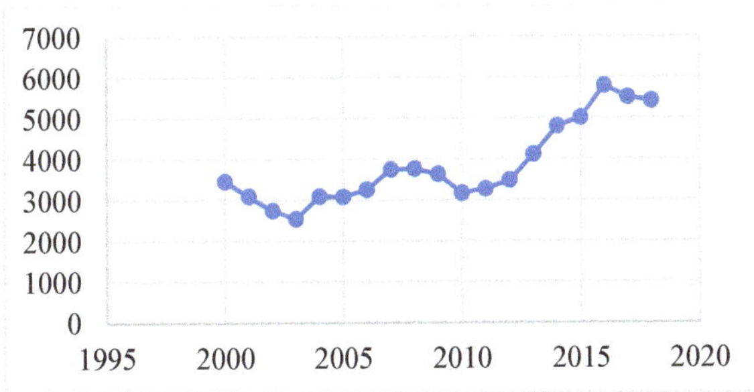

Figure 1. Increment of Domestic Violence Cases (2001–2018).

Table 1. Domestic Violence Cases in Malaysia [1].

Year Reports	Domestic Violence
2000	3468
2001	3107
2002	2755
2003	2555
2004	3101
2005	3093
2006	3264
2007	3756
2008	3769
2009	3643
2010	3173
2011	3277
2012	3488
2013	4123
2014	4807
2015	5014
2016	5796
2017	5513
2018	5421

[1] Source: Ministry of Women, Family and Society Development, 2020.

Since the COVID-19 pandemic in Malaysia, that started with first case on 25 January 2020, the Malaysian government imposed a Movement Control Order (MCO), to curtail the spread of the deadly virus. Malaysian citizens were forced to stay at home during this time, and this new scenario situated domestic violence victims under the same roof as their abusers, with only limited avenues of escape from these abusive actions. unsurprisingly, during the first MCO (2 October 2019–17 March 2020), the Malaysia Royal Police recorded 3258 domestic violence cases—a significant increase. During the MCO from 18 March 2020–31 October 2020, a further 3080 cases were recorded. The rate of cases of domestic violence reduced slightly after the government loosened the MCO restrictions during the recovery stage of the pandemic. Further waves of the pandemic spread again, with some localised MCOs extending into 2021, and the domestic violence issues remained.

Various police reports detail prolonged domestic violence that resulted in the murder of the victims [13]. Recently, the nation was shaken again after a boy was badly abused and brutally murdered under the care of his mother and stepfather, only three months after he was taken from his adopted family. Based on reports, there are many evident reasons why domestic violence increases during the MCOs, including:

- Abusers were always with the victim, spending more time together, increasing the chance of proximal conflict;
- Social isolation reduces opportunities for other outlets to channel or relieve stress;
- Financial problems and future uncertainties significantly increased conflicts and stress;

- A narrow or unconducive home environment causes drowning when at home;
- Abusers with drug addiction habits had difficulty obtaining supplies, both due to movement restrictions, and financial factors;
- Victims were unable to run away from the domestic violence situation due to movement restrictions;
- Absence of support from others, and no place to seek support, as many services had to close during MCO;
- Victim dependence on the abuser may have increased, as a result of restrictions, which exacerbates this abusive potential.

In summary, it was not the MCO that led to increased quarrels or abuse directly, but rather that domestic violence that was already happening before the MCO was now made worse when victims were unable to flee from their attackers.

3. Sufficiency of Initiatives and Law to Protect Domestic Violence Victims

In order to protect domestic violence victims, the Malaysian National Security Council (MKN) provided a helpline, Talian Kasih, together with the WAO helpline, to assist victims (beginning from the first MCO) [14]. Women's Aid Organisation reported that the helpline received 3.4 times as many calls related to domestic violence since the beginning of MCO in May 2019—highlighting the desperate need for more initiatives to support victims. MCOs were targeted to control the spread of the COVID-19 virus to keep the public safe, however in hindsight, the government should have taken into account that staying home was definitely not a safe place for women, children, the elderly, or those with disabilities, who are subject to abuse from family members. Furthermore, the MCOs limited the operations of shelters to receive domestic violence victims during the MCOs, and this situation trapped victims with their abusers, who controlled them physically, emotionally and socially.

In cases of domestic violence, first and foremost, the rules and regulations must be sufficient to protect the victims and bring abusers to justice. Currently, the Domestic [15] is enforced throughout Malaysia due to the increasing rates of domestic violence crime. The government believes that if it is not curtailed, it will result in a deterioration of the moral and social fabric of Malaysian society. The Domestic Violence Act (1994) is a law that pays special attention to domestic violence cases by providing several solutions for such cases [16]. Women's organisations in Malaysia faced nearly a decade of struggle fighting to protect domestic violence victims before the Domestic Violence Act was passed in 1994, under this law, domestic violence was legally declared a crime.

According to the act, where domestic violence occurs, third parties can report information to public welfare officers or police officers, in order to take action under the authority as enforcement officers. Those parties include the spouse, ex-spouse, daughter or son, mother or father, brother, sister, or any family members, as well as the lawyers of the victim. If the victim is a minor, or disabled adult, the caretaker or any person responsible for the child or disabled person can also report. The application to get a protection order can be made at the victim's residence, the residence where a victim is brought to live, a residence where the domestic violence has occurred, or to a temporary shelter for the victim's protection [17].

Concurrently, judicial actions can be taken under the Laws of Malaysia: Act 574: Penal Code Section 426: Punishment for Committing Mischief, whereby this section states that:

"Whoever commits mischief shall be punished with imprisonment or a term which may extend to *five years or with fine or with both."

According to [11], the Domestic Violence Act (1994) and the Penal Code 426 are sufficient to protect victims of domestic violence, but that there are some remaining issues that require further work to overcome problems, which include: (i) the length of time allocated for hearings under two different laws; (ii) the cost that victims must bear for pursuing a legal complaint; and (iii) the amount of compensation to be paid to the victims—particularly for families with reduced economic means.

Although both laws can be applied, one initial problem is that any domestic violence case is difficult to establish in order for justice to take place. Establishing concrete cases of domestic violence cannot be based on hearsay or general observations of victims—such as signs of physical injury or psychological illness resulting from the abuse. Certain experts need to be engaged to prove that domestic violence exists because victims are often psychologically and emotionally controlled by their abusers, and fear implicating their abuser. Victims who suffer domestic violence for extended periods, or from childhood, may be conditioned to believe that the behaviour is commonplace, or even socially acceptable, even if they are subject to severe injuries. Research by the Women and Gender Research Centre [18] shows that 9 percent of women in relationships suffer domestic violence at least once in their lifetime. According to [18], 35% of women worldwide have experienced some form of domestic violence by their intimate partner in their lifetime. Ref. [19] In the Women's Aid Organisation Annual Report 2020, it showed that 65 percent of girls reported that there was no action taken on the abuser.

4. How to Assist Domestic Violence Victims during MCOs?

Domestic violence may be unavoidable in any society or nation, however, it can be monitored and controlled with suitable initiatives—particularly during MCOs, when everyone is locked down at home with no escape from abusers. There are some significant initiatives that can be implemented, such as providing an appropriate number of shelters for victims, designing a first response system to assist victims immediately when domestic violence happens, and others. The existing laws must be reviewed to ensure that the victims do not have to undergo a long and painful journey to resolve their issues, in order to prevent escalating abuses, including murder.

During MCOs or other crucial situations, shelters must be made available to support victims. This is to ensure that the victims are not forced to remain in abusive situations, nor become homeless, or continue living in uncertainty and fear. After the MCO declarations, shelters for domestic violence were not included as essential services, and therefore were not allowed to operate. This was because the government believed that crowded shelters would facilitate the spread of disease without adequate precautions being taken. The only action that helplines could therefore suggest was for victims to be placed on a waiting list and advised them to get help and protection from other family members, neighbours, or other institutions that are able to assist [20].

According to [Figure 2], Malaysia is still far behind other nations in providing shelters for domestic violence victims, with only one family place available for every 70,000 people. The international standard is one family place for every 10,000 people.

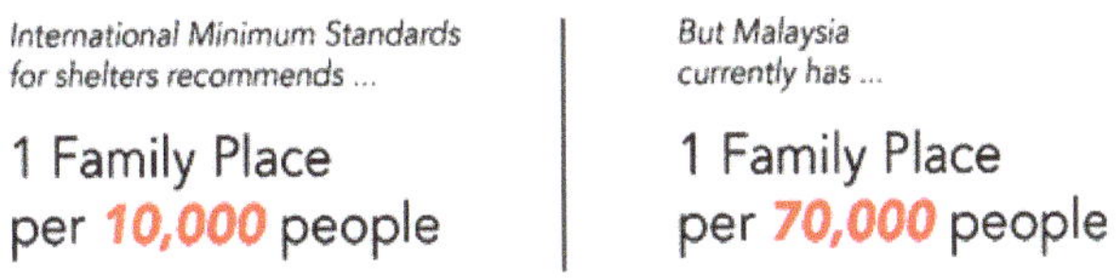

Figure 2. The Shelter Needs of Domestic Violence Survivors and the Availability and Accessibility and Related Services and Kuala Lumpur, Policy paper no. 1, 2019.

The government must take more proactive action in organising temporary shelters for domestic violence victims during pandemic crises. In France, for example, the government converted 20,000 hotel rooms to become shelters and "pop-up counselling centres" after cases of domestic violence increased during their MCO, to ensure the safety of victims of abusive domestic violence.

The effectiveness of Malaysia's first response system must be significantly improved when domestic violence cases increase during MCOs. First responders include crisis hotline operators, police officers, and social workers. Sufficient manpower resources must be operating to avoid them being overwhelmed as resources for victims to access legal protection and other support. Sufficient funds must be allocated for such services, as the number of cases has been shown to rapidly escalate during a pandemic/MCO.

According to Figure 3, as of December 2020, Malaysia has a very low ratio of 1:8756 citizens for the numbers of social workers when compared with countries such as the United States, the United Kingdom, Singapore, or Australia.

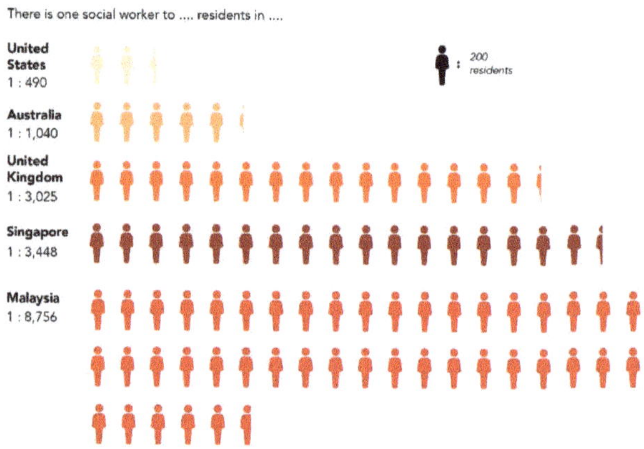

Figure 3. New Bill granting social workers professional status set for Dec tabling, Malay Mail Online, 2020.

Social workers not only cater to domestic violence cases, but also other issues such as flood relief, disaster relief, or other support services, often as volunteers. The quality of services provided to abuse victims is often low, as they are working under the pressure of new social norms, including social distancing and other measures for controlling the pandemic. To ensure an improved quality of support service for domestic violence victims, there is a need to focus on comprehensive coordination among various agencies or parties—particularly during a pandemic. Excessive bureaucratic red tape will only jeopardize the victim safety, and reduce the chance of them getting adequate protection.

Since this short paper only touches the tip of the iceberg that is domestic violence during the COVID-19 MCO in Malaysia, further research must be conducted to enable the country to continue to dig deeper and confront domestic violence while discovering the vulnerabilities of social support systems for protecting families from abuse.

Author Contributions: Conceptualization, T.A.T.Z.M., H.A.R.; methodology, R.N.; formal analysis, N.R.M.Z.; investigation, T.A.T.Z.M.; writing—original draft preparation, T.A.T.Z.M., H.A.R.; writing—review and editing, T.A.T.Z.M., H.A.R.; All authors have read and agreed to the published version of the manuscript.

Funding: This research received no external funding.

Institutional Review Board Statement: Not applicable.

Informed Consent Statement: Not applicable.

Data Availability Statement: Not applicable.

Conflicts of Interest: The authors declare no conflict of interest.

References

1. Abdullah, A. Themes and issues in research on social problems in the first decade of 21st century in Malaysia. *Akademia* **2010**, *78*, 3–14.
2. Adil, M.A.M. Remedi mangsa keganasan dalam rumahtangga mengikut undang-undang jenayah dan sivil. *J. Syariah* **1997**, *5*, 79–91.
3. Bruckert, C.; Law, T. *Women and Gendered Violence in Canada: An Intersectional Approach*; University of Toronto Press: Toronto, ON, Canada, 2018.

4. Daud, N.M.; Hussin, M.N.M.; Jamaludin, M.H. Keganasan rumahtangga di Malaysia: Satu analisis literatur: Domestic violence in Malaysia: Analysis of literature review. *J. Shariah Law Res.* **2021**, *6*, 273–296.
5. Erniwati, E.; Fitriani, W. Faktor-faktor penyebab orang tua melakukan kekerasan verbal pada anak usia dini, Yaa Bunayya. *J. Pendidik. Anak Usia Dini* **2020**, *4*, 1–8.
6. Firdaus, A.G. Mengapa Kes Keganasan Rumah Tangga Meningkat Semasa PKP? Astroawani. April 2020. Available online: https://www.astroawani.com/berita-malaysia/mengapa-kes-keganasan-rumah-tangga-meningkat-semasa-pkp-236528 (accessed on 1 December 2020).
7. Women's Centre for Change. *Statistics of Violence against Women in Malaysia Year 2000–2006*; Women's Centre for Change: Pulau Pinang, Malaysia, 2008.
8. Jafar, F. Hanya keberanian hapus keganasan. In *Berita Harian*; New Straits Times Press: Kuala Lumpur, Malaysia, 2000; p. 1.
9. Jal, Z.Y.M. *Jenayah Keganasan Rumah Tangga*; Universiti Malaya: Kuala Lumpur, Malaysia, 2004.
10. Lai, S.Y.; Maria, C.A.; Ong, J.L.; Wong, P.L. *The Rape Report: An Overview of Rape in Malaysia*; All Women's Action Society and Strategic Info Research (SIRD): Petaling Jaya, Malaysia, 2002.
11. Lee, W.S.; Masidi, Y. *The Shelter Needs of Domestic Violence Survivors and the Availability and Accessibility of Shelters and Related Services in Selangor and Kuala Lumpur, Policy Paper No. 1*; Women's Aid Organisation: Petaling Jaya, Malaysia, 2019.
12. Mukundan, S.R. *Domestic Violence Amid COVID-19 in Malaysia: Diving Deeper into the Heart of the Matter*; Malaysia, Singapore & Brunei Darussalam; UNDP: New York, NY, USA, 2020.
13. Nur, F.A.M. Laporan Keganasan Rumah Tangga Meningkat. Sinar Harian. June 2020. Available online: https://www.sinarharian.com.my/article/87263/BERITA/Nasional/Laporan-keganasan-rumah-tangga-meningkat (accessed on 1 December 2020).
14. Randawar, D.K.; Jayabalan, S. The definition of domestic violence in Malaysia: A cross-national comparison. *Akademika* **2018**, *88*, 77–89.
15. Randawar, D.K.; Jayabalan, S. Wife battering—A Malaysian perspective. In Proceedings of the Colloquium on Administrative Science and Technology, Kuching, Malaysia, 22 September 2016; pp. 275–283.
16. Ramha, R.R.; Sulaihha, M.I.; Zatul, I.M.F. Kita Cakna: Isu Keganasan Rumah Tangga. Buletin Akademik. University Teknologi Terengganu. September 2020. Available online: https://ir.uitm.edu.my/id/eprint/18820/1/BULETINAKADEMIKJANUARI-MAC2017[1].pdf (accessed on 1 December 2020).
17. Starsky, W.P. *Family LAw*, 5th ed.; West Thomson Learning: New York, NY, USA, 2001.
18. Minister: Violence against Malaysian Women Has Increased Alarmingly, The Sun. Available online: http://www.sun2surf.com/articlePrint.cfm (accessed on 12 December 2008).
19. University Sains Malaysia. *9% of Women Suffer from Domestic Violence. Apex—Laman Web Rasmi*; University Sains Malaysia: Penang, Malaysia, 2018.
20. World Health Organization. Responding to Intimate Partner Violence and Sexual Violence against Women. 2013. Available online: http://apps.who.int/rhl/guidelines/9789248595/en/ (accessed on 20 July 2022).

proceedings

MDPI

Proceeding Paper

Digital Skills Framework in Higher Education †

Sharipah Ruzaina Syed Aris [1,*], Sian Hoon Teoh [1], Sayang Mohd Deni [2], Fazyudi Ahmad Nadzri [1] and Siti Fairuz Dalim [1]

1 Faculty of Education, Universiti Teknologi MARA, Cawangan Selangor Kampus, Puncak Alam 42300, Selangor, Malaysia
2 Faculty of Computer and Mathematical Sciences, Universiti Teknologi MARA, Shah Alam 40450, Selangor, Malaysia
* Correspondence: sruzaina@uitm.edu.my
† Presented at the International Academic Symposium of Social Science 2022, Kota Bharu, Malaysia, 3 July 2022.

Abstract: Digital skills should be incorporated in the curriculum as the skills are essential attributes graduates should acquire before joining the workforce. Understanding the digital skills framework and abilities among learners may help educators create suitable learning outcomes, activities, and assessments to enable the learners to acquire these skills. This study shares expert judgements on a digital skills instrument (DSI) on content validation framework according to sufficiency, clarity, coherence, and relevance. A pre-test involving 218 students from various disciplines was conducted and Cronbach's Alpha was used to measure the validity and reliability of the DSI. The overall result indicated a strong reliability. Clarity was the characteristic garnering the highest value, followed by coherence, relevance, and sufficiency. The DSI was revised and 28 items encompassing five components were proposed following the feedback of: information literacy, computer and technology literacy, digital communication/collaboration skill, digital identity and well-being, and digital ethic. Based on the pre-test, all the components indicated high reliability, ranging from 0.847 to 0.958, and all of the items were retained to measure digital skills. Overall, the experts' evaluations and pre-tests suggested that the DSI and its five components as the digital skills framework can be used for measuring digital skills.

Keywords: digital skills; expert judgement; information literacy; computer and technology literacy; digital communication/collaboration skill; digital identity and well-being; digital ethic

Citation: Aris, S.R.S.; Teoh, S.H.; Deni, S.M.; Nadzri, F.A.; Dalim, S.F. Digital Skills Framework in Higher Education. *Proceedings* **2022**, *82*, 61. https://doi.org/10.3390/proceedings2022082061

Academic Editor: Mohamad Rahimi Mohamad Rosman

Published: 19 September 2022

1. Introduction

1.1. Digital Skills Proficiency: Influencing Factors

Having proficient digital skills is important in this era of modernity, as technology has become a necessity to keep the world functioning smoothly. Because of this, digital literacy has been an issue of concern regarding whether upcoming generations will be adequately equipped with digital knowledge. The ability to locate, evaluate, and produce clear information across a range of media platforms is referred to as digital literacy [1]. There are many factors that influence one's digital skills proficiency. Technical skills, critical understanding, and communicative abilities are some of the factors listed [1], with critical understanding being the most important. On the other hand, students' level of education also has an impact on the development of their digital skills [2]. Additionally, individuals who have a great critical understanding are also able to undertake a more in-depth analysis and evaluation of the information that they find in the media [1]. Another factor is technical skills, defined by Dewi et al. [1] as "the ability to access and operate media". As the factor implies, individuals possessing this skill are good at computer and internet skills, the use of media, and even advanced use of the internet. Another factor is one's ability to communicate. It has been suggested that individuals who have mastered the skill are able to interact with others and participate online, while also being able to produce media content. The other factor that affects digital skills proficiency that was discussed by Van Deursen et al. [3] is age. It has been postulated that a person's age may have an impact on their digital competency. On the other hand, digital skills have facilitated the growth of big data analysis. Because decision-making based on big data analysis is now widespread, there is a compelling need to improve digital skills [4].

1.2. Digital Skills as Graduate Attributes in the Curriculum Framework

Digital skills are among the future skill sets required of all graduates in order to embrace challenges and stay relevant [5]. They are known to provide the younger generation with the skills required in the 21st century, ensuring their survival in an unfamiliar environment. Therefore, those who have these skills are more employable, productive, and creative, and they will remain safe and secure in the new digital economy landscape. In other words, digital skills are needed as they encompass the ability to acquire and manage information through digital devices, communication applications, and networks [6]. The new skillset and the change in the education landscape also prompted the Malaysian Qualification Agency (MQA) to revise the 2010 qualification framework [7] so that it would emphasize the digital attribute. This was done to ensure that students at higher education institutions would be taught and trained in the skills, in addition to other skills such as numeracy, communication, interpersonal, and entrepreneurial skills. This new trajectory is aligned with the global agenda of preparing graduates to thrive in a volatile, uncertain, complex, and ambiguous (VUCA) environment. For this reason, since April 2019, the Malaysian Qualification Agency (MQA) has required all universities in Malaysia to comply with this new framework for their new curricula.

To meet this requirement, educators must not only expose their students to digital skills, but they must also determine the levels of their students' digital skills in the first place in order to help the former to design materials that can be used to equip the latter with the skills. This is important, as it can help them to develop appropriate assessments and approaches to achieve intended learning outcomes according to different disciplines or courses. Most importantly, the practice also helps them to understand their students better and to design a curriculum that meets their needs.

The Malaysian Qualification Framework (MQF) has established a general definition and attributes as the minimal criteria for higher education providers to develop their curricula. The MQF defines digital skills as the ability to use information/digital technology to support work and studies. The skills include gathering and storing information, processing data, using applications for problem solving and communication, and using digital skills ethically or responsibly. Four components were extracted from the general definition: information literacy, computer and technology literacy, visual literacy, and digital communication/collaboration skills.

The focus of this paper is to seek validation from experts based on their views on the digital skills instrument that contains the four components, and to establish the reliability of final items based on the pilot study. The aim is that the research outcome could guide curriculum developers and course designers to design appropriate content knowledge, delivery activities, and assessments that could instill and equip students with both skills that are aligned with the MQA description and the aspiration of the Malaysian Government who wishes to produce graduates with 21st-century skills.

2. Materials and Methods

This study is a cross-sectional study that uses the quantitative method to construct a digital skills framework. This study involves the development of questionnaire items, validation of the questionnaire process and testing the reliability of the questionnaire. The validation of the questionnaire involves expert content validation. For determining the reliability of the questionnaire, internal reliability was conducted and analyzed using Cronbach's Alpha.

2.1. Digital Skills Instrument (DSI)

There are 22 items that are divided according to the four components in the digital skills instrument (DSI), namely, information literacy, computer and technology literacy, visual literacy, and digital communication/collaboration skills. Table 1 shows a breakdown of the number of items for each of the components.

Table 1. Initial component of digital skills instrument (DSI).

Component	Number of Items
Information literacy	11
Computer and technology literacy	4
Visual literacy	3
Digital communication/ collaboration skills	4
Total	22

2.2. Expert Judgment Content Validation

Evaluation through expert judgements involves asking several individuals to make a judgement on an instrument or to express their opinions on a particular aspect (refer to Table 2 for the list of experts). The experts play a crucial role in explaining, adding, and/or modifying the required components. Initially, two experts in measurement and evaluation from a comprehensive and a research university were selected to evaluate the digital instrument. Then, feedback from eight experts was also taken into account in the refinement of the DSI. Overall, two digital content experts, three senior lecturers from science and technology, and three experts from social sciences were involved to ensure the validity of the DSI before its final version was administered to the actual sample.

Table 2. Judges and field of expertise.

Expert	Field of Expertise	Research/Teaching Experience (Years)
1	Language testing (social sciences)	18
2	Measurement and evaluation (social sciences)	20
3	Digital literacy (science and technology)	21
4	Artificial intelligence (science and technology)	15
5	Instructional communication and new media (social sciences)	16
6	Psychometric assessment (social sciences)	25
7	Pharmacy (health science)	20
8	Law and criminal justice (social sciences)	21

The volunteer experts came from a variety of disciplines, including science, technology, and social sciences (non-science and technology). The criteria for selection were based on either their experiences in curriculum development, in holding any academic post, or in having more than 10 years research or teaching experience. According to the standard and MQF, digital skills must be measured across disciplines and at all qualification levels. Therefore, experts from different fields should be more than able to provide better descriptions of appropriate digital attributes for different curriculum or academic programs.

All of the items were divided into the categories of sufficiency, clarity, coherence, and relevance (refer to Table 3 for the scales and indicators of the categories). All of the experts were required to rate each item using a scale ranging from 4 to 1. The experts were also encouraged to give comments in the open space in the instrument.

Table 3. Categories, indicators, and scale.

Category	Scale	Indicators
Sufficiency The items within the same dimension should suffice to measure this dimension.	4 3 2 1	The item is sufficient to measure the dimension The item measures some aspects of the dimension but does not represent the full dimension A few items must be added to fully assess the dimension The item is insufficient to measure the dimension
Clarity The items can be understood easily. (i.e., syntax and semantics are appropriate).	4 3 2 1	The item is sufficient to measure the dimension The item measures some aspects of the dimension but does not represent the full dimension A few items must be added to fully assess the dimension The item is insufficient to measure the dimension
Coherence The items are logically related to the dimension or indicator they are measuring.	4 3 2 1	The item is sufficient to measure the dimension The item measures some aspects of the dimension but does not represent the full dimension A few items must be added to fully assess the dimension The item is insufficient to measure the dimension
Relevance The items are essential or important. (i.e., they must be included).	4 3 2 1	The item is sufficient to measure the dimension The item measures some aspects of the dimension but does not represent the full dimension A few items must be added to fully assess the dimension The item is insufficient to measure the dimension

Source: adapted from Escobar-Pérez and Cuervo-Martínez [8].

2.3. Pilot Study

After the questionnaire was assessed by the experts, the resulting instrument was pilot tested with potential respondents to determine the reliability of each component and item in the final instrument. An online survey was done and randomly distributed to students from various faculties and fields.

3. Results

3.1. Content Validation by Expert Judgment

Cronbach's Alpha values were measured to check the data reliability and internal consistency, and values higher than 0.6 were considered to be acceptable. Values ranging from 0 to 1 with greater values indicate higher internal consistency measures on the opinions and perceptions among the respondents. The reliability of the instrument focusing on the experts' evaluations on the four constructs, namely sufficiency, clarity, coherence, and the relevance of the instrument were analyzed using Cronbach's Alpha. The overall result indicated strong reliability ($0.915 \leq$ Cronbach's $\alpha \leq 0.95$) for the 22 items. Table 4 presents the descriptive statistics (means and standard deviations) of the categories of the perceptions of digital skills. The means ranged from 3.40 to 3.56.

Table 4. Cronbach's Alpha and descriptive statistics for each of the categories.

Category	Cronbach's Alpha	Mean	Standard Deviation
Sufficiency	0.915	3.56	0.446
Clarity	0.944	3.40	0.532
Coherence	0.941	3.55	0.530
Relevance	0.934	3.52	0.529

Table 5 indicates the Cronbach's Alpha values for each component and characteristic based on the judgments by the experts.

Table 5. Cronbach's Alpha values for each of the components and characteristics of digital skills.

Construct	Characteristics	Characteristics	Cronbach's Alpha	Average
Digital Skills	C1—Information Literacy	Sufficiency	0.924	0.94
		Clarity	0.961	
		Coherence	0.949	
		Relevance	0.926	
	C2—Computer and Technology Literacy	Sufficiency	0.966	0.93
		Clarity	0.860	
		Coherence	0.934	
		Relevance	0.950	
	C3—Visual Literacy	Sufficiency	0.855	0.82
		Clarity	0.732	
		Coherence	0.812	
		Relevance	0.902	
	C4–Digital Communication/ Collaboration Skill	Sufficiency	0.736	0.76
		Clarity	0.650	
		Coherence	0.823	
		Relevance	0.837	

It can be seen that the Cronbach's Alpha values for information literacy ranged from 0.924 to 0.961 and the Cronbach's Alpha values for computer and technology literacy ranged from 0.860 to 0.966, indicating strong internal consistencies compared with the other two components, namely visual literacy, whose values ranged from 0.732 to 0.902, and digital communication/collaboration skills, whose values ranged from 0.650 to 0.837. It could be concluded that the experts' scores on their perceptions showed consistency towards digital skills on information literacy and computer and technology literacy. On the other hand, the respondents' perceptions of visual literacy showed a moderate level of consistency compared with digital communication/collaboration skills.

Further analysis on the enhancement of the questionnaires was conducted by considering the input given by the experts. For example, the questionnaire was improved by restructuring the order of the items, the sentence structure, and by adding two components, as well as removing the visual literacy component. Some of the items in the component were also merged with the other components. Consequently, 28 items representing five components in the DSI, namely information literacy, computer and technology literacy, digital communication/collaboration skills, digital identity and well-being, and digital ethics were developed.

3.2. Pilot Study

Distribution of Respondents

Figure 1 below shows the distribution of respondents based on college and faculties. Overall, 218 students from various academic programs participated in the online survey. Specifically, the highest percentage of the respondents came from the Faculty of Accountancy (44.04%), followed by students from the Faculty of Computer and Mathematical Sciences (26.61%) and from the Faculty of Film Theatre and Animation (12.84%). The results of the pilot study were analyzed using Cronbach's Alpha. Table 6 shows the proposed number of items by components, as well as the item-total statistics, which include the scale mean if an item is deleted, scale variance if an item is deleted, and corrected item-total correlation and Cronbach's Alpha if an item is deleted. As eliminating items from each component reduces the respective Cronbach's Alpha, most items were retained. Moreover, the overall Cronbach's Alpha analysis revealed a strong internal consistency among respondents for each component. The overall analysis and the value of Cronbach's Alpha if an item was deleted showed very high values with alpha values greater than 0.8.

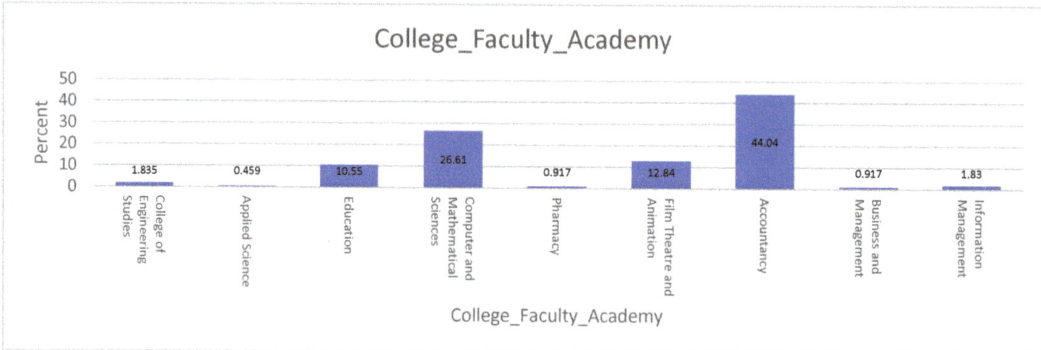

Figure 1. Distribution of respondents according to college and faculties.

Table 6. Item-total statistics: item and values in five components of perceptions on digital skills.

Component	Item	Scale Mean If Item Deleted	Scale Variance If Item Deleted	Corrected Item-Total Correlation	Cronbach's Alpha If Item Deleted	Cronbach's Alpha Overall
C1-Information Literacy	C1.1	110.49	235.726	0.687	0.966	0.957
	C1.2	110.52	235.264	0.738	0.966	
	C1.3	110.42	235.904	0.754	0.966	
	C1.4	110.60	232.683	0.804	0.965	
	C1.5	110.57	232.689	0.794	0.965	
	C1.6	110.71	234.522	0.736	0.966	
	C1.7	110.76	232.793	0.758	0.965	
	C1.8	110.69	233.006	0.790	0.965	
	C1.9	110.80	231.848	0.780	0.965	
	C1.10	110.59	234.732	0.717	0.966	
	C1.11	110.72	232.590	0.749	0.966	
	C1.12	111.07	235.193	0.607	0.967	
C2-Computer and Technology Literacy	C2.1	110.51	233.928	0.747	0.966	0.895
	C2.2	110.99	232.968	0.620	0.967	
	C2.3	110.69	233.790	0.741	0.966	
	C2.4	110.76	233.300	0.736	0.966	
	C2.5	11.13	234.324	0.572	0.967	
C3-Digital Communication/ Collaboration Skill	C3.1	110.69	232.702	0.746	0.966	0.919
	C3.2	110.67	233.016	0.718	0.966	
	C3.3	110.89	231.938	0.692	0.966	
	C3.4	110.61	234.082	0.701	0.966	
C4-Digital Identity and Well-Being	C4.1	110.31	236.879	0.616	0.966	0.847
	C4.2	110.63	233.745	0.654	0.966	
	C4.3	110.40	235.882	0.629	0.966	
C5-Digital Ethics	C5.1	110.39	234.846	0.688	0.966	0.942
	C5.2	110.59	233.691	0.714	0.966	
	C5.3	110.44	235.408	0.674	0.966	
	C5.4	110.43	235.167	0.665	0.966	

4. Discussion

The findings of this study led to the development of a digital skills framework. The construction of the digital skills framework in this study was subjected to a process of recurrent validation. There were two stages in the process. In the first stage, two experts provided extensive feedback on the content for four constructs (information literacy, computer and technology literacy, visual literacy, and digital communication/collaboration skills). The result of the first stage validation shows that two components, "visual literacy" and "digital communication/collaboration skills", had a moderate level of consistency. Only "visual literacy" was removed from the list of components. "Digital communication/collaboration skills" remained as a component as they are the skills that are currently needed. In the digital world, all learners or users must be able to collaborate with others in order to achieve the purpose and advantages of using digital technology. These qualities, notably communication and collaboration skills, are seen as crucial in the 21st century [9]. The second stage of validation involved eight experts. In the second stage, two important components were added based on the relevant items and comments from the experts. Their input recognized the need for practicing integrity in the digital world. Many recent studies have focused on the influence of digital technology use on well-being, particularly among adolescents [10]. Hence, components of the framework's five components are information literacy, computer and technology literacy, digital communication and collaboration skills, digital identity and well-being, and digital ethics. The importance of these primary components in the development of digital skills has been emphasized. Generally, these components incorporate basic awareness (such as managing information, full participation in the digital society, and showing responsibilities and appreciation), use (such as technical knowledge and the utilization of knowledge resources), and the creation of content with an effort towards adaptation to the digital society [11]. Figure 2 illustrates the analyzed components alongside the incorporated components, which have been highlighted by [11].

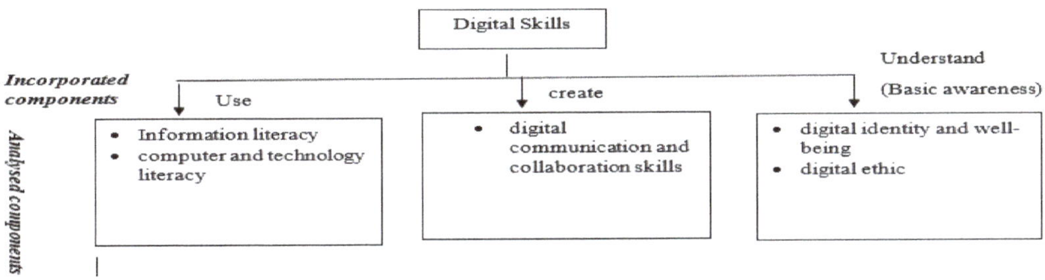

Figure 2. Comparison between the analyzed components (this study) and incorporated components [11].

As shown in Figure 2, anyone who wants to develop their digital skills must first learn how to obtain information and use computers or other digital devices, as shown in the incorporated component of "use". A healthy society emphasizes not only the application of digital technology and knowledge (which leads to information literacy, computer literacy, and technology literacy), but also the development of new knowledge through communication and collaboration. Additionally, everyone is drawn closer together via engagement and communication, as depicted in the incorporated component of "create". However, when there are several parties or individuals involved, communication becomes more complex. Thus, for ethical and well-being considerations, the use of knowledge exploration and social interaction in communication should be monitored. Hence, "understand" entails a socialization process, i.e., the individuals' responsibilities in interaction and communication. A guide on how to use digital tools wisely, such as social presence and emotional presence in social networks, is required for consideration in order for the development of digital skills. Additionally, a proper guide can assist individuals to develop their digital identities [12]. Eventually, the five components of the digital skills framework have been well-elucidated in the three aspects of "use", "create", and "understand" with basic awareness, as shown in Figure 2.

5. Conclusions

The aim of this study was to obtain the validity and reliability of the digital skills framework through expert judgements and a pilot test study. Consequently, information literacy, computer and technology literacy, digital communication and collaboration skills, digital identity and well-being, and digital ethics are the five agreed-upon components of the digital skills. Understanding the components of the digital skills is important to all educators before they can design their lessons, learning activities, and assessments during curriculum development. Another critical aspect is that educators must identify their students' digital literacy or ability in order to create significant and meaningful digital learning experiences in the classroom. The digital skills instrument (DSI) can be used as a tool to gain better insights into digital literacy among students. Finally, this tool can also serve as a digital needs analysis tool to investigate the strengths and weaknesses, and provide continuous quality improvement in curriculum development.

Author Contributions: Conceptualization, S.R.S.A.; methodology, S.R.S.A. and S.H.T.; software, S.M.D.; validation, S.R.S.A., S.H.T., and S.M.D.; formal analysis, S.R.S.A., S.H.T., and S.M.D.; investigation, S.R.S.A., S.H.T., and S.M.D.; resources, S.R.S.A.; data curation, S.R.S.A., S.H.T., and S.M.D.; writing—original draft preparation, S.R.S.A., and S.M.D.; writing—review and editing, F.A.N., and S.F.D.; visualization, S.R.S.A.; supervision, S.R.S.A.; project administration, S.R.S.A.; funding acquisition, S.R.S.A. All authors have read and agreed to the published version of the manuscript.

Funding: This research was funded by a special research grant from Universiti Teknologi Mara (600-RMC/GPK 5/3 (007/2020)).

Institutional Review Board Statement: Ethical review and approval for this study was obtained from the UITM Research Ethic Committee (REC/08/202) MR/713.

Informed Consent Statement: Informed consent was obtained from all subjects involved in the study.

Data Availability Statement: The data presented in this study are available upon request from the corresponding author. The data are not publicly available due to privacy restrictions.

Acknowledgments: The researchers would like to express their appreciation to all the experts who dedicated their time and expertise during the preparation of the survey in the study.

Conflicts of Interest: The authors declare no conflict of interest. The funders had no role in the design of the study; in the collection, analyses, or interpretation of data; in the writing of the manuscript; or in the decision to publish the results.

References

1. Dewi, R.S.; Hasanah, U.; Zuhri, M. Analysis Study of Factors Affecting Students' Digital Literacy Competency. *Ilkogr. Online* **2021**, *20*, 424–431.
2. Vodă, A.I.; Cautisanu, C.; Grădinaru, C.; Tănăsescu, C.; de Moraes, G.H.S.M. Exploring Digital Literacy Skills in Social Sciences and Humanities Students. *Sustainability* **2022**, *14*, 2483. [CrossRef]
3. Van Deursen, A.J.A.M.; Helsper, E.J.; Eynon, R. Measuring Digital Skills. From Digital Skills to Tangible Outcomes Project Report. 2014. Available online: https://www.oii.ox.ac.uk/research/projects/?id=112 (accessed on 20 January 2022).
4. Sivarajah, U.; Kamal, M.M.; Irani, Z.; Weerakkody, V. Critical analysis of Big Data challenges and analytical methods. *J. Bus. Res.* **2017**, *70*, 263–286. [CrossRef]
5. World Economic Forum. *Platform for Shaping the Future of the New Economy and Society. Schools of the Future: Defining New Models of Education for the Fourth Industrial Revolution*; World Economic Forum: Geneva, Switzerland, 2020.
6. UNESCO. *A Global Framework of Reference on Digital Literacy Skills for Indicators 4.4.2*; Information Paper No. 51; UNESCO Institute for Statistics: Montreal, QC, Canada, 2018; Available online: http://uis.unesco.org/sites/default/files/documents/ip51-global-framework-reference-digital-literacy-skills-2018-en.pdf (accessed on 20 January 2022).
7. Malaysian Qualifications Agency. *Malaysian Qualification Framework (MQF)*, 2nd ed.; Malaysian Qualification Agency: Kuala Lumpur, Malaysia, 2017.
8. Escobar-Pérez, J.; Cuervo-Martínez, Á. Content Validity and expert judgement: An approach to its use. *Av. Med.* **2008**, *6*, 27–36.
9. Van Laar, E.; Van Deursen, A.J.; Van Dijk, J.A.; De Haan, J. The relation between 21st-century skills and digital skills: A systematic literature review. *Comput. Hum. Behav.* **2017**, *72*, 577–588. [CrossRef]
10. Dienlin, T.; Johannes, N. The impact of digital technology use on adolescent well-being. *Dialogues Clin. Neurosci.* **2022**, *22*, 135–142. [CrossRef] [PubMed]

11. Belshaw, D.A. What Is' Digital Literacy'? A Pragmatic Investigation. Ph.D. Thesis, Durham University, Durham, UK, 2012.
12. Bozkurt, A.; Tu, C.H. Digital identity formation: Socially being real and present on digital networks. *Educ. Media Int.* **2016**, *53*, 153–167. [CrossRef]

proceedings

MDPI

Proceeding Paper

A Financial Literacy Model of Malaysian B40 Households: The Case of Financial Well-Being, Education Level and Socioeconomic Status †

Ananthan Munisamy, Sheerad Sahid * and Muhammad Hussin

Faculty of Education, Universiti Kebangsaan Malaysia, Bangi 43600, Malaysia
* Correspondence: sheerad@ukm.edu.my
† Presented at the International Academic Symposium of Social Science 2022, Kota Bharu, Malaysia, 3 July 2022.

Abstract: Financial literacy contributes to the economic well-being of every country. Specifically, the level of low-income households' financial literacy is crucial in the process of developing and building economic potential. Moreover, the knowledge, attitudes and behaviors of families are essential to the financial sustainability and money wisdom of low-income households. Thus, financial literacy and financial well-being are vital for low-income households to survive in highly competitive marketplaces and sustain their socioeconomic status. This conceptual paper relies on prior research and existing theory to focus on the impacts of financial literacy on socioeconomic status among low-income households. The paper proposes a model for future testing in which the results can provide support for enhancing and sustaining the socioeconomic status of low-income households. This proposed holistic model provides new insights, structure, and suggestions for more research on factors that lead to improving money wisdom and the socioeconomic sustainability of low-income households.

Keywords: socioeconomic status; financial literacy; financial well-being; financial sustainability; Malaysian B40 households

check for updates

Citation: Munisamy, A.; Sahid, S.; Hussin, M. A Financial Literacy Model of Malaysian B40 Households: The Case of Financial Well-Being, Education Level and Socioeconomic Status. *Proceedings* **2022**, *82*, 64. https://doi.org/10.3390/proceedings2022082064

Academic Editor: Mohamad Rahimi Mohamad Rosman

Published: 20 September 2022

Publisher's Note: MDPI stays neutral with regard to jurisdictional claims in published maps and institutional affiliations.

1. Introduction

By 2020, Malaysia aspires to become a developed country following a long-term development plan known as Vision 2020. This goal has recently been extended to 2025 as the government is aware of aspects that seem far from achievable at present. Household financial well-being and socioeconomic status (SES) are key elements in the national development of most countries [1]. Financial education enables every household to recognize its function as an economic actor and its various potentials [2]. B40 households need to gain economic balance by having access to information on personal financial management knowledge and financial well-being as these factors are key in measuring socioeconomic well-being [3–5]. Therefore, the Malaysian government through the theme "Tethering People's Growth" in the Eleventh Malaysia Plan centres on increasing the income and financial well-being of B40 households. This effort is implemented through the first strategic thrust of the 11MP in enhancing a just society [6–10]. In the Financial Education Network (FEN) report, financial education encompasses financial literacy that involves imparting 'money wise' knowledge to households to achieve financial well-being and the drive to improve their SES [11–13]. To produce a higher number of Malaysians who are financially literate especially among B40 households in Malaysia, various aspects such as poverty, financial management, financial education and financial literacy need to be considered [14–16]. It is common for common households that do not achieve financial well-being to have a stronger effect on their socioeconomic status (SES) [17–19]. For instance, 40% of Malaysian households earn an average monthly income of below MYR 4830, indicating that their socioeconomic status is still low [20,21]. Thus, as mentioned earlier, the components of the Eleventh Malaysian Plan focus on the B40 household group to improve their living standards alongside their financial well-being and SES [22–24].

2. Background of the Study

A pending and prominent issue with regards to the socioeconomic status (SES) of Malaysian B40 households is the low level of financial literacy in this population [25–27]. The overall income inequality is reported to continue to decline. The 2016 Family Well-Being Index that measures family well-being found that while well-being has improved, it remains at a moderate level. Despite the reported progress, challenges in strengthening household inclusion still exist including the low economic participation of B40 households, the limited ownership of Bumiputera financial and non-financial assets, the increased growth and population inequality between states/regions as well as between urban and rural areas. The Financial Education Network (FEN) outlined five major challenges faced by Malaysian B40 households with the first issue being the low level of confidence in financial knowledge among B40 households.

B40 households also face the rising costs of living and debt, prompting difficulties in managing finances [28,29]. Ethnic differences have emerged in financial vulnerability among Malaysians where Malays and Indians are more vulnerable when compared with Chinese since socioeconomic factors, financial attitudes and financial behaviours also influence their financial stability [30]. The third issue outlined relates to savings and budgeting: Bank Negara Malaysia and the Financial Education Network found that one in ten Malaysians believe that they are not disciplined enough in managing their finances. Approximately 84% of Malaysians claim that they only regularly save for the short term as they often withdraw their savings by the end of the month to cover their expenses, spending 43% of their income on health care and 39% on groceries [31–33].

Additionally, B40 households encounter a challenge that is considered a financial literacy factor, the readiness for unexpected life events [13]. Findings reported by [34–36]. align with the issues outlined in the Malaysian National Financial Literacy Strategy (2019–2023). As many as 52% of Malaysians find it difficult to come up with MYR 1000 in the event of an emergency, and only 24% can cover three months of living expenses should they lose their main source of income. Finally, the fifth issue involves ineffective retirement plans. Financial reports by [8,13] confirm that low-income households are less knowledgeable in planning for retirement, which can prevent them from enjoying financial well-being, affecting their SES.

These issues prompt the need for research that incorporates the elements of financial literacy suited to the Malaysian context, especially in measuring the financial literacy (FL) of the target group, B40 households. Furthermore, this necessity also involves the socioeconomic status (SES), the mediating role of financial well-being (FW) and the moderating variable of education level (LE) that should be proposed to stakeholders and households to enhance their financial management. Thus, an effective financial literacy model needs to be constructed as an alternative guide to B40 households in managing their finances. Financial literacy is an important aspect that can strengthen the SES status of families indirectly contributing to the progress of the country [37–40].

Literature Review

B40 households that intend to achieve a strong SES need to focus on effective financial planning, which is affected by factors such as poverty, financial management, financial education and also low financial literacy [41]. Financial literacy is deemed a major factor because households are fundamental decision-makers that are considered the country's economic growth drivers, highlighting the need for households to be well-equipped with basic financial knowledge [42]. The Malaysian National Strategy for Financial Literacy (2019–2023) implemented by the Financial Education Network consists of several agencies including the Ministry of Education Malaysia (MOE), Malaysian National Bank (MNB), Securities Commission Malaysia (PIDM), Employees Provident Fund (EPF), Malaysia Deposit Insurance Corporation, Credit Counselling and Management Agency (AKPK) and Limited National Settlement (PNB) and stressed the importance of household financial literacy in achieving the aspirations and direction of strong financial well-being and SES for families [43]. Financial literacy, i.e., solid financial knowledge, attitudes and behaviours, leads to strong SES [43]. Furthermore, the Financial Education Network, also further postulates that financially literate households can handle financial management and planning effectively in addition to making relevant financial decisions [43].

A designated set of financial education programmes and initiatives are readily available to increase the level of financial literacy among Malaysians at the national level. This set is in line with the five areas of focus proposed by the Malaysian government to improve the socio-economic status (SES) of B40 households in becoming a middle-class community (M40) instead. The lack of financial knowledge among B40 households triggers various financial issues preventing them from achieving financial well-being and affecting their SES [44]. The B40 group of households often run out of money before the next paycheque, possess no savings with increasing expenses and debt burden. They also tend to disregard proper retirement plans and thus require the need of financial assistance. These conditions and socioeconomic factors will also burden B40 households and affect their financial well-being [45]. The results of this study can go together with the Personal Finance Education Module issued by Bank Negara Malaysia as it involves every Malaysian household. The proposed model in this study can also serve as a baseline for MOE in integrating financial education elements in both the KSSR and KSSM curriculum by providing details to the MOE regarding the socioeconomic status of households that can affect student academic achievements.

This paper first gives an overview of B40 households and the issues they face. Following the outlined problem statement, a conceptual framework involving the main constructs is elaborated. These financial literacy constructs involve three aspects, namely, financial knowledge, financial attitudes and financial behaviour. In this study, financial literacy acts as an independent variable while the socioeconomic status of B40 households (SES) is a dependent variable mediated by two intermediary variables, namely, financial well-being (FW) as the mediator and education level (EL) as the moderator.

3. Developing the Model

3.1. Financial Literacy

Financial literacy is a skill that consumers should possess to operate in the increasingly complex financial landscape [23–30,45]. Therefore, governments nationwide are consistently experimenting with effective approaches to increase the levels of financial literacy among their citizens. The first step to improving financial literacy is measuring and understanding the constructs involved [45]. A financial literacy model was developed among university students which confirms the unique combination of the three dimensions: financial knowledge, financial attitudes and financial behaviour [39–45]. A review of past literature highlights the limitations in the testing models for literacy in Malaysia, especially when it comes to B40 households. Therefore, the current study aims to improvise the current financial liability model to meet the need for a rigorous literacy assessment that matches the growth of more complex financial products.

The Current State of Financial Literacy in Malaysia

The current level of financial literacy in Malaysia can be understood via five aspects namely, [1] level of financial knowledge, [2] savings and budgeting, [3] readiness in facing unforeseen situations, [4] planning for retirement and [5] awareness of investment risks and returns.

1. Level of Financial Knowledge

The Malaysian National Financial Literacy Strategy (2019–2023) reported that Malaysians have a low level of confidence in their own financial knowledge: one in three Malaysians perceived that they have low financial knowledge. This situation tends to hold true among low-income Malaysian households and in the B40 group [13–20,40–45]. Approximately 5% of Malaysians are aware of the rising living cost, but only 38% are able to relate the impact of investment on their purchasing power. Although 76% of Malaysians possess a budget, 2 out of 5 Malaysians find it difficult to keep up with their budget. Furthermore, 43% of Malaysians understand the concept of compounding money over time, while 22% believe that wealth can steadily increase. Furthermore, while 92% of Malaysians have deposit products, not many possess other types of investment products [13–20,45].

2. Cost of Living and Debt Burden

The high level of indebtedness among B40 households impacts their financial positions as they tend to take out personal loans to cover their daily expenses. In the year 2014, the indebtedness level of households earning MYR 3000 a month or less is seven times their annual income. Meanwhile, households earning MYR 5000 a month and below reported a level of indebtedness that is 3 to 4 times their annual income [16]. Additionally, the B40 household debt ratio to GDP was 89.1% in 2016 [6]. Economic factors such as financial well-being and social factors of education level are significant contributors that shape the households' attitudes towards debt [46].

3. Savings and Budgets

According to the Malaysia National Financial Literacy Strategy report, 2019–2023), one in ten Malaysians believe that they lack discipline in financial management. Despite 84% of Malaysians claiming that they practice regular savings, their savings are considered short term, as these savings are usually withdrawn by the end of the month to cover their daily expenses. Statistics also reported that one in five working Malaysian adults has not made savings in the previous six months, and three out of ten working Malaysian adults have to borrow money to buy basic necessities [46].

4. Willingness to Face Unexpected Events

The Financial Education Network also reported that 52% of Malaysians find it difficult to come up with MYR 1000 in the event of an emergency, and only 24% can afford to cover their living expenses for 3 months if they lose their main source of income (10% can last for more than 6 months). In terms of insurance and takaful coverage, only 17% of Malaysians subscribe to such plans.

5. Planning for Retirement

When it comes to retirement preparations, 41% of Malaysians rely on their EPF funds as their primary source of income, and almost 50% of Malaysians are unsure that they will have enough money to live on in their golden years. Since 68% of EPF members do not save the minimum amount of money per age group, 16% of Malaysians are extremely concerned about household expenses in old age. The suggested minimum savings amount for people 55 years and older is MYR 240,000, which allows for a withdrawal of MYR 1000 per month for 20 years following retirement. The majority of Malaysian workers—six out of ten—are self-employed and do not have access to any kind of institutional retirement or pension scheme [13–15,24–28,46].

3.2. *Financial Well-Being (Economic Factor)*

Financial well-being involves living a meaningful, comfortable and stable life that is free from prolonged financial problems in addition to being able to cope with unexpected financial problems and emergencies as well as changing living circumstances. It also refers to an individual's ability to meet current financial responsibilities and be secure about their financial future. The advent of increasingly sophisticated and accessible digital financial products and services further increases the importance of financial literacy. This prompts the need for Malaysians to understand the products, services and channels used to make the right financial decisions. Toward this aim, the National Financial Literacy Strategy (2019–2023) is a holistic five-year plan involving key strategies and comprehensive action plans that aims to provide guidance so that future efforts can produce a collective impact. The main goal of this strategy reinforces the role of financial education in improving financial literacy among Malaysians, as one of the prerequisites in strengthening their financial well-being [13,45,46].

3.3. Level of Education (Social Factor)

Education level is a sociodemographic factor that impacts financial literacy and reflects the socioeconomic status (SES) of households. A sufficient education level allows a household to adopt a positive attitude in making the right economic decisions [39]. In the current study, education level serves as a moderator between FL and SES B40 households. The second strategy report of the 11MP (2016–2020) and the 2010 Malaysia Millennium Development Goals (MDGs) Report state that the low level of educational attainment has affected the ability of B40 households to obtain high-paying jobs or venture into business activities [17,18,46].

3.4. Socioeconomic Status (SES) of B40 Households

Socioeconomic status is commonly associated with social science studies and refers to a deep understanding of how society should work. The common debate of this term among social scientists revolves around how the term indicates a person's position in the social hierarchy, how the hierarchy is structured, as well as a person's life options. In other words, SES refers to a person's access to a collectively desired resource be it material goods, money, power, friendship networks, health care, leisure, or education, alongside their access to these resources that enable them to function optimally in the social world.

To date, there is still no agreement on a proper definition for this term as it includes political ideologies regarding existing and desired social structures, and political and scientific ideologies often do not complement each other well. Over the past three decades, scholars have defined SES as a modest, measurable variable such as annual income. Some argue that race or ethnicity should be included in the definition, while others believe that financial status should also be part of the definition as these constructs are interrelated. For those who are not in the workforce, such as children, the elderly or those who intentionally drop out of school, SES can be a tricky construct to define [14]. A potential definition is proposed and described SES as a construct that reflects a person's access to collectively desired resources such as material goods, money, strength, friendship networks, health care, leisure, or educational opportunities [38,40–44,46].

3.5. Research Objectives

In general, this study examines the influence of Financial Literacy in relation to financial well-being and the education level of the socioeconomic status of B40 households. The objectives of the current study are as outlined:

(a) To examine the role of financial well-being (FW) in mediating the relationship between FL and SES of B40 households.
(b) To examine the role of the level of education (EL) as a moderator of the relationship between FL and SES of B40 households.
(c) To examine the effect of Financial Literacy (FL) on the socioeconomic status (SES) of B40 households.
(d) To examine the mean differences of FL, FW, SES of B40 households based on demographic factors.
(e) Proposed A Socioeconomic Status (SES) Sustainability Model of Malaysia's B40 households.

3.6. Research Hypotheses

The following six hypotheses were formulated to achieve the research objectives and test the proposed model:

H1. *FL has a significant impact and relationship with financial well-being (FW).*

H2. *FW has a significant relationship with the SES of B40 households.*

H3. *FW mediates the relationship between FL and SES of B40 households.*

H4. *FL has a significant effect and relationship with the SES of B40 households.*

H5. *Education Level (EL) has a moderating effect in the relationship between FL and SES of B40 households.*

H6. *Based on these demographic factors, there are differences in the mean scores of FL, FL, SES of B40 households.*

4. Conceptual Framework

Figure 1 presents the conceptual framework of the study where four variables or constructs namely financial literacy (FL), financial well-being (FW), level of education (LE) and socioeconomic status (SES) of B40 households are interrelated. Each of these constructs is developed where several dimensions can be tested. For the financial literacy construct, the following dimensions are focused upon: financial knowledge, financial attitudes and financial behaviour. The intermediary variable, financial well-being, includes the dimensions financial situation, future savings, financial resources, financial management, payment of bills and financial position. While the education level of the B40 household's (LE) serve as a moderating variable in this study, the socioeconomic status (SES) of B40 households served as a dependent variable where the researcher focused on monthly income, employment status, homeownership, car ownership and household composition (household size).

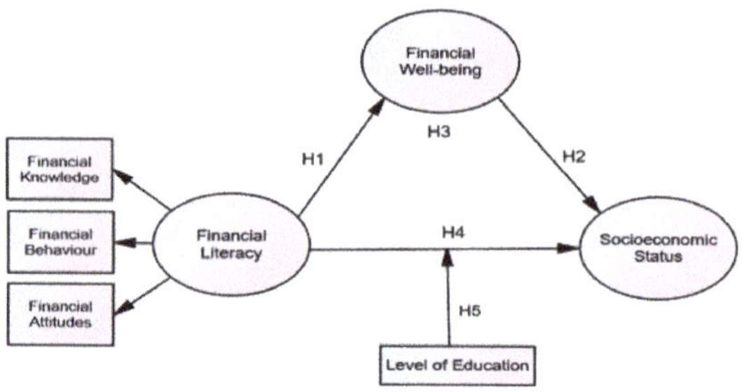

Figure 1. The Proposed Model for The Current Study.

5. Research Methodology

The current study employed a survey design in establishing a reliable model to measure the financial literacy (FL) construct among Malaysian B40 households as well as its effects on the SES of B40 households in Malaysia. Using a quantitative approach, the data of this study were obtained using a self-administered questionnaire where a detailed literature review was conducted to identify items that could measure the FL, FW, and SES constructs. The respondents for this study consisted of B40 households in the urban areas of Johor state who were selected using simple random sampling. According to Cooper and Schindler (2014), random probability sampling is where each population unit has an equal chance of being selected to ensure that the demographic and socioeconomic considerations are represented in the population. Random sampling also has fewer biased results [21,43–46] and enables study results to be generalised across the study population.

6. Research Instrument

The structured questionnaire used in this study consisted of 53 items rated on a 7-point Likert scale ranging from 1 = strongly disagree to 7 = strongly agree. Financial knowledge was measured using a 17-item scale where a sample item is "A budget is part of financial planning that I need to do". Financial behaviour was assessed using a 20-item measure, where a sample item for this construct was, "My purchase is according to ability". Similarly, financial attitude was assessed using a 17-item measure. All these scale items were adapted from [24,39–46]. Financial well-being was assessed using a 12-item measure outlined in [41–46], where a sample item is "I am satisfied with my current financial situation," while the socioeconomic status (SES) and level of education (LE) were measured using a 12-item scale that included monthly income, employment status, homeownership, car ownership and household composition (household size).

7. Discussion

This study aimed to test the relationship between the financial literacy model with financial well-being as a mediator and level of education as a moderator to test the impacts of the socioeconomic status of Malaysian B40 households. The exploratory factor analysis (EFA) methods, as well as the two-level structured equation modelling (SEM) approaches, were used to test the reliability, validity, and testing of hypotheses. B40 households face different financial problems over a long period and would need to possess financial literacy knowledge to manage their finances wisely and subsequently achieve financial well-being and improve their socioeconomic status. Thus, B40 households need to be educated to independently make wise financial decisions in addition to accessing relevant financial management information [36–38,46]. This also indicates the need to understand better the issues faced by B40 households in obtaining financial well-being and improving their socioeconomic status [39,40,46] The 11MP Mid-Term Review also highlights the need for further studies that account for government aspirations, economic challenges, and current global trends. Government efforts will be focused on stimulating economic growth for B40 households particular [41,44,46].

These findings highlight the importance of financial education based on financial literacy to encourage consumers to wisely manage their finances in addition to planning and managing the risks associated with financial problems. Meticulous financial planning can take place either through personal financial management or by appointing a qualified financial practitioner where personal financial planning is a more popular choice in Malaysia. Many still lack understanding of the importance of personal financial planning and various benefits that may arise from such planning [42–46]. Therefore, further study needs to be conducted to reveal the current financial literacy levels among Malaysians, especially among B40 households, to identify the impacts on SES of this group [47].

8. Conclusions

The findings of this study will give B40 households possibilities for enhancing their socioeconomic level. B40 households were able to pinpoint important social and economic elements in relation to financial literacy techniques to raise their socioeconomic status by improving the more reliable SES sustainability model (SES). Therefore, this research will provide future scholars with a new scholarly reference. With the help of this study, the knowledge and research gaps will be addressed, adding a wealth of important information to the field of scientific research and education in general. With the addition of information such as financial literacy, financial well-being, socioeconomic status (SES) and financial socialisation, knowledge sharing relating to these topics can continue. The outcomes of this study will be advantageous to multiple stakeholders. To ensure that every household is financially literate and can effectively manage their finances, the Ministry of Education (MOE) can take note of the findings of this study and implement some improvements, such as integrating financial education through KSSR and KSSM subjects in terms of financial literacy by strengthening the role of households in general and B40 in particular as economic actors.

The government's goals under the Malaysian Education Development Plan (PPPM) 2013–2025 are also met by these admirable initiatives and endeavours, which unmistakably encourage Malaysians to give financial literacy, financial well-being, financial socialisation and household SES their full attention. Because B40 parents' SES has a significant impact on children's academic achievement, the MOE can be asked to introduce a new B40 Household

SES Sustainability Model and initiatives as a means of achieving this goal in addition to the development of academic achievement among students from B40 family backgrounds. Scholarly references and research practise can be added by other stakeholders like students, researchers and academics. Additionally, this work creates a dynamic and contemporary reference area for future follow-up research.

Author Contributions: Conceptualization and writing—original draft preparation, A.M.; methodology, writing—review and editing, S.S. and M.H. All authors have read and agreed to the published version of the manuscript.

Funding: This research received no external funding.

Institutional Review Board Statement: Not applicable.

Informed Consent Statement: Not applicable.

Data Availability Statement: Not applicable.

Acknowledgments: Paper presented during a conference—organized by IASSC 2022—in the context of the conference, organized and hosted by Hafizah Mat Nawi, during the Parallel Session 6- online IASSC 2022.

Conflicts of Interest: The authors declare no conflict of interest.

References

1. Yusof, A. Ethnic disparity in financial fragility in Malaysia. *Int. J. Soc. Econ. Malays.* **2019**, *46*, 31–46. [CrossRef]
2. Ali, P.; Anderson, M.; McRae, C.; Ramsay, I. The Financial literacy of young people: Socio-Economic status, language background, and the rural-urban chasm. *Aust. Int. J. Rural. Educ.* **2016**, *26*, 54–66.
3. Aqmin, M.; Wahab, A.; Shahiri, H.; Mansur, M.; Azlan, M.; Zaidi, S. High cost of living in Malaysia: Slow Household income growth or rising standard of living? *Malays. Econ. J.* **2018**, *52*, 117–133. [CrossRef]
4. Arshat, Z.; Ismail, Z.; Mansor, M.; Madon, Z.; Omar, R. *Relationship between Economic Pressure and Family Functioning among B40 Family*; Faculty of Human Ecology, Universiti Putra, Malaysia: Serdang, Malaysia, 2018; pp. 8–11.
5. Azman, N.; Zakaria, R.H.; Marican, S. Household attitudes towards debt and indebtedness: A case study in the Klang Valley, Malaysia. *Malays. J. Soc. Adm.* **2017**, *9*, 59–86. [CrossRef]
6. Central Bank of Malaysia. *Financial Stability and Payment Systems Report*; Central Bank of Malaysia: Kuala Lumpur, Malaysia, 2019.
7. Central Bank of Malaysia. *Second Half Financial Stability Review*; Central Bank of Malaysia: Kuala Lumpur, Malaysia, 2019.
8. Central Bank of Malaysia. *Annual Report*; Central Bank of Malaysia: Kuala Lumpur, Malaysia, 2020.
9. Beckmann, E. *Household Savings in Central Eastern and Southeastern Europe: How Do Poorer Households Save?* World Bank Policy Research Working Paper 8751; The World Bank: Washington, DC, USA, 2019.
10. Blumberg, B.; Cooper, D.; Schindler, P. *Business Research Methods*, 12th ed.; McGraw Hill: New York, NY, USA, 2014.
11. Economic Planning Unit (EPU). Elevating B40 Households towards a Society. (Strategy Paper 2, Eleventh Malaysia Plan 2016-2020). Available online: http://rmk11.epu.gov.my/pdf/strategy-paper/StrategyPaper02.pdf (accessed on 1 January 2022).
12. Fazli, S.; Jariah, M.; Laily, P.; Amin, O. *A Study of Financial Resilience among Youth*; Malaysian Youth Development Research Institute: Putrajaya, Malaysia, 2007.
13. Financial Education Network, Report and Recommendations from the Task Force on National Financial Literacy Strategy, Malaysia. 2019. Available online: https://fenetwork.my/wp-content/uploads/2020/07/National-Strategy-ENG-F.pdf (accessed on 2 February 2022).
14. Gaur, K.L. Socio-Economic Status Measurement Scale: Thirst Area With Changing Concept For Socio-Economic Status. *Int. J. Innov. Res.* **2013**, 2, 139–145. Available online: http://www.ijird.com/index.php/ijird/article/view/37303 (accessed on 2 February 2022).
15. Department of Statistics. *Migration Investigation Report*; Department of Statistics Malaysia: Putrajaya, Malaysia, 2016.
16. Department of Statistics. *Household Income and Basic Amenities Survey Report*; Department of Statistics Malaysia: Putrajaya, Malaysia, 2016.
17. Department of Statistics Malaysia. *Press Release Household Expenditure Survey Report*; Department of Statistics Malaysia: Putrajaya, Malaysia, 2020.
18. Prime Minister's Office. *Malaysia Sustainable Development Goals Voluntary National Review 2017*; Economic Planning Unit: Putrajaya, Malaysia, 2017.
19. Kempson, E. Framework for the Development of Financial Literacy Baseline Surveys: A First International Comparative Analysis. OECD Working Papers on Finance, Insurance and Private Pensions. 2009. Available online: https://doi.org/10.1787/5kmddpz7m9zq-en (accessed on 3 February 2022).

20. Khalid, M.A. Climbing the Ladder. In *Socio-Economic Mobility in Malaysia*; Bloomsbury Publishing: Lodon, UK, 2016. [CrossRef]
21. Kothari, C.R. *Research Methodology, Methods & Techniques*; New Age International (P) Limited: New Delhi, India, 2006.
22. Kusairi, S.; Sanusi, N.A.; Muhamad, S.; Shukri, M.; Zamri, N. Financial households' efficacy, risk preference and saving behaviour: Lessons from lower-income households in Malaysia. *Econ. Sociol.* **2019**, *12*, 301–318. [CrossRef]
23. Lusardi, A.; Mitchell, O.S. The economic importance of financial literacy: Theory and evidence. *J. Econ. Lit.* **2014**, *52*, 5–44. [CrossRef]
24. Magesvari, R.; Kenayathulla, H.B.; Ghani, M.F.A. Socioeconomic factors influencing the financial literacy skills of secondary school students in Negeri Sembilan. *J. Educ. Leadersh.* **2018**, *5*, 10–33. Available online: http://e-jpurnal.um.edu.my/publish/JuPiDi/ (accessed on 5 February 2022).
25. Mahdzan, N.S.; Zainudin, R.; Sukor, M.E.A.; Zainir, F.; Wan Ahmad, W.M. Determinants of subjective financial well-being across three different household income groups in Malaysia. *Soc. Indic. Res.* **2019**, *146*, 699–726. [CrossRef]
26. Maki, M. *Financial Literacy and Household Performance*; The Department of Economics of Wisconsin-Milwaukee: Milwaukee, WI, USA, 2019. [CrossRef]
27. Mayan, S.N.A.; Nor, R.M.; Samat, N. Challenges To the Household Income Class B40 Increase in Developed Country towards 2020 Case Study: Penang. *Int. J. Environ. Soc. Space* **2017**, *5*, 35–41.
28. Mbarire, T.T.; Ali, A.I. Determinants Of Financial Literacy Levels Among Employees Of Kenya Ports Authority In Kenya. *Res. J. Financ. Account.* **2014**, *5*, 44–52. Available online: https://www.scribd.com/document/242812171/Determinants-of-Financial-Literacy-Levels-Among-Employees-of-Kenya-Ports-Authority-in-Kenya (accessed on 5 February 2022).
29. Ministry of Economic Affairs. *Executive Summary of the Semi-Final Review of the Eleventh Malaysia Plan 2016–2020: New Priorities and Emphasis*; Ministry of Economic Affairs Malaysia: Putrajaya, Malaysia, 2018.
30. Ministry of Economic Affairs Malaysia. *Twelfth Malaysia Plan*; Ministry of Economic Affairs Malaysia: Putrajaya, Malaysia, 2019; pp. 1–53. Available online: http://rmke12.mea.gov.my/about-us (accessed on 6 February 2022).
31. Mohamed, M.Z.; Xavier, J.A. Poverty Alleviation Strategies and New Economic Model in Malaysia. *Int. Acad. Res. J. Acad. Financ.* **2015**, *3*, 17–31.
32. Othman, M.A. *Study of Literacy, Socialization and Consumer Behavior Factors on Islamic Financial Planning Based on the Theory of Organized Behavior among Civil Servants in the Federal Territory*; University of Malaya: Putrajaya, Malaya, 2017.
33. Moreno-Herrero, D.; Salas-Velasco, M.; Sánchez-Campillo, J. Factors that influence the level of financial literacy among young people: The role of parental engagement and students' experiences with money matters. *Child. Youth Serv. Rev.* **2018**, *95*, 334–351. [CrossRef]
34. Mustapa, W.; Al Mamun, A.; Ibrahim, M. The Effect of Economic Vulnerability on the Participation in Development Programs and the Socio-Economic Well-Being of Low-Income Households. *Societies* **2018**, *8*, 60. [CrossRef]
35. Németh, E.; Zsótér, B. Anxious spenders: Background factors of financial vulnerability. *Econ. Sociol.* **2019**, *12*, 147–169. [CrossRef]
36. Nor, R.M. Inclusive Development for B40 Households Group: Malaysian Reality in Achieving Developed Nation Status by 2025. *Int. J. Acad. Res. Bus. Soc. Sci.* **2019**, *9*, 1035–1048. [CrossRef]
37. Norazam, N.S.; Othman, N. Quality of Financial Management Education With Financial Literacy Level. *J. Glob. Bus. Soc. Entrep.* **2017**, *1*, 183–193. Available online: http://gbse.com.my/v1no1jan17 (accessed on 8 February 2022).
38. Oakes, J.M.; Rossi, P.H. The measurement of SES in health research: Current practice and steps toward a new approach. *Soc. Sci. Med.* **2003**, *56*, 769–784. [CrossRef]
39. Potrich, A.C.G.; Vieira, K.M.; Mendes-Da-Silva, W.M. Management Research Review Development of a financial literacy model for university students For Authors Development of a financial literacy model for university students. *Manag. Res. Rev. Iss J. Risk Financ. Afr. J. Econ. Manag. Stud. Iss Int. J. Soc. Econ.* **2016**, *3944*, 356–376. [CrossRef]
40. Rahman, S.; Masih, M. *Munich Personal RePEc Archive Increasing Household Debts and Its Relation to GDP, Interest Rate and House Price: Malaysia's Perspective*. 2015. Available online: https://mpra.ub.uni-muenchen.de/62365/ (accessed on 2 February 2022).
41. Ramli, Z.; Awang, A.H.; Md Sum, S.M.; Haron, S.A.; Othman, M. Factors Determining the Financial Well-Being of Married Public Sector Employees in Malaysia. *Asian J. Univ. Educ. (AJUE)* **2018**, *8*, 169–177.
42. Rastogi, S.; Ragabiruntha, E. Financial inclusion and socioeconomic development: Gaps and solution. *Int. J. Soc. Econ.* **2018**, *45*, 1122–1140. [CrossRef]
43. Sekaran, U. *Research and Markets: Research Methods For Business—A Skill Building Approach*; John Wiley & Sons: Hoboken, NJ, USA, 2003.
44. Siwar, C.; Ismail, M.K.; Alias, N.A.; Zahari, S.Z. *The Lowest 40 Per Cent Income Group (B40) in Malaysia: Identifying Trends, Characteristics, Issues and Challenges. B40 Empowerment, Strategic Prosperity and Socioeconomic Implications*; Institut Alam Sekitar & Pembangunan (LESTARI) Universiti Kebangsaan Malaysia: Bangi, Malaysia, 2019; pp. 33–50.
45. Economic Planning Unit, Prime Minister's Department. Eleventh Malaysia Plan Strategy Paper 2: Improving the Status of B40 Households towards a Middle-Class Society. 2015. Available online: http://rmk11.epu.gov.my/pdf/kertas-strategi/Kertas%5CnStrategi%5Cn02.pdf (accessed on 12 February 2022).
46. Yusof, S.A.; Rokis, R.A.; Jusoh, W.J.W. Financial fragility of urban households in Malaysia. *Malays. Econ. J.* **2015**, *49*, 15–24. [CrossRef]
47. Arshat, Z.; Pai, F.H.; Ismail, Z. B40 Family: Stress and Strength B40 Family: Stress and Strength. *J. Adv. Res. Soc. Behav. Sci. J. Homepage* **2018**, *10*, 91–102. Available online: www.akademiabaru.com/arsbs.html (accessed on 15 February 2022).

proceedings

MDPI

Proceeding Paper

Domestic Travelers' Perceptions and Intention to Purchase Travel Insurance †

Shirelyzan Petrus [1], Ahmad Fareez Yahya [1,*] and Farashazillah Yahya [2]

1 Faculty of Hotel and Tourism Management, Universiti Teknologi MARA Sabah Branch, Kota Kinabalu 88997, Malaysia
2 Faculty of Computing and Informatics, Universiti Malaysia Sabah, Kota Kinabalu 88997, Malaysia
* Correspondence: ahmadfareez@uitm.edu.my
† Presented at the International Academic Symposium of Social Science 2022, Kota Bharu, Malaysia, 3 July 2022.

Abstract: Unfortunate events happening recently in increasing numbers, not least the outbreak of COVID-19, have caused drastic changes in tourist behaviours. Several studies have found that tourists' consumption patterns for travel products and services are changing due to threats of health hazards and travel risks at tourism destinations. Considering the increasing numbers of potential risks that might occur during traveling, it is crucial for travelers to have travel insurance that functions as a protection mechanism. This study investigates and determine travelers' awareness and perceptions of and their intentions towards travel insurance purchase in Malaysia. A non-probability sampling method and purposive sampling technique were applied in this study, with a total of 384 Malaysian domestic travelers over 18 years old used as respondents. Data were analysed using descriptive analysis and reliability testing for the pilot study. The study revealed that most of the respondents had positive perceptions regarding the variables and intended to purchase travel insurance for their future holidays.

Keywords: travel insurance; perceptions; intention; purchasing behaviours; Malaysian domestic travelers

Citation: Petrus, S.; Yahya, A.F.; Yahya, F. Domestic Travelers' Perceptions and Intention to Purchase Travel Insurance. *Proceedings* **2022**, *82*, 74. https://doi.org/10.3390/proceedings2022082074

Academic Editor: Mohamad Rahimi Mohamad Rosman

Published: 22 September 2022

1. Introduction

Insurance is one of the most complex financial products, and many types of insurance product are offered in the market [1]. The most common type of insurance related to the tourism industry is known as travel insurance. Travel insurance is one of the essential elements that must be involved in travel planning, especially if people intend to visit another country. As described by Lee et al. [2], people travelling abroad have high exposure risks, including injuries and infectious diseases. The main purpose of travel insurance is to cover unforeseen losses that travelers may be exposed to during their holidays. Through travel insurance, travelers can prepare for uncertain risks and unwanted events [3]. Several types of risks are potentially incurred during holidays, including equipment risk, financial risk, physical risk, psychological risk, satisfaction risk, social risk, and time risk [4].

Due to the ongoing COVID-19 outbreak, researchers have observed drastic changes in tourist behaviours. Tourist consumption patterns are changing, in terms of their travel and tourism product preferences, individual lifestyles, and daily behaviours. In the study by Gupta et al. [5], the global pandemic was found to effect travelers' levels of concern about risks associated with domestic and international traveling activities. Since COVID-19 was declared by the WHO as an infectious disease, peoples' fears and worries have increased because the virus can easily transmit through infected populations. Therefore, tourist attitudes have shown increasing tendencies towards avoiding risk in everyday life. After the outbreak, tourist preferences for travel products have changed, with more emphasis on health, personal safety, and hygiene elements in travel packages, products, or services that are offered to them [6]. As mentioned by Matiza [7], COVID-19 has influenced tourists' perceptions of travel risks; health and personal safety are the factors most likely to influence their decision-making processes in these uncertain times. It is believed that travel insurance is valued for its ability to reduce risk exposure during travel [8];considering the increased risk of health hazards during this pandemic it is crucial for travelers to have travel insurance, because most travel insurance provides coverage for medical expenses that occur during traveling [2]. Although travelers may have existing medical insurance, this will only cover medical expenses, while travel insurance covers a wider range of possible risks. As mentioned by Lee et al. [2], travel insurance policies provide coverage that include emergency, outpatient, and inpatient medical costs that occur during holidays, which are not usually covered by other medical insurance policies. Therefore, this study was conducted by the researchers to assess Malaysian domestic travelers' perceptions of travel insurance and their intention to purchase travel insurance, especially during this global pandemic.

The crisis we face because of this pandemic has clearly brought significant impact. Not only impacting the economic system, coronavirus has also impacted on tourist behaviours and caused drastic changes in consumption patterns of tourism and service products. Therefore, it is important to study and understand tourists' current behaviours, in order to further improve the quality of products and services. In this study, researchers focused on travel insurance products because tourists nowadays are more sensitive towards health risk and personal safety, as mentioned by Matiza [7]. According to Wen et al. [9], tourist behaviours and preferences in tourism marketing during the pandemic require more analysis in order to shape theories about consumer behaviours, marketing, and management, specifically in the tourism industry. Nevertheless, limited research has been published related to insurance, especially in the tourism and hospitality industries [2,4,8,10]. Moreover, most of the previous studies concentrated on other countries, whereas studies about travel insurance in Malaysia have been very scarce. Therefore, this research gathers new information related travel insurance, focusing on Malaysian travelers. This study aimed to investigate travelers' perceptions of and intention to purchase travel insurance in Malaysia. The specific objectives were to determine travelers' awareness of travel insurance, and their perceptions about purchasing travel insurance, and to identify travelers' intention to purchase travel insurance.

2. Travel Insurance

The most common insurance that related to the tourism industry, especially among travelers, is known as travel insurance. Travel insurance is an insurance policy that provides a safety net for travelers during trips, whether domestic or international [3]. For example, travel insurance policy in Malaysia by Persatuan Am Insurans Malaysia (PIAM) offers cover for accidents and losses that arise during a journey, whether by land, sea, or air. The coverage includes loss of luggage, medical and hospitalization expenses, death, permanent disablement, loss of money, loss of passport, repatriation, curtailment, and hijacking. In 2017, according to Olano [11] most Malaysian travelers were unwilling to purchase travel insurance despite their awareness about travel insurance; around 80% knew about travel insurance, but not more than 35% were willing to buy it. However, due to the Coronavirus outbreak, most insurers foresee rising demand for insurance products because Malaysian people's awareness of health risks has increased [12]. According to Li [6], COVID-19 pandemic, many countries have reportedly made travel insurance mandatory for travelers to destinations such as Singapore, Thailand, Vietnam, UAE, Portugal, and Turkey. Chin [13] reported that there was a growth and high demand for domestic tourism in Malaysia after the outbreak. Thus, this study focuses on domestic travelers since this group has become crucial in the Malaysian tourism market.

3. Purchasing Behaviours

According to Ramya and Ali [14], consumer purchasing behaviour refers to consumer activity towards the products, goods, or services; that activity includes selection, purchase, and consumption. Meanwhile, studies by Mashao and Sukdeo [15] defined consumer behaviours in purchasing products as practices that are used by individuals and groups, including organizations, to select, use, and dispose of products, services, or experiences, to fulfil their needs. Meanwhile studies by Auf et al. [16] stated that the purchase decisions of consumers are the results of many options, and purchasing patterns differ based on the goods or services involved. It was suggested by Qazzafi [17] that consumer behaviour is a broad area of activity that involves many different stages that have significant impact on purchase decisions for specific products, brands, or services. Based on these insights, this study examines the purchasing behaviours of domestic travelers in Malaysia, specifically towards travel insurance, based on their perceptions and intentions.

There are various factors, determinants, specificities, and characteristics that affect consumer purchase behaviours. The list includes consumer perceptions, which influence their final decisions towards products or services. Perception is a personal trait of an individual, insofar as consumers tend to form and interpret information related to the product or services based on their understanding [15]. Consumers acquire relevant information from their surroundings and form a perception about the products that they consider for purchase [17]. A recent study by Lim et al. [18] focused on peoples' perceptions towards life insurance purchase, and found that favourable perceptions of life insurance corresponded with a high tendency to purchase. Consequently, in the context of this current study, favourable perceptions of travel insurance are expected to be associated with a high intention to purchase travel insurance.

Lim et al. [18] used five indicators of perceptions to examine the association between peoples' perception and their intention to purchase life insurance. The five indicators assessed were objective knowledge, risk propensity, family influence, peer influence, and internet influence. These five indicators were confirmed to have significant linkage with peoples' behavioral intention to purchase life insurance in Malaysia. The research analysis showed more inclination towards life insurance purchase if people perceived the product positively. The indicators that were shown to have significant effect on peoples' perceptions were family, peer group, and internet influences; the study explained that if positive information about life insurance is available on the internet, potential consumers can self-educate themselves, which subsequently influences their attitudes towards these products and leads to better perceptions. Thus, this can be used as empirical evidence to support the linkages between perception and decision-making processes including intention to purchase certain products.

Therefore, the theoretical framework from Lim et al. [18] has been adapted in this study, because it provides a lens for exploring consumer perceptions' influences on individual intention to purchase. The indicators used for the current study were knowledge influences, perceived risk influences, social influences, and internet influences. There has been no specific research on domestic travelers' travel insurance purchases in Malaysia; the published information related to this topic is very limited. This study attempts to fill the literature gap related to this topic and contribute to an understanding of the changing patterns of travelers' behaviours. Figure 1 shows the research framework model for this study.

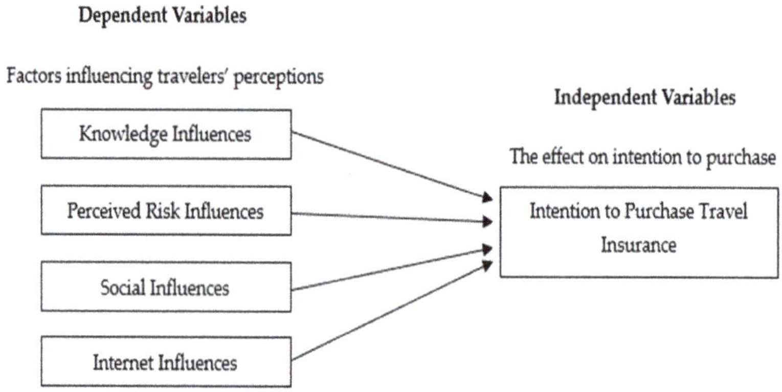

Figure 1. Research framework.

4. Materials and Methods

The study population for this research was Malaysian domestic travelers aged 18 years old and above. Based on the number of Malaysian domestic travelers, which is 131,700,000 [19], the number of samples estimated based on Krejcie and Morgan (1970) was 384 respondents. Since the probabilities of respondents being chosen were not related, the researcher used a non-probability sampling approach with purposive sampling for the data collection. The purposive sampling technique was used because the group of respondents had already been specified in advance. The questionnaire was designed using Google Forms and the hyperlink shared with the respondents through online platforms, i.e., Facebook, WhatsApp, and Telegram. The pilot test was conducted in December 2021 with 50 samples, in order to test the reliability of each item that was used in the questionnaire. The Cronbach Alpha values for independent variables were above 0.8 and the dependent variables were 0.6, which indicates that the items used were reliable. For data analysis the Statistical Package for Social Sciences (SPSS) software version 26.0 was used. The analysis carried out in this software included the reliability test which used Cronbach coefficient alpha, descriptive statistical methods, and inferential statistics methods with Pearson coefficient correlation to analyse the relationships between the variables in this study.

5. Results

5.1. Respondents Demograhic Profile

By gender, there were 72.1% female and 27.9% male respondents. The respondents in this study were from different range of ages, most were 22–30 years old (60.6%). The respondents' marital status was predominantly (82.8%) single, and the remainder were married (17.2%). Most respondents had a college or university level of education.

5.2. Perception and Intention to Purchase Travel Insurance

The descriptive analysis results for independent and dependent variables for each item are given in Table 1, with the percentages of Likert scale responses and mean values.

Table 1. Variables descriptive analysis (N = 355). The items were measured using five-point Likert scale:1—strongly disagree, 5—strongly agree.

Variables	1	2	3	4	5	Means
Knowledge Influences (KI)						
I acknowledge the benefits and types of travel insurance coverage.	5.9	12.4	21.4	42.0	18.3	2.54
I believe travel insurance will provide me with financial support during the travel period.	2.3	6.8	25.6	45.9	19.4	3.74
I think travel insurance is essential in travel planning.	0.8	4.8	13.2	45.1	36.1	4.11
I plan to purchase travel insurance with basic coverage only.	3.4	9.0	29.0	42.5	16.1	3.59
I plan to purchase travel insurance with comprehensive coverage.	2.0	6.2	21.7	38.9	31.3	3.91
Cronbach Alpha Value 0.803						
Perceived Risk Influences (PRI)						
Purchase travel insurance makes me feel safe from unforeseen risks during travel.	1.4	5.9	12.7	47.0	33.0	4.04
The outbreak of diseases influences me to purchase travel insurance.	2.3	6.2	20.3	43.9	27.3	3.88
I think medical expenses coverage is an essential feature of travel insurance.	0.8	3.7	13.8	44.5	37.2	4.14
I think personal accident coverage is an essential feature of travel insurance.	1.7	3.9	9.9	46.5	38.0	4.15
Coverage for travel inconveniences such as travel delays, loss, or damage of personal belongings are essential features of travel insurance.	1.1	5.6	13.5	44.8	34.9	4.07
I think trip cancellation coverage is an essential feature of travel insurance.	0.8	4.5	20.0	40.0	34.6	4.03
Cronbach Alpha 0.918						
Social Influences (SI)						
My family influences me to purchase travel insurance.	7.3	20.6	29.6	28.7	13.8	3.21
My friends influence me to purchase travel insurance.	7.6	18.3	25.9	33.2	14.9	3.30
An insurance agent influences me to purchase travel insurance.	9.3	15.5	27.9	32.7	14.6	3.28
A travel agent influences me to purchase travel insurance.	7.3	11.8	30.1	34.9	15.8	3.40
Social events influence me to purchase travel insurance.	7.0	12.1	28.7	37.2	14.9	3.41
Cronbach Alpha Value 0.819						
Internet Influences (II)						
I seek information online before purchasing travel insurance.	2.8	7.3	16.1	48.7	25.1	3.86
The insurance company website influences me to purchase travel insurance.	3.7	10.4	29.3	38.0	18.6	3.57
Social media influence me to purchase travel insurance.	3.4	10.1	25.4	43.4	17.1	3.62
Internet advertisements influence me to purchase travel insurance.	4.8	11.3	29.3	37.7	16.9	3.51
I prefer to purchase travel insurance online.	6.5	11.5	27.6	39.4	14.9	3.45
Cronbach Alpha Value 0.888						
Intention to Purchase Travel Insurance (INT)						
I will purchase travel insurance for my next domestic trip only.	7.6	15.5	33.2	32.1	11.5	3.25
I will purchase travel insurance for my next overseas trip only.	6.8	12.1	24.5	36.3	20.3	3.51
I will purchase travel insurance for my next domestic and overseas trips.	2.8	5.4	18.6	42.8	30.4	3.93
I will not purchase travel insurance for my next trip.	19.7	19.7	26.5	23.7	10.4	2.85
Cronbach Alpha 0.626						

6. Discussion

The study highlighted that more than half of respondents (52.1%) did not have any insurance coverage or policy. However, most of them were aware of the availability of travel insurance (73.25%). Domestic travelers ha a better understanding about insurance policy when they were consumers of insurance products. This result is aligned with the survey reported by Olano [11], indicating that most Malaysians were aware of travel insurance products, yet only a few of them were willing to purchase travel insurance because of their perception that they were not exposed to any potential risks during their trips. However, as reported by Azman [20], since the outbreak of COVID-19 there has been increasing awareness among Malaysians of the need to have life insurance. It was also stated by Malaysian RE [12] in their survey report that insurers predict is increasing demand for insurance products in Malaysia because of health risk awareness. Meanwhile, 43% of the respondents were aware about travel insurance through travel agents and 32% from social media; as mentioned by Akar and Dalgic [21] online networking is the medium where people can obtain information quickly and efficiently. These results supported the statements by previous studies [1] that insurance providers and marketers should have digitalization strategies for insurance products, to make consumers' purchasing processes easier.

The study revealed that almost half of the respondents acknowledged and understood the benefits of travel insurance during their journeys. Most respondents planned to purchase travel insurance including basic and comprehensive coverage. This result supports the study results of Tassiello and Tillotson [22], that subjective as well as objective knowledge influences travelers' decision making. This was additionally demonstrated by Kai et al. [23] in their studies on Chinese seniors, among whom having knowledge and information increased the intention to purchase travel insurance. They believed that travel insurance can make them feel safe, because it is a practical approach to reduce risk and loss occurred during travel [24], allowing them to enjoy their holiday without worrying too much. The respondents also agreed that the pandemic outbreak had the most significant impact on how they perceived risk in travel and tourism activity, and its influence on purchasing travel insurance, as claimed by previous researchers Tan and Caponecchia [25] and Matiza [7].

Many respondents agreed that social events influenced them to purchase travel insurance. However, these findings are contradictory to some previous research; Alt et al. [1] and Sarman et al. [4], indicated that people are more likely to purchase travel insurance based on advice from their friends and family.. The findings for this study showed that the mean scores for friends and family, respectively, were the lowest, indicating that the respondents' perceptions were not greatly influenced by their friends or family. Travelers tend to seek information related to travel insurance products before purchase, and the internet makes it easy for travelers to access that information at any time, as mentioned by Kian et al. [26]. Respondents also agreed that social media influenced their perceptions, leading them to buy products. Social media can change peoples' thoughts and opinions about products or services [21]. The analysis of traveler's intention to purchase travel insurance revealed that most of the respondents had an intention and future plan to purchase travel insurance for their holidays, whether domestic or overseas.

7. Conclusions

This study is one of very few relating to travel insurance that have been conducted in Malaysia. It examined domestic travelers' perceptions based on four factors that affect travelers' purchase decisions towards travel insurance. The four main factors found to affect travelers' perceptions were knowledge influences, perceived risk influences, social influences, and internet influences. Based on the findings, the dependent variables were found to affect the respondents' perceptions and their intentions towards travel insurance purchase in Malaysia; these findings align with previous studies by Lim et al. [18]. The information gathered from the present study contributes to current knowledge on tourists' consumption behaviours and the current demand for travel insurance in Malaysia among domestic travelers. Findings from this study can be used as a general outline and the information obtained can help related parties to develop effective strategies for the management and marketing of tourism products, especially after a public health crisis. From the results, perceived risk was found to effect travelers' perceptions significantly. Generally, travel insurance is very important in individuals' travel arrangements to ensure coverage for unforeseen risks and loss during holidays, including for illness such as COVID-19. A number of potential factors that influence tourists' insurance purchase decisions were not included in the present study. These should be researched further, since trends have changed and travel insurance has become a necessity for travel activities.

Author Contributions: Conceptualization, S.P. and A.F.Y.; methodology, S.P. and A.F.Y.; software, S.P.; validation, A.F.Y.; formal analysis, S.P.; investigation, S.P.; resources, S.P.; data curation, S.P.; writing—original draft preparation, S.P.; writing—review and editing, A.F.Y. and F.Y.; visualization, S.P. and A.F.Y.; supervision, A.F.Y.; project administration, S.P. and A.F.Y.; funding acquisition, A.F.Y. All authors have read and agreed to the published version of the manuscript.

Funding: This research received no external funding.

Institutional Review Board Statement: The study was approved by Branch Research Ethics Committee UiTM Sabah (BERC/12/2021/MR19 14 December 2021) in accordance with the-ICH Good Clinical Practices Guidelines, Malaysia Good Clinical Practice Guidelines and the Declaration of Helsinki.

Informed Consent Statement: Informed consent was obtained from all subjects involved in the study.

Data Availability Statement: Not applicable.

Acknowledgments: Paper presented during the International Academic Symposium of Social Science (IASSC) 2022, organized by the Faculty of Information Management, Universiti Teknologi MARA Kelantan Branch, Malaysia, in collaboration with Universitas Pembangunan Nasional Veteran Jakarta, Indonesia; Universitas Ngudi Waluyo, Indonesia; Camarines Sur Polytechnic Colleges, Philippines; Uttaranchal University, India; and UCSI University, Malaysia, on 3 July 2022 from Kelantan, Malaysia, in the context Advance in Digital Transformation Process and Mode, organized by Huda Hamidon and hosted by Dr Rudsada Kaewsaeng-On.

Conflicts of Interest: The authors declare no conflict of interest.

References

1. Alt, M.A.; Săplăcan, Z.; Benedek, B.; Nagy, B.Z. Digital touchpoints and multichannel segmentation approach in the life insurance industry. *Int. J. Retail. Distrib.* **2021**, *49*, 652–677. [CrossRef]
2. Lee, Y.H.; Lu, C.W.; Wu, P.Z.; Huang, H.L.; Wu, Y.C.; Huang, K.C. Attitudes and awareness of medical assistance while traveling abroad. *Glob. Health* **2018**, *14*, 67. [CrossRef] [PubMed]
3. Hasan, A.A.; Abdullah, N.C. Compulsory Travel Insurance in Malaysia: Some Regulatory Considerations. *Procedia Soc.* **2015**, *172*, 344–351. [CrossRef]
4. Sarman, I.; Curtale, R.; Hajibaba, H. Drivers of travel insurance purchase. *J. Travel Res.* **2020**, *59*, 545–558. [CrossRef]
5. Gupta, V.; Cahyanto, I.; Sajnani, M.; Shah, C. Changing Dynamics and Travel Evading: A Case of Indian Tourists Amidst the COVID-19 Pandemic. Emerald Insight. Available online: https://www.emerald.com/insight/content/doi/10.1108/JTF-04-2020-0061/full/html (accessed on 1 October 2021).
6. Li, K.S. Insurers See Rising Demand for Travel Insurance as More Countries Make It Mandatory. The Edge Malaysia 2021. Available online: https://www.theedgemarkets.com/article/insurers-see-rising-demand-travel-insurance-more-countries-make-it-mandatory (accessed on 9 October 2021).
7. Matiza, T. Post-COVID-19 crisis travel behaviour: Towards mitigating the effects of perceived risk. *J. Tour. Futures* **2020**, *8*, 99–108. [CrossRef]
8. Caponecchia, C.; Tan, D.T. Exploring the traveller underinsurance problem. *Ann. Tour. Res.* **2019**, *76*, 343–345. [CrossRef]
9. Wen, J.; Kozak, M.; Yang, S.; Liu, F. COVID-19: Potential effects on Chinese citizens' lifestyle and travel. *Tour. Rev.* **2020**, *76*, 74–87.
10. Allaberganov, A.; Preko, A.; Mohammed, I. Government commitment to tourism and hospitality sector during COVID-19 pandemic. *Tour. Crit.* **2021**, *2*, 153–169. [CrossRef]
11. Olano, G. More Than 60% of Malaysians Face Vacation Risk due to Lack of Travel Insurance. Insurance Business Asia. 2017. Available online: https://www.insurancebusinessmag.com/asia/news/breaking-news/more-than-60-of-malaysians-face-vacation-risk-due-to-lack-of-travel-insurance-78914.aspx (accessed on 1 February 2022).
12. Malaysian Insurance Highlights 2020. Malaysian Insurance Highlight. 2021. Available online: https://www.malaysian-re.com.my/our-insights/malaysian-insurance-highlight (accessed on 9 October 2021).
13. Chin, C. Domestic Travel Key to Tourism Recovery in Malaysia and Globally. The Star. 2021. Available online: https://www.thestar.com.my/lifestyle/travel/2021/12/28/domestic-travel-key-to-tourism-recovery-in-malaysia-and-globally (accessed on 9 October 2021).
14. Ramya, N.A.; Ali, S.M. Factors affecting consumer buying behavior. *Int. J. Appl. Res.* **2016**, *2*, 76–80.
15. Mashao, E.T.; Sukdeo, N. Factors that influence consumer behavior in the purchase of durable household products. In Proceedings of the International Conference on Industrial Engineering and Operations Management, Paris, France, 26–27 July 2018.
16. Auf, M.A.; Meddour, H.; Saoula, O.; Majid, A.H. Consumer buying behaviour: The roles of price, motivation, perceived culture importance, and religious orientation. *JBRMR* **2018**, *12*, 1. [CrossRef]
17. Qazzafi, S. Factor affecting consumer buying behavior: A conceptual study. *Int. J. Sci. Res.* **2020**, *8*, 1205–1208.
18. Lim, T.S.; Dzulkifli, D.Z.; Osman, Z.; Mohidin, R.; Jamal, A.A. Determinants of perception toward life insurance and its impact on intention to purchase. *LBIBF* **2020**, *15*, 16–26.
19. Department of Statistic Malaysia, Domestic Tourism Malaysia 2020. Department of Statistic Malaysia Official Portal 2020. Available online: https://www.dosm.gov.my/v1/index.php?r=column/cthemeByCat&cat=320&bul_id=eEl1dWtVVmdSRCtKNzJFUjJhWVQ5QT09&menu_id=b0pIV1E3RW40VWRTUkZocEhyZ1pLUT09 (accessed on 13 October 2021).
20. Azman, N.R. COVID-19 Increases Demand for Life Insurance. The Malaysian Reserve 2020. Available online: https://themalaysianreserve.com/2020/07/13/covid-19-increases-demand-for-life-insurance/ (accessed on 5 November 2021).
21. Akar, E.; Dalgic, T. Understanding online consumers' purchase intentions: A contribution from social network theory. *Behav. Inf. Technol.* **2018**, *37*, 473–487. [CrossRef]
22. Tassiello, V.; Tillotson, J.S. How subjective knowledge influences intention to travel. *Ann. Tour. Res.* **2020**, *80*, 102851. [CrossRef]
23. Kai, Y.; Zhujun, K.; Zhijie, C.; Xiaoting, S.; Wanyue, T. Social learning? Conformity? Or comparison?—An empirical study on the impact of peer effects on Chinese seniors' intention to purchase travel insurance. *Tour. Manag. Perspect.* **2021**, *38*, 100809. [CrossRef]
24. Liu, X.; Fu, X.; Hua, C.; Li, Z. Crisis information, communication strategies and customer complaint behaviours: The case of COVID-19. *Tour. Rev.* **2021**, *76*, 962–983. [CrossRef]

25. Tan, D.; Caponecchia, C. COVID-19 and the public perception of travel insurance. *Ann. Tour. Res.* **2021**, *90*, 103106. [CrossRef]
26. Kian, T.P.; Boonb, G.H.; Fongc, S.W.; Aid, Y.J. Factors that influence the consumer purchase intention in social media websites. *Int. J. Supply Chain Manag.* **2017**, *6*, 208–214.

proceedings

MDPI

Proceeding Paper

The Antecedents of Household Acceptance on Food Waste Innovation Products in Terengganu †

Noorazlin Ramli [1,*], Hayati Adilin Mohd Abd Majid [1], Wan Nazriah Wan Nawawi [1], Farah Adibah Che Ishak [2] and Syahida Maarof [3]

1 Faculty of Hotel and Tourism Management, Universiti Teknologi MARA Cawangan Terengganu, Kampus Dungun, Dungun 23000, Malaysia
2 Faculty of Food Science and Technology, Universiti Putra Malaysia, Serdang 43400, Malaysia
3 Food Science and Technology Research Center, Malaysian Agricultural and Development Research Institute (MARDI), Serdang 43400, Malaysia
* Correspondence: noora115@uitm.edu.my

† Presented at the International Academic Symposium of Social Science 2022, Kota Bharu, Malaysia, 3 July 2022.

Abstract: Food surpluses or waste materials may be transformed into new commercial products that aim toward food conservation. It is desirable to investigate the household level of knowledge and attitude towards acceptance of food waste innovations products. A total of 213 returned surveys are analyzed with the descriptive statistical and one-way ANOVA. It shows that knowledge and attitude are the antecedents of the acceptance on food waste innovation products. The findings might help minimize the amount of food waste produced on a daily basis by encouraging consumers and businesses to turn food surpluses or waste into useful items and a source of revenue.

Keywords: food surplus; food waste; food innovation products

Citation: Ramli, N.; Majid, H.A.M.A.; Nawawi, W.N.W.; Ishak, F.A.C.; Maarof, S. The Antecedents of Household Acceptance on Food Waste Innovation Products in Terengganu. *Proceedings* **2022**, *82*, 75. https://doi.org/10.3390/proceedings2022082075

Academic Editor: Mohamad Rahimi Mohamad Rosman

Published: 22 September 2022

1. Introduction

The worldwide waste crisis is becoming worse every day. According to a World Bank statement issued in 2019, there has been a 70% rise in worldwide urban solid waste, with emerging nations experiencing the greatest issues. The forecasted increase in waste volume from 2.01 billion tons per year currently to 3.40 billion tons per year by 2050 is expected to increase world expenses from USD 205 billion to USD 375 billion per year [1]. A surprising fact showed that around 1.3 billion tons of the world's entire food production is wasted each year [2]. In Malaysia, food wastes account for 32% of total waste, followed by paper and plastics, which account for 21% and 14% of waste, respectively [3]. Food waste production is predicted to rise dramatically in the future years, while landfill space is becoming scarce. Additionally, the findings from Solid Waste Management and Public Cleansing Corporation (SWCorp) presented that the total amount of food waste can definitely fill up the twin towers up to 16 times by 2020. Another fact was that Malaysians' waste is about 16,688 tons of food per day, which is an amount that can feed around 2.2 million people three times a day [4].

Food waste has become a world concern because it has had an impact on the environment as larger amounts of food are generated to meet the demand from an increasing and more affluent population [5]. Due to the increasing food waste, humans face serious environmental challenges such as global warming, increasing world population, overrun of solid waste materials and environmental pollution [2]. Food waste also contributes to the development of greenhouse gases in landfills. Food waste in landfills releases toxic substances in the soil that cause negative consequences for groundwater [6]. The act of shifting food waste from solid waste landfills to the recycling process is believed to contribute significantly in reducing greenhouse gases, making the earth a better place to live. Through near daily clean-up efforts around the city, the Kuala Terengganu City Council

(MBKT) gathers 300 tons of solid waste and transports it to a landfill in Sungai Ikan (Kuala Terengganu). Cleaning and solid waste collection operations in the city, which covers an area of 4291.31 hectares, are budgeted for MYR 7.75 million each year. Furthermore, cleaning and collection of solid waste on Pulau Redang costs MYR 557,280 per year, including solid waste transportation from the chalet to the mainland [7].

Furthermore, food-wasting is mainly high among consumers from an economic perspective. The data shown by [8] has stressed that Malaysian households of five have to spend an average of MYR 900 a month on food only. Meanwhile, a quarter of that food is wasted during preparation, cooking, and usage. This point has indicated that about MYR 225 goes into the bin every month, which works out to MYR 2700 a year per household. The data found in 2016 showed that food waste contributes to wastage, in which the Food Aid Foundation (Food Aid) documented that Malaysian food wastes almost 15,000 tones, including 3000 tons of edible food every day [9]. Besides this, food waste impacts social perspectives whereby it cannot be accepted as, in fact, there are still many people who live in hunger. An estimated 795 million hungry people out of 7.5 billion people living in this world suffer from starvation and chronic malnutrition [10]. Moreover, food security issues arise with the losses related to food access, such as food prices and purchasing power [11]. Food waste is a growing problem and its disposal is controversial, causing food prices to rise and access to food sources at high prices [12].

Food wastage mostly comes from the consumer level at surprising percentages and rates. Household waste, which includes food waste, paper, plastic, rags, metal, and glassware, is one of the most common sources of Municipal Solid Waste in populated neighborhoods [4]. Food waste from residential as well as commercial establishments such as restaurants, institutions, industries, school cafeterias, hospitals, and factories can actually be recycled [13]. In line with the above notion, several researchers are already engaged in the opportunities of using food waste to create new, value-added products that can be implemented to reduce food wastage. Furthermore, the importance of innovative advances in preventing or reducing food waste has been highlighted [14,15]. Food waste innovation is one of the solutions that can be made and accepted by society because of the flow of garbage that turns food waste into a reusable commodity. The extremely huge amount of wastage has drawn the attention of researchers to look for an alternative of utilizing food waste by-products [15]. Therefore, utilizing food waste innovation products not only helps the economics of residential and community, but can also increase food sustainability and reduce food insecurity, especially for underdeveloped countries [7,14–16]. In line with that, it is important for the study to identify the antecedents of acceptance on food waste innovation products which are knowledge and attitudes among Household in Kuala Terengganu.

2. Literature Review

2.1. *Food Waste among Household*

Malaysia is well known for being a foodie's paradise, with a wide variety of cuisine readily available at all times. Food is inextricably linked to Malaysia's identity and to the belief in warm hospitality among Malaysians. This stated declaration is most evident during the festive season such as the Hari Raya celebration, Chinese New Year, Deepavali and Gawai celebration, as people gather to celebrate with plenty of food. Unfortunately, this distinct culture and ideology has also been giving rise to household waste, since the amount given is frequently in excess of what is required. As a result, SWCorp discovered that, during the festive season, food waste might increase by up to 50% than usual [17]. Furthermore, according to Hayati Ismail, Director of the Food Aid Foundation, household waste is the leading source of food waste, followed by 'pasar malam' (night markets) and Ramadan bazaars, garbage from food courts, and finally the food and beverage [18].

Chronically, the average Malaysian discards 1.64 kg of waste each day, compared to 1.2 kg globally [8]. Households account for the majority of the country's daily food waste of 16,650 tones. Additionally, according to the data by SWCorp, households generated the most food waste (38%) compared to wet markets (24%), restaurants (23%), and hotels (23%). Overbuying food, poor food storage, and a lavish lifestyle are the primary causes of food waste [4]. The unconsumed food waste includes rotten fruits, expired bread, and eggs have doubled over the past three years and these could be a contributing factor to food waste issues in Malaysia [19]. Due to this, the use of food waste innovation products in the food industry seems to be a solution since it is not only caused by overabundance of waste among household level, but also by the depletion of renewable resources, the limitation of land available for agriculture, and the continual expansion of the world population [16].

2.2. *Antecedents of the Acceptance on Food Waste Innovation Products*

Knowledge may be characterized as a discrete outcome of acquiring and improving learning, with the knowledge being discrete until the individual uses it to complete a task, make a choice, or solve an issue [20]. Planning and purchasing routines, having or not having a stock overview, and not knowing if food may still be consumed are all factors that influence consumers' food management knowledge [20]. There are still 36% of consumers who are confused about the date labeling, which is the 'best before' date vs. the 'use-by date,' which might lead to food being thrown away unnecessarily [8]. Furthermore, consumers who have a strong understanding of the issues that would develop as a result of food waste are more inclined to avoid wasting food [13]. In keeping with this, when consumers have a high level of knowledge about the topic at hand, they are more likely easier to interpret the information and tend to accept something. As a result, having more knowledge has helped to better information processing, and it is a good preparation to accept innovations in food waste [7].

Attitudes are defined as "learned predispositions to respond consistently favorably or negatively toward a particular object" [21]. Consumers' acceptance of the effectiveness of innovation products were shown to be related to their degree of knowledge [16]. If education fails to enhance people's knowledge level, their attitudes do not alter or improve toward desired ones. However, attitude is a psychological evaluation that hinders the food waste reduction, and if some of the consumers have positive knowledge and attitudes, the acceptance for innovations will increase [7,13]. On the basis of previous elaboration in the literature review section, the proposed research framework is shown in Figure 1 below:

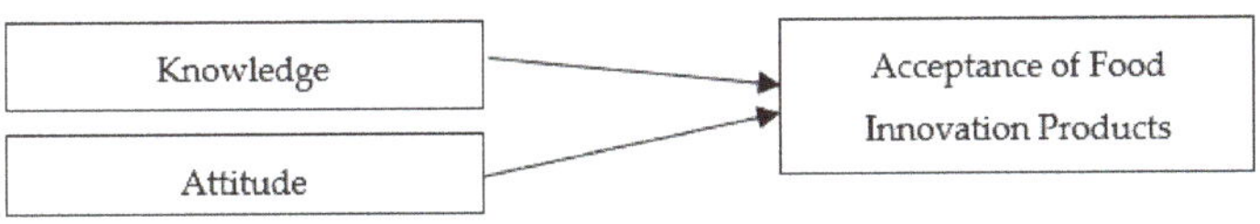

Figure 1. Research Framework.

3. Materials and Methods

This study is conducted to discuss the antecedents of acceptance on food waste innovation products among households. The convenience sampling approach was used in this study, with households randomly selected among the Kuala Terengganu population. The target population in this study is households in Kuala Terengganu, since this study seeks to determine public acceptance of food waste innovation products. Food waste is higher in urban areas due to a disparity in resident socioeconomic backgrounds. On the other hand, a household is defined as one or more people living in the same residential area, such as an apartment, a mobile home, or any single inhabited room [22]. All of the respondents are classed as household since they are assumed to reside in a house and produce municipal solid waste. The total population of Kuala Terengganu is estimated to be around 186,100 people, with a land area of 20,776 hectares [23]. However, due to time constraints and based on rule of thumb, only a total of 250 questionnaires were distributed to respondents. However, out of 250 questionnaires, only 213 surveys were completed and returned with accurate and significant results. PB Square, Plaza Paya Bunga, Pasar Payang, and Kuala Terengganu Express Bus Terminal were the selected areas in Kuala Terengganu.

The study calculated respondent information based on a questionnaire that has four sections: Section (A) demographic characteristics of respondents, Section (B) degree of knowledge, Section (C) consumer attitude, and Section (D) acceptance of food waste innovation products elements. Items in the questionnaire that show strongly disagree (1), disagree (2), neither agree nor disagree (3), agree (4), and highly agree (5) were interpreted using a 5-point Likert scale. Before the actual data collection, the questionnaire was subjected to a pre-test for content validity by experts. This test involved three experts who evaluated the statements' clarity, the questionnaire's presentation, and the acceptance of the statements among the respondents. Changes were made after receiving feedback from experts. After the questionnaire was amended, it was delivered to 50 randomly selected respondents in Malaysia for a pilot study. The researchers analyzed the information using IBM SPSS Statistics for Windows, version 26 (IBM Corp., Armonk, NY, USA) [24]. This study made use of descriptive statistics as well as correlation analysis.

4. Results and Analysis

The respondents' demographic factors are distributed into five categories. Most respondents are females with 64.3%, while the other 35.7% are males. Additionally, both categories' respondents are in the age groups 18–23 years and 24–54 years, and have the majority standing at 49.8% and 49.3% respectively, followed by the lowest respondents age groups of 55–64 years and 65 years and above, which are both at 0.5%. With respect to the respondents' income, it is as expected that the household income collected displayed by 51.2% of respondents is less than MYR 2500 as the majority, followed by 23.0% of respondents earning household income between MYR 2500 and MYR 4849, 16.9% of respondents earning household income between MYR 4850 and 10,959, and the lowest percentage of respondents' earnings of household income is MYR 10,960 with 8.9%. Furthermore, majority of the respondents with 43.7% are living in a household of six or more people, 18.8% of respondents are living in maximum of five people per household, 15.5% of respondents are living in maximum of four people per household, 13.6% of respondents are living in maximum of three people per household, 11.3% of respondents are living in maximum of two people per household, and finally, 6.1% of respondents are living alone. Therefore, this shows that majority respondents have families. In terms of educational level, it was found that 39.0% of respondents are Bachelor's degree holders, 31.5% of respondents are Diploma holders, 6.1% of respondents have graduated with a Master's degree, 1.4% of respondents have acquired a Ph.D., 0.9% of respondents are with PMR certificate, and finally, 0.5% of respondents have only acquired a UPSR certificate.

This section displays the mean scores from respondents on their knowledge and attitude on food waste innovation products. Tables 1 and 2 refer to the descriptive statistics for knowledge and attitude on food waste innovation products respectively. According to the findings, respondents had insufficient knowledge compared to their attitudes around food waste innovation items. Table 1 presents the highest mean of 1.4789, representing the knowledge that food waste can be composted into fertilizer and sold, followed by the knowledge that food waste can be innovated in Terengganu (M = 1.3944), food waste can be used as animal feed (M = 1.3615), the consumption of food waste can help reduce the increase in food waste disposal in the state of Terengganu. (M = 1.3427), and lastly, that wasted food can be recycled into innovative products that are beneficial to entrepreneurs or the people of Terengganu (M = 1.3286). The results can be seen as shown in Table 1.

In Table 2, the results provide the descriptive analysis of respondents' attitudes. The results show that the respondents have a high level of attitude on food waste innovation products. The results show the highest mean score of 4.4883 where respondents are concerned about how food surplus waste is collected, transported, and distributed. The lowest mean score is 3.0892, representing estimation of the food portions that can be consumed within a week.

Table 1. Descriptive Statistics of Knowledge.

	N	Mean	Std Deviation
I know food waste can be innovated in our state	213	1.3944	0.4897
I know that the consumption of food waste can help reduce the increase in food waste disposal in the state of Terengganu.	213	1.3427	0.5744
I know food waste can be composted into fertilizer and sold.	213	1.4789	0.5003
I know waste food can be recycled into innovative products that are beneficial to entrepreneurs or the people of Terengganu	213	1.3286	0.4702
I know waste food can be used as animal feed	213	1.3615	0.4817

Table 2. Descriptive Statistics of Attitude.

	N	Mean	Std Deviation
I immediately throw food waste into the trashcan after every meal.	213	4.2207	0.85404
I spend enough time to segregate food waste that can be reused from those that cannot.	213	3.8685	0.90679
I am concerned about how food surplus will be collected, transported, and distributed	213	4.4883	0.70450
I participate in food waste segregation if it's convenient to do it.	213	3.2958	1.16628
I estimate the food portions that can be consumed within a week.	213	3.0892	1.16414

One Way Analysis of Variance (ANOVA) is applied to compare more than two populations and analyze their equality. Moreover, the analysis is conducted to identify whether knowledge and attitude varied accordingly to demographic variables of education level and household income. In this research study, the population involved is divided in terms of respondents' education level, including UPSR, PMR, SPM/STPM, Diploma, Bachelor's degree, Master's degree, and Ph.D. Meanwhile, the group of household income is divided into four groups, namely, B40 (less than MYR 2500 and MYR 2500–4849), M40 (MYR 4850–RM10,959), and T20 (MYR 10,960 or higher). As mentioned in the previous Section 4 equal is defined as when the population shows a significant value greater than alpha value a = 0.05. Therefore, both "attitude and acceptance" factors are analyzed and displayed in Tables 3 and 4 below.

Table 3. Attitude and Acceptance by Education Level—ANOVA results.

		Sum of Squares	df	Mean Square	F	Sig.
Attitude	Between Groups	20.282	6	3.380	8.198	0.000
	Within Groups	84.939	206	0.412		
	Total	105.221	212			
Acceptance	Between Groups	16.021	6	2.670	2.733	0.014
	Within Groups	201.275	206	0.977		
	Total	217.296	212			

Table 4. Attitude and Acceptance by Household Income—ANOVA Results.

		Sum of Squares	df	Mean Square	F	Sig.
Attitude	Between Groups	3.330	3	1.110	2.277	0.081
	Within Groups	101.891	209	0.488		
	Total	105.221	212			
Acceptance	Between Groups	0.260	3	0.087	0.083	0.969
	Within Groups	217.036	209	1.038		
	Total	217.296	212			

Table 3 above shows that an "F" value represents the test statistic's value. The attitude variable shows a higher test statistic value (F = 8.198) with a significant *p*-value of 0.000. The "F" value calculated for the "acceptance" variable is (F = 2.733), and a significant *p*-value of acceptance is 0.014. Based on the analysis, both attitude and acceptance variables show a smaller significance value than the alpha value ($\alpha = 0.05$), which provides that both variables show a different mean population between different education levels.

Table 4 below shows that the "F" value represents the value of the test statistic. The attitude variable shows a higher test statistic value (F = 2.277) with a significant *p*-value of 0.081. Meanwhile, the F value calculated for the Acceptance variable is (F = 0.083), and a significant *p*-value of acceptance is 0.969. From the analysis, both attitude and acceptance variables show a bigger significant value than the alpha value (a = 0.05), hence proving that both variables show equal mean population between different groups of households' income.

5. Discussion and Conclusions

This study focused on investigating the antecedents of acceptance of household on food waste innovation products in Terengganu. Throughout the study, the researchers found that knowledge and attitude have an effect and significant relationship towards the acceptance of household on food waste innovation products, as the increases in those two factors will give a great improvement on how the household would accept the food waste innovation products. However, public knowledge of food waste recycling remains low, and the government should play a vital role in educating the public about the benefits of recycling food waste in preserving the environment and strengthening local and national economies. This study also emphasizes the significance of people's knowledge, as those with a strong understanding are more likely to have a positive attitude since they will think things through before acting. Consumers tend to pay more attention of their present predicament about how food surplus will be collected, transported, and distributed. This is due to the fact that most of the respondents are concerned with the management of food waste disposal (FWD), which can help protect the landfills and reduce wastage in Malaysia. Food waste is the most common type of municipal solid waste created in Terengganu, and if it could be effectively handled, the waste problem may be alleviated. The results are supported by [22], which agreed that people would improve household management behaviors when they realize that food waste innovation products could reduce waste. Aside from that, it can be concluded that the respondents have little knowledge on that food surplus can be composted into fertilizer and sold. Due to this fact, it is important for the community, especially at the household level, to be educated on how food waste can be made into fertilizer with composting methods. In line with this, the innovated waste products could be their business opportunity in Terengganu. In contrast, the households have little knowledge on that food waste can be recycled into innovative products that are beneficial to entrepreneurs or the people of Terengganu. With this fact, the communities and households can be involved to use culinary surpluses from fruits, vegetables, dairy products, or grains in daily activities and can generate their source of income and reduce the amount of wastage produced. Although the respondents who participated in this study do not represent the whole population, the results of this study have provided useful information regarding the elements that can impact a consumer's decision to not waste food.

Theoretically, the current study expands our understanding of households' knowledge and attitudes, which leads to food waste innovation products, notably among Kuala Terengganu's community and residence. Practically, the new study has significant implications for a variety of stakeholders, including entrepreneurs, society, and governments, in terms of developing a sustainability-oriented food waste by-products and innovation products plan. They may also be able to give more information on the importance of food waste prevention as well as ways for doing so. Practically, along with consumer needs, the food waste processing sectors that produce food waste innovative products will continue to expand across the world. Furthermore, in light of rising environmental concerns, much study is required to determine how food waste might be innovated for value addition and human consumption. This will benefit industry, the environment, and consumers to the greatest extent possible. Not only that, it will also assist the Kuala Terengganu City Council (MBKT) and the Solid Waste Management and Cleaning Corporation in dealing with massive amount food waste from residential houses and household by using innovative ways.

Author Contributions: Conceptualization, N.R. and W.N.W.N.; methodology, N.R.; software, H.A.M.A.M.; validation, F.A.C.I. and S.M.; formal analysis, N.R.; investigation, H.A.M.A.M.; resources, W.N.W.N.; data curation, H.A.M.A.M.; writing—original draft preparation, N.R.; writing—review and editing, N.R.; visualization, S.M.; supervision, F.A.C.I.; project administration, W.N.W.N.; funding acquisition, N.R. All authors have read and agreed to the published version of the manuscript.

Funding: This research was funded by Research Collaboration Fund (RCF) 2021/1, RMU, PJI UiTM Cawangan Terengganu.

Institutional Review Board Statement: Not applicable.

Informed Consent Statement: Not applicable.

Data Availability Statement: Not applicable.

Acknowledgments: The authors are grateful for the collaboration with the members from Universiti Putra Malaysia (UPM) and Malaysian Agricultural Research & Development Centre (MARDI). This paper was presented at the International Academic Symposium of Social Science (IASSC) 2022.

Conflicts of Interest: The authors declare no conflict of interest.

References

1. World Bank. Solid Waste Management. Available online: https://www.worldbank.org/en/topic/urbandevelopment/brief/solid-waste-management (accessed on 8 February 2021).
2. FAO. Global Food Losses and Food Waste—Extent, Causes and Prevention. Available online: http://www.fao.org/3/i2697e/i2697e.pdf (accessed on 27 January 2021).
3. Ramdzan, S.N.; Kadir, A.A.; Kamil, M.N.A.F.; Shahabuddin, N.A.S.; Mizad, M. Implementation of Food Waste Composting in Malaysia. In *Water and Environmental Issues*; Micropollutant Research Centre (MPRC) UTHM: Johor, Malaysia, 2018; Volume 2, p. 35. ISBN 978-967-2183-94-5.
4. SWCorp Malaysia. *Kompendium Pengurusan Sisa Pepejal Malaysia 2019*; Solid Waste Corporation Malaysia: Cyberjaya, Malaysia, 2019.
5. Aktas, E.; Sahin, H.; Topaloglu, Z.; Oledinma, A.; Huda, S.; Irani, Z.; Sharif, A.M.; Kamrava, M.; van't Wout, T. A consumer behavioral approach to food waste. *J. Enterp. Inf. Manag.* **2018**, *31*, 658–673. [CrossRef]
6. Pappalardo, G.; Cerroni, S.; Nayga, R.J., Jr.; Yang, W. Impact of COVID-19 on Household Food Waste: The Case of Italy. *Front. Nutr.* **2020**, *7*, 585090. [CrossRef] [PubMed]
7. Nawawi, W.N.W.; Khalid, S.A.; Ramli, N.; Daud, N.M. Attitude and expectation: Food waste recycling as a business opportunity in Terengganu. *Soc. Sci. Humanit.* **2017**, *25*, 65–74.
8. Jarjusey, F.; Chamhuri, N. Consumers' awareness and knowledge about food waste in Selangor, Malaysia. *Int. J. Bus. Econ. Aff.* **2017**, *2*, 91–97.
9. Sulaiman, N.F.A.R.; Ahmad, A. Save The Food for A Better Future: A Discussion on Food Wastage in Malaysia. *Int. J. Law Gov. Commun.* **2018**, *3*, 12–21.

10. Jamaludin, A.; Norazam, S.; Kamal, S.M. Food Wastage Awareness Among Restaurant Consumers in Subang Jaya. *Int. J. Sci. Technol. Res.* **2020**, *9*, 1232–1236.
11. Stangherlin, I.D.C.; Barcellos, M.D.D. Drivers and barriers to food waste reduction. *Br. Food J.* **2018**, *120*, 2364–2387. [CrossRef]
12. Talwar, S.; Kaur, P.; Yadav, R.; Sharma, R.; Dhir, A. Food waste and out-of-home-dining: Antecedents and consequents of the decision to take away leftovers after dining at restaurants. *J. Sustain. Tour.* **2021**, 1–26. [CrossRef]
13. Ramli, N.; Rosli, N.S.; Wahap, F.A.; Nawawi, W.N.W.; Majid, H.A.A. Acceptance of food waste recycling products among public toward sustainable food waste management. In *Charting a Sustainable Future of ASEAN in Business and Social Sciences*; Springer Nature Singapore Pte Ltd.: Singapore, 2020; Volume 1, pp. 391–401.
14. Martin-Rios, C.; Hofmann, A.; Mackenzie, N. Sustainability-Oriented Innovations in Food Waste Management Technology. *Sustainability* **2021**, *13*, 210. [CrossRef]
15. Lau, K.Q.; Sabran, M.R.; Shafie, S.R. Utilization of Vegetable and Fruit By-products as Functional Ingredient and Food. *Front. Nutr.* **2021**, *8*, 661693. [CrossRef] [PubMed]
16. Torres-León, C.; Ramírez-Guzman, N.; Londoño-Hernandez, L.; Martinez-Medina, G.A.; Díaz-Herrera, R.; Navarro-Macias, V.; Alvarez-Pérez, O.B.; Picazo, B.; Villarreal-Vázquez, M.; Ascacio-Valdes, J.; et al. Food Waste and Byproducts: An Opportunity to Minimize Malnutrition and Hunger in Developing Countries. *Front. Sustain. Food Syst.* **2018**, *2*, 52. [CrossRef]
17. Bashir, M.; Majid, A.A.; Alden, H.N.A.; Hussin, S.M.S.; Zahari, M.S.M. The Role of Environmental Knowledge and Mediating Effect of Pro-Environmental Attitude towards Food Waste Reduction. *Int. J. Acad. Res. Bus. Soc. Sci.* **2018**, *8*, 60–72.
18. Naidu, S. What a Waste: Malaysia's Struggle with Excess Food. 2017. Available online: https://www.channelnewassia.com/news/asia/what-a-Waste-Malaysia-s-struggle-with-excess-food-8735458 (accessed on 2 January 2019).
19. Razak, A.; Ghafar, S.a.A.; Padzil, S.W.a.M.; Kamaruddin, N.A.a.; Zin, A.a.M.; Saim, N.a.; Suhaimi, M.a.M.S.; Husna, A. Household food wastage prevention in Malaysia: An issue processes model perspective. *Econ. Technol. Manag. Rev.* **2018**, *13*, 51–62.
20. Alattar, M.; DeLaney, J.; Morse, J.; Pincus, M.N. Food waste knowledge, attitudes, and behavioral intentions among university students. *J. Agric. Food Syst. Community Dev.* **2020**, *9*, 1–16. [CrossRef]
21. Viachaslau, F.; Coşkun, A.; Derqui, B.; Matute, J. Restaurant management and food waste reduction: Factors affecting attitudes and intentions in restaurants of Spain. *Int. J. Contemp. Hosp. Manag.* **2022**, *34*, 1177–1203.
22. Jereme, I.A.; Siwar, C.; Begumand, R.A.; Abdul, B. Food waste and food security: The case of Malaysia. *Int. J. Adv. Appl. Sci.* **2017**, *4*, 6–13. [CrossRef]
23. Department of Statistics. *Household Expenditure Survey Report by State and Administrative District Terengganu 2019*; Department of Statistics: Putrajaya, Malaysia, 2020.
24. IBM Corp. *Released 2019. IBM SPSS Statistics for Windows*; Version 26.0; IBM Corp.: Armonk, NY, USA, 2019.

proceedings

MDPI

Proceeding Paper

Evaluation of Reinventing Government Policy on Financial Management Performance of Education Public Service Agencies on Ministry of Transportation †

Eka Sari Prihartanti [1,*], Muhammad Zilal Hamzah [2] and Eleonora Sofilda [2]

1 Flight Calibration Internal Audit, Ministry of Transportation, Tangerang 15820, Indonesia
2 Faculty of Economic and Business, Trisakti University, Jakarta 11440, Indonesia
* Correspondence: ekasariprihartanti@gmail.com

† Presented at the International Academic Symposium of Social Science 2022, Kota Bharu, Malaysia, 3 July 2022.

Abstract: The objectives of this research are (1) to analyze and assess the level of interest or expectations of the government, the management of the Public Service Agency (PSA), and stakeholders on the implementation of bureaucratic entrepreneurship, and (2) to analyze and assess the perception of the government, the management of the Public Service Agency, and stakeholders on current problems. This research uses a combination of quantitative and qualitative descriptive research approaches, where the data are sourced from primary data and secondary data through focus group discussions and processed with the NVivo program. The object of this research is 23 work units within the patterns of the financial management section of the PSA at the Ministry of Transportation.

Keywords: policy; reinventing government; performance

Citation: Prihartanti, E.S.; Hamzah, M.Z.; Sofilda, E. Evaluation of Reinventing Government Policy on Financial Management Performance of Education Public Service Agencies on Ministry of Transportation. *Proceedings* **2022**, *82*, 76. https://doi.org/10.3390/proceedings2022082076

Academic Editor: Mohamad Rahimi Mohamad Rosman

Published: 22 September 2022

1. Introduction

1.1. Preface

Provisions on budgeting are set in Law Number 17 of 2003 concerning State Finances, and in Government Regulation Number 23 of 2005 concerning the Financial Management of Public Service Agencies. The shift from the traditional financial management system to a performance-based financial management system is then accommodated through a legal basis, namely in Law Number 1 of 2004 concerning the Treasury. Furthermore, both Article 68 and Article 69 opened a new corridor for the implementation of the performance base in the government environment, and government agencies whose main tasks and functions are providing services to the community can apply a flexible financial management pattern by prioritizing productivity, efficiency, and effectiveness

Reinventing government is often referred to as bureaucratic entrepreneurship. Entrepreneurial bureaucracy means transforming the entrepreneurial spirit into the public sector. The entrepreneurship of bureaucracy comprises a fundamental change in government systems and organizations to improve the effectiveness, efficiency, and ability of government agencies to innovate. Innovation must always be updated because organizational sustainability means that the company must continuously change itself over a certain period of time [1]. A semi-autonomous agency has three key features: (1) it is structurally disaggregated from a ministry, (2) it carries out public tasks, and (3) it operates under more business-like conditions than traditional government bureaucracies [2].

A PSA is expected to be more independent in that reliance on financing from the State Revenue and Expenditure Budget (APBN), often referred to as the pure rupiah (RM), should decrease. This is in line with the initial expectation of the establishment of the PSA, namely organizational independence. The target and realization of the Central Government's PSA revenue are described as always increasing beyond the set target, and having a positive influence on the performance of the APBN and fiscal policy making of the government. This coincides with the targeting and realization of PSA revenue from the Human Resources Development Agency (BPSDM). The Ministry of Transportation is one of the government agencies that implements the PSA in several of its work units. To improve organizational performance, the implementation of the PSA is deemed necessary to be applied to government agencies in Indonesia, including the Ministry of Transportation. There are ten main factors/dimensions that determine service quality. The existing dimensions can be summarized into five main dimensions. Therefore, as well as measuring financial performance, it is also necessary to measure service performance [3]. The growth of performance audits continues but is inconsistent over time [4]. In order to be able to dig deeper into the extent to which the concept of bureaucratic entrepreneurship policy is applied to the Educational PSA of the Ministry of Transportation, a related study is needed; this is conducted by parties relevant to the financial management of the PSA. One of these related studies involves conducting a focus group discussion (FGD). The use of the FGD method requires combination with other data collection tools to increase the richness of the data and make the resulting data more valuable and more informative to answer research problems [5]. The PSA of the Ministry of Transportation is a government agency or government organization that provides and maximizes public services without prioritizing profit, with the characteristics of managing its finances independently with efficient and productive principles as well as private organizations. The achievement of public financial management can be seen from the performance of the organization through performance appraisals in public sector organizations. So, the formulation of the problem in this study is (1) What is the level of interest or expectation of the government, the management of the PSA, and stakeholders on the implementation of bureaucratic entrepreneurship? (2) What are the perceptions of the government, the management of the PSA, and stakeholders on the current problems?

1.2. Research Purpose

The research objectives that were achieved are:

1. Analyze and assess the level of interest or expectations of the government, management of the Public Service Agency, and stakeholders on the implementation of bureaucratic entrepreneurship.
2. Analyze and assess the perception of the government, management of the Public Service Agency, and stakeholders on current problems.

1.3. Research Gap

To be able to delve deeper into the extent to which the concept of bureaucratic entrepreneurship policy is applied to the Educational PSA of the Ministry of Transportation, a study related to this is needed by parties with an interest in PSA financial management, one of which must be conducted through a focus group discussion.

1.4. Literature Review

Broadly, public policy can be defined as the relationship between a government unit and its environment [6]. Public policy is defined as whatever the government chooses to do and not do [7]. Reinventing government, or entrepreneurship bureaucracy, is a fundamental change in government systems and organizations to increase the effectiveness, efficiency, and ability of government agencies to innovate [8]. Entrepreneurial bureaucracy has ten principles, namely: (1) Catalyst governance; (2) The government belongs to the community; (3) Competitive government; (4) The government is mission-oriented; (5) Results-oriented government; (6) Customer-oriented government; (7) Entrepreneurial government; (8) Anticipated government; (9) Decentralized governance; and (10) The government is market-oriented.

Agencies are part of the complex New Public Management (NPM) reforms [9]. These reforms were based on the expectation that breaking up government departments by delegating executive and regulatory tasks to single-purpose agencies and introducing business-like instruments would boost performance. Government organizations as policy-makers are separated from implementing policies or government organizations with more specific tasks and organizing and in direct contact with public services and emphasizing how public institutions should be run in a private-like manner [10]. The PSA is an agency within the Government that was formed to provide services to the community in the form of providing goods and/or services that are sold without prioritizing seeking profit and in carrying out their activities focus on the principles of efficiency and productivity. Educational institutions with limited budgets have the ability to apply NPM concepts that make the schools able to compete with other private schools while implementing NPM [11]. The successful implementation of NPM for developing countries is affected by the level of economic progress, the existence of a formal market the economy, the rule of law, the advanced administration of infrastructure, and the efficiency of the state for the success of NPM-oriented reforms, and there is still a greater role of the state in socio-economic transformation in general and in the implementation of market-oriented reforms in particular [12].

NPM is applied not only in countries with a high level of prosperity but can also be applied in countries with conditions similar to those of Indonesia [13]. Apart from the shortcomings in the implementation of regional autonomy, the implementation of NPM in the management of local governments in Indonesia has had a positive impact on several aspects, namely the accountability performance of government institutions and the moratorium and early retirement policies for civil servants who do not meet the requirements as an effort to increase the efficiency and productivity of local government performance, which will improve the quality of public services and the creation of partnerships between the public and private sectors or public-private partnership (PPP), has now become a standard concept in the local government environment.

2. Research Methodology

This study uses data collected through secondary data and focus group discussions followed by data analysis using NVivo 12. The characteristics of the FGD method, namely using semi-structured interviews with a group of individuals with a moderator leading the discussion in an informal setting, and aims to collect data or information on a particular issue topic [14]. Collaboration and coordination from the views of agencies related to the development of the PSA for Financial Management (Ministry of Finance, PSA Management and Supervisory Board, Ministry of Transportation, and other stakeholders) based on several existing regulations, will analyze through this FGD. The results of the FGD were processed using the NVivo analysis method to obtain policy recommendations for the Ministry of Transportation's Educational PK-PSA Implementation Model Based on the concept of Entrepreneurship Bureaucracy. The framework of this research is described in the diagram as shown in Figure 1 below.

NVivo data analysis was carried out using the coding method, which involves coding field notes, observations, and archival materials [15]. Data analysis using NVivo 12 will be analyzed by Matrix Coding Query, Word Frequency Query, Analytical Map, and Framework Matrices. NVivo has various advantages and may significantly improve the quality of research. The analysis of qualitative data has become easier than ever before and yields more professional results. The software greatly reduces manual tasks and gives the researcher more time to discover tendencies, recognize themes, and derive conclusions [16].

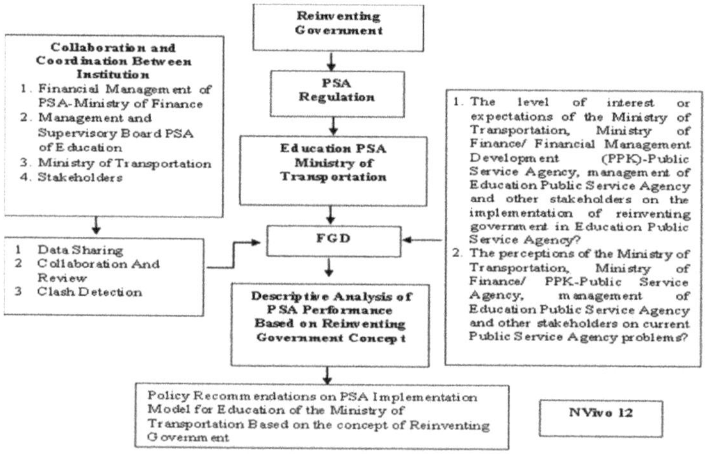

Figure 1. Framework.

3. Research Result and Discussion

3.1. Performance

There are 9 PSA revenue performance achievements that do not reach 100%, namely: BP3IP Jakarta, PPI Curug, Poltekpel Banten, PTDI STTD, BP2TL Jakarta, Poltrada Bali, PPI Madiun, Poltekbang Jayapura and BP3 Curug. In terms of revenue performance, asset optimization reached 100%, only 4 PSA, namely: STIP Jakarta, Poltekpel Banten, Poltekpel Sorong and Poltekbang Palembang. For the management modernization performance, achievements that did not reach 100% were: Poltekbang Surabaya, Poltekpel Barombong, PPI Curug, Poltekpel Malahayati, Poltekbang Medan, Poltrada Bali, Poltekbang Jayapura, BP3 Curug and Poltekbang Semarang. For the establishment training performance achievement, which reached 100%, only 3 PSAs out of 20 PSAs, namely BP3IP Jakarta, PIP Semarang, and PIP Makassar, while for the technical training performance achievement which reached 100% only 3 out of 23 PSAs, namely: Poltekpel Malahayati, Poltran SDP Palembang, and BP3 Curug. The achievement of graduate absorption performance reached 100% in only 4 out of 21 PSAs, namely: BP3IP Jakarta, Poltekpel Barombong, Poltekpel Banten, and API Banyuwangi and for lecturer performance achievement, which reached 100%, there are 2 PSAs out of 23 existing, namely: BP3IP Jakarta and BP3P Curug. For research performance achievements which reached 100%, 9 out of 23 existing PSAs were: BP3 Curug, Poltekbang Palembang, Poltekpel Surabaya, Poltekbang Surabaya, API Banyuwangi, Poltekpel Mahalayati, Poltekbang Makassar, Poltekbang Medan, and PPI Madiun, and for the achievement of community service performance, which did not reach 100%, there are 4 of the 23 existing PSAs, namely: PKTJ Tegal, Poltekpel Malahayati, PTDI-STTD, and BP2TL Jakarta.

For the performance achievement of the Community Satisfaction Index which reached 100%, 6 of the 23 existing PSAs are: PIP Makassar, Poltekpel Surabaya, Poltekbang Surabaya, Poltekbang Makassar, Poltekbang Medan, and Poltekbang Palembang. For the performance achievements of cadets from low-income communities that did not reach 100%, 7 out of 21 PSAs are Poltekpel Barombong, Poltekbang Makassar, Poltekbang Medan, PTDI-STTD, BP2TL Jakarta, Poltrada Bali, and PPI Madiun. For the achievement of relevance performance that reached 100%, 8 of the 21 existing PSAs are: STIP Jakarta, PIP Makassar, Malahayati Poltekpel, Surabaya Poltekpel, Barombong Poltekpel, Medan Poltekbang, Bali Poltrada, and PPI Madiun and for the achievement of the assignment program performance that did not reach 100% in 4 of the 20 existing PSAs, namely: STIP Jakarta, Poltekpel Malahayati, Poltekbang Medan, PTDI STTD, Poltrada Bali, and PPI Madiun.

3.2. Continuity

At the Center for Refresher Education and Improvement of Shipping Science (BP3IP) Jakarta and the Indonesian Aviation Polytechnic Curug (PPIC), sustainability is maintained through various innovations. The sustainability issue at BP3IP is mainly related to market changes where there are now many other schools that provide the same services as those offered by BP3IP, whereas BP3IP was initially the only provider. Therefore, several breakthroughs were made as a way to optimize revenue and maintain BP3IP sustainability, including increasing the types of education and training offered, picking up registration balls/revalidating seafarers' certificates, collaborating with operational work units for shipping safety and security as well as drafting regulations related to education and training costs online and revising the targets set in the previous strategic plan. Meanwhile, at PPIC, the concept of sustainability is maintained by means of program diversification as a form of self-development during the pandemic. Additional services such as the IDUKA program (Industry and the World of Work) in addition to the main PPIC services, can be a source of income boost during a pandemic. Poltekpel Surabaya also carries out education and training activities as its main service. Financial achievements also experience ups and downs, especially during a pandemic. Therefore, the sustainability of the institution is maintained by conducting long-distance learning. Service diversification in the context of sustainability is also implemented in other Educational PSAs of the Ministry of Transportation. The stimulus program that is handed down to business entities as stakeholders is also believed to be one way to maintain the sustainability of the organization. Where the government must be present in supporting aviation safety, upgrading the skills or knowledge possessed by employees of the Business Entity, in this case, Perum LPPNPI or other Business Entities related to the world of aviation, by sending them to schools in the Ministry of Transportation.

3.3. The Result of FGD

In this case, matrix coding is divided into 2 categories, which are single and cross the border. The single category refers to only one aspect of the part. For example, the Regulator part (in the row) will also be matched with the Regulator part itself (in the column), where this is carried out to see if there is a relationship or connection, either implicitly or explicitly, on each node in the Regulator part. While the cross-border category applies to the equivalent between parts, for example, between the PSA and Regulator, and so on. It aims to be able to see the relationship or connection nodes/statements in each category of informants.

- a. Matrix coding in a single category. Starting at the Regulator, where there is a relationship between one node with another node. For example, "Integration between Stakeholders" with "Efficiency". The integration of stakeholders (the Technical Ministry, the Ministry of Finance, the PSA, Supervisory Board, Society, and market) is one of the conditions for the creation of an efficient PSA, particularly in optimizing revenue. Unfortunately, there are still misunderstandings in the perceptions of these stakeholders. The disparity in perception must be supported by socialization regarding PSAs, when there is an understanding, the integration process between fellow stakeholders will be relatively better. As mentioned above, the market is one of the stakeholders included in this research. In fact, it was found that there was one PSA (one of the oldest Poltekpel in Indonesia, BP3IP) that had difficulty achieving its revenue target. This is because major training activities (formerly) were centered in the PSA. The relationship between the "PSA Sustainability" nodes and "PSA Efficiency" on the PSA section, relates to the more efficient the PSA (the better the performance), at least having the hope of relatively better business sustainability against the less efficient PSA. If referring to the constraints that occur, especially in BP3IP, the difficulty in optimizing revenue affects how the sustainability of the PSA is indirectly affected. While the relationship between one node with another node. For example, there is a relationship between the "Business Portfolio Diversification" and "Market Dynamics" nodes. This refers to a PSA's efforts to diversify its business activities to deal with market dynamics, especially during the current pandemic. The diversification seems to be carried out by all existing PSAs (all informants who represent PSAs in FGD activities), for example through the formation of non-regular education and training to optimize their assets. In the Supervisory Board section, where there is a relationship between one node and another node. For example, there is a relationship between the "Moral Hazard" and "Target Revenue Manipulation" nodes. The use of the term "Moral Hazard" does seem excessive, but the manipulations carried out give rewards to PSA officials, but do not provide optimum service to the community. So there needs to be "honesty" in planning income targets. The occurrence of this is triggered by the provision of rewards if the realization of income exceeds the target. Furthermore, there is a relationship between the nodes "Professionalism of PSA Officials" and "Manipulation of Revenue Targets". The manipulation of revenue targets by PSA officials, apart from being a matter of honesty, is clearly a matter of the professionalism and credibility of PSA officials. This can be related to the statement of Informant 5 where the elected PSA officials are usually only experts in their technical fields but do not have the ability to act like an entrepreneur. Furthermore, in the Business Entity section, where there is a relationship between one node and the other nodes. For example, there are a relationship between the "Airline Difficulty" nodes and with "Increased Intensity of Accounts Receivable" nodes. This shows the relationship between a significant decrease in traffic movement on all types of flights that occurred on Airnav Indonesia, resulting in a decrease in revenue. From this, efforts were made to maintain business continuity, one of which was through cash optimization, namely the collection of receivables.

- b. Cross Border, it will be shown in the matrix coding table in a cross-border manner, starting from the Regulator and the Supervisory Board (the matrix can be seen in the NVivo–Output-Matrix Coding-Cross Border-Regulator vs. Supervisory Board-Matrix folder. There is a relationship between "Fit and Proper" nodes with "PSA Efficiency". A PSA, in optimizing its income, needs to be accompanied by knowledge and competence in the economic field, not only from technical aspects. This makes the competence of HR in a PSA, especially officials, to be Fit and Proper. From this, it is possible to carry out a kind of Fit and Proper test to ensure that the PSA leader is truly an expert in both technical and economic knowledge. Apart from that, a relationship was found between the "Discipline" nodes with PSA Efficiency". Still, in the same context, this refers to the need for the role of PSA officials themselves to discipline their human resources, especially themselves, in issues of innovation, creativity, and encouraging knowledge. The relationship between the nodes "Supervisory Board Role" and "Urgency of Socialization". This is related to preventing negative perceptions from the PSA towards the Supervisory Board regarding its role as supervisor. For Cross Border, it will be shown in the matrix coding table in a cross-border manner, starting from the Regulator and the PSA that there is a relationship between the "Changes in Market Structure" nodes with "PSA Sustainability". Namely, changes in market structure will have an impact on the sustainability of the PSA. This is due to people's preferences in making requests. One of the PSAs in the spotlight here is BP3IP, which used to monopolize the service and now has to spread it out. If the pandemic, will add to the difficulties of the PSA, it is proven that its income has decreased for several periods. For Cross Border, it will be shown in the matrix coding table in a cross-border manner, starting from the Regulator with the Business Entity that the relationship between the "Link and Match Industry-Education" nodes with "Delay in HR Acceptance" is related to the impact of the pandemic that made Airnav Indonesia (as the agency businesses that absorb labor) delay the absorption of human resources. As mentioned earlier, this is due to a decrease in traffic movement from domestic, international, and overflying flights. This is indirectly related to the incompatibility of the absorption capacity of graduates with their competencies. Cross Border Supervisory Board with the PSA that the relationship between the nodes "Market Dynamics" and "Community Preferences" is related to the problem of BP3IP experiencing difficulties due to changes in market structure that occurred. This has previously been explained in the previous section that with the influence of preference, the market becomes more dynamic. This is part of consumer behavior. Then there is a relationship between the "Business Portfolio Diversification" nodes with "Community Preferences".

The following table describes the aggregate number of references (including sub-nodes) of each of these nodes, as follows:

These results indicate that the 20 nodes above have the largest contribution to the overall hierarchy, as shown in Figure 2 below. This indicates, overall (4 Categories of Informants), both implicitly and explicitly, alluding to the problem of PSA results of the FGD, explained that there are several problems that are often expressed by the speakers and can be a priority for improvement, namely PSA performance as much as 14.29%, Optimizing revenue by 11.01%, Integration between stakeholders as much as 10.42% and improvement of PSA regulation as much as 5.65%, as shown in Figure 3 below.

Figure 2. Hierarchy Aggregate.

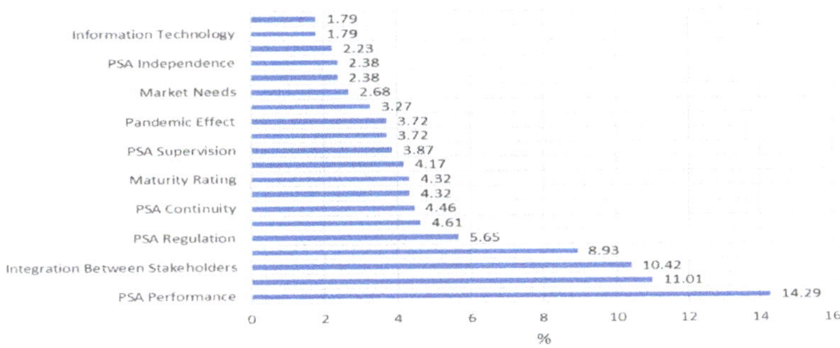

Figure 3. Reference Nodes.

4. Conclusions and Recommendations

4.1. Conclusions

For the first objective (the level of interest or expectations of the governments, the management of the PSA, and stakeholders on the implementation of bureaucratic entrepreneurship) the relevant aspects are: (a) To increase the POBO ratio (Operational Expenses Operating Income), independence in investment financing, dependence on the pure rupiah, increase the PNBP performance, and to optimize the assets in the Education PSA of the Ministry Transportation; (b) PSAs in the Ministry of Transportation can finance their own activities, either through loans or conducting KSOs; (c) To cooperate with external parties, especially in terms of optimizing fixed assets in order to get the additional income that will definitely an impact on increasing capacity (monetarily) to finance PSA activities; (d) To increase the integration between stakeholders (among others between the Technical Ministry, the Ministry of Finance, the PSA, the supervisory board, and market); and (e) To improve service quality of PSA Education without being oriented towards profit.

The second objective (the perception of the government, the management of the PSA, and stakeholders on current problems) stated that the main problems related to performance realization, target achievement, and impact are on PSA performance (as much as 14.29%), the optimization of revenue (11.01%), integration between stakeholders (10.42%), and the improvement of PSA regulations (5.65%). These five main problems can be seen in the current conditions, namely (a) PSA education that is not yet independent and very much depends on APBN budget allocations (pure rupiah), (b) Inequality in perception must be supported by socialization regarding PSAs, (c) The level of cooperation between PSAs and external parties in terms of cooperation in asset utilization is still low. Assets that should be used as income for PSAs are not maximized, (d) PSA services are not optimal, one of which occurs due to manipulation of planning and revenue targets, (e) PSA income is still highly dependent on the implementation of offline training, and (f) Changes in the market structure that are no longer part of amonopoly, where several new players have emerged in providing the same service, thus providing several alternatives to the market in making purchases. There are many new schools that offer the same type of education and training as the existing schools.

4.2. Recommendation

Recommendations that can be given regarding the results of this research are: (1) More innovation is needed in the services provided, the methods of service development, main services, or additional services, as well as cooperation in asset utilization to reduce dependence on pure rupiah; (2) Shared perceptions and integration among stakeholders are necessary so that the situation is relatively improved. The more efficient the performance of the PSA, the higher the expectation of business sustainability will be. This is especially true for PSAs that are less efficient, by reducing expenditure items that are deemed less important; (3) PSA management must be more innovative regarding cooperation in the use of assets to reduce dependence on pure rupiah; (4) A review of the government's policy regarding the provision of rewards for PSAs that can achieve the target is needed, followed by a review of the procedure for selecting PSA officials at the education work unit of the Ministry of Transportation. A leader of the PSA must have knowledge in the technical field that he is involved in, as well as be professional and have the spirit of business management (entrepreneurship) from the PSA itself as well as the principles of entrepreneurship bureaucracy; (5) Optimizing the digitalization/technology capabilities of his human resources, especially in terms of providing online education and/or training; and (6) Propose to the government to regulate the market structure and/or provide clear boundaries regarding the establishment and management of schools in the Ministry of Transportation, especially during the current pandemic.

Author Contributions: Conceptualization, E.S.P.; methodology, E.S.P. and M.Z.H.; software, E.S.P.; validation, E.S.P., E.S. and M.Z.H.; formal analysis, E.S.P.; investigation, E.S.P.; resources, M.Z.H.; data curation, E.S.P.; writing—original draft preparation, E.S.P.; writing—review and editing, E.S. and M.Z.H.; visualization, E.S.P.; supervision, E.S. and M.Z.H.; project administration, E.S.P.; funding acquisition, E.S.P. All authors have read and agreed to the published version of the manuscript.

Funding: This research received no external funding.

Institutional Review Board Statement: Not applicable.

Informed Consent Statement: Not applicable.

Data Availability Statement: Not applicable.

Acknowledgments: The data collection method in this research was carried out through focus group discussions, therefore the authors would like to thank the focus group discussion informants. The author does not forget to thank the Ministry of Transportation, which has provided valuable data for this research.

Conflicts of Interest: The authors declare no conflict of interest.

References

1. Buchanan, D.; Fitzgerald, L.; Ketley, D.; Gollop, R.; Jones, J.L.; Lamont, S.S.; Neath, A.; Whitby, E. No going back: A review of the literature on sustaining organizational change. *Int. J. Manag. Rev.* **2005**, *7*, 189–205. [CrossRef]
2. Overman, S.; van Thiel, S. Agencification and Public Sector Performance: A Systematic Comparison in 20 Countries. *Public Manag. Rev.* **2015**, *18*, 611–635. [CrossRef]
3. Parasuraman, A.; Zeithaml, V.A.; Berry, L. SERVQUAL: A Multiple-Item Scale for Measuring Consumer Perceptions of Service Quality. *J. Retail.* **1988**, *64*, 12–37.
4. Parker, L.D.; Jacobs, K.; Schmitz, J. New Public Management and The Rise of Public Sector Performance Audit Evidence From the Australian Case. *Account. Audit. Account. J.* **2019**, *32*, 280–306. [CrossRef]
5. Lambert, S.D.; Loiselle, C.G. Combining Individual Interviews and Focus Groups to Enhance Data Richness. *J. Adv. Nurs.* **2008**, *62*, 228–237. [CrossRef] [PubMed]
6. Eyestone, R. *The Thread of Policy: A Study in Policy Leadership*; Bobbs-Merril: Indianapolis, IN, USA, 1971.
7. Dye, T.R. *Understanding Public Policy*; Pearson Education Inc.: Saddle River, NJ, USA, 2005.
8. Osborne, D.; Plastrik, P. *Memangkas Birokrasi, Lima Strategi Menuju Pemerintahan Wirausaha*; Victory Jaya Abadi: Jakarta, Indonesia, 2000.
9. Bach, T.; van Thiel, S.; Hammerschmid, G.; Steiner, R. Administrative Tradition and Management Reforms: A Comparison of Agency Chief Executive Accountability in Four Continental Rechsstaat Countries. *Public Manag. Rev.* **2016**, *19*, 765–784. [CrossRef]
10. Lukman, M. *Badan Layanan Umum Dari Birokrasi Menuju Korporasi*; Bumi Aksara: Jakarta, Indonesia, 2015.
11. Amalia, A.P. Education Budget Management, New Public Management and the COVID-19 Pandemic. *Din. Pendidik.* **2021**, *16*, 1–11. [CrossRef]
12. Sarker, A.E. New Public Management in Developing Countries. *Int. J. Public Sect. Manag.* **2006**, *19*, 180–203. [CrossRef]
13. Pratama, P.Y.; Kalalinggi, R. Application of New Public Management in Indonesia in the Field of Transportation. *J. Gov. Public Policy* **2019**, *6*, 126–147.
14. Dick, W.; Carey, L. *The Systematic Design of Instruction*; Harper Collins Publishers: New York, NY, USA, 1996.
15. Miles, M.B.; Huberman, A.M. *Analisis Data Kualitatif Tentang Metode-Metode Baru*; UIP: Jakarta, Indonesia, 1992.
16. Hilal, A.H.; Alabri, S.S. Using NVivo for Data Analysis in Qualitative Research. *Int. Interdiscip. J. Educ.* **2013**, *2*, 1.

proceedings

Proceeding Paper

The Reality of Happiness According to Scholars' Viewpoints: A Systematic Literature Review (SLR) Analysis †

Noraini Junoh [1,*], Zanirah Mustafa@Busu [1], Ahmad Murshidi Mustapha [1], Abdul Manam Mohamad [2] and Nurhidayah Muhammad Hashim [3]

1 Academy of Contemporary Islamic Studies (ACIS), Universiti Teknologi MARA (UiTM) Cawangan Kelantan, Machang 18500, Kelantan, Malaysia
2 Pusat Pengajian Teras (PPT), Universiti Sultan Zainal Abidin, Kuala Nerus 21300, Terengganu, Malaysia
3 Academy of Contemporary Islamic Studies (ACIS), Universiti Teknologi MARA (UiTM), Shah Alam 40450, Selangor, Malaysia
* Correspondence: norainijunoh@uitm.edu.my; Tel.: +60-139842483
† Presented at the International Academic Symposium of Social Science 2022, Kota Bharu, Malaysia, 3 July 2022.

Abstract: Happiness is one of the main topics that dominates the discourse of ethical philosophy. The concept of happiness began to be discussed from the emergence of philosophy in classical Greece to the present day in both Western and Islamic perspectives. This study aims to analyze the views of scholars in debating the topic of happiness. This study is a descriptive study using a qualitative systematic literature review method. The main source of data collection for this study is by highlighting previous works obtained from Scopus and Web of Science (WoS) online databases and it is also supported by data obtained from Google Scholar. The analysis was done based on Preferred Reporting Items for Systematic Reviews and Meta-Analyzes (PRISMA). The results of the SLR analysis identified fifteen (15) articles featuring discussions on the topic of happiness among scholars, among them Plato, Aristotle, Thomas Aquinas, Maimonides, and Jeremy Bentham representing the West and Ibn Sina, al-Ghazali, al-Farabi, Ibn Miskawayh, HAMKA, al-Attas, and Said Nursi representing Islamic scholars. The findings also highlighted three main elements discussing the topic of happiness, namely the spiritual, intellectual, and material elements. These three elements are not discussed in a balanced way in Western civilization causing confusion in uniting various views, but in Islam, although scholars' opinions differ, they are however united in the principle of revelation that balances the three elements to explain the topic of true happiness. This study is important from the point of view of understanding the topic of happiness traced from the history of thought of scholars.

Keywords: happiness; sa'adah; eternal; eudaimonia; scholar; SLR

check for updates

Citation: Junoh, N.; Mustafa@Busu, Z.; Mustapha, A.M.; Mohamad, A.M.; Hashim, N.M. The Reality of Happiness According to Scholars' Viewpoints: A Systematic Literature Review (SLR) Analysis. *Proceedings* **2022**, *82*, 77. https://doi.org/10.3390/proceedings2022082077

Academic Editor: Mohamad Rahimi Mohamad Rosman

Published: 23 September 2022

Publisher's Note: MDPI stays neutral with regard to jurisdictional claims in published maps and institutional affiliations.

1. Introduction

Things which are always associated with the highest goal of human life such as health, wealth, money, and knowledge are often misunderstood as happiness. In fact, all these criteria are not the end goal of one's life, but rather things needed in order to achieve something else known as happiness, that is the highest goal of one's life. Humans always want to achieve success, happiness, and blessings in their lives. Different approaches are taken by people to acquire happiness as the perception and definition of happiness is different among human beings. In daily life, humans would feel happy or blissful if they manage to gain good position, possess a lot of money and wealth, have children, earning high salary, earn a lot of profits in business and valued by others. Unfortunately, happiness and success for some people are based on material measurements alone. By using this kind of measurement, various problems arise when human beings are fighting to achieve luxury and power but neglecting manners and morals.

Many scholars agree that since the Age of Enlightenment, especially in the West, the pursuit of happiness has been sought primarily through material consumption. Although global capitalism has succeeded in generating more and more wealth, many studies have revealed that gaining more wealth from a material point of view does not increase human happiness. The failure to achieve this happiness is due to adherence to hedonic and materialist values at the expense of spirituality and pure values [1].

In this modern society, happiness index is considered as one of the measures used in assessing personal and community well-being. It is believed that the index reflects the sustainability of the well-being of society which will form a peaceful and harmonious nation. As far as our country is concerned, Malaysia was ranked 80th among the happiest countries in the world according to the World Happiness Report 2019 [2]. In 2021, the happiness index of Kuala Lumpur, the main city in Malaysia was 78.6 percent which was classified as "moderately happy" [3]. The Happiness Index of a society will be measured from time to time by taking into consideration several new factors. For instance, issues of depression, mental illness of the population, stress level and the ability to cope with stressful life as well as economic burden have been accepted as crucial problems today.

In a study conducted by N. A. A. Pauzi and R. Koris [4], it is claimed that in 2020, depression would be the focal mental health issue in Malaysia. This is due this is due to very high work pressure, family issues and the ability to cope with pressure. Pressure at work as well as working extra hour to complete the task given before deadlines would affect work productivity and work performance and it might also give direct impact to the happiness index of the population [5]. When looking into the life of an employee, research shows that there are five factors that could lead to happiness; positive relationship with other staff, working performance, workplace environment, compensation received from the employer, and health condition of the employee [4]. Sources from the Royal Malaysian Police (PDRM) found that there was an increase in suicide cases reported in 2020, with 631 cases compared to 609 cases in 2019. As of March 2021, a total of 336 suicide cases were reported [6].

In the Islamic perspective, the most important factor that needs to be considered in finding happiness is the religious factor. The connection between religion and happiness is something which can be said as subjective because it is difficult to measure religion through empirical observation. This depends on one's practice and obedience in following all religious obligations [7].

Discussion on the concept of happiness has been a major focus among philosophers and scholars either from both Western or Islamic worlds which had started since the Greek era. What is the meaning of happiness? How to gain happiness in life? What is the connection between happiness and the ultimate goal in life? These questions have been debated widely by them in searching for the answer. Because of that, various perspectives and concept of happiness have been developed by various scholars in the world. By referring to literature relating to this area, the objective of this paper is to analyze the views of scholars from Islamic and West in debating the concept of happiness using a qualitative systematic literature review method.

2. Research Methodology

2.1. Highlights of Systematic Works Based on the PRISMA Method

The highlights of this systematic work were guided by Preferred Reporting Items for Systematic Reviews and Meta-Analyzes (PRISMA). This method is a quality literature review guideline because its relatively rigorous and detailed process involves large quantities of data. The advantage of PRISMA is that it can show elements of transparency, consistency, and high standards to produce qualitative research reports through certain processes. This process is referred to as a systematic process, a clear procedure of how it is produced and a comprehensive scope covering all relevant and reusable materials (replicate) other researchers with a similar approach have used for discussing a topic [8]. The advantage of using a systematic highlighting method with the PRISMA method in the field of social sciences is that the works obtained are from quality data sources and recognized by all researchers and that it can explain the limitations of the study with the help of keywords. These guidelines also help authors not waste time and wonder if the highlights of their work are enough. PRISMA helps the author to locate the right literature according to the objectives of the study through the following four processes: identification, screening, eligibility, and inclusion [9].

The discussion in this study uses two (2) online databases in research, namely, a main database consisting of Scopus and Web of Science (WoS) databases and a support database, Google Scholar. The main database contains scientific and empirical studies across various fields such as social sciences, engineering, environment, economics, accounting and development, medicine, law, as well as business and marketing [10]. Supporting databases are needed primarily to obtain additional resources such as journals without indexing and any articles missing from the leading database. A support database is needed because Y. Xiao and M. Watson [11] claims that there is no perfect database and the sensitivity of the database to the keywords used to find relevant articles does not reach 100 percent. The process begins by performing a series of searches (search strings) for keywords using the available search facilities and exclusive commands that have been set in the Scopus and WoS databases, respectively. In this process, the authors used keyword matches found in the thesaurus, and were assisted by Boolean operator functions in both databases (Table 1). The keywords and search strings used could be repeated to get the latest material and future research. While searching for these keywords, the author found 13,275 articles (79 articles through Scopus, 96 articles in WoS, and 13,100 articles in Google Scholar). A total of eight (8) overlapping articles in the three databases were sorted with the help of Endnotes software. This makes a total of 13,267 articles to go through a screening and eligibility process.

Table 1. Database resources and search series (search string).

Database	Keyword Order Sequences and Boolean Operators
Scopus	TITLE-ABS KEY ((((true* OR eternal* OR reality*) AND (happiness* OR sa'adah* OR eudaimonia* OR well-being* OR felicity* OR prosperity* OR success*) AND (scholar* OR philosopher*) AND (Islamic* OR eastern* OR western*))))
Web of Science (WoS)	TS = ((((true* OR eternal* OR reality*) AND (happiness* OR sa'adah* OR eudaimonia* OR well-being* OR felicity* OR prosperity* OR success*) AND (scholar* OR philosopher*) AND (Islamic* OR eastern* OR western*))))
Google Scholar	TS = ((((true* OR eternal* OR reality*) AND (happiness* OR sa'adah* OR eudaimonia* OR well-being* OR felicity*) AND (scholar* OR philosopher*) AND (Islamic* OR eastern* OR western*))))

This stage is the phase where the authors screened with certain criteria so that the selected articles were more specific and led to the research question. First, the authors chose journal articles in the form of quantitative studies only, while articles consisting of chapters in books, seminar articles, and proceedings, as well as literature review articles were excluded. Second, the authors selected only English-language articles as well as those related to the fields of religion, philosophy, art and humanities, social science, and multidisciplinary. Third, in terms of the period of time, the authors did not limit themselves to a specific period for the searching in the main database and limited the search from 2018 to 2022 in the supporting database. This is because when the initial screening was done, the authors found that writings related to this field started around 1998 worldwide (see Table 2). After a screening process, out of 13,267 articles, the authors excluded 370 articles, selecting only 45 articles. Although the excluded articles were not included in the analysis of the study, some articles were used as additional references.

Table 2. Eligibility and elimination criteria.

Criteria	Eligibility	Elimination
Types of literature	Journal articles (quantitative and qualitative)	Literature review articles, books, chapters in books, seminar articles and proceedings
Language	English	In addition to English
Area	Religion, philosophy, art and humanities, social science and multidisciplinary humanities	History, management, ecology, hospitality, leisure, sport, tourism, geochemistry, geography, multidisciplinary geosciences, nursing, earth and planet, agricultural, economics, finance, business
Duration	All the time (Scopus and WoS) 2018–2022 (Google Scholar)	None 2017 and below
National coverage	Whole world	

2.2. Eligibility and Inclusion Process

At this stage, the authors only focused on articles that discuss happiness from the point of view of Western and Islamic scholars. All 45 articles were analyzed by making in-depth readings starting from the abstract and the entire article to identify the views of the scholars. As a result of reading, only fifteen (15) articles were found to lead to discussions on true happiness based on scholars. All the article selection processes for the purpose of systematic literature review (SLR) are shown and explained through the PRISMA flow chart in Figure 1 below.

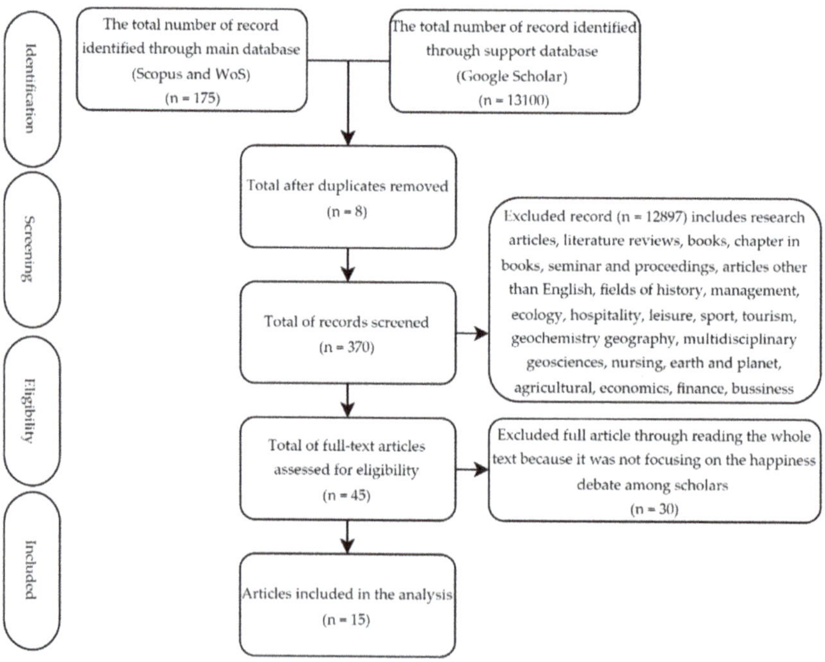

Figure 1. PRISMA Flow Chart. Source: Adapted from Moher et al. [8] and Shafrill et al. [10].

3. Findings

This study details the analysis studies on the topic of happiness based on the views of scholars according to the descriptive analysis and thematic analysis resulting from the highlights from the selected articles.

3.1. Descriptive Analysis

Based on the search, only two (2) articles were quantitative, while thirteen (13) articles were qualitative. All these articles focused on the views of scholars on the concept of happiness ranging from classical to modern figures either from the Western or Islamic point of view. The search results showed that the happiness debate was discussed in 1998 in various journal publications by the following researchers: M. A. Wahab [12], M. Ebrahimi, A. Gholami and K. Yusoff [13], T. Lomas et al. [14], A. Nur, S. K. Yuzar, M. F. A. Mohd Sa'ad [15], A. R. Alebouyeh [16], N. Aydin and H. Khan [17], M. N. Omar et al. [18], M. Ulfa and E. D. A. Puspita [19], A. R. Ateerah and Z. M. Lukman [20], D. Eidukienė [21], R. R. Shah [22], N. Aydin and E. Manusov [1], C. M. Neria [23], M. Joshanloo [24], and S. Stroumsa [25]. This descriptive analysis is summarized in Table 3 below. For the country of origin, it was found that Western countries dominate the discussion related to the concept of happiness by Islamic and Western scholars, namely the article written by M. A. Wahab [12], D. Eidukienė [21], R. R. Shah [22], C. M. Neria [23], N. Aydin and H. Khan [17], T. Lomas et al. [14], A. R. Alebouyeh [16], M. Joshanloo [24], and S. Stroumsa [25]. Meanwhile, in the Malaysian context, from the four articles, it was found that a quantitative study was used in conceptualizing happiness based on the views of Islamic scholars. This shows that Malaysia has shown good skills in applying the ideas of Islamic scholars to the practice of society for achieving true happiness.

Table 3. General information of articles and data sources from selected journal.

Author	Year of Publication	Study Location	Journal of Publication	Research Design	Data Source
M. A. Wahab [12]	2022	United States	Journal of Religion and Health	Qualitative	NVivo
M. Ebrahimi, A. Gholami and K. Yusoff [13]	2021	Malaysia	International Journal of Islamic Thought	Qualitative	Peer-reviewed and refereed journals
T. Lomas et al. [14]	2021	New Zealand	International Journal of Wellbeing	Qualitative	Primer and secondary data
A. Nur, S. K. Yuzar, M. F. A. Mohd Sa'ad [15]	2021	Indonesia	Mashdar: Jurnal Studi Al-Quran dan Hadis	Qualitative	Library research
A. R. Alebouyeh [16]	2021	Iran	Theosophia Islamica	Qualitative	Content analysis
N. Aydin and H. Khan [17]	2021	Switzerland	Sustainability	Qualitative	Multidimensional human nature mode
M. N. Omar et al. [18]	2021	Malaysia	International Journal of Islamic Thought	Qualitative	Content analysis
M. Ulfa and E. D. A. Puspita [19]	2020	Indonesia	TASFIYAH Jurnal Pemikiran Islam	Qualitative	Content analysis
A. R. Ateerah and Z. M. Lukman [20]	2019	Malaysia	International Journal of Research and Innovation in Social Science (IJRISS)	Quantitative	PLS-SEM
D. Eidukienė [21]	2017	Lithuania	Filosofija, Sociologija	Qualitative	Primer and secondary data

Table 3. *Cont.*

Author	Year of Publication	Study Location	Journal of Publication	Research Design	Data Source
R. R. Shah [22]	2016	United States	International Journal of Religion and Spirituality in Society	Qualitative	Content analysis
N. Aydin and E. Manusov [1]	2014	Malaysia	Al-Shajarah	Quantitative	Questionnaire
C. M. Neria [23]	2013	United Kingdom	Arabic Sciences and Philosophy	Qualitative	Content analysis
M. Joshanloo [24]	2013	Netherlands	Journal of Happiness Studies	Qualitative	Primer and secondary data
S. Stroumsa [25]	1998	Netherlands	Medieval Encounters	Qualitative	Content analysis

3.2. Thematic Analysis

After analyzing the general information of the articles, an analysis was made based on the findings of the study of each article. This analysis found six (6) articles that made comparisons between Muslim and Western figures on the topic of happiness and nine (9) articles which focused on specific scholars as summarized in Table 4. Next, some elements of happiness were extracted from the scholars' writings which were summarized into three elements, namely, spiritual, intellectual, and material (see Table 5).

Table 4. Figures that highlight the debate on the topic of happiness discussed by selected authors.

Author (Years)	Comparison of Islamic and Western Scholars	The Views of Islamic Scholars	Western View
M. A. Wahab (2022) [12]		al-Ghazali, al-Farabi, Ibn Miskawayh, al-Attas	Aristotle
M. Ebrahimi, A. Gholami and K. Yusoff (2021) [13]		al-Ghazali	
T. Lomas et al. (2021) [14]	14 different eras and spanning a range of global regions		
A. Nur, S. K. Yuzar, M. F. A. Mohd Sa'ad (2021) [15]		HAMKA	
A. R. Alebouyeh (2021) [16]			Jeremy Bentham
N. Aydin and H. Khan (2021) [17]	Christianity, Islam, and Judaism		
M. N. Omar et al. (2021) [18]		Ibn Miskawayh	
M. Ulfa and E. D. A. Puspita (2020) [19]		HAMKA	
A. R. Ateerah and Z. M. Lukman (2019) [20]		al-Ghazali	
D. Eidukienė (2017) [21]			Jeremy Bentham
R. R. Shah (2016) [22]	Thomas Aquinas and al-Ghazali		
N. Aydin and E. Manusov (2014) [1]		Said Nursi	
C. M. Neria (2013) [23]		al-Farabi	
M. Joshanloo (2013) [24]	Islamic and Western	Sufism	Plato and Aristotle
S. Stroumsa (1998) [25]	Avicenna and Maimonides		

Table 5. Elements of happiness in the context of scholarly debate presented by the selected authors.

Author (years)	Element		
	Spiritual	Intellectual	Material
M. A. Wahab (2022) [12]	✔	✔	✔
M. Ebrahimi, A. Gholami and K. Yusoff (2021) [13]	✔		
T. Lomas et al. (2021) [14]	✔	✔	✔
A. Nur, S. K. Yuzar, M. F. A. Mohd Sa'ad (2021) [15]		✔	✔
A. R. Alebouyeh (2021) [16]			✔
N. Aydin and H. Khan (2021) [17]	✔	✔	✔
M. N. Omar et al. (2021) [18]	✔		
M. Ulfa and E. D. A. Puspita (2020) [19]	✔	✔	
A. R. Ateerah and Z. M. Lukman (2019) [20]	✔		
D. Eidukienė (2017) [21]			✔
R. R. Shah (2016) [22]	✔		✔
N. Aydin and E. Manusov (2014) [1]	✔		
C. M. Neria (2013) [23]		✔	
M. Joshanloo (2013) [24]	✔		✔
S. Stroumsa (1998) [25]		✔	✔

4. Discussion

Based on the above analysis, there were two themes that emerged regarding the scholars who discussed the topic of happiness and the main elements related to the topic of happiness.

4.1. Scholars Who Discussed the Reality of Happiness

Referring to the analysis of all the articles, it was found that the discussion of happiness started since classical Greek times about 529 A.D. through two famous figures, namely, Plato and Aristotle. Happiness was one of the main topics that dominated the discourse of ethical philosophy at the time. The question of happiness arose when philosophers tried to find answers about the ultimate and highest goal of human life. Various concepts and meanings about happiness were highlighted to find a solution to the question [26,27]. Plato and Aristotle pioneered the tradition of eudaimonia, which relied on the ethics of goodness to determine happiness and how to achieve it. The main ethical questions such as "how should I live? or "what do I need to be like?" provoked a deep debate resulting in the philosophical stream of happiness of that time [24].

In the study of M. A. Wahab [12], the philosophers' discussion of happiness has opened up a space of debate among Islamic philosophers. Al-Farabi, Ibn Miskawayh, Ibn Sina, and Imam al-Ghazali are among the Islamic scholars who discussed the question of happiness in detail. This philosopher's discussions can be seen in the study of M. Ebrahimi, A. Gholami and K. Yusoff [13], D. Eidukienė [21], R. R. Shah [22], C. M. Neria [23], M. Joshanloo [24], and S. Stroumsa [25] covering almost the entire articles which were analyzed. Based on the findings, happiness is a major topic discussed in all religions and ethical systems. However, the discussion of happiness is not the same for all systems of thought, as the concept varies between the ethical systems of classical Greek philosophy and Islam. In the tradition of classical Greek society and philosophers, the discussion of happiness is very important because it is at the core of every ethical discussion among them. The discussion is based on the question of "what is the ultimate purpose in human life". The happiness discussed by every Greek philosopher is like a riddle that needs to be put together because of the diversity of views regarding happiness. For example, Plato associated the ultimate goal of happiness with justice, Aristotle considered happiness closely related to good behavior, while Epicureanism and Cynicism considered pleasure as the source of happiness. Yet the discussion of happiness by Aristotle can be said to be the culmination and complement of the classical Greek discussions of happiness. This is because Aristotle was a classical Greek philosopher who discussed about happiness completely and systematically in his writings Nicomachean Ethics [12,28].

The debate on happiness presented by the Greek philosopher is seen to influence the Islamic moral philosophical figures. The debate on happiness has also been a topic of discussion traditionally in the discussion of Islamic moral philosophy. However, most of the discussions of Islamic philosophers especially in the field of morality are entirely theological in nature, i.e., linking the discussion of happiness with religion, for example in [28].

In the Western world, the discussion of happiness did not stop in Greek times but continued into the Middle Ages, which can be seen in the studies of S. Stroumsa [25] and D. Eidukienė [21], who, respectively, studied the figures of Maimonides and Thomas Aquinas. Maimonides (ca. 1135–1204) was a Jewish philosopher and astronomer who became one of the most prolific and influential Torah scholars and physicians, while St. Thomas Aquinas (1225–1274 AD) was a philosopher and theologian, who became Doctor of the Church in 1323. His system harmonized Aristotelianism and Catholic theology in his Summa Theologica. These two figures when compared to medieval Islam are represented by Ibn Sina and Imam al-Ghazali. For this reason, both the studies of S. Stroumsa [25] and D. Eidukienė [21] have associated a great deal of relationship and influence between the figures comparing between Ibn Sina and Maimonides and Thomas Aquinas and al-Ghazali.

Among the most famous Western figures in modern times is Jeremy Bentham (1748–1832), who was a British philosopher, jurist, and social reformer. He is considered the founder of modern utilitarianism and a Western hedonistic figure. Contemporary figures who also discussed happiness were Michel de Montaigne (1533–1592), Arthur Schopenhauer (1788–1860), James Stuart Mill (1806–1873), and Sidgwick (1838–1900) [29,30]. The ideas of happiness are seen as an effect of the influence of Greek philosophy which outlined pleasure as the main foundation of life. This ideology which was first introduced in Greece subsequently spread to Europe and eventually spread throughout the world according to Western secular philosophy. This understanding has changed into a culture and way of life for the society in that era [30].

In the modern era of Islam, scholars have also discussed the question of happiness systematically through the production of excellent works such as Said Nursi (1877–1960), HAMKA, and Syed Muhammad Naquib al-Attas. Said Nursi (1877–1960) was one of the key figures for the development of spirituality in modern-era Turkey [1]. HAMKA or Haji Abdul Malik Karim Amrullah was one of the great Muslim scholars of Indonesia from a philosophy and Sufism perspective, at once directly lived in modern times. It is seen that the paradigm of his thoughts on happiness is the design between Sufism and philosophy [19]. Syed Muhammad Naquib al-Attas, on the other hand, is the First Holder of the Honorable Chair of Abu Hamid al-Ghazzali. This shows that the ideas are very dynamic and of high quality as a result of ideas from previous Islamic figures, especially Imam al-Ghazali [27,31].

Based on the figures mentioned above, it shows that happiness is an important aspect in every human's life. Discussions about happiness have existed from time to time. Such scholars have played an important role in finding and determining the direction for the topic of happiness. Their ideas have had a great influence in shaping the thoughts during their time. There is also a relationship and connection between the views expressed by various scholars despite them coming from different times and beliefs.

4.2. *Key Elements in the Reality of Happiness*

Based on Table 5, the second theme analyzed was related to the three main elements that are often debated by scholars on the topic of happiness, namely the spiritual, intellectual, and material elements. These three elements are important elements that underlie the life of every human being. Spirituality, according to Islam, is an inner aspect of human beings that cannot be revealed by the human senses [32]. It is unique because it relates to the divine dimension [27]. To summarize the meaning from various perspectives, Emie

Sylviana [33] defined the spiritual element as the inner part of human beings who feel and want, higher abilities, the inner qualities of a person, existing in the body naturally, and a tool to acquire knowledge for self-perfecting and knowing Allah SWT. While the intellectual element that is rooted in the word intellect refers to reason, thinking, or 'aql in Arabic. In general, an intellectual means someone who has good intellectual power and is free from ignorance. In other words, a person who is educated and has high abilities in one branch or several branches of knowledge [34] or a group of thinkers who speak only about ideas and questions involving nonobjects [35]. In Islam, apart from the spiritual aspect, human beings are created from something of a material form, either from soil or semen which is then interpreted as a body or corpse [36]. This material element leads to matter and wealth that are synonymous with the material realm or the realm of this world.

Thus, these three elements become the main points in the debate of scholars in the direction of true happiness. In general, most people accept happiness as the ultimate goal in human life. However, the assessment and measurement of happiness for each individual are different because happiness is relative to each individual. If traced in the thoughts of Greek scholars, Plato and Aristotle disagreed on the topic of happiness, known as eudaimonia. Starting from the view of the sophist, who placed lust as a key element in achieving happiness, Plato and Aristotle discussed in detail the concept of happiness. The essence of Plato's commentary on this discussion focused on human happiness from the point of view of the soul through the mastery of four virtues, namely al-hikmah, al-syaja'ah, al-'iffah, and al-'adl. In Aristotle's view, the ultimate goal in human life is to achieve happiness in which he made deliciousness as one of the most important conditions for achieving happiness [37]. Aristotle also emphasized three things that should be prioritized by someone who wants to be happy. First, perform one's function well, which means that people can achieve the goal by pursuing bodily pleasures and avoiding suffering (hedonistic/materialistic). Second, practice moderation in life by choosing good deeds moderately and collectively such as engaging in politics. Third, maximize the use of reason in a philosophical way. These three things need to be combined to achieve deliciousness and happiness in life [38]. From the viewpoint of Aristotle, the deliciousness of the five senses is a compatible feeling and the deliciousness of the mind is the perfection that reaches its essence that evokes feelings. If the perfection of its substance, whether its reality or its glory, is revealed, it can evoke deliciousness and even produce a sense of security and prosperity. The emphasis on the psychic or spiritual element can be seen in Plato's view as a high moral formation for obtaining happiness [24]. Meanwhile, Aristotle saw that the intellectual and material elements needed to be balanced to obtain happiness [17,38].

These views influenced the thinking of subsequent Western scholars. Maimonides revealed that the last thing to be achieved in life was intellectual, based on happiness [25]. However, Thomas Aquinas states that happiness can be achieved by cultivating some intellectual and moral values, which enable one to understand the nature of happiness and motivate one to seek happiness in a reliable and consistent manner. Yet, one will never be able to find the greatest happiness in this life because ultimate happiness consists of a supernatural union with God. Thus, human happiness does not consist of property, status, pleasure, or any created goodness at all [17]. Most goods do not have the necessary relation to happiness because the main object of human will can only be found in God, who is the source of all goodness. This view of Thomas Aquinas saw the existence of a spiritual relationship with divinity in the context of religion [22].

The emphasis on the material element can be seen in the thinking of modern scholars, especially Jeremy Bentham. As a leader of utilitarianism, he argued that the most moral action was the one which resulted in the highest amount of utility, where utility was defined as the achievement of pleasure after deducting the suffering of all involved in any action [16]. Thus, happiness is the experience of pleasure and the absence of pain. Acts that do not promote the greatest happiness are morally wrong, such as being ascetic. Bentham also raised the principle of hedonism which links human nature with the factors of pain and joy. Only this factor will affect human beings in an action that is to show what must be done in addition to determining an action that should be done. Jeremy Bentham also stated that the only moral fact is that humans seek pleasure and avoid pain. His principle underlines that due to the factors of pleasure and pain, humans prescribe an action that should be done. The concept of goodness and happiness brought about by hedonism is based solely on worldly pleasures as well as its focus on materialism alone [38].

In the view of Islamic scholars, as a figure who mastered the thought of Plato and Aristotle, al-Farabi [39] examined happiness from an intellectual point of view by stating that happiness was the ultimate achievement of human beings who had reached the highest level of intellect (al-'aql al -mustafad). The path to achieve this happiness must go through four priorities, namely the priority of knowledge, the priority of thought, the priority of morality, and the priority of practice. Moral precedence becomes a condition to the precedence of thought especially cultural thought [40]. In this regard, al-Farabi [39] opined that intellectuality played a major role in ensuring someone achieved happiness.

Ibn Miskawayh [41] also acknowledged that the topic of happiness was the highest goal of every human being, and each had different forms of happiness based on how to obtain it. The most important happiness or absolute goodness is that one can attain the rank of an angel and then know Allah SWT [42]. It can be achieved by using the mind fully through contemplation of all the existence of beings on earth. Thus, one's soul is also filled with truth and clean from all impurities [12,18]. By also combining the ideology of Plato and Aristotle, human happiness according to Ibn Miskawayh [43] includes two elements, namely body and soul. Bodily happiness is not permanent and contains pain and remorse. It also prevents the soul from reaching Allah SWT. In contrast to the happiness of the soul, which is the most perfect happiness, it is obtained by prioritizing the rational soul (al-nafs al-natiqah) over al-nafs al-ghadabiyyah and al-nafs al-bahimiyyah [44]. These two types of soul, if not controlled can cause suffering later because the happiness obtained from them is only temporary. Thus, to ensure that one can balance the three spiritual elements, one should practice the values of goodness, such as al-ḥikmah (wisdom), al-'iffah (simplicity), al-shajā'ah (courage), and al-'adālah (justice) [12].

Similar to Aristotle, Ibn Sina [45] also discussed the aspect of deliciousness in achieving happiness. It is the ultimate goal of every human being. The delicacies or pleasures that people often crave for are divided into three types, namely the delicacies of lust (al-lazzah al-hissiyyah) such as food, luxury, and sexuality, spiritual delicacies (al-lazzah al-batinah) such as self-dignity and glory, and intellectual delicacies (al-lazzah al-'aqliyyah). Of the three forms of delicacy, Ibn Sina gave more attention to the intellectual delicacy to which the nature of the body is subject. He also focused on individual happiness as opposed to political happiness [46]. This contrasts with Aristotle who was inclined to materialism and emphasized the influence of the body on the intellect [40].

Ibn Sina [45] also divided human beings into several categories according to the way human beings acquired happiness. First, the group of al-'arifun, which consists of philosophers. This group always fills their time with contemplation (tafakkur) on God while setting aside all the physical elements that are external and material. When a person is free from all that, then the soul becomes clean and can reach the pure realm (alam al-quds), which subsequently brings the person to the situation of high knowledge so that it can experience eternal happiness. Second, al-nafs al-salimah, known as the Sufi group, is pure by nature and not contaminated by material elements. When the Sufi group is given spiritual remembrance that signals things in the divine realm, then they can enjoy a deep preoccupation without feeling confused and terrible [24]. This would not be the case if there was no chastity. The Sufi group also manages to enjoy the delights of life just as the al-'arifun group above. Third, the normal public who can taste happiness according to their own suitability. What is important is that this group must strive to purify the soul (tazkiyah al-nafs) by performing various acts of worship, whether obligatory or sunnah [46].

Meanwhile, the topic of happiness according to al-Ghazali [47,48] refers to the highest and essential happiness as a result of humans' ability to acquire the moral virtues and purity of their soul. The real happiness is to be able to remember Allah SWT and then know

Allah SWT by having the virtues of the soul and character. Priority and purity of soul can distinguish between the path of happiness and misery. While the goodness of morals or character is obtained by continuous deeds of worship until the disappearance of vile and ugly habits. This is the main way for human beings to obtain blessings and enjoyment in this world and the hereafter.

This means that Imam al-Ghazali emphasized the happiness of the hereafter compared to the happiness of the world. This is in line with the main purpose of why the Apostles were sent among us which was to guide people to achieve true happiness. Glory in the judgment of Allah SWT also lies in the pursuit of happiness in the hereafter and whoever fails to obtain it is more despicable than animals. Such group of people will even face misery in the hereafter. Accordingly, the characteristics of the happiness of the hereafter according to al-Ghazali [49] are eternal happiness without any ruin, deliciousness without hardship, joy without hardship, wealth without poverty, perfection without flaw and glory without disgrace. This defines the topic of happiness by obtaining the blessings of heaven. With this, it is understandable that the topic of happiness emphasized by Imam al-Ghazali is the unification of knowledge and deeds, spiritual and physical [27]. All these aspects have their respective roles in determining the direction of human life in this world and the hereafter.

This view of Imam al-Ghazali is also in line with al-Attas who emphasized the balance between the demands of the world and the hereafter. He explained that the material element had an important contribution in the intellectual and spiritual development of a human being. This is due to the body or wealth and the environment, in which humans are able to obtain certain information and data about the world and its experience [50]. Health and wealth are happiness in itself for human beings, but it must be remembered that in Muslim society, happiness must remain based on Islamic values. Al-Attas [51] explained that to get the happiness of the hereafter, the world should be used as a tool. Human being is not only considered as a subject, but also as an object of science. In the problem of happiness, human beings have the potential to be objects as well as subjects who feel and study about happiness. Every activity must be based on the Islamic worldview, so that the happiness obtained is the real happiness. HAMKA's view of happiness also emphasizes the balance of life in this world and the hereafter, but the main purpose of human life is to achieve the pleasure of Allah SWT. Humans will not achieve happiness if they focus on material pleasures only [15,19]. N. Aydin, and E. Manusov [1] also presented Said Nursi's views which are in line with the above scholars. Elements of spiritual growth, positive emotions, psychology, behavior, desire for self-improvement and humanity, living devotion to God, compassionate love, service to others, and high spiritual achievement become essential elements in the pursuit of happiness.

It can be seen in the above discussion that all Islamic scholars agreed on true happiness, which is the happiness of the hereafter when one's soul resembles an angel and then tastes the blessings of heaven and sees Allah SWT. However, from the point of view of the way human beings obtain happiness, it can be seen in two aspects. For philosophers such as al-Farabi and Ibn Sina, true happiness is obtained through the path of reason. However, they also accepted the practices of spiritual purification that are done practically and continuously [12]. However, they emphasized the intellect by understanding the theoretical sciences first, then it can be practiced through action [52]. This means that philosophers prioritized theory or reason based on analysis and continuous learning in achieving happiness, whereas the practices of worship that are practiced are derived from the study of the mind [53].

Although Greek philosophy is seen to be influential in the views of Islamic philosophical figures, what is discussed is in line with the Qur'an and Sunnah [54]. S. H. Nasr [55] also asserted that the Qur'an and Sunnah are the main references in Islamic philosophy. According to I. Madkur [53], in general, the tendency toward Sufism is an important aspect that distinguishes between Islamic philosophy and other philosophies. Accordingly, the thoughts of Islamic philosophers in the field of ethics cannot be separated from the influence of the teachings of Sufism. Meanwhile, mystics emphasize the ability of the human soul beyond the intellect. Ibn Miskawayh and Imam al-Ghazali are among the figures of Sufism who strongly emphasized the ability of the soul to achieve happiness. However, they also discussed the relationship of the mind with the human soul in which the mind plays an important role in understanding the states of one's soul. In the meantime, the discussion of Islamic philosophers especially in the field of morality is entirely in the form of theology [56]. Surely, the Sufis put the Qur'an and Sunnah as the main tenets in their life. Each issue discussed is accompanied by strong arguments from the text of the Qur'an and Hadith. For this reason, the understanding of the fact of happiness in particular is in line with Islamic law in contrast to the Western view which relies on intellect and lust alone [24]. Thus, the spiritual, intellectual, and material elements in Islam need to be balanced for obtaining true happiness. The discussion in the Western view has no end and is seen as misguided because there is no alignment in the relevant elements.

5. Conclusions

The concept of happiness discussed by Western and Islamic scholars is closely related to the philosophy held by them. The main goal they want to achieve is happiness in the life they live. The emphasis on the three basic elements in happiness is focused on the spiritual, intellectual, and material. However, the context of happiness according to Islamic scholars is more towards the goal of real life that is to be achieved according to Islam and it is not only happiness in this world, but also in the hereafter. The foundations for achieving such happiness are based on Islamic epistemology from the source of arguments and propositions of revelation, as opposed to the views of Western scholars who rely on reason and lust alone. The emphasis on ethics and morals becomes a balanced consideration especially in understanding and achieving happiness according to Islam. This is because Islamic figures such as Ibn Sina, al-Ghazali, Ibn Miskawayh, HAMKA, and al-Attas still place the value of moral purity and the pleasure of God as the culmination of happiness. Material is only a tool and a link to achieve that happiness.

Although the study of happiness from Western scholars has influenced Islamic scholars and researchers, it has not become a major policy that set aside the true spiritual aspects of Islam. In the context of contemporary happiness studies, only eight studies have been conducted effectively for explaining happiness according to Islam as well as the West. However, with only two quantitative studies, this shows that studies in the aspect of happiness can be studied more comprehensively, especially when applying it according to the current situation and reality. This is because all religions basically discuss the aspect of happiness specifically to be achieved by each of its followers. Islam basically makes happiness as one of the measuring lines of human evaluation in achieving God's pleasure as well as gaining worldly and hereafter success.

Author Contributions: Conceptualization, N.J. and Z.M.; methodology, A.M.M. (Ahmad Murshidi Mustapha); software, N.J.; validation, A.M.M. (Abdul Manam Mohamad), N.M.H., and N.J.; formal analysis, A.M.M. (Ahmad Murshidi Mustapha); investigation, Z.M.; resources, N.J.; data curation, A.M.M. (Ahmad Murshidi Mustapha); writing—original draft preparation, N.J.; writing—review and editing, N.J.; visualization, Z.M.; supervision, A.M.M. (Abdul Manam Mohamad) and N.M.H.; project administration, A.M.M. (Abdul Manam Mohamad); funding acquisition, N.J. and Z.M. All authors have read and agreed to the published version of the manuscript.

Funding: This research was funded by Research Management Center (RMC), grant number 600-RMC/GPM LPHD 5/3 (173/2021).

Institutional Review Board Statement: Not applicable.

Informed Consent Statement: Not applicable.

Data Availability Statement: Not applicable.

Acknowledgments: A word of appreciation is extended to the Research Management Center (RMC) for funding the research grant entitled "Shah Wali Allah al-Dihlawi Happiness Index: An Analysis of Academic Happiness Rates in the COVID-19 Pandemic Era Sustainable Work Culture" 600-RMC/GPM LPHD 5/3 (173/2021).

Conflicts of Interest: The authors declare no conflict of interest.

References

1. Aydin, N.; Manusov, E. Materialism, hedonism, spirituality, and subjective well-being: An empirical study of risale-I nur (RN) readers. *Al-Shajarah* **2014**, *19*, 251–284. [CrossRef]
2. Berita Harian. Ukuran Indeks Kebahagiaan Tanda Aras Kualiti PBT. Available online: https://www.bharian.com.my/berita/nasional/2022/04/941847/ukuran-indeks-kebahagiaan-tanda-aras-kualiti-pbt-pakar (accessed on 17 July 2022).
3. Malaysiadateline. Malaysia Jatuh ke Tangga 80 dalam Laporan Kebahagiaan Sedunia 2019. Available online: https://malaysiadateline.com/malaysia-jatuh-ke-tangga-80-dalam-laporan-kebahagiaan-sedunia-2019 (accessed on 17 July 2022).
4. Pauzi, N.A.A.; Koris, R. Tekanan Kerja Dan Kos Kesihatan Mempengaruhi Kebahagaian Staf Universiti Malaysia Terengganu. *J. Undergrad. Res.* **2020**, *2*, 101–114.
5. Wulandari, S.; Widyastuti, A. Faktor-faktor kebahagiaan di tempat kerja. *J. Psikol.* **2014**, *10*, 41–52.
6. Kementerian Kesihatan Malaysia. Kesihatan Mental dan Sokongan Psikososial Ketika Pandemic COVID-19. Available online: https://covid-19.moh.gov.my/semasa-kkm/2021/06/mhpss-kesihatan-mental-dan-sokongan-psikososial-ketika-pandemik-covid-19 (accessed on 17 July 2022).
7. Abdul Aziz, A.R. Konsep Dan Perspektif Tauhidik Dalam Meningkatkan Kebahagiaan Kendiri. *Int. J. Humanit. Technol. Civiliz. (IJHTC)* **2021**, *10*, 64–79.
8. Moher, D.; Liberati, A.; Tetzlaff, A.; Altman, D.G. Group preferred reporting items for systematic reviews and meta-analyses: The PRISMA statement. *BMJ* **2009**, *6*, 1–7.
9. Jamaludin, N.I.; Shahimi, S.; Musafar Hameed, L.B. Kepercayaan Sebagai Teras Sumbangan Melalui Pendanaan Awam Bertujuan Kebajikan. *Asian J. Account. Gov.* **2020**, *13*, 55–65.
10. Shafrill, H.A.M.; Krauss, S.E.; Samsuddin, S.F. A systematic review on Asian's farmers' adaptation practices towards climate change. *Sci. Total Environ.* **2018**, *644*, 683–695. [CrossRef]
11. Xiao, Y.; Watson, M. guidance on conducting a systematic literature review. *J. Plan. Educ. Res.* **2017**, *39*, 93112. [CrossRef]
12. Wahab, M.A. Islamic spiritual and emotional intelligence and its relationship to eternal happiness: A conceptual paper. *J. Relig. Health* **2022**, *4*, 1–24.
13. Ebrahimi, M.; Gholami, A.; Yusoff, K. Al-Ghazali's Ma'rifah and Mahabbah's relations. *Int. J. Islamic Thought* **2021**, *20*, 65–76. [CrossRef]
14. Lomas, T.; Case, B.W.; Cratty, F.; Vander Weele, T.J. A global history of happiness. *Int. J. Wellbeing* **2021**, *11*, 68–87. [CrossRef]
15. Nur, A.; Yuzar, S.K.; Mohd Sa'ad, M.F.A. The understanding of Al-Adabiy Al-Ijtima'iy (A Study of the Verses of Happiness in The Book of Tafsir Al-Azhar Buya Hamka). *Mashdar J. Studi Al-Quran Dan Hadis* **2021**, *3*, 97–124.
16. Alebouyeh, A.R. Compatibility or Incompatibility of Bentham's Utilitarianism with the Quran's Doctrine. *J. Theosoph. Islamica* **2021**, *1*, 48–74. [CrossRef]
17. Aydin, N.; Khan, H. G-Donic happiness: An alternative to Hedonic and Eudemonic happiness for sustainable consumption. *Sustainability* **2021**, *13*, 6096. [CrossRef]
18. Omar, M.N.; Long, A.S.; Mahmood, A.R.; Rahman, Z.A. Islamic notion of happiness (Al-Sa'ada): An analysis of Miskawayh's Thought. *Int. J. Islamic Thought* **2021**, *19*, 49–57. [CrossRef]
19. Ulfa, M.; Puspita, E.D.A. Pursuing happiness in modern era: Study on Hamka's Perspective. *TASFIYAH J. Pemikir. Islam* **2020**, *4*, 1–26. [CrossRef]
20. Ateerah, A.R.; Lukman, Z.M. Happiness instrument development. *Int. J. Res. Innov. Soc. Sci. (IJRISS)* **2019**, *3*, 288–292.
21. Eidukienė, D. Jeremy Bentham: The ideal of the greatest happiness for the greatest number of people as a modus vivendi. *Filos. Sociol.* **2017**, *28*, 29–37.
22. RShah, R.R. nSaint Thomas Aquinas and Imam al-Ghazālī on the attainment of happiness. *Int. J. Relig. Spiritual. Soc.* **2016**, *6*, 15–29.
23. Neria, C.M. Al-fārābī's lost commentary on the ethics: New textual evidence. *Arab. Sci. Philos.* **2013**, *23*, 69–99. [CrossRef]
24. Joshanloo, M. A comparison of western and Islamic conceptions of happiness. *J. Happiness Stud.* **2013**, *14*, 1857–1874. [CrossRef]
25. Stroumsa, S. True felicity': Paradise in the thought of avicenna and Maimonides. *Mediev. Encount.* **1998**, *4*, 51–77. [CrossRef]
26. Mustain, M. Etika dan ajaran moral filsafat Islam: Pemikiran para filosof muslim tentang kebahagiaan. *Ulumuna* **2013**, *17*, 191–212. [CrossRef]
27. Ayob, M.A.S. Pemikiran kebahagiaan dalam tamadun Yunani klasik 470 S.M-529 M.: Satu analisis ringkas. *Jurnal Peradaban* **2019**, *12*, 1–25. [CrossRef]
28. Ayob, M.A.S.; Zaini, M.N.M.; Mohd Sairi, F.; Md Soh, N.S. Pemikiran kebahagiaan menurut Aristotle dan Ibnu Miskawayh: Satu analisis awalan. *MALIM J. Pengaj. Umum Asia Tenggara* **2020**, *21*, 55–67.

29. Wikipedia. Philosophy of Happiness. Available online: https://en.wikipedia.org/wiki/Philosophy_of_happiness (accessed on 2 June 2022).
30. Hamzah, H.; Sabjan, M.A.; Mat Akhir, N.S. Konsep budaya Hedonisme dan latar belakangnya dari perspektif ahli falsafah Yunani dan barat moden. *J. Al-Tamaddun* **2016**, *11*, 49–58. [CrossRef]
31. Wan Daud, W.M.N. *Falsafah dan Amalan Pendidikan Islam Syed M. Naquib al-Attas: Satu Huraian Konsep Asli Islamisasi*; Universiti Malaya: Kuala Lumpur, Malaysia, 2005.
32. Abdul Aziz, A.; Mamat, M.N.; Salleh, D.M.; Syed Abdullah, S.F.; Nordin, M.N. Analysis of literature review on spiritual concepts according to the perspectives of the Al-Quran, hadith and Islamic scholars. *Turk. J. Comput. Math. Educ.* **2021**, *12*, 3152–3159.
33. Mohd Zahid, E.S. Pembangunan spiritual: Konsep dan pendekatan dari perspektif Islam. *E-J. Islamic Thought Underst.* **2019**, *2*, 64–87.
34. Ahmad, A. Aplikasi konsep ulama' dan intelektual: Suatu penilaian kritis. *J. Usuluddin* **2000**, *11*, 13–32.
35. Al-Attas, S.M.N. *Faham Agama dan Asas Akhlak*; Angkatan Belia Islam Malaysia: Kuala Lumpur, Malaysia, 1977.
36. Stapa, Z. Manusia pembina tamadun: Perspektif pemikiran Islam. *J. Hadhari* **2009**, *1*, 31–44.
37. Saliba, J. *Al-Mu'jam Falsafi bi Alfaz al-'Arabiyyah wa al-Faransiyyah wa Inkliziyyah wa al-Latiniyyah*; Dar al-Kitab al-Lubnani: Beirut, Lebanon, 1982.
38. Ayob, M.A.S. Konsep Kebahagiaan Aristotle Menurut Perspektif Falsafah Akhlak Islam. Doctoral Thesis, Jabatan Akidah dan Pemikiran Islam, Akademi Pengajian Islam, Universiti Malaya, Kuala Lumpur, Malaysia, 2016.
39. al-Farabi, A.N.M. *Ara' Ahl al-Madinah al-Fadilah*; Matba'ah al-Sa'adah: Misr, Egypt, 1906.
40. Nasution, H. *Filsafat Islam*; Penerbit Gaya Media Pratama: Jakarta, Indonesia, 1999.
41. Ibn Miskawayh, A.A.A. *Kitab al-Fawz al-Asghar*; Dar Maktabah al-Hayah: Beirut, Lebanon, 1970.
42. Fakhry, A. *A History of Muslim Philosophy*; Trans. Dewan Bahasa dan Pustaka: Kuala Lumpur, Malaysia, 1993.
43. Ibn Miskawayh, A.A.A. *Tahdhib al-Akhlaq*; Mansyurat al-Jamal: Beirut, Lebanon, 2012.
44. Musa, M.Y. *Falsafah al-Akhlaq fi al-Islam*; Mua'assasah al-Khariji: Kaherah, Egypt, 1963.
45. Ibn Sina, A.A.H. *al-Isharat wa al-Tanbihat*; Dar al-Ma'arif: Kaherah, Egypt, 1968.
46. Zakaria, I. Ketuhanan Kenabian Dan Kebahagiaan Menurut Ibn Sina. *Islamiyyat* **2010**, *32*, 135–156.
47. Al-Ghazali, A.H.M. *Ihya' 'Ulum al-Din*; Dar al-Minhaj li al-Nashr wa al-Tawzi': Jeddah, Saudi Arabia, 2011.
48. Al-Ghazali, A.H.M. *Ma'arij al-Quds fi Madarij Ma'rifah al-Nafs*; Dar al-Kutub al-Ilmiyyah: Beirut, Lebanon, 1988.
49. Al-Ghazali, A.H.M. *Mizan al-'Amal*; Dar al-Ma'arif: Kaherah, Egypt, 1964.
50. Wan Daud, W.M.N. *The Educational Philosophy and Practice of Syed Muhammad Naquib Al-Attas*; ISTAC: Kuala Lumpur, Malaysia, 1998.
51. Al-Attas, S.M.N. Islam religion and morality. In *Prolegomena to The Metaphysics of Islam: An Exposition of the Fundamental Elements of the Worldview of Islam*; ISTAC: Kuala Lumpur, Malaysia, 1995.
52. Leaman, O. *An Introduction to Medieval Islamic Philisophy*; Abdullah, T.M.A., Ed.; Rajawali: Jakarta, Indonesia, 1989.
53. Madkur, I. *Fi al-Falsafah al-Islamiyyah*; Rajawali Press: Jakarta, Indonesia, 1988.
54. Bakar, O. *Al-Farabi: Kehidupan, Hasil Karyanya dan Peri Penting Tokoh*; Dewan Bahasa dan Pustaka: Kuala Lumpur, Malaysia, 1991.
55. Nasr, S.H. The Qur'an and Hadith as source and inspiration of Islamic Philosophy. In *History of Islamic Philosophy Part 1*; Nasr, S.H., Leaman, O., Eds.; Routledge Taylor & Francis Group: London, UK, 1996.
56. Ansari, M.A.H. Miskawayh's conception of Saadah. *Islamic Stud.* **1963**, *2*, 317–335.

proceedings

MDPI

Proceeding Paper

Exploring an Alternative Material of Holy Book Stand Holder through Malaysian Muslim Perspectives on the Design of Rehal †

Xiao Hao Gu and Hwee Ling Siek *

De Institute of Creative Arts and Design, UCSI University, Kuala Lumpur 56000, Malaysia
* Correspondence: perlinesiek@ucsiuniversity.edu.my

† Presented at the International Academic Symposium of Social Science 2022, Kota Bharu, Malaysia, 3 July 2022.

Abstract: The existing holy book stand holder (rehal) made from wood or plastic is a product commonly used by Muslims. Its purpose is to hold and preserve the sanctity of the holy book during recitation. For this study, the researcher formed an understanding of the meaning and cultural value of the rehal from a Malaysian Muslim's perspective. The materials used to make rehals, their optimum height, as well as acceptable colours, designs, and ornamentations were all analysed. The data collected serves as a guideline by integrating craftsmanship, eco-friendly materials, and relevant design ornaments to create new ceramic rehals that can provide a comfortable user experience.

Keywords: rehal; ceramic; Malaysian Muslims; design

Citation: Gu, X.H.; Siek, H.L. Exploring an Alternative Material of Holy Book Stand Holder through Malaysian Muslim Perspectives on the Design of Rehal. *Proceedings* **2022**, *82*, 79. https://doi.org/10.3390/proceedings2022082079

Academic Editor: Mohamad Rahimi Mohamad Rosman

Published: 23 September 2022

1. Introduction

The rehal is a holy book holder. It is essentially a small stand used by Muslims, Hindus, Sikhs, and Christians to hold and ensure respect for religious texts during recitation. In Malaysia, rehals are normally used by the Muslim majority, who place their Quran on top while they recite the surahs. Traditionally, rehals are made from wood, and assembled by interlocking two similarly sized pieces of wood. This means the rehal can be easily dismantled for storage when not in use [1]. The surface of the rehal is usually decorated with Islamic calligraphy, abstract flowers, or Arabic motifs [2]. Rehals are used by Muslim people worldwide, regardless of race, age, status, or social class [3].

Most of the rehals found on the market nowadays are made out of wood or plastic. However, these materials present different sustainability issues. Under moist conditions, wooden rehals provide a breeding ground for pathogenic microorganisms, which pose a threat to human health. Meanwhile, the plastics used for rehals are derived from non-renewable fossil fuels, which will eventually be depleted and are therefore not sustainable. This study explores the Malaysian Muslim perspective on prospective new rehal designs that use eco-friendly sustainable alternative materials because the Quran says that human beings should have self-discipline and respect all life with kindness, and protect and maintain the balance of the natural ecosystem [4]. Ahmad et al. suggested adjusting the height of commercial rehals, which ranges from 20 to 25 cm [5], to achieve optimum reading height [6] because ease of reading allows users to focus on their studies. Thus, a good rehal design is crucial. Since traditional rehal ornamentation and decorations have remained largely the same throughout the years, this study also investigates the possibility of creating rehals with more modern decorations, patterns, or ornamentation.

The objectives of this study are to (1) create an environmentally friendly rehal by using sustainable material, (2) optimize the height of rehal based on human ergonomics, and (3) design modern rehal embellishments that are acceptable in the eyes of Islam. The data collected from this study will serve as a design guideline that enables the creation of a well-received Malaysian rehal.

The research scope includes the standard rehal as the research object and aims to design an environmentally friendly and sanitary rehal by improving the material, height, and decorations based on the theory and data collected from the pilot test. From the literature review, the researcher learned that modern rehals may come with extra features, such as a box or modifications that allow it to be transformed into a bag [1]. To streamline the initial research, only traditional designs were examined in this study. However, new rehal ideas should be explored via innovation and design in the future.

2. Literature Review

Learning the Quran is an important act in Islam. The Quran is a sacred text that requires a proper holder, which in this case is called rehal. Rehal's craftsmanship is mainly manifested in shape, form, and decoration, which indirectly reflects the artistic identity of the culture. Rehal is one of the treasures of Islamic art; its historical origin and art culture have spread to China, Indonesia, Turkey, India, Japan, Nepal, Africa, and all over the world [3].

2.1. The Types of Rehals

The rehal is an X-shaped book stand that can be collapsed into a flat form (Figure 1). It is used to hold sacred texts during recitation by keeping them elevated off the floor. There are various types of rehals in the market today, which are mainly made of wood, plastic, and metal.

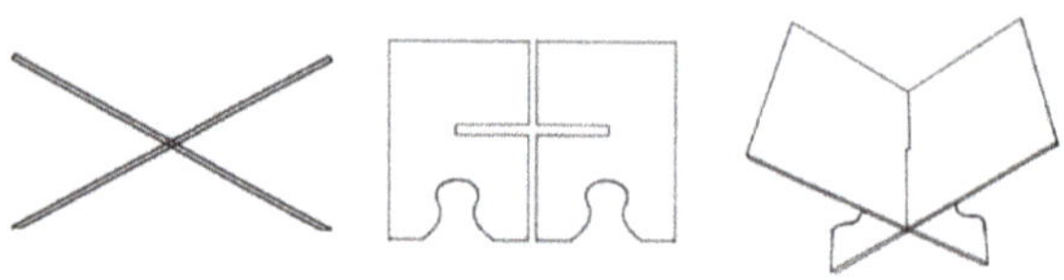

Figure 1. Visual sketch of the rehal, illustrated by researcher.

There are two rehals on display in the Islamic Arts Museum of Malaysia in Kuala Lumpur. The one from India has inlaid brass and exquisitely crafted floral motifs on its surface (Figure 2a). The second, which is from Egypt, is decorated with carvings, inlaid with bones, pearls, shells, and aggregates (Figure 2b). The rehal in (Figure 2c), is a Mongolian Empire Quran shelf displayed in the National Palace Museum, Taipei, Taiwan [7]. This rehal is made of jade and covered with flower and leaf patterns. The edge is decorated with a column of crown lace, which is full of Islamic style and imbued with cultural value.

(a) (b) (c)

Figure 2. Rehal from (**a**) India, (**b**) Egypt, and (**c**) Mongolia, respectively (Photos source: (**a**,**b**), taken by the researcher in Islamic Arts Museum Malaysia, KL).

2.2. Dimensions and Height of Rehal

According to the architect's handbook of ergonomic guidelines, the average adult male sitting in a cross-legged position will place their hands at about 300 mm off the ground [8]. However, commercial rehals have a height that ranges from 200 to 250 mm (Figure 3). Hence, the height of the existing rehal can be further optimised (Ahmad et al., 2021).

Figure 3. The difference between the height of a standard rehal and the height where the hands of an average adult male would rest.

2.3. Islamic Wooden Rehal

Rehals are a sacred tool that Muslims use during Quran recitations. It preserves the sanctity of the Quran by keeping it off the floor while placing it within an accessible distance and angle. However, the usage of wooden rehal raises a few issues. First of all, wooden rehals require the cutting down of trees, which is not ideal. Secondly, wooden rehals are prone to discolouration with the passing of time. They may fade or darken, losing their original lustre. In addition, improperly prepared wood may crack due to moisture, leading to the growth of microorganisms that can cause health issues. The combination of these issues require a wooden rehal to be replaced eventually. As the Quran has stated: *"I created all things by measure"* (54:49), wasting resources is strictly forbidden. The Sharia stipulates that even if one does their cleaning by the river, they should not use an excessive amount of water. Similarly, Islam advocates for environmental protection and the optimisation of the living environment. The Prophet Muhammad said: *"A Muslim planted a tree and let people and livestock enjoy it, that is his giving, a good deed."* Planting a tree is not only a social virtue but an act of closeness to the Lord [9].

2.4. Material Requirements for Product Design

Proper research should be conducted before the creation of a new rehal. The requirements for the new product include: resources should not be wasted, and the material should be eco-friendly [10]. Since the product's function, size, shape, manufacturing process, and aesthetics all affect consumer acceptance and decisions, only a product that tackles all these aspects deserves mass production.

3. Research Methods

3.1. Content Analysis

Two research methods were used in this study: Content analysis and in-depth interviews. Content analysis methods are aimed at providing knowledge and understanding of the phenomenon via studies and textual data [11–13]. The researcher applied a directed approach to content analysis commencing coding with predetermined codes analysed from literature reviews, and analysing in-depth interviews with Malaysian Muslims, thus obtaining a consensus regarding rehal requirements in terms of material, height, colour, design, and ornamentation.

3.2. In-Depth Interview

An in-depth interview is a research method that conducts intensive interviews with a small group of interviewees to explore and obtain their views and opinions on a particular topic. It provides detailed information and personal thoughts on the topic and issues discussed [14]. The interview methods can be grouped into three types: unstructured interviews, semi-structured interviews, and structured interviews.

3.3. Sample Size

This study explored the Malaysian Muslim's perspective on present rehal materials [15], optimum rehal height for comfortable reading while in a cross-legged position, and saisfaction level regarding rehal aesthetics. Purposive sampling was used to select seven adult Malaysian Muslims respondents. The seven respondents include three students —one postgraduate student and two undergraduate students, two businessmen, one Arabic language teacher, and one administrative officer. Before conducting the interviews, a check was performed to ensure that each respondent either owned, or had close proximity to a family member who owned at least one rehal to qualify as a sample. These interviews, each 20-30 minutes long, were conducted in Kuala Lumpur, on a weekday in April 2022 for the first four respondents, and on a weekend in early May 2022 for the remaining three respondents (see Table 1). The communications were audio recorded with no interference from the interviewees' thoughts during the interview process.

The interview questions are as below:

- The material of their present rehal and whether they are able to accept other eco-material.
- The extension of the height of rehal.
- The colour, design, and ornaments of their present Rehal and are they able to accept new design.

Table 1. Compilation of interviewees' information.

Interviewee	Gender	Occupation	Do You Have a Rehal	No. of Rehal at Home	Users Of Rehal
R1	M	PhD Student	Family members using	1	Mother, Wife
R2	F	Bachelor's Degree Student	Family members using	2	Mother, Grandma
R3	F	Bachelor's Degree Student	Family members using	2	Mother, Sister
R4	F	Arabic Language Teacher	Yes	1	Mother, Self
R5	F	Administrative Officer	Yes	5	Father, Mother, Brother
R6	M	Businessman	Family members using	2	Mother, Nieces, Self
R7	M	Businessman	Yes	3	Self, Wife

3.4. Analysis of Interview Content

ATLAS.ti is a qualitative research tool for deductive coding, analysing transcripts, and creating network diagrams [16]. It is rigorous in the analysis of the data and can produce credible results [17]. Primary data was collected via in-depth interviews and three network diagrams were created by sorting out the vocabulary of the seven respondents.

Q1: What is their current rehal made of and can they accept rehals made from alternative eco-friendly materials?

In terms of material, all the interviewees responded that they had either owned, or seen wood rehals. Although R1, R4, and R5 have come across plastic rehals (57%), their family members prefer wood rehals (100%). In terms of rehal care, only R6 uses a dry brush to clean and preserve their rehal. The rest of the respondents use a wet cloth to wipe it (86%). When asked about problems encountered during rehal care, R1 and R6 mentioned deformation (29%), R2 talked about discoloration, R3 and R4 shared about how the wood was splitting apart (29%), and R7 said that their rehal became mouldy (29%) (Figure 4, Table 2)

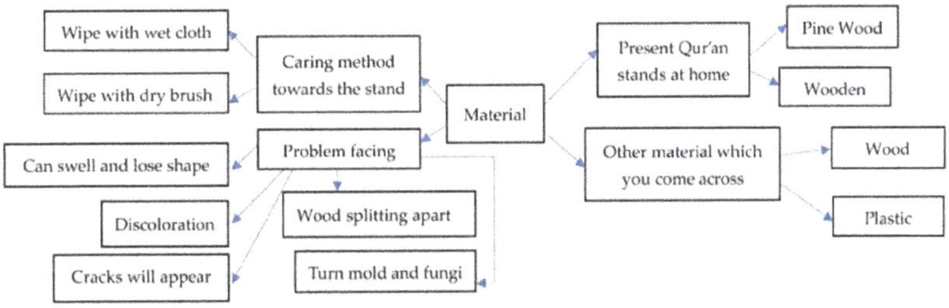

Figure 4. Atlas-ti network of rehal materials.

Table 2. Analysis of interview content of Question 1.

Respondents	Rehal Materials		Method of Caring		Problems Faced			
	Plastic	Wood	Wet Cloth	Dry Brush	Deformation	Discolouration	Wood Splitting	Mouldiness
R1	✓	✓	✓		✓			
R2	✓	✓	✓			✓		
R3		✓	✓				✓	
R4	✓	✓	✓				✓	
R5	✓	✓	✓					✓
R6		✓		✓	✓			
R7		✓	✓					✓
Total	4/7	7/7	6/7	1/7	2/7	1/7	2/7	2/7
Percentage	57%	100%	86%	-	29%	-	29%	29%

Q2: What is the optimum height of the rehal?

All the respondents replied saying that rehals should be well kept and cared for properly. R1, R2, R4, and R6 mentioned that rehals should be placed in high places (57%). In particular, R3 mentioned that rehals should be wrapped with clean material after usage. Meanwhile, R5 and R6 mentioned that the book holders should be placed on a bookshelf or inside a book cabinet. Regarding the height of rehal, all the respondents have rehals with different sizes and heights; R1 and R4 use 21 cm × 29 cm rehals, R2 uses a 18 cm × 26 cm rehal, R3 uses a 10 cm × 15 cm rehal, R5 uses a 20 cm × 40 cm rehal, R6 uses a 15 cm × 30 cm rehal, and R7 uses a 15 cm × 21 cm rehal. The interviewees revealed that they felt that using a rehal helped them to read. Specifically, R6 mentioned that using a rehal increased their comfort while reading (71%). Although all respondents expressed satisfaction with the height and size of their current rehals, they also thought that adjusting the height was preferred (100%) (Figure 5, Table 3).

Table 3. Analysis of interview content of Question 2.

Respondents	Storage Place for the Rehal		Height of the Current Rehal						Reasons for Stand Usage		
	High Place	Inside a Book Cabinet	21 × 29 cm	18 × 26 cm	10 × 15 cm	20 × 40 cm	15 × 30 cm	26 × 21 cm	Helps Reading, Comfortable	Preserves Sanctity	Fine
R1	✓		✓						✓		✓
R2	✓			✓						✓	✓
R3					✓				✓		✓
R4	✓		✓							✓	✓
R5		✓				✓			✓		✓
R6	✓						✓		✓		✓
R7		✓						✓	✓		✓
Total	4/7	2/7	2/7	1/7	1/7	1/7	1/7	1/7	5/7	2/7	7/7
Percentage	57%	29%	29%	-	-	-	-	-	71%	29%	100%

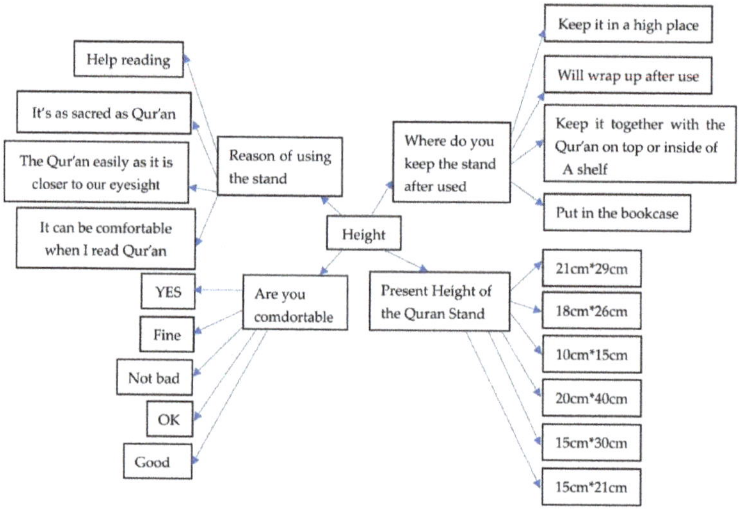

Figure 5. Atlas-ti network view of the height of the rehal.

Q3: What are the designs and ornaments that adorn their current rehal and are different designs acceptable?

When it came to rehal decorations and designs, respondents R1 and R5 stated that their rehals featured traditional floral patterns. Meanwhile, R3 and R6 had Arabic calligraphy patterns (29%) on their rehals. R6's rehal also had arabesque and geometric designs. R4 and R7 mentioned that their rehals were engraved with patterns (29%). While all the respondents felt that rehals are suitable as gifts, the appropriate occasion differed. According to R1, rehals are a suitable all-occasion gift. Meanwhile, R2 considered it more appropriate for VIPs. Respondents R3 and R4 mentioned that it could be gifted to the elderly. On the other hand, R5 considered it a wedding gift and R6 considered it a good housewarming gift. Lastly, R7 expressed that it was a decent present for a coming-of-age ceremony (100%). Respondents were willing to consider purchasing rehal with new or different designs. R1 and R6 indicated that they would accept newer designs while R2 stressed that there should be no animal motifs used (29%). R3 suggested green and yellow colour schemes and R4 recommended Jawi calligraphy. In contrast, R5 and R7 preferred contemporary minimalistic designs (29%) (See Figure 6, Table 4).

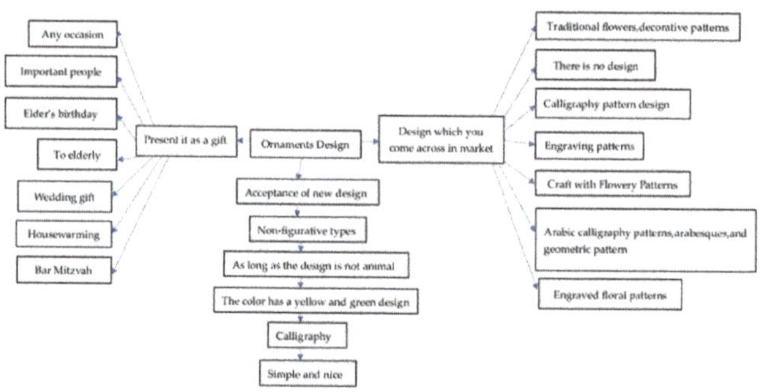

Figure 6. Atlas-ti network view of the surface design of the rehal.

Table 4. Analysis of interview content of Question 3.

Respondents	Design in the Market				As a Gift		Acceptance of New Design			
	Floral	Arabic Calligraphy	Geometric	Engraving	All Occasion	Non-Figurative	No Animal	Colours (Yellow and Green)	Calligraphy	Minimalistic
R1	✓				✓	✓				
R2					✓		✓			
R3		✓			✓			✓		
R4				✓	✓				✓	
R5	✓				✓					✓
R6		✓	✓		✓	✓				
R7				✓	✓					✓
Total	2/7	2/7	1/7	2/7	7/7	2/7	1/7	1/7	1/7	2/7
Percentage	29%	29%	-	29%	100%	29%	-	-	-	29%

4. Conclusions

The interviews suggest that almost every Muslim family has a rehal. Wood and plastic are the two common materials used to make the rehal, and the most common cleaning method is to wipe the item with a wet cloth. Wooden rehals in particular are prone to moldiness, deformation, or discolouration. The intricate wood carvings may also present a challenge during the cleaning process. Respondents believe the rehal to be sacred. As such, they prefer to keep their rehals in locations far off the ground when not in use. While the current height of commercial rehals are sufficient, comfort can be increased if the height is increased because it will provide an illuminance level suitable for human visual comfort while sitting in a cross-legged position on the floor. At present, commercial rehals are mainly decorated with floral motifs and patterns. However, new designs can be used if they follow Islamic laws such as using non-figurative designs, and adding calligraphy.

4.1. Proposed Ecological Material for Rehal

After proper consideration of available materials, the researcher would like to propose ceramic as a new alternative eco-friendly material for rehals. Ceramic does not retain moisture. Hence, it does not face the same problems as wood such as discolouration, deformation, and health issues. Moreover, it fits the current aesthetic trends in modern society. Thus, it is a sustainable choice for new product designs.

Ceramic rehals have the following three advantages:

(1) It is made from clay—a material that can significantly reduce environmental pollution.
(2) The ceramic rehal can be cleaned arbitrarily. The process of producing ceramic requires firing glazed clay in a kiln at a high temperature. The glaze and the high density of the ceramic reduces concerns regarding moisture, microbial growth, and regular wear-and-tear.
(3) Various patterns and designs can be freely added to the clay before firing. Consumers can even choose to custom-make designs as a gift for their friends.

4.2. Ceramic Rehal Exploration

The usage of environmentally friendly material is crucial to rehal design. The researcher, who is also a product designer, considered a range of eco-materials such as paper boards, cement, metal, gypsum, and glass, but finally chose ceramic. Ceramic is made by shaping and firing clay at a high temperature. The process renders it resistant to heat, corrosion, and oxidation. The present bisque-fired prototype is a semi-finished product (Figure 7, the unglazed version). Future rehals will be made in different sizes in accordance with different Quran sizes, then glazed to achieve the ideal eco-friendly and sustainable rehal.

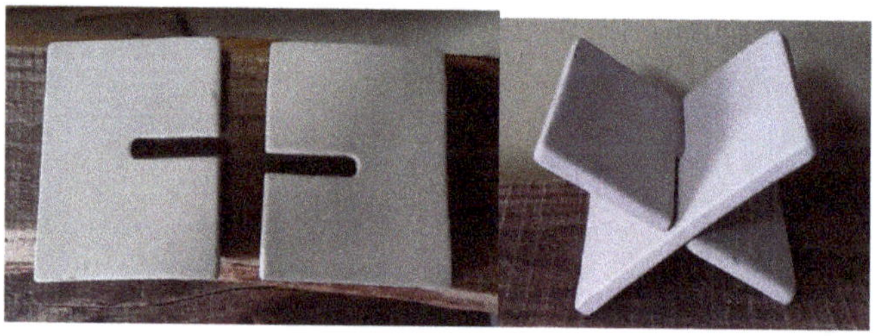

Figure 7. Prototype of ceramic rehal.

5. Discussion

The interviews allowed for a deeper understanding of the feasibility of alternative rehal materials, and laid a foundation for in-depth research on designing new rehals. This study reflects the cultural value of these book holders, which is helpful for designers. The data collected serves as a guideline for the creation of a new range of ceramic rehals by integrating eco-materials, relevant designs, and different heights to improve user experience and comfort.

The cultural value of a product is particularly important in the context of globalisation, which embraces cultural recognition. Designing an eco-friendly product that is also capable of capturing cultural connotations and enhancing user identity by creating a sense of national belonging allows for better overall user experience and acceptance. In future studies, the researcher shall add on a detachable ceramic groove to optimise the height of a standard rehal for the comfortable reading then refine the patterns and designs on the surface with glazing in accordance with Islamic doctrine.

The researcher applied their crafting skills to produce a ceramic rehal that took into consideration physical characteristics and user habits. The most difficult part of the production process was achieving a precise size and thickness for the clay slabs (Figure 8a). More improvements to the production process are required to ensure the two uniform sides of the clay tablets (Figure 8b) interlock perfectly (Figure 8c) to provide a sense of solidity and reliability. After trial and error, the researcher obtained a better understanding of the process and gained valuable experience. The product is thus presented as a Malaysian design.

(a) (b) (c)

Figure 8. (**a**–**c**) Experimentation by trial and error.

Author Contributions: Conceptualization, X.H.G. and H.L.S.; methodology, X.H.G. and H.L.S.; software, ATLAS.ti; formal analysis, X.H.G.; investigation, X.H.G.; resources, X.H.G.; data curation, X.H.G.; writing—original draft preparation, X.H.G.; writing—review and editing, H.L.S.; visualisation, X.H.G.; supervision, H.L.S. All authors have read and agreed to the published version of the manuscript.

Funding: This research received no external funding.

Institutional Review Board Statement: Not applicable.

Informed Consent Statement: Informed consent was obtained from all subjects involved in the study.

Data Availability Statement: Not applicable.

Conflicts of Interest: The authors declare no conflict of interest.

References

1. Samad, M.A.A.; Ramli, Z.; Wahab, M.R.A.; Zakaria, R.M.A. Rehal Tradisional Di Pantai Timur Semenanjung Malaysia: Aralisis Bahan Teknik Dan Ukuran. *J. Arkeol. Malays.* **2018**, *31*, 35–48.
2. Hagedorn, Annette, "Qur'an stand" in Discover Islamic Art, Place: Museum with No Frontiers. 2015. Available online: http://www.discoverislamicart.org/database_item.php?id=object;ISL;de;Mus01;22;en (accessed on 10 April 2022).
3. Kurz, O. Folding chairs and Koran stands. In *Islamic Art in the Metropolitan Museum of Art*; Ettinghausen, R., Ed.; Metropolitan Museum of Art: New York, NY, USA, 1972.
4. *Islamic Environmental Protection Concepts: A Perspective of the Quran*; Minzu University of China, the School of Philosophy and Religious Studies: Beijing, China, 2019.
5. Husini, R.N. A review Influence of national day Lighting on Quranic memorisation (Hafazan) Learning Task. In Proceedings of the 1st International Conference on Immersive Educational Technology (ICIET), Bangi, Malaysia, 6–7 August 2018.
6. Ahmad, N.; Ghoni, E.; Mohamad, D.; Harun, M.; Naser, M.; Jaafar, J.A.; Yusoff, A.M. Influence of rehal design on user comfort. In Proceedings of the 8th International Conference on Advanced Materials Engineering & Technology (ICAMT 2020), Langkawi, Malaysia, 26–27 November 2021.
7. National Palace Museum, Quran Holy Book Stands. Available online: https://theme.npm.edu.tw/opendata/DigitImageSets.aspx?sNo=04027009 (accessed on 2 April 2022).
8. Littlefield, D. *Metric handbook: Planning and Design Data*; Architectural: New York, NY, USA, 2012.
9. Qi, X.Y. *Bu Hali Shengxun Shilu Quanji (Buhar I Hadit the Complete Record)*; The Commercial Press: Beijing, China, 2018.
10. Yu, Z.J. Analysis of material application in product design. *Ind. Des. Des. Theory* **2019**, *5*, 59–60.
11. Hsieh, H.F.; Shannon, S. Three approached to qualitative content analysis. *Qual. Health Res.* **2005**, *15*, 1277–1288. [CrossRef] [PubMed]
12. Downe-Wamboldt, B. Content analysis: Method, applications, and issues. *Health Care Women Int.* **1992**, *13*, 313–321. [CrossRef] [PubMed]
13. Xiao, H.H. *Juece Kexue Cidian (Dictionary of Decision Sciences)*; People's Publishing House: Beijing, China, 1995.
14. Boyce, C. *Conducting In-Depth Interview: A Guide For Designing And Conducting In-Depth Interviews For Evaluation Input*; Pathfinder International: Watertown, MA, USA, 2006.
15. Siek, H.L.; Lee, C.E. The influence of national cultural attributes on locally produced designs: Case study of Malaysian design, IAFOR. *J. Cult. Stud.* **2021**, *6*, 33–43. [CrossRef]
16. Archer, E.; Vuuren, H.; Walt, H. *Introduction to ATLAS.ti*; Research Rescue: Pretoria, South Africa, 2017.
17. Skjott Linneberg, M.; Korsgaard, S. Coding qualitative data: A synthesis guiding the novice. *Qual. Res. J.* **2019**, *19*, 259–270. [CrossRef]

proceedings

Proceeding Paper

Telecommuting and Employee Productivity: Mediating Role of Work-Family Conflict and Autonomy †

Noor Aini Jaafar [1] and Ramita Abdul Rahim [2,*]

1 Proposal Management Unit, Mesiniaga Berhad, Menara Mesiniaga, Subang Jaya 47500, Malaysia
2 Department of Technology and Supply Chain Management Studies, Faculty of Business and Management, Universiti Teknologi MARA, Puncak Alam 42300, Malaysia
* Correspondence: ramita@uitm.edu.my
† Presented at the International Academic Symposium of Social Science 2022, Kota Bharu, Malaysia, 3 July 2022.

Abstract: A flexible work arrangement known as telecommuting enables employees to work remotely from their homes or other locations. Telecommuting work arrangements demonstrate that work is no longer a place but rather what you do, and this new working model is expected to persist and become more popular. Even with inevitable quibbles, their similarities outweigh their differences and provide more benefits to both firms and employees. Despite its significance, the implementation of telecommuting has caused some firms to question how effective their employees are at work. Here, the researchers use two mediating variables to identify the association and interactions of autonomy and work-family conflict with employee productivity. This study used a quantitative research design, convenience sampling technique, and sample size of 155, and data were analyzed with Smart PLS. The key inside this paper is indicated that autonomy mediates and has a crucial influence on employee productivity. The telecommuters were to have more control over their schedules, making it easier to balance work and personal obligations and possibly reducing stress and improving work-life balance, which contributes to better performance.

Keywords: telecommuting; work-family conflict; autonomy; employee productivity

Citation: Jaafar, N.A.; Rahim, R.A. Telecommuting and Employee Productivity: Mediating Role of Work-Family Conflict and Autonomy. *Proceedings* **2022**, *82*, 84. https://doi.org/10.3390/proceedings2022082084

Academic Editor: Mohamad Rahimi Mohamad Rosman

Published: 23 September 2022

1. Introduction

The telecommuting work arrangement was first applied way back in the 1970s by [1]. Ref. [2] define telecommuting as working from anywhere that employees feel comfortable working using tools and devices such as smartphones, tablets, desktop computers, and notebooks while online remotely with their workplaces, organizations, clients, and others. As authors, Ref. [3] noted that telecommuting is another operational mode that requires technology. Telecommuting is a flexible work arrangement that permits work to be done remotely, from an employee's home or other location, thereby benefiting both parties; employees and organizations [4]. Despite some arguable variances, their similarities outweigh their differences. They all refer to outside-the-office work that uses ICT to connect. Telecommuting, telework, and remote work are now synonymous.

Telecommuting reduces traffic and parking concerns, especially in major cities. A recent study found that companies that enable telecommuting have higher productivity, a better work atmosphere, more accessible access to talent pools, and lower absenteeism, attrition, and costs [5]. Teleworking improves well-being, work-life balance, performance, motivation, satisfaction, and autonomy and reduces stress and work-family conflict [6]. However, too many telework hours might impair employee productivity and stress levels. In addition, the pandemic that devastated global health systems and the potential for future illnesses may motivate teleworkers [7]. This new phenomenon opens new possibilities for further experimentation and, eventually, the implementation of telecommuting. The organization's concerns in health and safety, supply chain, labor force, cash flow, business demand, marketing, and public health are reshaping the world's sustainable future. Sustainability, according to [8], sustainability achieves economic balance and responsibility. Telecommuting provides a workaround for the organization to operate for its sustainability and development despite lockdowns and movement control orders, according to [7].

According to previous research, the introduction of telecommuting has provided some insight into employees' productivity levels. Although previous studies strongly suggested that when employees telecommute or work remotely, it contributes to many other benefits besides increasing employee productivity, there is some gap the researchers wanted to discover in this study. For instance, authors [9] in their study, found that employees that work in a virtual office concept or telecommuting are those who can choose where and when to execute their work, significantly emulate a higher level of job satisfaction, and contribute to higher employee productivity. This finding was discovered earlier by [10]. They added that besides contributing to higher employee productivity, job satisfaction also translates to benefits for the employer with lower employee turnover. The employees have a higher organizational commitment and a lower or weaker possibility of resigning and leaving the organization.

There are numerous reasons for implementing telecommuting, including reducing the time spent traveling to and from work, coping with rising travel costs, reducing the risk of hazards and diseases, implementing a business continuity strategy, and assisting employees with work-life balance in the workplace. These factors have been supported in the studies done by [11] and emphasize that employees can fulfill the demands of work-life balance with flexible work scheduling. Another study by [12], found that one could achieve a work-life balance when fully involved, engaged, and committed to work, social, and family demands. There is no conclusive definition for it. The work-life balance depends on an employee's capacity to meet work and life's multiple demands and responsibilities. With telecommuting, there is the opportunity to work hours flexibly and fulfill the work-life balance.

However, today's employees are more concerned about the conflicting demands of work and family duties. Telecommuting with a flexible work schedule is likely to be viewed as a desirable perk that can assist in achieving a healthier work-life balance. Even though telecommuting is a modern notion, and some organizations are experimenting with it, it is not acceptable for the outlined reasons discussed above. Employers and organizations have acknowledged the significant flexibility in working environments where employees have a sense of autonomy and perform at their best.

2. Literature Review and Hypothesis Development

2.1. *Telecommuting and Employee Productivity*

In [13] stated that employees' ability to meet or exceed agreed-upon job tasks with the provided tools, technologies, and procedures is employee productivity. In this context, telecommuting is a procedure to achieve employee productivity. Authors [14] found that the call center employees in a large Chinese travel agency were randomly selected to work from home most days and only came to the office to attend training. The authors stated that after nine months of telecommuted increased, their productivity by 13% compared to their colleagues who worked from the office. Staff retention improved by half among telecommuters.

They believe call center employees' jobs are particularly well-suited for telecommuting as they do not involve collaboration or face-to-face time [14]. Another study done by [15] stated that managers in 73% of the Dutch teleworking organizations said that telecommuting boosted their productivity, and 60% of the Dutch managers said that higher work motivation was too due to credit for their increased production. Using a natural experiment with a public sector organization in the United Kingdom [16], discovered that productivity is higher when teammates are in the same room and that the effect is more substantial for urgent and complex work. Teleworking, they argue, is unsuited for tasks that necessitate face-to-face communication. According to [17], based on an experimental approach, telecommuting has a good impact on the productivity of creative work but a negative impact on the productivity of boring tasks.

Telecommuting has increased employee productivity and performance [18], since employees can choose how and when to complete tasks, improving and increasing productivity. The authors concluded that telecommuting impacts employees, families, and organizations better. Hence, it benefits employees to improve how they do the work, contributing to best organizations practices. Telecommuting arrangements in working from anywhere (WFA) could be even better for productivity than working from home, depending on the type of work. According to [19], work-from-home arrangements presuppose that employees reside close enough to the office to commute a few days a week or as needed. They asserted that work-from-anywhere agreements enable employees to work remotely and physically away from their firm's offices.

The authors' research in [20] reveals the impact of job stability as the most significant and critical element of perceived teleworking productivity. Employees perceived that telecommuting does not jeopardize their job performance, and employers acknowledged the fact. Hence, the hypothesized:

H1. *Telecommuting will have a significant influence on employee productivity.*

2.2. The Mediating Effect of Work-Family Conflict (WFC)

Flexible work schedules are one of the characteristics of telecommuting, as employees can adjust their time between completing their work and attending to family demands, respectively. In their study paper, Telecommuting's Differential Impact on Work-family Conflict [21], surveyed 454 respondents who telecommute and work in the office. The authors focused on the impact of working from home or telecommuting on work-to-family conflict (WFC) and family-to-work conflict (FWC). The WFC defines working while interfering with family responsibilities, while FWC is the interring of a family with work duties.

Other authors [22], have examined the temporal nature of telecommuting and its impact on work overload, exhaustion, and work-family conflict. After studying 600 Certified Practicing Accountants (CPA) Australia participants, the researcher found that telecommuting during office hours improved coping with work overload, reduced after-office hours work, and decreased work-family conflict. Work-family conflict is the failure to reconcile work and responsibility roles with family [23]. It happens when work and family responsibilities are incompatible. According to [24], it is causing incompatible pressures from work and family roles.

In the year 2020, Ref. [25] concluded that work-family conflict happens due to the interference of one role: work/family or family/work; and complying with expectations. It's identified as weakening employee responsibility's different roles (work and family). According to [26], from the findings of the study, telecommuting after work hours is related to both work-family conflicts, i.e., work interference with family (WIF) and family interference with work (FIW). They stated that telecommuting from home after work hours significantly benefited employees through cost savings.

As stated earlier, flexible work schedules are a component of telecommuting, favoring employees in managing their personal needs. Thus, employees assume that organizations support them in managing and coping with work-family demands [27]. They were known as the signaling theory: flexible work schedules contribute to employees' positive perceptions and increase their job satisfaction and commitment, leading to increased productivity. In researcher study [28], used signal theory to explain how this results in increased job satisfaction and commitment among employees and increased productivity. According to [29], the availability of flexible work schedules may increase employees' feelings of control over work-family issues and benefit their ability to cope with work-family conflict. The studies regarding work-family conflict do not support a causal relationship. WFC makes work hours a significant aspect; as a result, telecommuting is proposed and mediated by WFC. Hence, the hypothesized:

H2. *Work-Family Conflict (WFC) will mediate the relationship between telecommuting and employee productivity.*

2.3. The Mediating Effect of Autonomy

Employee satisfaction is strongly correlated with a sense of control over one's work environment, and this study found that telecommuters reported higher levels of autonomy in their jobs. Employees stay with the company without motivation to leave, de-stress, improved work-family balance, and higher employee productivity ratings by their reporting manager and supervisor [30]. According to [18], telecommuting allows employees to carry out their job obligations outside of the workplace, uncontrolled by their supervisor. Telecommuting gives individuals autonomy and the capacity to manage the allocated work. They could schedule their work according to their preferences without jeopardizing its completion. Ref. [31], on the other hand, found that the level of autonomy could depend on the telecommuting job, the processes and procedures in place, the equipment used, and the type of work done at home or elsewhere [6].

In the 1970s, authors [32] defined autonomy as the degree to which employees have control and discretion over how they carry out their tasks, consistent with self-determination theory (SDT). This macro theory identifies the origins and consequences of human agentic action [33]. This SDT explains the interaction between need satisfaction and intrinsic motivation and the management of extrinsic motivation. More precisely, autonomy is defined [34] as the degree to which an individual considers his or her acts as the outcome of his or her own free choice, independent of external influence in a particular context. Thus, employees become intrinsically driven when they believe their decisions and actions are independent.

A meta-analysis by [30] involving 12,883 employees found that telecommuting is beneficial for psychological autonomy and work-family conflict. Additionally, Ref. [35] found an association between telecommuting and higher job performance evaluations from managers. These studies supported [36]. They explored how telecommuting design influences the work environments and outcomes through its effects on their social system, autonomy and self-management opportunities and needs, and role limitations [36]. Most recent studies suggest that autonomy as an intangible element, besides salary and pay, will be a motivating factor in increasing and changes in productivity [37]. The authors concluded that perceived autonomy, rather than total autonomy, influenced work determination and quality and that organizations must use appropriate interference to achieve higher productivity levels. They also noted that telecommuting employees had increased autonomy and that autonomy influenced their behaviors.

Understanding autonomy is significantly more vital with the increase in telecommuting. Thus, the following hypothesis was made:

H3. *Autonomy will mediate the relationship between telecommuting and employee productivity.*

3. Materials and Methods

3.1. Sample and Data Collection

To test the hypotheses, the researchers used online self-administered questionnaires to collect the data. Our procedure for delivering the questionnaire was based on [38]. The questionnaire was validated by an industry expert and two representatives from UiTM academicians.

A total of 175 were distributed to the headquarter of the private organization in the Klang Valley through Google form that integrated with WhatsApp and corporate email. A total of 155 responses were gathered and analyzed during the collection period. Therefore, the total response rate for collected questionnaires is 89%. Table 1 demonstrates the demographic information of the respondents. The sample consisted of 56.8% females and 43.2% males. In most of the sample, 53.5% identified themselves as married, 43.9% as single, and 2.6% as single parents. About 11.0% of the respondents were newly employed with the organization, with the working years less than a year, followed by 19.4% between

1–2 years, 16.1% between 3–4 years, 19.4% between 5–10 years, and highest 34.2% having been with the organizations for more than 10 years. Most respondents were permanent employees with 81.9% and contract positions reported 18.1%. The participants varied in organizational position within the organization, with 20.0% as non-executive, 53.5% as executive, 5.8% assistant manager, 14.2% manager, 3.9% senior manager, 0.6% general manager and 1.9% director level.

Table 1. Respondent Profile.

Profile	Frequency	Percent (%)
Gender		
Female	88	56.8
Male	67	43.2
Marital Status		
Single	68	43.9
Married	83	53.5
Single Parent	4	2.6
Working Years		
Less than a year	17	11.0
1–2 years	30	19.4
3–4 years	25	16.1
5–10 years	30	19.4
More than 10 years	53	34.2
Employment Status		
Permanent	127	81.9
Contract	28	18.1
Current Position		
Non-executive	31	20.0
Executive	83	53.5
Assistant Manager	9	5.8
Manager	22	14.2
Senior Manager	6	3.9
General Manager	1	0.6
Director	3	1.9

3.2. Measures

Telecommuting was measured using four items, "I could adopt telecommuting arrangement", "I have demonstrated interest in telecommuting", "Implementing telecommuting brings a positive impact on my organization", and "I agree that my organization has a proper design and planning in implementing telecommuting". The items were measured from 1 (strongly disagree) to 5 (strongly agree).

Work-Family Conflict (WFC) construct was assessed through a five-item scale using five items, "The demands of telecommuting interfere with my home/family life", "The amount of time telecommuting takes up makes it difficult to fulfill home/family responsibilities", "Things that I want to do at home do not get done because of the telecommuting demands", "The telecommuting produces strain that makes it difficult to switch off and enjoy home/family life", and "Due to the telecommuting related, I often make changes to my plans for home/family".

Autonomy was measured using five items, "When I am telecommuting, my manager allows me to make my own decisions about how to schedule my work, to make decisions on approaches use to complete my work, to plan how I execute my work, to make my own decision at many of the time" and "offers me a significant chance for freedom in how I do and complete the work".

Employee productivity construct was measured and assessed through a five-item scale [39], "My manager evaluated me as an effective employee", "consistently satisfied with my work output", "agreed that I work at a very steady pace", "I am a highly productive employee", and "believe that I am competent worker".

3.3. Normality

According to [40], skewness measures the normal distribution's symmetry, or more precisely, its lack of balance. Kurtosis is a measure of a distribution's peaked ness. If the skewness or kurtosis of the distribution results in a z value in the range of 1.96 for a small sample size (n 50), the distribution is approximately normal. However, it has been found that the sample distribution is normal for medium-sized samples (50 n 300), resulting in an absolute z value of 3.29. Table 2 indicates the normality of skewness and kurtosis for the independent, dependent, and mediating variables is listed below.

Table 2. Descriptive Statistics.

Constructs	Skewness Statistic	Kurtosis Statistic
TELE12	−1.123	1.485
TELE13	−1.114	1.161
TELE14	−0.813	0.542
TELE15	−0.608	−0.335
WFC16	−0.470	−0.695
WFC17	−0.405	−0.789
WFC18	−0.149	−1.027
WFC19	−0.374	−0.792
WFC20	−0.492	−0.696
AUTO21	−0.948	0.464
AUTO22	−1.085	1.270
AUTO23	−0.729	0.332
AUTO24	−1.042	1.060
AUTO25	−1.071	0.904
EP31	−0.697	0.586
EP32	−0.625	0.257
EP33	−0.413	−0.510
EP34	−0.478	0.130
EP35	−0.646	0.629

TELE = Telecommuting, WFC = Work-family Conflict, AUTO = Autonomy, EP = Employee. Productivity

3.4. Reliability

Cronbach Alpha was done to analyze the reliability of the data, and that seemed to be similar to the literature. Table 3 demonstrates the scale used for measuring the constructs of Telecommuting, Work-family conflict, Autonomy, and Employee productivity, was taken from the research. Cronbach's Alpha coefficient of the studied variables was more significant than 0.7; hence it can be concluded that the data is reliable.

Table 3. Reliability of the study.

Variables	Cronbach's Alpha for Pilot Study (n = 30)	Cronbach's Alpha for Actual Study (n = 155)
Telecommuting	0.857	0.862
Work-family conflict	0.869	0.926
Autonomy	0.985	0.960
Employee Productivity	0.957	0.945

3.5. Data Analysis

The data were analyzed by employing statistical software to assess the psychometrics of the measurement model and estimate the parameters of the structural model. The validity and reliability of the measurement model are evaluated by assessing the internal consistency reliability, indicator reliability, convergent validity, and discriminant validity.

3.6. Measurement Model

Internal consistency reliability (CR) is higher than 0.7 for all constructs. Outer loadings for all indicators that have been used in this study were above the recommended stringent threshold loading value of more than 07 except for TELE 0.630 however the AVE for it was more than 0.5. Convergent validity for all AVE values is more than 0.5

Fornell and Larcker's square root of each construct's AVE should have a greater value than the correlations with other latent constructs and the cross-loadings criterion which each indicator load high on its own constructs, and this indicates discriminant validity is achieved. The HTMT values should be below a strict criterion of 0.85 for this paper, the value of HTMT was all lover than the stricter criterion mentioned which concluded that the respondents understood that all constructs are distinct.

3.7. Structural Model

The coefficient of determination (R^2) value should be equal to or greater than 0.1 and this result indicated that R^2 for employee productivity is at 0.358 and R^2 adjusted is 0.340, hence, the predictive capability is established.

4. The Findings

Table 4 indicates the hypothesis testing for this paper. The hypothesis (H1) revealed no significant positive relationship between telecommuting and employee productivity. As for the hypothesis-specific indirect effects (H2 and H3), using the Smart PLS 3, run the bootstrapping consistently. The researcher can access specific indirect effects, which is relatively convenient as this paper involves multiple mediating variables. The H2 is not significant, which means that the influence of IV and DV is not passing thru the mediator. Only H3 is significant and accepted, meaning autonomy mediated the IV on DV. Hence, the employees who perform telecommuting have more autonomy in their work, making them more productive. Autonomy impacted the determination and quality of work done, achieving a higher employee productivity level.

Table 4. Hypothesis Testing Direct and Specific Indirect Effects.

Path	Effects	β	t	*p*	Confidence Intervals
H1 (Telecommuting- > Employee Productivity)	Total Indirect	0.364	5.090	0.855	(−0.185, 0.151)
H2 (Telecommuting- > Work-family Conflict- > Employee Productivity)	Specific Indirect Effect	0.020	0.595	0.552	(−0.044, 0.094)
H3 (Telecommuting- > Autonomy- > Employee Productivity)	Specific Indirect Effect	0.254	4.926	0.000	(0.165, 0.362)

5. Discussion and Recommendation

The current study looked at how telecommuting affects employee productivity, as well as how it affects work-family conflict and autonomy. According to the studies, telecommuting does not considerably boost employee productivity. This scenario is said to be due to a typical blunder made by employers that force employees to work during set hours rather than allowing them to work when it is most convenient for them when telecommuting. To achieve employee productivity, employers or organizations must integrate the flexibility aspect [6] when introducing or supporting telecommuting for their employees [18]. Employees had more flexibility over their time and more options because of flexible work arrangements [41]. Distraction is, in fact, another critical issue that arises because of telecommuting. According to this study, employees telecommute face 51% of the time. The claim was supported based on the previous study by [42], which indicated that distractions negatively influence telecommuters' productivity.

The location, whether at home or elsewhere, determines the productivity-enhancing effectiveness of telecommuting [19]. When an organization implements telecommuting, the critical points to consider are the location and the employee's family size. If the employees need to share space with other family members in a tiny house with a large family, they will not have adequate placement or workstation. Without a designated workstation, it is difficult for the employee to draw a line between household and business responsibilities [43].

Aside from distraction and internet bandwidth, according to this report, one of the most significant obstacles for employees working from home is a lack of a designated workspace (39%) and a lack of infrastructure for video conferencing (32%). Organizations need to invest in or give one-off payments to their employees to purchase the furniture and equipment for their home offices to give them a better telework experience. Ref. [44] discovered that this study confirmed that work-family conflict increases when telecommuting. Work-family conflict exacerbates when a failure to set clear lines and boundaries between work and personal life results in work interfering with home life and vice versa [44]. Working women, married or single parents with children under 18, and the elderly, who wear numerous hats at home: worker, wife, mother, and daughter [45], should pay special attention to this factor. Women make up 56.8% of those who responded to the survey.

Employers could develop practical instructions for employees to set boundaries and examine their expectations of telecommuters to address the problem [10,30]. Having a daycare arrangement for youngsters or the elderly and not attempting to multitask while working, which could lead to work-family conflict, is one example of such a barrier [6,46]. Employers should also provide guidelines, proper equipment, and training to assist employees in dealing with the challenges of telecommuting [43]. These might help to reduce work-family friction.

Autonomy has a favorable association with employee productivity, according to our research. Employees who telecommute do so to have more control over how they execute their job. In essence, telecommuting is supposed to give people more control over their schedules, making it easier to balance work and personal obligations, reducing stress, and improving work-life balance.

Employees who telecommute did not connected to a centralized office have more autonomy and flexibility in organizing their work and personal commitments, which benefits work performance and productivity. Employees have freedom in how and when they complete their work without physical monitoring. They might also alter their work schedules, take time off, and attend to personal affairs without jeopardizing their yearly leave or job quality [47].

6. Conclusions

The study's main goal was to investigate how telecommuting impacts employees' productivity levels. The findings signify that employees' productivity level increases when employees telecommute, mediated by the autonomy variable. The employees who telecommute have greater control and autonomy, which leads them to become more content with their work. Every employee has different productive times; therefore, telecommuting will allow employees to take advantage of their most productive time by performing their work the time. Overall, we believe this research has significantly contributed to assisting the organization and perhaps help others to decide whether to permanently implement telecommuting mode of work and invest in its enhancement.

Author Contributions: Conceptualization, N.A.J. and R.A.R.; methodology: N.A.J. and R.A.R.; writing—original draft preparation: N.A.J.; data collection, N.A.J.; data analysis, N.A.J.; review and editing, N.A.J. and R.A.R.; validation, R.A.R.; Supervision, R.A.R. All authors have read and agreed to the published version of the manuscript.

Funding: This research received no external funding.

Institutional Review Board Statement: Not applicable.

Informed Consent Statement: Not applicable.

Data Availability Statement: Not applicable.

Acknowledgments: The researchers would like to thank the Faculty of Business Management, Universiti Teknologi MARA (UiTM) for supporting and funding the publication of this paper.

Conflicts of Interest: The authors declare no conflict of interest.

References

1. Nilles, J.M.; Carlson, F.R.; Gray, P.; Hanneman, G. Telecommuting—An Alternative to Urban Transportation Congestion. *IEEE Trans. Syst. Man Cybern.* **1976**, *SMC-6*, 77–84. [CrossRef]
2. Ahmed, A.; Ishaque, A.; Nawaz, T.; Ali, Y.; Hayat, F. Telecommuting: Impact on productivity of telecommuters. In Proceedings of the 2014 IEEE International Conference on Management of Innovation and Technology, Singapore, 23–25 September 2014; pp. 187–192. [CrossRef]
3. Feldman, D.C.; Gainey, T.W. Patterns of telecommuting and their consequences: Framing the research agenda. *Hum. Resour. Manag. Rev.* **1997**, *7*, 369–388. [CrossRef]
4. Nieminen, L.R.G.; Nicklin, J.M.; McClure, T.K.; Chakrabarti, M. Meta-analytic Decisions and Reliability: A Serendipitous Case of Three Independent Telecommuting Meta-analyses. *J. Bus. Psychol.* **2011**, *26*, 105–121. [CrossRef]
5. Brunelle, E.; Fortin, J.-A. Distance Makes the Heart Grow Fonder: An Examination of Teleworkers' and Office Workers' Job Satisfaction through the Lens of Self-Determination Theory. *SAGE Open* **2021**, *11*, 2158244020985516. [CrossRef]
6. Allen, T.D.; Golden, T.D.; Shockley, K.M. How Effective Is Telecommuting? Assessing the Status of Our Scientific Findings. *Psychol. Sci. Public Interest* **2015**, *16*, 40–68. [CrossRef]
7. Belzunegui-Eraso, A.; Erro-Garcés, A. Teleworking in the Context of the Covid-19 Crisis. *Sustainability* **2020**, *12*, 3662. [CrossRef]
8. Hakovirta, M.; Denuwara, N. How COVID-19 Redefines the Concept of Sustainability. *Sustainability* **2020**, *12*, 3727. [CrossRef]
9. Beauregard, T.A.; Basile, K.A. Telework: Outcomes and Facilitators for Employees. In *The Cambridge Handbook of Technology and Employee Behavior*; Cambridge University Press: Cambridge, UK, 2019. [CrossRef]
10. Golden, T.D.; Veiga, J.F. The Impact of Extent of Telecommuting on Job Satisfaction: Resolving Inconsistent Findings. *J. Manag.* **2005**, *31*, 301–318. [CrossRef]
11. Beauregard, T.A.; Henry, L.C. Making the link between work-life balance practices and organizational performance. *Hum. Resour. Manag. Rev.* **2009**, *19*, 9–22. [CrossRef]
12. Dragani, J. Work/Life Balance in the 21st Century. *Way Ahead* **2012**, *8*, 23–24. [CrossRef]
13. Campbell, K.M. Flexible Work Schedules, Virtual Work Programs, and Employee Productivity. Doctoral Dissertation, Walden University, Minneapolis, MN, USA, 2015. Available online: https://scholarworks.waldenu.edu/dissertations/509 (accessed on 2 October 2021).
14. Bloom, N.; Liang, J.; Roberts, J.; Ying, Z.J. Does Working from Home Work? Evidence from a Chinese Experiment. *Q. J. Econ.* **2014**, *130*, 165–218. [CrossRef]
15. Hoornweg, N.; Peters, P.; van der Heijden, B. Finding the Optimal Mix between Telework and Office Hours to Enhance Employee Productivity: A Study into the Relationship between Telework Intensity and Individual Productivity, with Mediation of Intrinsic Motivation and Moderation of Office Hours. *Adv. Ser. Manag.* **2016**, *16*, 1–28. [CrossRef]
16. Battiston, D.; Blanes i Vidal, J.; Kirchmaier, T. Is Distance Dead? Face-to-Face Communication and Productivity in Teams (March 2017). CEPR Discussion Paper No. DP11924. Available online: https://ssrn.com/abstract=2941509 (accessed on 20 July 2020).
17. Glenn Dutcher, E. The effects of telecommuting on productivity: An experimental examination. The role of dull and creative tasks. *J. Econ. Behav. Organ.* **2012**, *84*, 355–363. [CrossRef]
18. Fazrida, F.; Khan, P.; Mohammed, N.; Hafizah, N.; Harith, M. The Relationship Between the Impacts of Telecommuting Engagement and Employee Performance in Oil and Gas Industry in Kuantan, Pahang. *Malays. J. Soc. Sci. Humanit.* **2018**, *3*, 1–9. [CrossRef]
19. Choudhury, P.; Foroughi, C.; Larson, B. Work-from-anywhere: The productivity effects of geographic flexibility. *Strateg. Manag. J.* **2020**, *42*, 655–683. [CrossRef]
20. Aboelmaged, M.G.; El Subbaugh, S.M. Factors influencing perceived productivity of Egyptian teleworkers: An empirical study. *Meas. Bus. Excel.* **2012**, *16*, 3–22. [CrossRef]
21. Golden, T.D.; Veiga, J.F.; Simsek, Z. Telecommuting's differential impact on work-family conflict: Is there no place like home? *J. Appl. Psychol.* **2006**, *91*, 1340–1350. [CrossRef]
22. Campbell, J.; Boell, S.; Keating, B.; Cecez-Kecmanovic, D. Temporal aspects of telework and its impact on work-family conflict. In Proceedings of the ACIS 2013, Niigata, Japan, 16 June 2013; p. 90. Available online: https://aisel.aisnet.org/acis2013/90 (accessed on 23 July 2020).
23. Duxbury, N.; Hosagrahar, J.; Pascual, J. Why must culture be at the heart of sustainable urban development? *Agenda 21 for Culture*. 2016. Available online: http://hdl.handle.net/10316/46481 (accessed on 22 September 2020).
24. Greenhaus, J.H.; Beutell, N.J. Sources of Conflict Between Work and Family Roles. *Acad. Manag. Rev.* **1985**, *10*, 76–88. [CrossRef]
25. Jerg-Bretzke, L.; Limbrecht-Ecklundt, K.; Walter, S.; Spohrs, J.; Beschoner, P. Correlations of the "Work–Family Conflict" With Occupational Stress—A Cross-Sectional Study Among University Employees. *Front. Psychiatry* **2020**, *11*, 134. [CrossRef]
26. Jostell, D.; Hemlin, S. After hours teleworking and boundary management: Effects on work-family conflict. *Work* **2018**, *60*, 475–483. [CrossRef]

27. Gardiner, J.; Tomlinson, J. Organisational approaches to flexible working. *Equal Oppor. Int.* **2009**, *28*, 671–686. [CrossRef]
28. Grover, S.L.; Crooker, K.J. Who appreciates family-responsive human resource policies: The impact of family-friendly policies on the organizational attachment of parents and non-parents. *Pers. Psychol.* **1995**, *48*, 271–288. [CrossRef]
29. Wayne, J.H.; Randel, A.E.; Stevens, J. The role of identity and work–family support in work–family enrichment and its work-related consequences. *J. Vocat. Behav.* **2006**, *69*, 445–461. [CrossRef]
30. Gajendran, R.S.; Harrison, D.A. The good, the bad, and the unknown about telecommuting: Meta-analysis of psychological mediators and individual consequences. *J. Appl. Psychol.* **2007**, *92*, 1524–1541. [CrossRef] [PubMed]
31. Gagné, M.; Bhave, D. Autonomy in the Workplace: An Essential Ingredient to Employee Engagement and Well-Being in Every Culture. In *Human Autonomy in Cross-Cultural*; Springer: Dordrecht, The Netherlands, 2010; pp. 163–187. [CrossRef]
32. Hackman, J.R.; Oldham, G.R. Motivation through the design of work: Test of a theory. *Organ. Behav. Hum. Perform.* **1976**, *16*, 250–279. [CrossRef]
33. Howard, A.G.; Zhu, M.; Chen, B.; Kalenichenko, D.; Wang, W.; Weyand, T.; Andreetto, M.; Adam, H. MobileNets: Efficient Convolutional Neural Networks for Mobile Vision Applications. *arXiv* **2017**, arXiv:1704.04861.
34. Jung, Y. Understanding the Role of Sense of Presence and Perceived Autonomy in Users' Continued Use of Social Virtual Worlds. *J. Comput. Commun.* **2011**, *16*, 492–510. [CrossRef]
35. Gajendran, R.S.; Harrison, D.A.; Delaney-Klinger, K. Are Telecommuters Remotely Good Citizens? Unpacking Telecommuting's Effects on Performance Via I-Deals and Job Resources. *Pers. Psychol.* **2015**, *68*, 353–393. [CrossRef]
36. Allen, D.G.; Renn, R.W.; Griffeth, R.W. The impact of telecommuting design on social systems, self-regulation, and role boundaries. In *Research in Personnel and Human Resources Management*; Emerald Group Publishing Limited: Bingley, UK, 2003; Volume 22, pp. 125–163. [CrossRef]
37. Johannsen, R.; Zak, P.J. Autonomy Raises Productivity: An Experiment Measuring Neurophysiology. *Front. Psychol.* **2020**, *11*, 963. [CrossRef] [PubMed]
38. Evans, J.R.; Mathur, A. The value of online surveys: A look back and a look ahead. *Internet Res.* **2018**, *28*, 854–887. [CrossRef]
39. Staples, D.S.; Hulland, J.S.; Higgins, C.A. A Self-Efficacy Theory Explanation for the Management of Remote Workers in Virtual Organizations. *Organ. Sci.* **1999**, *10*, 758–776. [CrossRef]
40. Mishra, P.; Pandey, C.M.; Singh, U.; Gupta, A.; Sahu, C.; Keshri, A. Descriptive statistics and normality tests for statistical data. *Ann. Card. Anaesth.* **2019**, *22*, 67–72. [CrossRef] [PubMed]
41. Wheatley, D. Employee satisfaction and use of flexible working arrangements. *Work. Employ. Soc.* **2017**, *31*, 567–585. [CrossRef]
42. Greer, T.W.; Payne, S.C. Overcoming telework challenges: Outcomes of successful telework strategies. *Psychol. J.* **2014**, *17*, 87–111. [CrossRef]
43. Seva, R.R.; Tejero, L.M.S.; Fadrilan-Camacho, V.F.F. Barriers and facilitators of productivity while working from home during pandemic. *J. Occup. Health* **2021**, *63*, e12242. [CrossRef]
44. Sarbu, M. The role of telecommuting for work-family conflict among German employees. *Res. Transp. Econ.* **2018**, *70*, 37–51. [CrossRef]
45. Van der Lippe, T.; Lippényi, Z. Co-workers working from home and individual and team performance. *New Technol. Work Employ.* **2020**, *35*, 60–79. [CrossRef]
46. Lautsch, B.A.; Kossek, E.E.; Eaton, S.C. Supervisory approaches and paradoxes in managing telecommuting implementation. *Hum. Relat.* **2009**, *62*, 795–827. [CrossRef]
47. Ramaswamy, M.; Viswanathan, R.; Sarathlal, N.; Byju, K. Teleworking with work autonomy enhances productivity of Indian IT professionals during COVID-19 pandemic. *PalArch's J. Archaeol. Egypt/Egyptol.* **2020**, *17*, 6758–6770.

proceedings

Proceeding Paper

Maritime Security Policy for Increasing National Economic Growth in Archipelagic Country †

Ahmad Muhdar *, Muhammad Zilal Hamzah and Eleonora Sofilda

Faculty of Economic and Business, Trisakti University, Daerah Khusus Ibukota Jakarta 11440, Indonesia
* Correspondence: ahmad1703@dephub.go.id

† Presented at the International Academic Symposium of Social Science 2022, Kota Bharu, Malaysia, 3 July 2022.

Abstract: Transport security is a leading strategic issue in an archipelago country with many back entrances to secure. Both ports and shipping lanes are significant for multinational investment activities. This paper aims to determine the impact of maritime transport security and the expectations of stakeholders on Indonesia's economic growth. This paper used quantitative and qualitative approaches. The analytical technique in the quantitative method used Input-Output analysis, and for the qualitative method, Focus Group Discussions (FGD) activities were held and analyzed by NVivo. The results of the quantitative approach showed an impact of changes in GDP by 21.99 billion IDR and an increase in the income of workers by 0.13 billion IDR. The other 53 sectors changed the output by 36.96 trillion IDR, the workforce by 79,072 people, and labor income by 860.31 billion IDR. The qualitative method explained that the Stakeholder Synergy has the most significant contribution among stakeholders to creating a better maritime transport security ecosystem. The multiplier impact on all sectors, income, and employment, will always appear in a public policy.

Keywords: transportation security policy; national economic growth; input–output analysis

Citation: Muhdar, A.; Hamzah, M.Z.; Sofilda, E. Maritime Security Policy for Increasing National Economic Growth in Archipelagic Country. *Proceedings* **2022**, *82*, 86. https://doi.org/10.3390/proceedings2022082086

Academic Editor: Mohamad Rahimi Mohamad Rosman

Published: 27 September 2022

1. Introduction

The concept of an archipelagic state is regulated in the third United Nations Convention on the Law of the Sea, 1982 (UNCLOS, 1982). It explains that an archipelagic country consists of one or more islands, and includes a group of islands, the waters of which are made as a natural manifestation of a unified economic and political geography [1]. Indonesia declared itself to be an archipelago country through the Djuanda Declaration on 13 December 1957. The Indonesian seas became one unit with the territory (land) of the Unitary Republic of Indonesia, including the sea around, between, and within the archipelago. Furthermore, Indonesia is also located in a strategic cross position between two oceans and two continents, so it merits being called an archipelagic country on the maritime axis [2].

In terms of business, Indonesia is at the crossroads of international trade routes where 90% of world trade is transported using sea transportation and 40% is through the three straits in Indonesia. Furthermore, Indonesia is a world natural tourist destination [3,4]. Additionally, in terms of productivity, port security is still a house of cards [5]. As a member of the International Maritime Organization (IMO) board since 1973, Indonesia has ratified international cooperation conventions in maritime safety and security, including marine environment protection. As an archipelagic country with many back doors that must be secured, transportation security is a major strategic issue; both ports and shipping lanes are essential for multinational investment activities [6].

Shipping security practices in Indonesia are carried out by the Maritime Port Authority (Directorate General of Sea Transportation-DJPL) according to the International Ship and Port Facility Security Code (ISPS-Code) based on guidance from the IMO and the Safety of Life at Sea (SOLAS) chapter XI-2. Both ships and port facilities have reported a decrease in the number of incidents of theft and accidents in security-restricted areas. In addition, it was reported that there was a significant decline in stowaway cases at US ports in the first six months after the introduction of the ISPS Code [7]. Meanwhile, the DJPL oversees its implementation and compliance with its implementation in Indonesia. For the Standards of Training, Certification, and Watchkeeping for Seafarers (STCW), seafarers and related security personnel have received the IMO model course training as required.

A conducive climate is needed to create competitive ports so that all activities at the port can run smoothly without interruption. Port security is an important issue and it is the main requirement in determining the location of a business in multinational investment activities [8]. Furthermore, a port is also required to provide security both from the water and shore sides because any security disturbance can significantly impact port performance and the economy [9]. This paper aims to determine the impact of sea transportation security at ports, and the expectations of all parties with maritime security interests on Indonesia's economic growth.

2. Research Methodology

A mixed method (quantitative and qualitative) is used in this study. The data consisted of two kinds: secondary data and primary data. For primary data, this paper uses transcripts of eight informants involved in a Focus Group Discussion (FGD). The quantitative analysis uses Input—Output analysis techniques, while the qualitative method uses the NVivo tools analysis.

2.1. Input–Output Quantitative Analysis

The Input–Output (IO) analysis shows that the economy contains relationships and interdependencies in the industrial sector. The inputs of one industry are the outputs of other industries, and vice versa. Eventually, the interrelationships between them lead to an equilibrium between supply and demand in the economy. The initial purpose of developing this model is to determine the interdependence between industries. Today, the basic concepts created by Leontief are critical components of many types of economic analysis. The main objective of Leontief's Input-Output model, which was developed in the 1930s, is to study the interdependence of various sectors in the economy [10]. Several previous researchers have conducted studies using the I–O model with several variations. For example, Sener's research (2013) combining the I–O model with time-series regression states that the integrated I–O model can also be extended to include other indexes modeled with a time-series approach, such as sector indexes. For example, one can analyze changes in the NASDAQ or the DOW stock market indexes on the economy [11].

According to Firmansyah et al. (2018), the huge potency of fisheries and the fish manufacturing industry has been made significant by the Indonesian government by employing the Input-Output analysis approach [12]. The article aimed to elaborate on and evaluate the role of the sea transportation sector to support other sectors with an up-to-date I–O RAS matrix.

The novelty of this research is the combination of sectoral Input–Output with the up-to-date Input–Output RAS matrix in the transportation sector, and the improvement of stakeholder expectation with the NVivo method.

As a quantitative model, the I–O table will provide a comprehensive picture of: (i) The structure of the national/regional economy, which includes the output structure and added value of each sector; (ii) The structure of intermediate inputs, namely the use of various goods and services by the production sectors; (iii) The structure of the supply of goods and services, both in the form of domestic production and goods originating from imports; and (iv) The structure of demand for goods and services, both intermediate demand by the production sectors and final demand for consumption, investment, and exports [13].

Sectors within the production system are called endogenous variables, while variables outside the production system, namely those in the second, third, and fourth quadrants, are called exogenous variables. Thus, it can be clearly seen that the I–O model differentiates endogenous variables from exogenous variables. I–O is also used outside the production system in the form of final requests. The I–O result table consists of input coefficient variables called the input coefficient matrix, a variable or multiplier matrix, an attractiveness index variable, a driving force index, various supporting variables, and other analytical variables depending on the breadth of the field to be discussed. The I–O table format can be seen in Table 1 below [14].

Table 1. Input–output table.

Input \ Output		Production Sector 1	2	....	N	Final Demand	Total Output
Production Sector	1	Z11	Z12	....	Z1N	F1	X1
	2	Z21	Z22	....	Z2N	F2	X2
	3	Z31	Z32	....	Z3N	F3	X3
	.....			....		....	....
	N	ZN1	ZN2	....	ZINN	FN	XN
Primary Input		V1	V2	....	VAN		
Total Input		X1	X2	.....	XN		

Source: [14].

Table 1 represents the distribution of output, both from domestic and abroad. At the same time, when viewed vertically, these figures are also the composition of inputs of a sector obtained from other sectors. Table 1 also shows the arrangement of numbers in the form of a matrix between the sectors in an economy. There are interrelated relationships among several sectors. Sector 1, the output of which is Z1, is allocated in a row, namely Z11, Z12, and Z13, respectively, to sectors 1,2,3 as an intermediate demand of F1 to meet the final demand. In general, the I–O equation is formulated below (1).

$$\sum_{i=1}^{n} Z_{ij} + F_i = X_i \tag{1}$$

where i is $1, 2, 3, \ldots, n$; Z_{ij} is the number of sectors i outputs used as sector j inputs; F_j is the final demand for the transportation sector (i); X_j is the total output of sector i. The general equation that expresses the relationship between the columns is denoted in Equation (2).

$$\sum_{j=1}^{n} Z_{ij} + V_j = X_j \tag{2}$$

where j is $1, 2, 3, \ldots, n$; Z_{ij} is the number of sectors i outputs used as sector j inputs; V_j is the final demand for the transportation sector (j); X_j is the total output of sector i.

2.2. Input–Output Linkage Analysis

Linkage analysis can provide an overview of inter-sectoral linkages. Linkage consists of forwarding linkage and backward linkage directly or indirectly. The forward linkage itself provides information on the degree of linkage between sectors that produce outputs used as inputs from other sectors. In comparison, the backward linkage is used to see the linkage of a sector that supplies inputs to the sector under study [13].

2.3. Impact Analysis of Input–Output Spread

The direct and indirect linkages forward and backward are not sufficient to be used as a basis for selecting key sectors. These indicators cannot be compared between sectors because the role of final demand for each sector is not the same. Therefore, it must be normalized by comparing the average impact of all sectors. This analysis is the impact of the spread that can determine whether a sector can grow its upstream or downstream sector.

2.4. Input–Output Multiplier Analysis

Multiplier analysis is one type of I–O analysis, and there are two types of multipliers, namely, Type I Multipliers and Type II Multipliers. The Type I Multiplier is obtained from further processing of the open Leontief inverse matrix, while the Type II Multiplier is obtained from the closed Leontief inverse matrix. The Type I Multiplier and the Type II Multiplier result from an impact mechanism process consisting of an initial effect, a first-round effect, an indirect impact (industrial support effect), and a consumption-induced effect. Based on the Leontief inverse matrix for both the open and closed model, the values of the output multiplier can be determined.

2.5. RAS Method Input–Output Matrix Update

The creation of an I–O table for this region cannot be achieved every year. Each region can only compile I–O tables with a very long interval period of up to 10 years. To overcome the problem of creating regional I–O matrices, we can use various approaches to update the data. The non-survey method for compiling or updating the regional I–O matrix can use the RAS method [13].

2.6. Qualitative Analysis (NVivo)

The NVivo method is used as an analytical model because a standard systematic data analysis process is used in this study, namely coding analysis. The qualitative data management process in NVivo can assist researchers in better and more efficient data management. The most important thing to note in its use is the existence of coding and nodes. The coding process is filling the nodes with information related to the categories of concepts (codes) that have been formed in the node system. Before carrying out data processing and analysis, it is necessary to understand the flow of qualitative research analysis by Jozef Raco in Kalambo et al. (2019) [15].

Interview data was captured by conducting interviews with eight informants who were grouped into three groups, namely those who act as regulators (policymakers), operators (service users), and observers of maritime transportation. The research data obtained from interviews were then processed using the Nvivo 12 Plus software tools. The NVivo 12 Plus software is used in coding from data from interviews with selected sources. The primary purpose of coding is to form main categories based on various data sources that researchers have collected.

3. Research Result and Discussion

3.1. Indonesian Input–Output Analysis

Based on Indonesia's I–O 2019 table of 53 sectors, the highest contribution of each sector to final demand are: (i) Building sector at 22.55%; (ii) Food and Beverage Industry sector at 12.73%; (iii) Wholesale and Retail Trade sector and Non-Cars and Motorcycles with 8.16%; and (iv) The Sea Transportation and Port Security sector was 0.47%. The contribution of the above sectors to the 2019 National GDP were: (i) The building sector by 11.00%; (ii) The Wholesale and Retail Trade sector, and Non-Cars and Motorcycles by 10.72%; (iii) The Food and Beverage Industry sector by 6.56%; and (iv) The Sea Transportation and Port Security by 0.34%.

3.2. Sector Forward and Backward Linkage Analysis

The value of the direct forward linkage of the Sea Transportation and Port Security sector is 0.129. The indirect is 1.576, which means if there is an increase in final demand of one million rupiahs. The Sea Transportation and Port Security sector's output is directly allocated to other sectors, including the construction sector itself, which will experience an increase of IDR 129,000. Meanwhile, the value of the forward linkage directly means that if there is an increase in final demand by one million rupiahs, the construction sector's output is sold or allocated either directly or indirectly to other sectors, including the construction sector itself, which will increase by IDR 1,576,000.

The value of the direct backward linkage of the Sea Transportation and Port Security sector is 0.154 and the indirectly backward is 1.734. The value of this direct backward linkage can be interpreted as if there is an increase in final demand by one million rupiahs, the construction sector will directly increase demand for other sectors, including the construction sector itself, by IDR 154,000. As for the value of direct and indirect backward linkages, if there is an increase in the final demand of one million rupiahs, the construction sector will increase its input demand to other sectors, either directly or indirectly by IDR 1,576,000.

3.3. Multiplier Impact Analysis

Multiplier impact analysis of the Sea Transportation and Port Security sector output was 1.734, the labor multiplier value was 0.067, and the income value was 0.001. It can be interpreted that if there is an increase in the final demand for the sea transportation sector by one million rupiahs, then the average output in other sectors will increase by IDR 1,734,000.00; labor increase by one person, and income increases by IDR 10,000.

3.4. Maritime Security Policy Linkages with Other Sectors

The relationship between Maritime Security Policies mostly relates to the Port Security sector with other sectors based on the analysis results. The Maritime Security Policy outputs through the Port Security sector, the building sector, the shipping industry sector, and the metal, computer, electronic, optical, and electrical goods industrial sector, which are directly sold or allocated to other sectors, including the sector itself worth IDR 1.28 trillion. The Maritime Security Policy outputs through the Port Security sector, the building sector, the shipping industry sector, and the metal, computer, electronic, optical, and electrical goods industrial sector, which are not directly sold or allocated to other sectors, including the sector itself, amounting to IDR 3.33 trillion.

The requests for Maritime Security Policy input through the Port Security sector, the building sector, the shipping industry sector, and the metal, computer, electronic, optical, and electrical goods industry sector, which are directly sold or allocated to other sectors, including the sector itself to the amount of IDR 12.38 trillion. The requests for Maritime Security Policy input through the Port Security sector, the building sector, the shipping industry sector, and the metal, computer, electronic, optical, and electrical goods industry sector, which are indirectly sold or allocated to other sectors, including the sector itself in the amount of IDR 142.02 trillion.

The Maritime Security Policy in 2019, through the Port Security Sector, impacted other sectors. The most significant change is the output in the building sector of IDR 3.92 trillion, the Wholesale and Retail Trade sector, Non-Cars and Motorcycles of IDR 2.87 trillion, and the Port Security sector itself amounted to IDR 2.35 trillion as seen in Figure 1.

The most extensive labor absorption is in the food crop agriculture sector with as many as 56,519 people, the Food and Beverage Industry sector with 5361 people, the Plantation sector with 3382 people, and the Chemical, Pharmaceutical, and Drug Industry with 2575 people as seen in Figure 2. Meanwhile, in labor income, the food crop agriculture sector was IDR 614.93 billion, the Food and Beverage industry sector was IDR 58.32 billion, and the Wholesale and Retail Trade sector was IDR 40.24 billion.

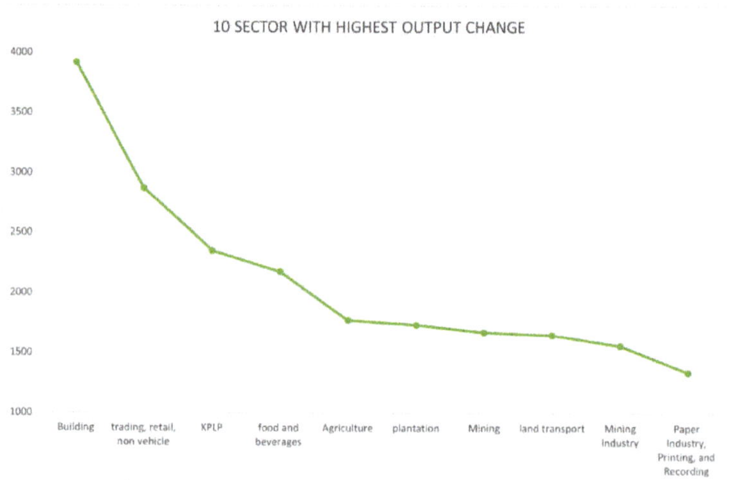

Figure 1. Sectors Affected by the Implementation of Maritime Security Policy in 2019 (Source: [16]).

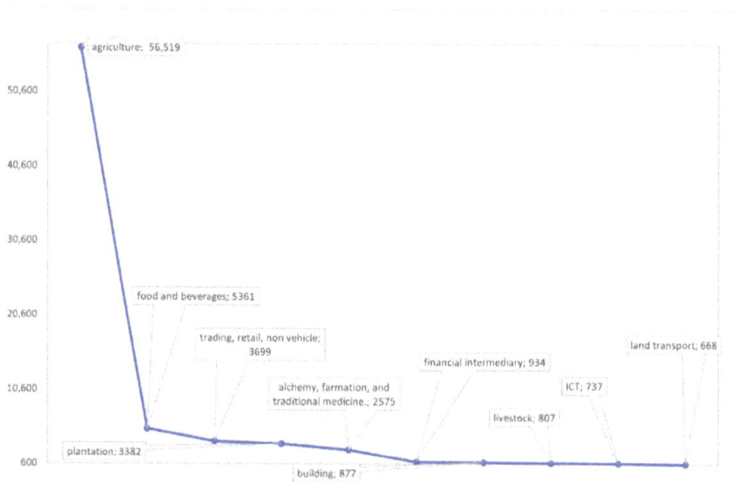

Figure 2. Sectors Affected by Implementation of Maritime Security Policy from total Labor (Source: [16]).

3.5. Impact of Maritime Security Policy Implementation on Economic Growth

In 2019, the Ministry of Transportation allocated a budget of IDR 5.75 trillion to implement maritime security policies, especially the implementation of the ISPS Code in Indonesia. This simulation is to determine the economic impact of maritime security policies, especially the implementation of the ISPS Code worth IDR 5.75 trillion in 2019. The results obtained are a change in output of IDR 39.52 trillion, an increase in GDP of IDR 21.99 billion, and an increase in worker income of IDR 860 billion. The number of workers absorbed is 79,084.

3.6. NVivo Qualitative Analysis

The results of the FGD underlined several items as being part of the Maritime Security Policy Implementation, namely: Stakeholder Synergy, Shipping Safety and Security, Maritime Environment Protection, Connectivity, Dangerous Goods and Solid Bulk Handling, Role of Surveyors, Verification and Technical Investigation, HR Competence, Information and Communication Technology, Maritime Concepts, Data Integration, Dangerous Goods and Solid Bulk, ISPS Code, Data Compliance, Maritime Sector Potential, Implementation Commitments, Information Technology and Port Infrastructure.

After grouping, several references are obtained in the aggregate (20 nodes) as seen in Table 2, which have the most significant contribution to the overall hierarchy. It is indicated, as a whole (3 categories of informants); both implicitly and explicitly, alluding to the issue of stakeholder synergy. In this case, the stakeholder synergy nodes contain statements of informants from all categories that discuss the need for stakeholder involvement in marine resource management, HR competence, shipping administration, sailing safety and security, logistics business performance, etc.

Table 2. NVivo node reference.

No.	Nodes	Reference	%	Condition
1	Stakeholder Synergy	61	13%	Sufficient
2	Shipping Safety & Security	57	12%	Sufficient
3	Maritime Environment Protection	44	9%	Not Sufficient
4	Connectivity	34	7%	Not Sufficient
5	Economic Aspect	34	7%	Good
6	Hazardous Goods & Solid Bulk Handling	30	6%	Not Sufficient
7	Surveyor's Role	26	6%	Not Sufficient
8	Technical Verification & Tracing	23	5%	Good
9	HR Competence	20	4%	Sufficient
10	Information and Communication Technology	17	4%	Good
11	Maritime Concept	16	3%	Not Sufficient
12	Data Integration	14	3%	Good
13	Dangerous Goods & Solid Bulk	14	3%	Not Sufficient
14	ISPS Code	13	3%	Good
15	Data Compatibility	13	3%	Good
16	Price Disparity	12	3%	Good
17	Maritime Sector Potential	11	2%	Good
18	Implementation Commitment	9	2%	Sufficient
19	Information Technology	9	2%	Good
20	Port Infrastructure	9	2%	Good
	Total number	466		

(Source: Data Processed).

After discussions between experts, six categories of nodes were deemed to be in poor condition and required more attention, namely: the protection of the maritime environment, connectivity, handling of dangerous goods and solid bulk, the role of surveyors, understanding of maritime concepts, and implementation commitments.

4. Conclusions

The linkage of Maritime Security Policy implementation with other sectors in the Indonesian Economy, namely, the Output of Maritime Security Policy implementation through sectors: KPLP, building sector, shipping industry sector, industrial sector of metal goods, computers, electronics, optical and electrical equipment directly sold or allocated to other sectors including the sector itself is 0.007% of the total national final demand. Meanwhile, what was not directly sold or allocated to other sectors, including the sector itself, was 0.019% of the total national final demand. The requests for input for the implementation of the Maritime Security Policy through the following sectors: KPLP, the building sector, the shipping industry sector, the metal goods industry, computers, electronic goods, optics, and electrical equipment which are directly sold or allocated to other sectors including the sector itself are of 0.042% of the total national input. Meanwhile, indirectly sold or allocated to other sectors, including the sector itself, amounted to 0.48% of the total national input.

The implementation of the Maritime Security Policy in 2019 through the KPLP sector was IDR 5.75 trillion. It impacts other sectors, where the largest is in the building sector, which is 0.013% of the total national output, then the Wholesale and Retail Trade sector, the Non-car and Motorcycle sector by 0.01%, and the KPLP sector itself by 0.008%. The most significant employment absorption of the national workforce is in the Food Crop Agriculture sector, which is 0.043%; Next are the Food and Beverage Industry sector (0.004%), the Wholesale and Retail Trade, the Non-car and Motorcycle sectors (0.003%), and the Plantation sector (0.003%).

Meanwhile, in labor income, the largest sector was the Food Crop Agriculture sector (0.001%), followed by the Food and Beverage Industry sector (0.006%), and Wholesale and Retail Trade sector (0.006%). The impact of implementing the Maritime Security Policy in 2019 on the increase in GDP is 0.14% of the National GDP. This increase absorbs the workforce by 0.06% of the total national workforce, and increases workers' income in all sectors by 0.014% of the national workforce income.

The stakeholder expectations for the implementation of Maritime Security policies on a priority scale were: (i) 13% of stakeholders expect Synergy between Stakeholders; (ii) 12% expect the realization of Shipping Safety and Security; (iii) 9% expect the realization of Maritime Environment Protection; (iv) 7% expect the creation of Connectivity, and (v) 6% expect the implementation of the Handling of Dangerous Goods and Solid Bulk.

Author Contributions: All authors contributed to the study. Conceptualization, A.M. and M.Z.H.; methodology, A.M. and E.S.; software, A.M. and E.S.; validation, A.M., E.S. and M.Z.H.; formal analysis, A.M.; investigation, A.M.; resources, A.M.; data curation, A.M. and M.Z.H.; writing—original draft preparation, A.M.; writing—review and editing, A.M. and E.S.; visualization, A.M. and E.S.; supervision, M.Z.H.; project administration, A.M.; funding acquisition, A.M. All authors have read and agreed to the published version of the manuscript.

Funding: This research received no external funding.

Institutional Review Board Statement: Not applicable.

Informed Consent Statement: Informed consent was obtained from all subjects involved in the study.

Data Availability Statement: In this study, the data collection method was carried out through the Focus Group Discussion and the Indonesian Statistic Agency.

Acknowledgments: Thank you to the Trisakti University Economics Program for supporting the implementation of this study.

Conflicts of Interest: The authors declare no conflict of interest.

References

1. Lasabuda, R. Pembangunan Wilayah Pesisir dan Lautan Dalam Perspektif Negara Kepulauan Republik Indonesia. *J. Ilm. Platax* **2013**, *1*, 92–101. [CrossRef]
2. Rustam, I. Tantangan ALKI dalam Mewujudkan Cita-cita Indonesia sebagai Poros Maritim Dunia. *Indones. Perspect.* **2016**, *1*, 1–21. [CrossRef]
3. UNCTAD. *World Seaborne Trade*; United Nations Publication: New York, NY, USA, 2019.
4. UNWTO. *Compendium of Tourism Statistics, Data 2013–2017*; World Tourism Organization (UNWTO): Madrid, Spain, 2019.
5. Flynn, S.E. Port Security Is Still A House of Cards. *Far East. Econ. Rev.* **2006**, *169*, 5–11.
6. Hutagalung, S.M. Penetapan Alur Laut Kepulauan Indonesia (ALKI): Manfaatnya dan Ancaman Bagi Keamanan Pelayaran di Wilayah Perairan Indonesia. *J. Asia Pac. Stud.* **2017**, *1*, 75–91. [CrossRef]
7. IMO. *Guide to Maritime Security and the ISPS Code*, 2012th ed.; Inter-Governmental Maritime: London, UK, 2012.
8. Ruswandi, A.; Gartika, D. Strategi Pengembangan Investasi di Sekitar Pelabuhan Perikanan Tipe B di Jawa Barat. *J. Akuatika* **2013**, *4*, 89–101.
9. Adam, L.; Dwiastuti, I. Membangun Poros Maritim Melalui Pelabuhan. *J. Masy. Indones.* **2015**, *41*, 163–176.
10. Miller, R.E.; Blair, P.D. *Input-Output Analysis, Foundations and Extensions*, 2nd ed.; Cambridge University Press: New York, NY, USA, 2009.

11. Sener, S. *A Time-Series Integration of Confidence Indices with Input-Output Model*; The George Washington University ProQuest Dissertations Publishing: Michigan, MI, USA, 2013.
12. Firmansyah; Oktavilia, S.; Sugiyanto, F.X.; Hamzah, I.N. Impact of Fishery Policy on Fishery Manufacture Output, Economy and Welfare in Indonesia. In Proceedings of the IOP Conference Series: Earth and Environmental Science, Banda Aceh, Indonesia, 26–27 September 2018.
13. Daryanto, A.; Hafizrianda, Y. *Analisis Input-Output & Socia l Accounting Matrix untuk Pembangunan Ekonomi Daerah*; IPB Press: Bogor, Indonesia, 2010.
14. Amir, H.; Nazara, S. Analisis Perubahan Struktur Ekonomi (Economic Landscape) dan Kebijakan Strategi Pembangunan Jawa Timur Tahun 1994 dan 2000: Analisis Input-Output. *J. Ekon. Dan Pembang. Indones.* **2005**, *5*, 37–55. [CrossRef]
15. Kalambo, Y.L.; Susmoro, H.; Riono, R. Peningkatan Kemampuan Organisasi Tim Unit Survei Tanggap Segera (USTS) Pusat Hidrografi Dan Oseanografi Angkatan Laut (Pushidrosal) Dalam Operasi Penanggulangan Bencana Alam dan SAR di Laut. *J. Studi Manaj. Organ.* **2019**, *16*, 18–26.
16. Ahmad. Analisis Keterkaitan Kebijakan Keamanan Maritim Dengan Pertumbuhan Perekonomian Nasional. Ph.D. Dissertation, Trisakti University, Jakarta, Indonesia, 2022.

proceedings

Proceeding Paper

Non-Traditional Security Issues in Southeast Asia during COVID-19: Implications and Mitigation Strategies by ASEAN †

Nur Ain Yaacob [1,*], Muhammad Izmer Yusof [2], Suzei Mat Nuruddin [1], Zarina Mohd Zain [1] and Nuramira Akmal Mustapa [3]

1 Faculty of Administrative Science and Policy Studies, Universiti Teknologi Mara, UiTM Seremban, Seremban 70300, Malaysia
2 Pusat Sains Liberal, Universiti Malaysia Perlis, Arau 02600, Malaysia
3 Counsellor at High Commission of Malaysian in New Delhi, Kementerian Luar Negeri, Putrajaya 62100, Malaysia
* Correspondence: nurainyaacob@uitm.edu.my
† Presented at the International Academic Symposium of Social Science 2022, Kota Bharu, Malaysia, 3 July 2022.

Abstract: Since the end of the Cold War in the 1990s, problems of security have become increasingly important. The contemporary notion is that the security of the state is no longer threatened just by conflict, but also by other circumstances and challenges that continue to endanger the well-being of citizens and the state. The objective of this concept paper is to examine ASEAN's response towards selected non-traditional security issues during the COVID-19 pandemic, such as infectious disease (COVID-19), regular and irregular migration, food security, cyber security, and the implications faced by ASEAN in such matters. The library research method was used by the author in this article. Several keyword search strings were used, including "nontraditional security issues", "COVID-19 in Southeast Asia", and "mitigation measures by ASEAN". Thus, this study offers new insights and suggestions by focusing on the implications and prospects of ASEAN solutions to NTS concerns caused by COVID-19 as the virus spread throughout Southeast Asia. The findings also indicate that regional cooperation is an integral feature of coordinated efforts by governments and non-state entities, such as non-governmental organizations (NGOs), in combating these security concerns. The blueprint for NTS also can be suggested to help further understand these issues.

Keywords: ASEAN; implications; non-traditional security issues; policy response

check for updates

Citation: Yaacob, N.A.; Yusof, M.I.; Nuruddin, S.M.; Zain, Z.M.; Mustapa, N.A. Non-Traditional Security Issues in Southeast Asia during COVID-19: Implications and Mitigation Strategies by ASEAN. *Proceedings* **2022**, *82*, 90. https://doi.org/10.3390/proceedings2022082090

Academic Editor: Mohamad Rahimi Mohamad Rosman

Published: 30 September 2022

1. Introduction

Southeast Asia has seen multiple crises in the last decade, all of which have had a substantial impact on state security agendas. In 1997, there was an Asian financial crisis, which was followed by the SARS pandemic in 2003, avian influenza in 2005, Typhoon Haiyan in 2013, the escalation of the Rohingya refugees' issue in 2017 and, currently, the region is dealing with the COVID-19 pandemic. All of these were nonmilitary, yet they were discovered to pose a serious threat to the existence and well-being of the nations and societies [1]. Indeed, NTS challenges continue to threaten the well-being of nations and communities throughout Asia and the rest of the world. Threat issues are no longer stipulated only the threat of war, but have expanded to include health, food, cyberspace, the economy, smuggling, human trafficking, drugs, and many other issues. Non-traditional security is regarded as a non-military source that endangers the well-being of nations, societies, and communities [2]. These NTS issues are progressively becoming the highest priority for most governments and policymakers in Asia, as well as in other parts of the world. In summary, the proponents of NTS ought to be extended in terms of security so that those vulnerable, namely non-state actors, political entities, organizations, and human individuals, can have a bigger part in providing or ensuring their security. Transnational issues can cause societal and political instability and, hence, become a threat to society.

Thus, it is vital to have multilateral and regional cooperation, as national solutions are not adequate to solve these issues.

The COVID-19 pandemic has posed several security risks to Southeast Asian countries, and has had a devastating impact on their politics, economies, and social stability. As an infectious disease, COVID-19 causes a lot of damage to the sustainable growth of the state. Due to restrictions on people's movement, other concerns arise, such as regular and irregular migration, which has a spillover effect into neighboring Southeast Asian countries, such as the issue of "Rohingya refugees". Furthermore, it prompted another issue that posed a threat to people's health, namely food security challenges, such as food availability, people's access to food, and adequate food preparation to maintain sanitation and maximum nutrition. Finally, as a result of increased connectivity due to the COVID-19 shutdown and online learning, cyber security issues have increased, resulting in a major increase in cybercrime issues, such as the spread of malware, ransomware, and potential victims of online child sexual exploitation. Therefore, this paper is important as it provides a brief explanation of way of addressing the NTS issues faced by ASEAN due to COVID-19, as well as highlighting the mitigation measures that were implemented to deal with such issues. It will also give guidance to all policymakers, including state and non-state actors, for them to recognize the risk connected with COVID-19 and other NTS risks.

2. Methods

The discussion of this conceptual paper was created using a scoping study of the literature and a secondary data collection approach. In this study, data was gathered using the library research approach. It is based on the findings of previous research on this topic by other researchers. This theoretical research can be used to develop a conceptual framework or hypothesis that reflects the researcher's overall writings [3]. Therefore, by utilizing this approach, the researcher would collect significant data on material relating to the researcher's topic from books, journals, documents, manuscripts, papers, conferences, and online sources, and evaluate it using content analysis. Several keyword search phrases were utilized in this paper, which are "nontraditional security issues", "COVID-19 in Southeast Asia", and "mitigation measures by ASEAN". The literature search was carried out on many websites, including Scopus and ScienceDirect. The initial search yielded 25 (SCOPUS) and 10 (Science Direct) items. However, 20 articles were eliminated owing to their premature conclusions and anecdotes, or because they did not address the NTS concerns during COVID-19 and mitigation measures. Articles that are incomplete, or have a broken connection and overlap were also found during the search. As a result, the final work will be scrutinized until it is reduced to no more than 15 articles, each supported by a variety of publications, hearings, documents, and other reports.

3. Implications of Non-Traditional Security Issues in Southeast Asia during COVID-19

3.1. *Infectious Disease (COVID-19)*

One of the most serious global non-traditional security risks is the COVID-19 pandemic. This worldwide health crisis has resulted in an unacceptably high number of deaths and a significant economic impact. This year, we have seen that the coronavirus disease 2019 (COVID-19) has been evolving into different variants. Even though vaccines are being distributed, the pandemic continues to have a devastating effect on the world. The multiple consequences of COVID-19 have worsened existing socioeconomic inequalities while creating new vulnerabilities and raising the overall degree of risk throughout the world [4]. In Southeast Asia, as of March 2022, ASEAN member countries have confirmed at least 265,744 reported cases and 3919 death cases. Vietnam has the most daily cases, with around 164,596 cases recorded, while Cambodia has the greatest mortality rate, with approximately 3043 death cases recorded, despite having the lowest incidence of daily cases among ASEAN member nations [5].

Before the COVID-19 global recession that started in the second half of 2020, Southeast Asia was one of the regions with the highest rates of industrialization and urbanization, as well as one of the regions with the highest rates of population growth. It was one of the first regions to be affected when the World Health Organization (WHO) declared the coronavirus outbreak in the People's Republic of China. This was due to its close geographic proximity to China, as well as its business, tourism, and supply chain ties to the country. Most ASEAN countries have taken a variety of steps to combat the spread of COVID-19, including forming a special task force on the disease, enacting movement control orders, closing borders, restricting travel, and shutting non-essential businesses, schools, and public spaces [6].

The ASEAN market is expected to lose US$400 billion (RM1.62 trillion) of GDP growth in 2020 and 2021 because of the pandemic. The dramatic decrease in oil prices, as well as the reduction in exports in several Southeast Asian economies, notably Thailand and Vietnam, has contributed to the economic slowdown. This will have a substantial influence on the business activity of several countries in the area, including the United States. Aside from that, socially disadvantaged groups in ASEAN countries, particularly those in the informal sector, have been adversely affected by COVID-19, as well as by the containment measures taken by governments. They did not have enough access to healthcare, and many had constrained living environments that increased their chances of being exposed to the disease. However, the community quarantine also places them in a more vulnerable situation with no income, loss of job, no savings to restock goods for families, and no access to the social safety net.

3.2. Regular and Irregular Migration

The effects of COVID-19 have impacted almost all migrants, whether they are migrating domestically or internationally, forcibly or voluntarily, and on a regular or irregular basis [7]. It caused further difficulties for migrants, particularly those who were illegal. Regular migration can become irregular due to changes in national laws and policies, and vice versa. Since the deployment of strengthened border controls and limits on the freedom of movement around the region due to COVID-19, there has been an increase in countries refusing entry to abandoned ships carrying Rohingya migrants, prompting worries of a repeat of the 2015 "boat people crisis". Because there was no coordinated response to the crisis at the time, most of the countries involved agreed to the 2016 Bali Declaration, which outlined action to avoid its repetition. Even though at a recent meeting of the Bali Process Task Force on Planning and Preparedness, countries, such as Indonesia and Malaysia, stressed the importance of "saving lives at sea and not putting people's lives and safety at risk when responding to irregular maritime migration", this seems to have not been maintained in the face of COVID-19. Malaysia has previously allowed boats, although, on an ad hoc basis, and its position has hardened in recent months [8].

Many Rohingya Muslims have fled Myanmar and are now living in overcrowded refugee camps in Bangladesh, with some being trafficked throughout Southeast Asia. They have a far higher risk of infections of the coronavirus because they lack basic healthcare and precautions [9]. It is estimated that the majority of the seven million migrant workers in and from ASEAN are undocumented. As undocumented workers, they are not entitled to any form of social protection. Apart from the risk to their health, this also jeopardizes the host country's restricting measures. After losing their employment in Thailand owing to travel restrictions enforced from March to May, about 150,000 migrant workers from Myanmar, 50,000 from Cambodia, and 60,000 from Laos went home. Indeed, COVID-19's impact on the labor market left many of them trapped and with terrible employment prospects. Due to this, the number of people living in poverty and hardship will increase [10].

Furthermore, the pandemic also put them at risk due to uneven refugee protections across Southeast Asia countries. Only a few states have ratified the 1951 United Nations Refugee Convention, and the rest have not agreed-upon principles or standards for protection within ASEAN. This allows the countries to react in whichever manner they see fit. So far only Cambodia, the Philippines, and Timor-Leste in Southeast Asia have agreed to sign the United Nations' 1951 Convention on Refugees and the 1967 Protocol on Refugees'

Status [11]. The UNHCR Director for Asia and the Pacific has criticized the Bali Process, claiming that there is no concerted reaction from ASEAN member nations, and that worries of COVID-19 are being exploited to send boats out to sea. As a result of this issue, at least 30 refugees died during the journey after being stranded on the sea for about seven months.

3.3. Food Security

There are around 61 million people in Southeast Asia who are malnourished, and this number may continue to rise as a result of the COVID-19 outbreak [12]. Indeed, COVID-19 has exacerbated food security issues in some parts of the world, particularly in the Asia–Pacific region, where stringent quarantine regulations and export prohibitions on essential foods have affected every stage of the food supply chain [13]. The logistics of the food value chain, including transportation, warehousing, procurement, packaging, and inventory management, have been affected, which has an effect not only on the amount of food that is available but also on its quality, freshness, safety, market access, and cost. As a result, it leads to an increase in hunger and the number of malnourished children under five that suffer from poor growth and development globally due to COVID-19. Indonesia is one of the countries hardest hit by the COVID-19 virus in Southeast Asia. Overburdened healthcare facilities, disrupted food-supply networks, and income loss as a result of the pandemic might all contribute to a significant increase in the number of malnourished children in the archipelago. Similar to Cambodia, the rate of poverty remains high, and there is lack of food supply, and rise in the prices of food due to the crisis. Singapore also faces major challenges in the food supply chain, including food imports, local manufacturing, retail grocery shops, food and beverage services, and in terms of vulnerable people's care [14].

The pandemic poses a threat to food security in terms of the production and availability of food, physical and market access, and impact from imported products. Due to lockdown, all farmers are unable to get agricultural products required for the cropping process, such as fertilizer, seeds, and pesticides. If there is not enough fertilizer, agricultural output will decrease. If there are not enough seed and pesticides, crops will be more sensitive to natural stresses, such as droughts and floods, as well as pests and diseases. Thailand and Vietnam were affected by this problem, since they are among Asia's top rice exporters. The prices of food will also be increased when physical access is restricted and food is scarce. When this happens, it affects poor people the most. The poorest households were unable to purchase enough nutritious food as prices increased, and most of them reduced their food intake to save more money. In Malaysia, 63.76% of respondents from the Bottom 40% (B40) families have reduced their food intake to save more money [15]. As in Vietnam and Laos, individuals with low incomes were forced to eat less nutritious meals owing to increased food costs.

3.4. Cyber Security

The COVID-19 pandemic has increased the pace of digital transformation, compelling both the public and private sectors to embrace it, transforming the work, study, retail, and business environments. Digital transformation has also opened up new opportunities for cybercriminals to attack the computer networks and systems of people, companies, and even multinational corporations [16]. As refer to the Figure 1 below, it shows that Southeast Asia has a 66% Internet penetration rate. On the higher end of the spectrum, Brunei has a 95% of Internet penetration rate followed by 88% in Singapore, 83% in Malaysia, 75% in Thailand, 70% in Vietnam, 67% in the Philippines, and 64% in Indonesia. At the lower end of the spectrum, in Laos and Myanmar, the penetration rate is 43% and 41%, respectively.

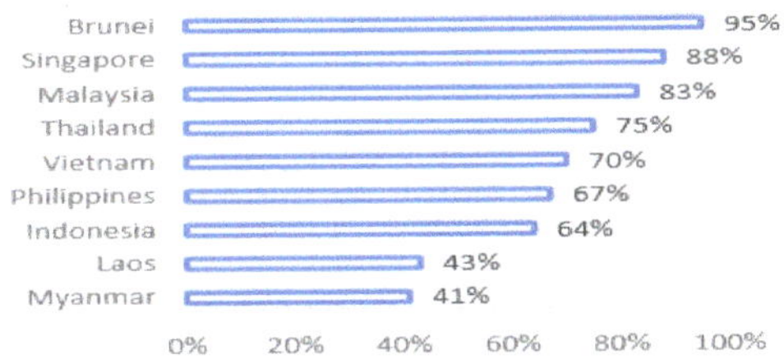

Figure 1. Internet penetration (%) in Southeast Asia Countries [17].

Thus, as people become more reliant on the Internet, a multitude of new security threats have emerged, all of which have the potential to cause significant harm. As a consequence, digital economic trust and resilience will be diminished, preventing the region from achieving its full digital potential if no action is taken to make it secure. Malicious emails, fraud, phishing attacks, and malware have all escalated in the last 18 months as a result of the pandemic. Small and medium-sized businesses (SMEs) have become popular targets for cybercriminals, with many being hacked or losing data. Most of these companies are unaware of how to protect their data from cyberattacks. Small and medium-sized enterprises in the region should prioritize cyber security in light of recent cyberattacks on the systems of several ASEAN nations, including Indonesia and Malaysia [18].

Furthermore, as Internet access and mobile device penetration have increased during the COVID-19 lockdown, children have spent more time online than ever before in recent years. Due to this, there has been a significant increase in the number of potential victims of online child sexual exploitation. It also stated that COVID-19 has resulted in a major surge in child exploitation in many parts of the world. Lockdowns are driving children from low socioeconomic situations to engage in Internet prostitution due to financial concerns. It is also becoming increasingly difficult for these children to receive adequate supervision because of limited access to childcare and other support services.

4. Asean Mitigation Strategies towards Selected Non-Traditional Security Issues during Pandemic COVID-19

4.1. The COVID-19 Pandemic

Southeast Asia has proved that regional cooperation which is through ASEAN is essential for assisting the AMSs to mitigate the COVID pandemic without working at the global level. Several efforts and initiatives were introduced and implemented by ASEAN in order to deal with this public health crisis. It can be seen that, on 10 March 2020 in Da Nang, Vietnam, the 26th ASEAN Economic Ministers (AEM) conference purposely stressed the importance to have a coordinated effort to strengthen the economy in response to the outbreak of COVID-19. It also emphasized the use of technology, especially in economic activity, such as e-commerce and trade facilitation in the digital economy platforms, such as the ASEAN Single Window, which assists the SMEs to continue operating during COVID-19. These SMEs were being given an opportunity to upgrade and enhance their supply chain connectivity, as well as allow them to operate the business as usual [5].

Furthermore, instead of having regional cooperation, ASEAN is also collaborating with its Development and Dialogue Partners. This started in March 2020, whereby ASEAN and the EU collaborated through a Ministerial Video Conference to discuss several challenges in mitigating the COVID-19. The important outcome of the meeting was that ASEAN needs to invest more in scientific research and maintain the supply chain so that it will be able to minimize the negative impact of COVID, particularly in the economy and social sector. Besides this, ASEAN continues to work with another dialogue partner, the US. In April 2020, the two powers purposely discussed improving a public health cooperation in a more progressive manner. Furthermore, ASEAN also continues to work together with the ASEAN

Plus Three (South Korea, Japan, China) through video conferencing by sharing and learning from South Korea how to best manage the pandemic, because of the successful effort carried out by South Korea to flatten the curve without imposing movement restrictions or an economically damaging lockdown. Additionally, the establishment of the ASEAN Respond Fund also was discussed during the Special ASEAN Summit in order to ensure every proposal of the meeting could be implemented, and that the goal to mitigate the pandemic could be easily achieved.

Additionally, Figure 2 below show that ASEAN has also created several important apparatuses to combat this pandemic, such as the ASEAN Emergency Operating Centre Network for Public Health Emergency, and the ASEAN BioDiaspora Virtual Centre. These two important mechanisms were established in order to facilitate and provide accurate information and technical inputs related to this disease. It is also responsible for initiating health protocols, such as quarantine measures, testing techniques, tracing close contacts, and isolation that must be followed by AMSs. Apart from that, ASEAN also strives to combat false news and misinformation about COVID-19 issues by establishing an ASEAN Risk Assessment and Risk Communication Centre. They will help to provide a control measure and disseminate relevant prevention actions in such matters. The ASEAN also has created a Regional Public Health Laboratories Network purposely to give support and expertise, especially on scientific research and technical support among AMSs.

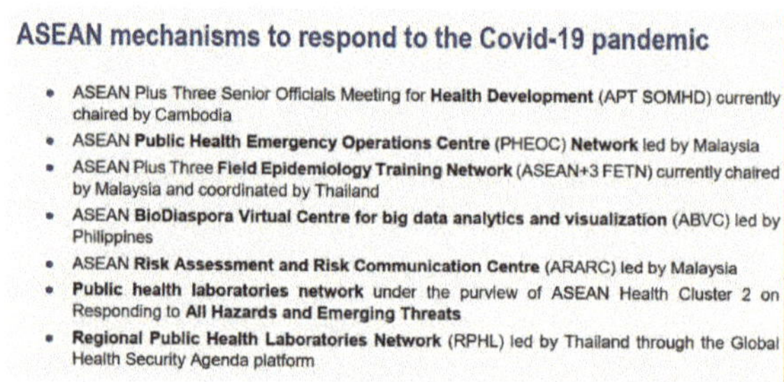

Figure 2. The ASEAN mechanism to respond to the COVID-19 pandemic [5].

4.2. *The Regular and Irregular Migration during COVID-19*

The ASEAN has also addressed several best approach and initiatives, especially among employees, and particularly in regard to the migrant workers that were badly affected due to the COVID-19 pandemic. One of the measures taken is to provide free COVID-19 testing and healthcare to the migrant workers in order to deal with the process of automatic visa extension. Furthermore, the process of registration, management of logistics for organizing returns, providing support and temporary income support for returnee migrants have also been implemented. Additionally, efforts have been made at the national level, whereby most AMSs, such as Philippines and Thailand, provided the affected migrant workers with a one-off cash transfer and assisted their return home through chartered flights. In addition, Thailand as a receiving country also gave support to migrant workers by granting permission to stay and work to those from Myanmar and Cambodia. Not only that, but depending on their contributions and terms of service, they were also eligible for unemployment insurance and severance money. Furthermore, Singapore also provides migrant workers with counseling and covers the cost of COVID-19 testing and treatment, whilst Malaysia offers migrant employees COVID-19 screening subsidies.

4.3. *Food Security Issues during COVID-19*

The ASEAN has developed a number of initiatives and measures to ensure food security in Southeast Asia as a result of COVID-19. It began on 15 April 2020, when ASEAN published a joint statement during the ASEAN Ministers of Agriculture and Forestry (AMAF) meeting, aimed at strengthening the ASEAN countries' commitment to ensure food security, food safety and food nutrition during pandemic. It focused on minimizing the disruption of food supply chains, ensuring the markets remained open, and facilitating the shipping process for agriculture and food product. Additionally, ASEAN also seeks to provide a sustainable supply of adequate, safe, and nutritious food that fits the dietary requirements of ASEAN communities during and after the COVID-19 pandemic. The ASEAN social protection programs also was expanded, especially to the smallholder farmers of micro, small, and medium enterprises inorder to assure countries food security and promoting food production [19].

Due to the high level of food self production and the active supply chain among the AMSs, Intra-ASEAN commerce also were vitalized by ASEAN, especially in terms of food activity, such as transmitting the raw materials from producer to the processor and towards the end to the consumer. It essentially took critical steps to preserve various activities, such as harvesting and delivering food by ensuring labor availability and establishing agri-food as a critical service [20]. Additionally, ASEAN is currently operating on a new set of digital agriculture rules and procedures as digitalization becomes important due to the post COVID-19 era. It is also known as knowledge-driven agriculture, which is focused on improving labor and land productivity. The use of the Internet and technology has increased due to COVID-19 and, therefore, it is urged that ASEAN countries move into digitalization of their agriculture sectors.

For instance, ASEAN has implemented a Green Revolution Technologies purposely to enhance crop varieties by using a conventional breeding methods. Other than that, Smart Farming also was introduced in order to assist and support the farmer through the use of laser guided tractors, GPS and also ploughing equipment. FINTECH, on the other hand, was made available to all small farmers in various ASEAN countries via credit and loan services, allowing them to trade in both local and national markets [21].

4.4. *Cyber Security Issues during COVID-19*

Throughout COVID-19 in Southeast Asia, ASEAN made numerous efforts to address cyber security concerns. To begin, a special webinar was held to raise public awareness about cyber security risks. The ASEAN Foundation, with the support of Microsoft, launched the ASEAN Cybersecurity Skilling Programme (ASEANCSP) during the "Cybersecurity in ASEAN: Lessons for Youth and How the COVID-19 is Shaping the Every-Evolving Digital Landscape" on 24 February 2022 [22]. The ASEANCSP was designed to help people in ASEAN have a better understanding of cyber security. These master trainers will then impart cyber security knowledge to 30,000 underserved young people across ASEAN member countries. This will contribute to the development of a secure digital ecosystem in the region. In addition, The ASEAN–Singapore Cybersecurity Centre of Excellence (ASCCE) also aims to improve the creation, legislation, and research capacities of member states' cyber security strategies. It is critical to address cyber security challenges in ASEAN countries through technical training, information sharing on cyber risks and attacks, and best practices across these teams. A cyber think tank is also part of the center's job. It carries out research and training in several areas, such as international law, cyber strategy, cyber law, cyber norms, and other cyber security policy issues [23].

In order to ensure that children received a significant benefit from online opportunities, ASEAN Member States have realized that there is an urgent need for collective action, as well as to improve laws at national and regional levels. One of the collaborations that have been made between ASEAN and UNICEF is through the Safe Online Initiative. This program explicitly promotes internet safety through better legislation and provides support and guidance, investment, and also the building of partnerships at the regional and national levels. The ASEAN region also has committed over US$ 10 million to it. Other than that, ASEAN has also developed a regional plan of action, the ASEAN-ICT Forum, to occur every year in order to assist private sector companies, especially financial institutions, Internet services providers, and social media by working together to keep children safe online. It is crucial and shows a timely commitment by the AMS to ensure that children and young people can use the Internet safely without being exploited and abused.

5. Discussion

Since the outbreak of the COVID-19 pandemic, the Association of Southeast Asian Nations (ASEAN) has made numerous efforts to mitigate the resulting non-traditional security issues. It is a global health problem that should be best addressed through a coordinated regional response. A regional reaction is more essential than ever to support national measures. The COVID-19 Vaccine for All that was implemented by ASEAN is one of the initiatives under the ASEAN socio-cultural community pillar that should be one of the significant security improvements in combating the threat of COVID-19. It can be shown by the declaration of endemic status in several AMSs, such as Brunei on 15 December 2021, Malaysia in April 2022, and Thailand by July 2022. The number of positive and death cases also reduced, as the new variant, Omicron, is 75% less likely to lead to the development of serious illness or death than in those infected with the Delta variant. As a result, the political and economic issues can be slowly resolved and mitigated once the virus has been thwarted.

Moreover, it is recommended that the migrant workers should receive social protection in proportion to their economic contributions to their country. Additionally, such a system should strengthen cross-border social protection for a highly mobile population. If they are unable to return to their home country, they should be allowed to obtain emergency social assistance in the country they are being sent to. This will help not only the migrant workers' health and economic well-being, but also the host country. Aside from that, the government must take aggressive steps to address people's most fundamental need, which is food, by giving financial help and support to keep small farmers competitive in production while ensuring that impoverished customers receive appropriate food. The government may utilize a variety of measures, including subsidies, financial support, price regulation, and import licenses.

On the other hand, it is important for AMSs to take immediate collective action and enhance legislation at the national and regional levels in order to prevent and respond to online exploitation and abuse while ensuring that children have no barriers to participating in online opportunities. The ASEAN also should continue to strengthen existing cooperation and identify existing gaps to improve and enhance the regional cyber security agenda. This cooperation will further strengthen ties between AMSs and provide a more general understanding of how to deal with cyber-attacks targeted at the Southeast Asian region.

As a result, during COVID-19, non-traditional security concerns cannot be dismissed as a minor matter, but rather must be viewed as serious security risks. Government must pay attention to socioeconomic shifts and broaden their awareness of security risks beyond the military. More importantly, it must also be recognised that many of these problems, such as pandemics and the environment, affect more than one country. This means that states cannot solve these problems on their own. To deal with these security threats, it is important for states and non-state actors, such as non-governmental organizations (NGOs), to work together on a regional level. Only by pooling their resources, knowledge, and skills will these players be able to effectively handle the ever-increasing complexity of non-traditional security. Thus, ASEAN's unity is essential to overcome challenges due to the global financial crisis, climate change, natural disasters and, now, the COVID-19 pandemic.

6. Conclusions

In conclusion, the emergence of COVID-19 has resulted in major disruptions to the AMSs, most notably to the economy, as well as to the social and political stability of the states. This may be seen through the quarantine or lockdown measures that have been enforced by AMS, which have contributed to an economic slowdown because they decrease export activity and reduce the supply of food. A limited food supply will indirectly contribute to growing hunger, particularly among children with stunted growth and those from low socioeconomic backgrounds. Loss of jobs, lack of resources, and lack of access to a social safety net all contribute to an increase in poverty. Furthermore, cybercriminal assaults targeting SMEs in Southeast Asia will detract from ASEAN's goals to adopt a Go Digital ASEAN policy as part of digital economy activities.

In addition, it is essential for the AMSs to address the requirements and to take immediate policy action in order to deal with all of the NTS dangers that have arisen as a result of the COVID-19 pandemic. In order to effectively combat the COVID-19 pandemic, it is necessary to give it not just more serious attention at the national level but also at the regional and global levels. This is because the pandemic has an influence that cuts across multiple levels. Furthermore, strong support and a response from every AMS to mitigate the NTS issues are needed to ensure that all the programs, policies, joint action, and other initiatives can be successfully implemented, as well as to achieve their intended goals.

Author Contributions: Conceptualization, N.A.Y. and M.I.Y.; methodology, N.A.Y.; software, N.A.Y., Z.M.Z. and S.M.N.; validation, N.A.Y.; formal analysis, N.A.Y. and M.I.Y.; investigation, N.A.Y. and M.I.Y.; resources, N.A.Y. and N.A.M.; data curation, N.A.Y. and Z.M.Z.; writing—original draft preparation, N.A.Y.; writing—review and editing, N.A.Y. and S.M.N.; visualization, N.A.Y.; supervision, N.A.Y., S.M.N. and Z.M.Z.; project administration, N.A.Y.; funding acquisition, N.A.Y. All authors have read and agreed to the published version of the manuscript.

Funding: This study was supported by the FSPPP Internal Research Grant (FIRG), Faculty of Administrative Science and Policy Studies, Universiti Teknologi Mara (UiTM), Malaysia with the collaboration of Kementerian Luar Negeri, Putrajaya (project code:600-icaen/firg-01/2021). The study also has been granted with the approval from UiTM Research Ethics Committee (REC).

Institutional Review Board Statement: Not applicable.

Informed Consent Statement: Not applicable.

Data Availability Statement: Not applicable.

Acknowledgments: Not applicable.

Conflicts of Interest: The authors declare no conflict of interest.

References

1. Caballero-Anthony, M. *Negotiating Governance on Non-Traditional Security In Southeast Asia And Beyond*; Columbia University Press: New York, NY, USA, 2016.
2. NoorOsman, D.; Hassan, F.M.; Tajudin, A.A.; Awang@Ali, M.N. COVID-19 as a National Security Issue in Malaysia: A Comparison with the Italian and Australian Perspectives. *Acad. J. Interdiscip. Stud.* **2021**, *10*, 261.
3. Tesch, R. Qualitative Research: Analysis Types and Software. Available online: https://books.google.com.my/books?isbn=1134077378 (accessed on 12 April 2022).
4. Organization for Economic Co-operation and Development. COVID-19 Crisis Response in the ASEAN Member States. 2020. Available online: https://www.oecd.org/coronavirus/policy-responses/covid-19-crisis-response-in-asean-member-states-02f828a2/ (accessed on 16 April 2022).
5. ASEAN BioDiaspora Virtual Center. COVID-19 Situational Report in the ASEAN Region. 2022. Available online: https://asean.org/wp-content/uploads/2022/03/COVID-19_Situational-Report_ASEAN-BioDiaspora-Regional-Virtual-Center_4Mar2022.pdf (accessed on 13 April 2022).
6. Djalante, R.; Nurhidayah, L.; Van Minh, H.; Phuong, N.; Mahendradhata, Y.; Trias, A.; Lassa, J.; Miller, M.A. COVID-19 and ASEAN responses: Comparative policy analysis. *Prog. Disaster Sci.* **2020**, *8*, 100129. [CrossRef] [PubMed]
7. IOM. IOM Ireland Rapid Policy Survey Report: The Irish Government's Policy Response to COVID-19 from a Migration Governance Lens. 2020. Available online: https://publications.iom.int/books/iom-ireland-rapid-policy-survey-reportthe-irish-governments-policy-response-covid-19 (accessed on 16 April 2022).
8. Nanthini, S. In Limbo in ASEAN: Pandemic and Irregular Migration. 2020. Available online: https://www.rsis.edu.sg/ (accessed on 16 April 2022).
9. Tan, N.A. Plight of Rohingyas under COVID-19 Spotlights ASEAN's Failure. The Diplomat. 2020. Available online: https://thediplomat.com/2020/06/plight-of-rohingyas-under-covid-19-spotlights-aseans-failure/ (accessed on 20 April 2022).
10. Ducanes, G.; Dutta, P.; Nixon, N. Displaced, Stranded, Unprotected: ASEAN's Migrant Workers. 2020. The Diplomat. Available online: https://thediplomat.com/2020/12/displaced-stranded-unprotected-aseans-migrant-workers/ (accessed on 20 April 2022).
11. A Guide to International Refugee Protection and Building State Asylum Systems. Inter-Parliamentary Union and the United Nations High Commissioner for Refugees. 2017. Available online: https://www.unhcr.org/3d4aba564.pdf (accessed on 18 April 2022).

12. Food and Agriculture Organization (FAO). Mitigating Impacts of COVID-19 on Food Trade and Markets. Joint Statement by QU Dongyu, Tedros Adhanom Ghebreyesus, and Roberto Azevedo, Directors General of FAO, WHO, and WTO. 2020. Available online: http://www.fao.org/news/story/en/item/1268719/icode/ (accessed on 18 April 2022).
13. Nortajudin, A. Starving In a Pandemic. The ASEAN Post. 2020. Available online: https://theaseanpost.com/article/starving-pandemic (accessed on 18 April 2022).
14. Mardones, F.O.; Rich, K.M.; Boden, L.A.; Moreno-Switt, A.I.; Caipo, M.L.; Zimin-Veselkoff, N.; Alateeqi, A.M.; Baltenweck, I. The COVID-19 Pandemic and Global Food Security. *Front. Veterinar. Sci.* **2020**, *7*, 578508. [CrossRef] [PubMed]
15. Ibrahim, A.Z.; Othman, Z. COVID-19: Coping Strategies among B40 Households in Malaysia to Achieve Food Security during Movement Control Order (MCO). *Eur. J. Mol. Clin. Med.* **2020**, *7*, 1513–1524.
16. INTERPOL. Asean Cyberthreat Assessment. 2021. Available online: https://www.interpol.int/content/download/16106/file/ASEAN%20Cyberthreat%20Assessment%202021%20-%20final.pdf (accessed on 16 April 2022).
17. Digital Global Report (2020). Internet Usage in Southeast Asia—Statistics & Facts. Available online: https://www.statista.com/topics/9093/internet-usage-in-southeast-asia/#topicHeader__wrapper (accessed on 18 April 2022).
18. Rebecca, O. Cybersecurity Is Still Challenging for ASEAN Businesses. 2021. Available online: https://techwireasia.com/author/becs/ (accessed on 18 April 2022).
19. ASEAN Secretariat. ASEAN Pledges to Ensure Food Security during COVID-19 Outbreak. Available online: https://asean.org/asean-pledges-to-ensure-food-security-during-covid-19-outbreak/ (accessed on 16 April 2022).
20. Teng, P. COVID-19: Maintaining Food Security in Asia Pacific. Opinion Editorial published by Inter Press Service News Agency, FAO, UN, Rome. 2020. Available online: https://www.ipsnews.net/2020/04/covid-19-maintaining-food-security-asia-pacific/ (accessed on 21 March 2022).
21. Teng, P.P.S.; Caballero-Anthony, M.; Montesclaros, J.M.L. ASEAN responses to COVID-19 for assuring food security. *Adv. Food Secur. Sustain.* **2021**, *6*, 83–118.
22. Octaviano, A. ASEAN Cybersecurity Skilling Programme Launched by ASEAN Foundation with Microsoft. 2022. Available online: https://www.aseanfoundation.org/asean_cybersecurity_skilling_programme_launched_by_asean_foundation_with_microsoft (accessed on 16 April 2022).
23. Kenny, C. Asean-S'pore Centre for Training National Cyber-Security Teams Opens New Campus. 2021. Available online: https://www.straitstimes.com/tech/tech-news/asean-spore-centre-for-training-national-cyber-security-teams-opens-new-campus (accessed on 12 April 2022).

proceedings

Proceeding Paper

Examining a Model of Entrepreneurial Success Factors among Undergraduate Students: A Quantitative Study †

Kay Hooi Keoy [1,*], Yung Jing Koh [1], Japos Genaro [2], Su Mon Chit [1], Luqman Lee [3] and Meng Chuan Ho [4]

1 Institute of Computer Science and Digital Innovation, UCSI University, Kuala Lumpur 56000, Malaysia
2 School of Graduate Studies, North Eastern Mindanao State University, Tandag 8300, Philippines
3 De Institute of Creative Arts and Design, UCSI University, Kuala Lumpur 56000, Malaysia
4 Faculty of Social Sciences and Liberal Arts, UCSI University, Kuala Lumpur 56000, Malaysia
* Correspondence: keoykh@ucsiuniversity.edu.my; Tel.: +603-9101-8880 (ext. 2459)
† Presented at the International Academic Symposium of Social Science 2022, Kota Bharu, Malaysia, 3 July 2022.

Abstract: Promoting entrepreneurship is important to develop competitive advantages, as it is one of the determinants of a country's economy and development. Students are the future pillar of society, so stimulating entrepreneurship from education is the best way to shape their entrepreneurial mindset and attitudes. Empirical studies have demonstrated the important relationship between entrepreneurship education and its success among higher education students. However, the impact of technology enablement on perceived entrepreneurial outcomes has received minimal focus. In total, 304 data were collected from universities in the Philippines showing that, among entrepreneurial success factors, technology enablement has the greatest impact. Our results demonstrated that related stakeholders should understand the importance of and need to enhance the use of technology in relation to entrepreneurship.

Keywords: entrepreneurial education mechanism; perceived entrepreneurial outcome; technology enablement

Citation: Keoy, K.H.; Koh, Y.J.; Genaro, J.; Chit, S.M.; Lee, L.; Ho, M.C. Examining a Model of Entrepreneurial Success Factors among Undergraduate Students: A Quantitative Study. *Proceedings* **2022**, *82*, 95. https://doi.org/10.3390/proceedings2022082095

Academic Editor: Mohamad Rahimi Mohamad Rosman

Published: 10 October 2022

1. Introduction

The effect of entrepreneurship has been widely studied in various respects such as innovation, economic growth, and country development [1]. It has been proved that entrepreneurship education positively impacts stimulating entrepreneurial activities by providing practical skills such as problem-solving skills and critical thinking [2]. However, technology enablement's influence on entrepreneurship is yet to be discussed and investigated widely [3]. It is well known that technology has a huge impact on people's daily life in different areas, making life easier. Hence, there is a need for scholars and the public to study how technology can be a push factor for entrepreneurship. This study aims to provide a comparative study to evaluate perceived entrepreneurial success and adopt technology enablement so as to incorporate these finding into the conventional education system. With the proposed hypotheses, this study hopes to provide results that demonstrate that technological enablement can have a higher and more significant influence on perceived success. This result will guide institutions and the government to implement sound strategies to focus on the adoption and usage of technologies on entrepreneurial activities.

2. Literature Review

2.1. Definition of Entrepreneurship

Entrepreneurship acts as a link between economic growth and self-satisfaction. This can be shown in a country's development, as entrepreneurship both creates job opportunities and pursues innovative ideas to meet the market needs [4]. Entrepreneurship is an achievement that reaches something tangible and new, such as the introduction of goods and services to society [5]. In this case, entrepreneurs need to have a sense of unerring market understanding to recognise market opportunities by using technology such as social media and blogs. Consumers nowadays may utilise social media to share their experiences on a particular product or service, enabling entrepreneurs to search for unmet demands. Entrepreneurship is a challenging journey as it is risky and uncertain [6]. Therefore, promoting entrepreneurship is important for economic growth.

2.2. Entrepreneurship in Education

Entrepreneurship education aims to provide syllabi and programmes that can shape students' entrepreneurial mindsets with skilled-based learning outcomes [7]. This refers to their willingness to be involved directly or indirectly in entrepreneurship activities by using the skills learned, such as technology-related tools and knowledge, to compete in a dynamic business environment. Hence, entrepreneurship education can provide sufficient and diverse learning approaches that shape entrepreneurial attitudes. Additionally, higher education institutions adapt to the rapid changes in the education system by improving teaching content and entrepreneurship-related events for students to learn how to start a new business [8]. For example, these might include teaching content improvements, including designed-based thinking skills, as an experiential learning process can deliver entrepreneurs' intention and business basics. Hence, institutions and educators need to promote the entrepreneurial ecosystem in order to offer interesting and practical learning experiences to develop students' entrepreneurial intentions.

2.3. Entrepreneurship Education in Philippines

The Philippines' economy is dominated by SMEs (99.6%), just like other ASEAN countries [9], forcing the Philippines to focus on entrepreneurship stimulation. It is well known that the United States influenced the Philippines' education system; as such, English is the language of delivery. The education system focuses on delivering practical knowledge and training for students to start a new business. However, the sustainability of entrepreneurship growth is not focused [10]. Therefore, there is urgent attention needed to be aware of entrepreneurship sustainability by combining technology adoption in the learning process. Moreover, institutions play a key role in creating opportunities and growth for students by offering, enhancing and stimulating proper training to develop students' entrepreneurial intentions and initiation.

2.4. Entrepreneurship Education Mechanism (EEM)

An entrepreneurship education mechanism refers to programmes that are effective for entrepreneurship skills' enhancement [11]. Some studies pointed out that higher institutions' entrepreneurial environment, such as providing related events and competition, can motivate students' learning process and performance [12]. For example, business plan competitions that can provide students with real-life experiences, teamwork, and problem-solving skills to help them understand the process of entrepreneurship. Moreover, some entrepreneurship-related events will assign an instructor to each team to maximise educational outcomes by providing supervision for venture creation's theories and practical skills [13]. In short, an entrepreneurial education mechanism can provide valuable mentoring that can enhance and develop students' intentions by removing institutional obstacles to venture creation.

2.5. Technological Enablement (TE)

Technology enablement in entrepreneurship refers to the streamlining of business internal operation processes through the adoption of technology. Technology implementation in entrepreneurship has been widely investigated [14]. Technology's capabilities are proven to improve overall business efficiency. For instance, increased speed allows businesses to enhance process systems and overall business performance. It is claimed that entrepreneurship success depends on the adoption of technology [15]. With the adoption of technology, a business can gain competitive advantages in terms of enhancing customer relationships, increasing sales, and increasing its market share. It is believed that technological enablement can greatly impact entrepreneurial success.

2.6. Perceived Entrepreneurial Outcome (PEO)

There are different driving forces behind venture creation. The most popular perceived outcome is a financial reward that will strongly motivate an individual to achieve their goals [16]. Self-satisfaction by pursuing original ideas to the market by gaining a loyal customer group and market share is another one of the most popular drivers in entrepreneurship [17]. Moreover, individuals seeking a work–life balance between family and work, and the positive return between the time and money invested, will have a higher chance of achieving entrepreneurial success [18]. In short, this study considers various outcome indicators to determine potential entrepreneurial success.

3. Research Methodology

3.1. Research Rationale

This study investigates the direct and indirect impacts of technology enablement to boost higher education students' perceived entrepreneurial outcomes. The recognition and adoption of digital transformation is a new trend that can help in improving socio-economic development and improve overall business efficiency, thus developing competitiveness to achieve short- and long-term goals [19]. The result generated should provide a brief understanding of the importance of technology enablement in relation to students' perceived entrepreneurial success. Therefore, related agencies such as policymakers and universities can implement supportive strategies to ensure the success of entrepreneurial mindset enhancements.

3.2. Research Questions

1. What are the effects of the variables 'education mechanism' and 'technological enablement' on higher education students' perceived entrepreneurial success?
2. Is the impact of technological enablement more critical than the conventional educational system?

3.3. Research Framework and Hypothesis

The importance of education mechanisms and technology enablement was covered in the previous sections. Both factors have proved to have a positive impact on entrepreneurial outcomes among higher education students. Therefore, the factors are grouped into the success elements for perceived entrepreneurial outcomes. Figure 1 shows the research framework of the study.

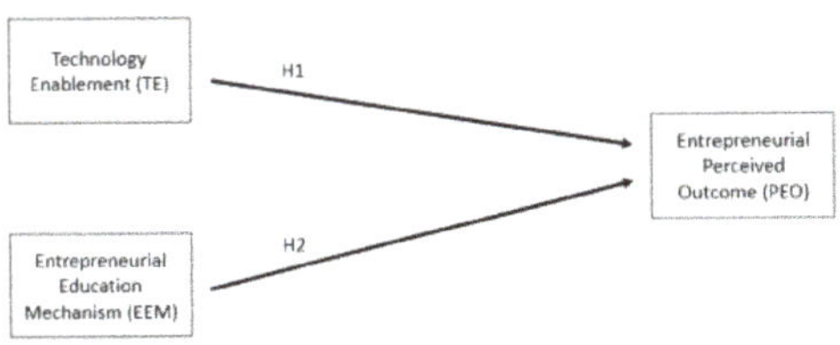

Figure 1. Research framework.

The hypothesises are proposed as follows:

H_1. *Technology Enablement (TE) will positively impact Perceived Entrepreneurial Outcome (PEO).*

H_2. *Entrepreneurial Education Mechanism (EEM) will positively influence Perceived Entrepreneurial Outcome (PEO).*

3.4. Research Procedure

This study uses the Philippines as example and proposes to identify the critical elements that will positively impact perceived entrepreneurial outcomes. The study first goes through a literature review, collects information for the designed questionnaire, and analyses the findings. The questionnaire was randomly sent to the intended respondents in universities in the Philippines via Google Form. A 5-Likert scale was used to represent the respondents' level of agreement, from (5) Strongly Agree to (1) Strongly Disagree. In order to reduce common method bias (CMB) in the research, a clear instruction was included in the Google Form to provide a basic understanding of the research. The data were collected separately from different higher institutions to minimise the CMB. As a result, 304 respondents from Philippines were surveyed. Moreover, partial least squares structural equation modelling (PLS-SEM) was conducted, which the measurement model refers to as the relationship between latent variables and observed data [20]. Hence, SEM-PLS with Smart PLS 3.0 software was implemented to analyse the proposed research model.

4. Findings

4.1. Descriptive Analysis

Table 1 indicates the respondents' demographic information. There are 104 male and 200 female respondents. Most of the students are postgraduates (52.96%), followed by undergraduates (38.82%), diploma (5.92%) and foundation (2.3%). Moreover, 150 respondents had prior experience or involvement in entrepreneurial fields, and we able to answer based on personal experiences. In contrast, the remaining 154 students were surveyed based on knowledge and opinions, as they had not previously participated in entrepreneurial pursuits.

Table 1. Demographic analysis.

Demographic Characteristics	Items	Philippines Respondents	%
Gender	Male	104	34.21
	Female	200	65.79
Education level	Foundation	7	2.3
	Diploma	18	5.92
	Undergraduate	118	38.82
	Postgraduate	161	52.96
Ventured into entrepreneurial activities either directly or indirectly	Yes	150	49.34
	No	154	50.66

4.2. Measurement Model

4.2.1. Construct Validity and Reliability Test

Before the test, the outer loadings for each item were defined in order to measure the relationship. The acceptable value for outer loadings was equal to or bigger than 0.5, and 0.7 or above was considered as highly satisfactory [21]. Construct validity was measured by composite reliability (CR) and average variance extracted (AVE); any CR values greater than 0.7 and AVE values larger than 0.5 were acceptable [22]. Construct validity is important for research that cannot be measured or observed directly; in this case, it refers to perceived entrepreneurial outcome. Table 2 shows the result of construct validity; PEO3 was removed due to a higher VIF. The values of CR and AVE are all greater than 0.7 and 0.5, indicating the satisfactory nature of both construct validity and reliability.

Table 2. Construct validity and reliability.

Construct	Items	Outer Loadings	Composite Reliability	Average Variance Extracted (AVE)
EEM	EEM1	0.901	0.944	0.809
	EEM2	0.907		
	EEM3	0.879		
	EEM4	0.91		
PEO	PEO1	0.893	0.953	0.835
	PEO2	0.923		
	PEO4	0.914		
	PEO5	0.925		
TE	TE1	0.799	0.944	0.740
	TE2	0.869		
	TE3	0.898		
	TE4	0.906		
	TE5	0.882		
	TE6	0.800		

4.2.2. Discriminant Validity Test

A discriminant validity test is needed to ensure there is no overlapping between the factors [23]. The Fornell–Larcker criterion and HTMT value are used to determine the ability to differentiate between two constructs in a model. The acceptable value of the Fornell–Larcker criterion is any value greater than the other correlation, while the HTMT value should be smaller than 0.9 [24]. Table 3 refers to the value of the Fornell–Larcker criterion of this model; all the values are greater than other correlations. Table 4 indicates the HTMT values, all of which are smaller than 0.9. Both of the results show the existence of discriminant validity in the model.

Table 3. Fornell–Larcker Criterion.

Construct	EEM	PEO	TE
EEM	0.899		
PEO	0.508	0.914	
TE	0.546	0.705	0.860

Table 4. HTMT Value.

Construct	EEM	PEO	TE
EEM			
PEO	0.542		
TE	0.590	0.755	

4.3. Structural Model

Multicollinearity Test

Multicollinearity occurs when there is a substantial intercorrelation between independent variables. The existence of multicollinearity will lead to misleading conclusions [25]. The acceptable VIF value is smaller than 5; PEO3 was removed due to the higher VIF and the model was then run again to obtain better results. Table 5 shows that all the VIF values are below 5, indicating no multicollinearity in the model.

Table 5. Variance inflation factor (VIF) value.

Construct	Items	VIF
EEM	EEM1	3.176
	EEM2	3.498
	EEM3	2.885
	EEM4	3.234
PEO	PEO1	3.002
	PEO2	3.954
	PEO4	3.662
	PEO5	4.043
TE	TE1	2.08
	TE2	3.148
	TE3	3.89
	TE4	4.011
	TE5	3.098
	TE6	2.282

4.4. Hypothesis Testing

The hypothesis was tested based on two-tailed test at a 95% confidence level to prove the null hypothesis's plausibility [26]. Table 6 indicates the path coefficient and the *p*-value for each relationship. All *p*-values are smaller than 0.05, which supports the null hypothesis. In other words, the TE has a positive impact on PEO (β = 0.610, $p < 0.05$). In addition, the EEM is critical in relation to influencing PEO in entrepreneurship (β = 0.175, $p < 0.05$). Lastly, the research hypothesises and relationship are depicted in Figure 2.

Table 6. Hypothesis testing.

Hypothesis	Beta	SE	Standard Deviation (STDEV)	T Statistics (|O/STDEV|)	*p* Values	Decision
H_1: TE → PEO	0.610	0.617	0.049	12.471	0	Accepted
H_2: EEM → PEO	0.175	0.170	0.053	3.300	0.001	Accepted

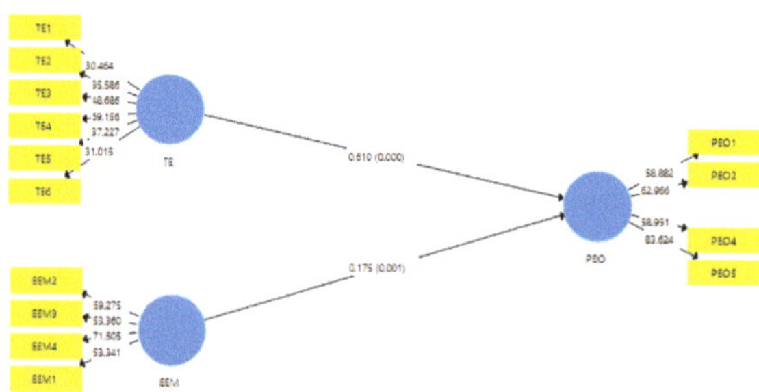

Figure 2. Research model.

5. Discussion

This research uses the Philippines as an example and aims to identify the impact of technology enablement and conventional education systems on perceived entrepreneurial outcomes. As a result, H_1 is supported, as it is believed that the greater the technology support, the better the achievement of students' perceived entrepreneurial outcome. Technology adoption enables business to quickly adapt to dynamic business environments that may encounter various market possibilities and difficulties [27]. For example, customer needs are changing dynamically with the growth of different digital platforms. Technology can also help students to recognise market opportunities and discover innovative solutions that can develop sustainable business models [28]. Therefore, the adoption of technology will greatly increase the likelihood of achieving achieve entrepreneurial outcomes. Moreover, H_2 showed statistically significant results. The results demonstrate the importance of entrepreneurial education, as it believed that the stronger the support from education mechanisms, the greater the perceived entrepreneurial outcomes among higher education students [29]. This is because supportive education mechanisms, such as faculty support, professional lecturers, financial support and many others, are able to help in shaping students' attitude and mindsets, thus encouraging them to be involved in entrepreneurship [30]. With the skills and mindset gained from educational support, students can understand the best and most efficient ways to achieve their goals. In short, the results generated refer to a higher impact of TE than the conventional education mechanism. This may be due to the fast-changing pace of technology, which can help businesses operate effectively and better encourage students' intentions due to the greater outcomes achieved.

6. Conclusions

This study has pointed out the importance of the influence of critical elements on perceived entrepreneurial success. Moreover, the results generated show that the impact of technology enablement is greater than effect of an entrepreneurial education mechanism on the perceived outcomes. This is a relatively novel result, as the majority of prior studies have not measured the importance of technology enablement in relation to entrepreneurship. Therefore, major stakeholder in the Philippines need to pay attention to the implementation of technology in higher education systems in order to enhance and increase the likelihood of students achieving perceived entrepreneurial outcomes. In short, there is a need for the government and institutions to be aware of the use of technology in entrepreneurship in order to develop its economic goals.

7. Limitation

There were only 304 data collected from a few universities in Philippines, which is a small sample size for analysing the impact of critical factors. Moreover, there were only a few items included in each factor, meaning that the result generated may not be accurate enough. Lastly, the impact of the mediator factor, i.e., technology enablement, is not specific enough, as the research could have included more aspects to generate a more accurate result.

8. Future Study

It is suggested that the future studies should collect more respondents from more universities in order to have a larger sample size. This is because the standard error and bias will be reduced and minimised due to the larger sample size. More items should be included in each factor to better interpret the impact of the critical elements on perceived success. Lastly, the effect of technology as a mediating factor should be analysed in a more diverse manner to provide a broader picture for readers.

Author Contributions: Conceptualization, K.H.K. and M.C.H.; formal analysis, Y.J.K.; writing—original draft preparation, J.G.; writing—review and editing, S.M.C.; project administration, L.L. All authors have read and agreed to the published version of the manuscript.

Funding: This research received is funded from UCSI University, Research External and Innovation Grant.

Institutional Review Board Statement: Not applicable.

Informed Consent Statement: Not applicable.

Data Availability Statement: Not applicable.

Conflicts of Interest: The authors declare no conflict of interest.

References

1. Muhhammad Shafiu, A.; Abdul Manaf, H.; Muslim, S. Utilization entrepreneurship for job creation, poverty reduction and national development. *J. Soc. Sci. Res.* **2020**, *6*, 97–102. [CrossRef]
2. Hameed, I.; Irfan, Z. Entrepreneurship education: A review of challenges, characteristics and opportunities. *Entrep. Educ.* **2019**, *2*, 135–148. [CrossRef]
3. Wang, W.; Pan, C.; Ray, P.K. Technology entrepreneurship in developing countries: Role of telepresence robots in healthcare. *IEEE Eng. Manag. Rev.* **2021**, *49*, 20–26. [CrossRef]
4. Ali, A.; Kelley, D.J.; Levie, J. Market-driven entrepreneurship and institutions. *J. Bus. Res.* **2020**, *113*, 117–128. [CrossRef]
5. Mehmood, T.; Al-Gasaymeh, A. Schumpeterian entrepreneurship theory: Evolution and relevance. *Acad. Entrep. J.* **2019**, *25*, 1–10.
6. Nuhu, S.; Abddulraheem, N.; Abubakar, H.; Abdurauf, A.; Sulaiman, K. Entrepreneur Skills as Determinants to Job Opportunities among LIS Postgraduate Students in University of Ilorin. Available online: https://digitalcommons.unl.edu/libphilprac/5876 (accessed on 20 March 2022).
7. Ratten, V.; Usmanij, P. Entrepreneurship education: Time for a change in research direction? *Int. J. Manag. Educ.* **2021**, *19*, 100367. [CrossRef]
8. Wang, S.Y.; Wu, X.L.; Xu, M.; Chen, Q.X.; Gu, Y.J. The evaluation of synergy between university entrepreneurship education ecosystem and university students' entrepreneurship performance. *Math. Probl. Eng.* **2021**, 3878378. [CrossRef]
9. Velasco, A.L. Entrepreneurship education in the philippines. *DLSU Bus. Econ. Rev.* **2013**, *22*, 1–14.
10. Mubanga, P.; Hock, O.Y.; Karim, A.M.; Mulenga, I.M. Methods of financing technical and vocational education and training, and entrepreneurship education to support skills development in Lusaka Province, Zambia. *Int. J. Res. Sci. Innov.* **2019**, *6*, 96–107.
11. Hynes, B.; Richardson, I. Entrepreneurship education: A mechanism for engaging and exchanging with the small business sector. *Educ. Train.* **2007**, *49*, 732–744. [CrossRef]
12. Watson, K.; McGowan, P. Rethinking competition-based entrepreneurship education in higher education institutions: Towards an effectuation-informed coopetition mode. *Educ. Train.* **2020**, *62*, 31. [CrossRef]
13. Indarti, S. The effects of education and training, management supervision on development of entrepreneurship attitude and growth of small and micro enterprise. *Int. J. Organ. Anal.* **2021**, *29*, 16–34. [CrossRef]
14. Leong, C.; Tan, F.T.C.; Tan, B.; Faisal, F. The emancipatory potential of digital entrepreneurship: A study of financial technology-driven inclusive growth. *Inf. Manag.* **2022**, *59*, 103384. [CrossRef]
15. Cunningham, J.A.; Lehmann, E.E.; Menter, M.; Seitz, N. The impact of university focused technology transfer policies on regional innovation and entrepreneurship. *J. Technol. Transf.* **2019**, *44*, 1451–1475. [CrossRef]
16. Shepherd, D.A.; Patzelt, H. Motivation and entrepreneurial cognition. *Entrep. Cogn.* **2018**, 51–103. [CrossRef]
17. Ojo, S. Black African perceptions of entrepreneurial outcomes in the UK. *Soc. Bus. Rev.* **2021**, *16*, 278–305. [CrossRef]
18. Hassan, S.; Shamsudin, M.F. Measuring the effect of service quality and corporate image on student satisfaction and loyalty in higher learning institutes of technical and vocational education and training. *Int. J. Eng. Adv. Technol.* **2019**, *8*, 533–538. [CrossRef]
19. Rodrigues, M.; Franco, M. Composite index to measure cities' creative performance: An empirical study in the Portuguese context. *Sustainability* **2019**, *11*, 774. [CrossRef]
20. Shmueli, G. Predictive model assessment in PLS-SEM: Guidelines for using PLSpredict. *Eur. J. Mark.* **2019**, *53*, 2322–2347. [CrossRef]
21. Taylor, T.; Geldenhuys, S. Using partial least squares to measure tourism students' satisfaction with work-integrated learning. In *Tourism-Perspectives and Practices*; Sabah, S., Ed.; Intechopen: London, UK, 2019.
22. Shrestha, N. Factor analysis as a tool for survey analysis. *Am. J. Appl. Math. Stat.* **2021**, *9*, 4–11. [CrossRef]
23. Bertea, P.; Zait, A. Methods for testing discriminant validity. *Ed. Univ. Craiova* **2011**, *2*, 217–224.
24. Franke, G.; Sarstedt, M. Heuristics versus statistics in discriminant validity testing: A comparison of four procedures. *Internet Res.* **2019**, *29*, 430–447. [CrossRef]
25. Gujarati, D. *Econometrics by Example*; Palgrave Macmillan: New York, NY, USA, 2011.
26. Emmert-Streib, F.; Dehmer, M. Understanding statistical hypothesis testing: The logic of statistical inference. *Mach. Learn. Knowl. Extr.* **2019**, *1*, 945–961. [CrossRef]
27. Ndofirepi, T.M. Relationship between entrepreneurship education and entrepreneurial goal intentions: Psychological traits as mediators. *J. Innov. Entrep.* **2020**, *9*, 2. [CrossRef]
28. Evans, D.; McKee, J. *Social Marketing*; AG Printing & Publishing: Nairobi, Kenya, 2021.
29. Chen, L.; Ifenthaler, D.; Yau, J.Y.-K. Online and blended entrepreneurship education: A systematic review of applied educational technologies. *Entrep. Educ.* **2021**, *4*, 191–232. [CrossRef]
30. Shen, T.; Osorio, A.E.; Settles, A. Does family support matter? The influence of support factors on entrepreneurial attitudes and intentions of college students. *Acad. Entrep. J.* **2017**, *23*, 24–43. [CrossRef]

proceedings

MDPI

Proceeding Paper

Sociodemographic Factors, Parental Stress, and Depressive Symptoms among Mothers of Children with ADHD †

Qarem Mohamed Mustafa [1,*], Mimi Iznita Mohamed Iqbal [2], Pramesti Pradna Paramita [3] and Trias Mahmudiono [4]

1 Universiti Teknologi MARA Cawangan, Pulau Pinang 13500, Malaysia
2 Department of Psychology, International Islamic University Malaysia, Gombak 53100, Malaysia
3 Faculty of Psychology, Universitas Airlangga, Jawa Timur 60115, Indonesia
4 Faculty of Public Health, Universitas Airlangga, Jawa Timur 60115, Indonesia
* Correspondence: qarem@uitm.edu.my; Tel.: +604-562-3511

† Presented at the International Academic Symposium of Social Science 2022, Kota Bharu, Malaysia, 3 July 2022.

Abstract: The increased prevalence of children with ADHD signals the increase in mothers who are caring for those children. This will then put the mothers at risk of various mental health issues, consequently impacting their ability to care for their children who need their support. Thus, the objectives of this study are to identify the parental stress experienced by mothers of children with ADHD and examine the relationship between sociodemographic variables and parental stress with the presence of depressive symptoms based on the transactional model of stress. The findings highlight the importance of ensuring parental stress experienced by mothers through appropriate support.

Keywords: strain; maternal; disability

check for updates

Citation: Mustafa, Q.M.; Iqbal, M.I.M.; Paramita, P.P.; Mahmudiono, T. Sociodemographic Factors, Parental Stress, and Depressive Symptoms among Mothers of Children with ADHD. *Proceedings* **2022**, *82*, 96. https://doi.org/10.3390/proceedings2022082096

Academic Editor: Mohamad Rahimi Mohamad Rosman

Published: 12 October 2022

1. Introduction

The global prevalence of ADHD is between 3% and 7%, averaging at 5%. This number is expected to increase based on the comparison of data from previous years [1]. In the United States for example, the prevalence of children diagnosed with ADHD was at 2.5% in 2010 [2], while the current available data show a prevalence of 9.4% [3]. A similar pattern can also be seen in Malaysia, in which the prevalence of children with ADHD, specifically those who are predominantly hyperactive and impulsive, was found to be at 0.32% in 2011 [4], while the National Health and Morbidity Survey conducted in 2015 found a prevalence of 4.6% [5]. This can be considered relatively high when compared to the prevalence of ADHD in other countries in Southeast Asia, such as Japan with a prevalence of 1.65% [6], the Philippines with a prevalence of 3% [7], or Taiwan with a prevalence of 4.2% [8]. ADHD is typically diagnosed during childhood because the symptoms tend to surface in early childhood [9], and parent ratings of the symptom's presentation are necessary for the process of diagnosis [1]. As the number of children with ADHD increases, the number of parents who have to care for those children would also increase at an equal rate.

Caring for children with ADHD may not be the same as caring for typically developed children. This is because children with ADHD may appear to be stubborn with a low frustration threshold, causing them to have difficulty obeying instructions and rules set by the parents [10]. Additionally, they are also found to be more susceptible to maladjustment, oppositional behaviours, inattention–disorganization behaviours, and angry outbursts, due to the nature of the disorder [10,11]. The parents may need more resources, as compared to other parents of typically developed children. The lack of resources to meet the demands of the children with ADHD, creating an aversive psychological reaction, is known parental stress [12]. Past studies also found that parenting children with ADHD is more stressful compared to parenting children who are HIV-affected, have asthma, learning disabilities [13], and have autism spectrum disorder [12]. This may be due to how the needs of children with ADHD may exceed the demands of children with other illness.

High levels of parental stress are found to contribute to the development of negative mental health outcomes across numerous populations [11,12,14]. Specifically, parents with high levels of stress were found to also have high symptoms of psychological distress [12,14], anxiety [11], and depression [11,12]. When the parents have poor mental health, their children are exposed to a greater risk of socioemotional impairments, poor cognitive and academic abilities, and to some extent, their well-being as they grow older [15]. This highlights the importance of safeguarding the parents' mental health, which can be done by identifying significant predictors to not only the mental health, but also the parental stress. There are many possible factors that are suggested to be associated with the mental health of the parents, including parenting style, coping strategies, and also parental stress [16]. Children's characteristics, such as age, gender, atypical behaviours, sensory processing difficulties, and parents' socioeconomical status, were found to be associated with the parental stress of the parents, among others [16].

Due to the different nature of parental responsibility in Malaysia, the current study will also focus only on mothers, as opposed to both mothers and fathers. This is due to the nature of parenting in Malaysia [17]. Although there are studies that suggest there is not a significant difference between the stress experienced by mothers and fathers [18], there are also studies that suggest otherwise [17], which makes the comparison inconclusive. This is where a cultural component may play an important role. In Malaysia, for example, mothers are known to be have dual roles, while the father is the main breadwinner of the family. Mothers are expected to not only work, but also to manage household chores once they completed their work, which may not be expected of fathers [17]. The need to meet the demand of working and household chores, added with the need to meet the demands of children with ADHD, may impact the mental health of the mothers significantly. According to the transactional model of stress [19], when mothers are exposed to an event that is perceived to be stressful, they will appraise whether the event is something threating to their well-being. They will also identify resources that will be useful to mitigate the stressful event, and failure to do so is suggested to put the mothers at risk of negative outcomes, such as negative emotions and reducing the ability to function [19]. Therefore, the current study will focus specifically on the parental stress experienced by mothers of children with ADHD, taking into account their socioeconomic status and determining how these variables are associated to their mental health.

2. Methodology

2.1. Study Design and Sample

A quantitative cross-sectional design was used, and a questionnaire was distributed for data collection. The population of interest is mothers of children with ADHD, in which purposive sampling was applied due to the specific characteristics of the population that is knowledgeable about the variables of interest [20]. A total of 104 mothers participated in the study, but eight questionnaires were removed because the information provided did not match the inclusion criteria or have significant missing data. The sample size was determined by using G*Power software version 3.1, with the statistical test being correlation, two-tailed, 0.3 effect size, 0.05 probability of error, and 0.80 power. The recommended sample size is 82, but was increased to 100 to take into account the 10% dropout rate [21]. Therefore, only data from 94 mothers were used as the sample, with a response rate of 92.3%. The inclusion criteria to be counted as part of the sample are:

- Having a child between the age of 6 years old and 12 years old with ADHD.
- The child has no other comorbidities other than ADHD.
- The ADHD diagnosis was provided by a medical doctor, specialist, or clinical psychologist.

These inclusion criteria were decided to reduce the possibility of confounding variables. Therefore, mothers who have children with ADHD that are 12 years old and above are excluded. Moreover, if the child only shows symptoms but are yet to receive a formal diagnosis, the data provided by the mother will not be included. Mothers who are unable to read or understand English and Malay or have any cognitive impairment were also excluded from participating in answering the questionnaire.

2.2. Data Collection

A knowledge sharing program was organized with a psychology center to provide knowledge on how to care for children with ADHD. A psychiatrist, occupational therapist, and clinical psychologist were appointed to share their knowledge based on their respective field. The program was promoted in the Facebook group ADHD Malaysia, which was made available for all to participate. At the end of the program, the questionnaires were distributed only to the mothers, since some fathers were also present. The completed questionnaires were retrieved, and additional data were gathered by sharing the version of the same questionnaire in a Google Form in the same Facebook group for other mothers who were interested in participating.

2.3. Measurement Tools

The questionnaire consists of sections for sociodemographic information, which also includes socioeconomic status, the parental stress inventory-short form (PSI-SF), and the Center of Epidemiologic Studies—Depression (CES-D). The PSI-SF was developed by Abidin [22] to measure parental stress. There are 36 items with the Likert scale as the response format. Participants have to select between the range of 1 and 5 with 1 representing "Strongly Agree" and 5 representing "Strongly Disagree". The items can be further divided into four domains, mainly parental sacrifices, parental expectations, parental emotion, and behavioral stress. The Cronbach alpha of each domain for the current study was found to be within the range of 0.90 to 0.74, which is an acceptable range for internal consistency [23]. Radloff [24] developed the CES-D to measure depression within the community and inpatient samples. The scale consists of 20 items, in which the participants will have to select from the Likert scale between 1, representing "Not at all or less than 1 day last week", and 5, representing "Nearly every day for two weeks". High scores represent a high presence of depressive symptomatology. The internal consistency of the scale for the current study is within the acceptable range (Cronbach alpha = 0.84).

2.4. Statistical Analysis

The data were analyzed using the Statistical Packages for Social Science (SPSS) software version 25. The data were cleaned, removing outliers before further analysis was conducted. Descriptive statistical analysis was done to obtain mean and standard deviation for each variable of interest. Assumptions of normality, multicollinearity, and independence of errors were tested to decide upon parametric or non-parametric inferential statistics. These were then followed with the Pearson product-moment correlation to obtain the correlation coefficient and multiple regression analysis to identify the variance contributed by socioeconomic status and parental stress on mental health among mothers of children with ADHD.

2.5. Ethical Approval

The Postgraduate Committee and examiners appointed by the Department of Psychology Postgraduate Thesis Committee ensures that the current research follows the ethical standards of the Department of Psychology, International Islamic University Malaysia (Reference: IIUM/301/DPGS/13/12/01). Informed consent was given to all the participants, and they were made aware of the risk and benefits of participating in the research. All information was also kept private and confidential, used only by the researchers.

3. Result and Findings

The sociodemographic information, which also includes the socioeconomic status of the parents, can be seen in Table 1. The mean age of the participants is 39.57 (SD = 6.07), with all the participants being Malays. Most of the mothers have a bachelor's degree (40.6%), while only a few of their highest education certificates are the Sijil Pelajaran Malaysia (13.5%). A total of 46 participants have a monthly household income between RM4001 and RM6000, which constitutes 47.9%, while only 2% have a monthly household income of RM1001 to RM3000. Most of the children were diagnosed at the age of 8 years old (n = 34; 34.5%), while only 7 children were diagnosed at the age of 11 years old (7.3%).

Table 1. Sociodemographic information of participants.

Demographic	Total	
	Mean	**SD**
Age	39.57	6.07
	n	%
Level of education		
SPM	21	21.9
STPM	12	12.5
Diploma	40	41.7
Bachelor's degree	17	17.7
Master's degree	6	6.3
Monthly household income		
RM1001–RM3000	3	3.1
RM3001–RM4000	36	37.5
RM4001–RM6000	50	52.1
RM6001 and above	7	7.3
Age of child when diagnosed		
6 years old	8	8.3
7 years old	15	15.6
8 years old	34	35.4
9 years old	15	15.6
10 years old	8	8.3
11 years old	7	7.3
12 years old	9	9.4

The Pearson product–moment correlation was used to identify the correlation between the variables of interest, in which a p-value of 0.05 and below indicates a statistically significant relationship. A significant negative relationship was found between the age of the mothers and parental expectations ($r = -0.267$, $p < 0.01$), while a significant positive relationship was found between level of education and behavioural expression ($r = 0.255$, $p < 0.05$). No significant association was found between socioeconomic variables and depressive symptomatology. However, parental stress in the form of parental expectations was found to be significantly associated with depressive symptomatology, such that an increase in parental expectations is suggested to significantly increase depressive symptomatology ($r = 0.412$, $p < 0.01$). Table 2 shows the result for other variables of interest.

Table 2. Correlation coefficient of sociodemographic information, parental stress, and depressive symptomatology.

	Variable	Mean	SD	1	2	3	4	5	6	7	8
1	Age	39.67	6.06								
2	Level of education	4.74	1.17	**0.210** *							
3	Average household income	3.64	0.67	**0.293** **	0.120						
4	Number of children	3.38	1.47	0.177	**−0.224** *	−0.042					
5	Emotional expression	37.85	8.77	−0.048	0.139	−0.079	0.073				
6	Parental sacrifices	27.26	6.25	−0.150	−0.049	−0.057	−0.133	**0.239** *			
7	Parental expectations	13.11	3.34	**−0.267** **	−0.006	−0.076	−0.035	−0.038	0.144		
8	Behavioural expression	12.5	3.26	−0.092	**0.255** *	−0.167	−0.068	**0.440** **	**0.504** **	−0.070	
9	DepressionSymptomatology	19.14	7.88	0.119	−0.165	0.007	0.029	0.154	**0.412** **	0.143	0.068

* $p < 0.05$; ** $p < 0.01$.

Further analysis used multiple linear regression, in which the assumptions of normality, multicollinearity, and independence of errors were met. The independent variables are sociodemographic information and parental stress, while the dependent variable is depressive symptomatology. The analysis found a significant interaction between parental expectations and depressive symptomatology (B = 0.466, $p < 0.01$). This means that the depressive symptomatology experienced by parents of children with ADHD is predicted to increase by 0.466 standard deviation units with every 7.88 increase in CES-D score. Overall, parental stress was found to contribute a 23.2% variance in explaining depressive symptomatology experienced by the participants (F [7,95] = 48.03, $p < 0.001$). More information can be seen in Table 3.

Table 3. Multiple regressions of sociodemographic information, parental stress, and depressive symptomatology.

Predictors	R^2 Change	β	SE B
Step 1	0.052		
Age		0.165	0.140
Level of education		−0.198	0.699
Average household income		−0.017	10.257
Step 2	**0.232** ***		
Emotional expression		0.153	0.090
Parental sacrifices		**0.466** **	0.138
Parental expectations		0.137	0.228
Behavioural expression		−0.161	0.296

** $p < 0.01$; *** $p < 0.001$.

4. Discussion

The current research aims to identify the parental stress experienced by mothers of children with ADHD in addition to identifying the status of their mental health. This is achieved by measuring their stress level based on four main domains and identifying depressive symptomatology experienced by the mothers. Socioeconomic status was also considered as a variable of interest.

The present study found high parental stress experienced by mothers of children with ADHD, specifically within the domain of parental sacrifices, which puts their mental health at risk. This is consistent with past studies [11,12,14], in which parents of children with ADHD have higher levels of stress as compared to typically developed children or children with other disorders [12,13]. This can be explained by the symptoms of ADHD, in which the children tend to be more active than other children [10,11]. Additionally, some children with ADHD tend to be impulsive and inattentive, resulting in behaviours that are performed without considering consequences, as well as difficulty following rules and instructions [10,11]. Mothers are then expected to make sacrifices to manage these behaviours, so much so that their resources can be depleted faster compared to other mothers. Examples of sacrifices based on the measurements used are in terms of sleep schedules, inability to perform leisure activities, inability to enjoy attending gatherings, and many others [20]. When such sacrifices are made and the behaviours are still present, it is not surprising that a sense of hopelessness can start to exist. This is due to the nature of ADHD, which is suggested to be a lifelong disorder, which means that the symptoms can exist throughout childhood, adolescence, and even adulthood [6]. This could explain the significant positive interaction between parental sacrifices and depressive symptomatology.

In relation to parental stress, the age of the mothers and level of education was found to be significantly associated with parental stress. Results show that as the mothers' age increases, their stress due to parental expectations decreases. This is consistent with a past study, which found that mothers who are more experienced are able to manage their stress better [16]. Furthermore, the added experience that comes with age helps mothers to better regulate their expectations to their children's development. Expectation management is beneficial because they will then have more realistic expectations toward the children. The realistic expectations are possibly a reflection of acceptance of the children's condition that they will not behave similar to typically developed children. A study found that one reason for parents to hesitate to seek help from a professional is because they first need come to accept that the problem is beyond what they can handle themselves [25]. Once the mothers are able to accept the situation that they are in, they can then focus on problem solving and identify the best method for their family to adapt to the demands of the child that has ADHD. In addition to age, level of education was found to increase parental stress due to the behavioural expression of ADHD. This can be explained due to the atypical progress of the children's behavioural expression in relation to their development. The discrepancies between the knowledge that the parents possess due to their level of education does not coincide with the behaviours shown by the children creating stress. It is also suggested that those with high cognitive functioning tend to ruminate and worry more [26], which means that parents with a high level of education tend to be affected more by the behavioural manifestation of the ADHD symptoms. This suggests the benefit of cognitive behavioural therapy for parents so that they can manage the way they think, feel, and behave with regard to the condition of their children. Therefore, intervention can be beneficial for the parents as well as the children.

5. Conclusions and Recommendation

The findings of the current study provide empirical support of the relationship between parental stress and the mental health among mothers of children with ADHD. This highlights the importance of caring for the mothers as they are caring for their children. Although no socioeconomic status was found to be a significant predictor to mental health, some sociodemographic factors were found to be associated with the parental stress. The empirical information can then be used to expand on policies to provide further assistance to mothers of children with ADHD. Since socioeconomic status is not a significant predictor of parental stress and the mental health of the mothers, other types of resources may be important to the mothers, which need to be identified by future studies. Moreover, future studies can expand on the limited number of participants of the current study, in addition to diversifying the race of the participants, since the samples of the current research were only among Malays. It is also suggested that future studies focus on qualitative methods to have an in-depth understanding of the stress experienced by the mothers and perform an analysis of needs regarding the resources required by the mothers to manage their parental stress.

Author Contributions: Conceptualization: Q.M.M., M.I.M.I., P.P.P. and T.M.; methodology: Q.M.M. and M.I.M.I.; formal analysis: Q.M.M. and M.I.M.I.; data curation: Q.M.M.; visualization: Q.M.M.; validation: Q.M.M.; writing—original draft preparation: Q.M.M.; writing—review and editing: Q.M.M., M.I.M.I., P.P.P. and T.M.; supervision: M.I.M.I.; project administration: Q.M.M., M.I.M.I., P.P.P. and T.M. All authors have read and agreed to the published version of the manuscript.

Funding: This research received no external funding.

Institutional Review Board Statement: Not applicable.

Informed Consent Statement: Not applicable.

Data Availability Statement: Not applicable.

Acknowledgments: This research was made possible with the support and guidance of allahyarhamah Azlin binti Hj. Alwi. Additionally, the support from JiwaDamai Psychology Center to organize the program for the participants to gather and gain knowledge and the Facebook group Persatuan ADHD Malaysia made it possible for data collection. Finally, the support from Universiti Teknologi MARA (UiTM) Pulau Pinang is also essential for the publication of the article.

Conflicts of Interest: The authors declare no conflict of interest.

References

1. Sayal, K.; Prasad, V.; Daley, D.; Ford, T.; Coghill, D. ADHD in children and young people: Prevalence, care pathways, and service provision. *Lancet Psychiatry* **2018**, *5*, 175–186. [CrossRef]
2. Tian, Y.; Frazee, S.G.; Henderson, R.R.; Iyengar, R. Geographic variation in diagnosis, medication use and associated costs of attention deficit disorder (ADD). *Value Health* **2013**, *16*, A68. [CrossRef]
3. Danielson, M.L.; Bitsko, R.H.; Ghandour, R.M.; Holbrook, J.R.; Kogan, M.D.; Blumberg, S.J. Prevalence of parent-reported ADHD diagnosis and associated treatment among U.S. children and adolescents 2016. *J. Clin. Child Adolesc. Psychol.* **2018**, *47*, 199–212. [CrossRef] [PubMed]
4. Gomez, R.; Hafetz, N. DSM-IV ADHD: Prevalence based on parent and teacher ratings of Malaysian primary school children. *Asian J. Psychiatry* **2011**, *4*, 41–44. [CrossRef]
5. *National Health and Morbidity Survey, Non-Communicable Diseases, Risk Factor & Other Health Problems*; Institute of Public Health, Ministry of Health: Shah Alam, Malaysia, 2015; Volume 2.
6. Aoki, Y.; Tsuboi, T.; Furuno, T.; Watanabe, K.; Kayama, M. The experiences of receiving a diagnosis of attention deficit hyperactivity disorder during adulthood in Japan: A qualitative study. *BMC Psychiatry* **2022**, *20*, 1–8. [CrossRef]
7. Zabala Jr, B.A.; Manabat, J.A.S.A. Prevalence of behavioral symptoms leading to attention deficit hyperactivity disorder (ADHD) among grade I learners. *Psychol. Behav. Sci.* **2017**, *6*, 70–78. [CrossRef]
8. Liu, A.; Xu, Y.; Yan, Q.; Tong, L. The prevalence of attention deficit/hyperactivity disorder among Chinese children and adolescents. *Sci. Rep.* **2018**, *8*, 1–15. [CrossRef]
9. Nomanbhoy, A.; Hawkins, R. Parenting practices of Singaporean mothers of children with ADHD. *Proc. Singap. Healthc.* **2018**, *27*, 33–40. [CrossRef]
10. Ben-Naim, S.; Gill, N.; Laslo-Roth, R.; Einav, M. Parental stress and parental self-efficacy as mediators of the association between children's ADHD and marital satisfaction. *J. Atten. Disord.* **2019**, *23*, 506–516. [CrossRef]
11. Mustafa, Q.M.; Alias, I.A.; Alwi, A.H. Stress experiences of mothers of children With ADHD: Significant coping strategies that safeguard their mental health. *Malays. J. Med. Health Sci.* **2021**, *17*, 14–21.
12. Martin, C.A.; Papadopoulos, N.; Chellew, T.; Rinehart, N.J.; Sciberras, E. Associations between parenting stress, parent mental health and child sleep problems for children with ADHD and ASD: Systematic review. *Res. Dev. Disabil.* **2019**, *93*, 103463. [CrossRef] [PubMed]
13. Gupta, N.D.; Simonsen, M. Non-cognitive child outcomes and universal high quality child care. *J. Public Econ.* **2010**, *94*, 30–43. [CrossRef]
14. Bernardo, J.; Rent, S.; Arias-Shah, A.; Hoge, M.K.; Shaw, R.J. Parental stress and mental health symptoms in the NICU: Recognition and interventions. *NeoReviews* **2021**, *22*, e496–e505. [CrossRef]
15. Haines, L.; Grimes, A. What matters for the wellbeing of mothers and children in material hardship? Application of a modified indicator framework. *Soc. Indic. Res.* **2022**, *162*, 449–474. [CrossRef]
16. Enea, V.; Rusu, D.M. Raising a child with autism spectrum disorder: A systematic review of the literature investigating parenting stress. *J. Ment. Health Res. Intellect. Disabil.* **2020**, *13*, 283–321. [CrossRef]
17. Narkunam, N.; Hashim, A.H.; Sachdev, M.K.; Pillai, S.K.; Ng, C.G. Stress among parents of children with attention deficit hyperactivity disorder, a Malaysian experience. *Asia-Pac. Psychiatry* **2014**, *6*, 207–216. [CrossRef]
18. Theule, J.; Wiener, J.; Tannock, R.; Jenkins, J.M. Parenting stress in families of children with ADHD: A meta-analysis. *J. Emot. Behav. Disord.* **2013**, *21*, 3–17. [CrossRef]
19. Lazarus, R.S. *Stress and Emotion*; Springer: New York, NY, USA, 2006.
20. Palinkas, L.A.; Horwitz, S.M.; Green, C.A.; Wisdom, J.P.; Duan, N.; Hoagwood, K. Purposeful sampling for qualitative data collection and analysis in mixed method implementation research. *Adm. Policy Ment. Health* **2016**, *42*, 533–544. [CrossRef] [PubMed]
21. Kadam, P.; Bhalerao, S. Sample size calculation. *Int. J. Ayurveda Res.* **2010**, *1*, 55–57.
22. Abidin, R.R. *Parenting Stress Index, Professional Manual*, 3rd ed.; Psychological Assessment Resources Inc.: Lutz, FL, USA, 1995.
23. Bolarinwa, O.A. Principles and methods of validity and reliability testing of questionnaires used in social and health science researches. *Niger. Postgrad. Med. J.* **2015**, *22*, 195–207. [CrossRef]
24. Radloff, L.S. The CES-D scale a self-report depression scale for research in the general population. *Appl. Psychol. Meas.* **1977**, *1*, 385–401. [CrossRef]
25. Hansen, A.S.; Telléus, G.K.; Mohr-Jensen, C.; Lauritsen, M.B. Parent-perceived barriers to accessing services for their child's mental health problems. *Child Adolesc. Psychiatry Ment. Health* **2021**, *15*, 4. [CrossRef] [PubMed]
26. Karpinski, R.I.; Kolb, A.M.K.; Tetreault, N.A.; Borowski, T.B. High intelligence: A risk factor for psychological and physiological overexcitabilities. *Intelligence* **2018**, *66*, 8–23. [CrossRef]

proceedings

Proceeding Paper

The Psychosemantic Realm of Social Media Warfare Affecting Public Diplomacy †

Rus Shafinas Shafie

Centre of Media and Information Warfare Studies, Faculty of Communication and Media Studies, Universiti Teknologi MARA (UiTM), Shah Alam 40450, Malaysia; russhafinas@gmail.com

† Presented at the International Academic Symposium of Social Science 2022, Kota Bharu, Malaysia, 3 July 2022.

Abstract: The psychosemantic realm of social media warfare is an emerging field that relates to psycholinguistics in warfare. The paucity of literature relating to the subject paved the interest of this study. Moreover, it aims to address fake news, misinformation, and disinformation, with the essence of understanding meaning-making used as a tactic to influence both digital society and public diplomacy. Therefore, the present study intends to explore the realm of psycholinguistics in social media warfare, focusing on how meaning is inferred and how it influences public diplomacy. The study amalgamates behaviorism and constructivism in understanding the essence of meaning-making. A sample is taken from a Facebook wall post of state actors representing the Malaysian government and the Malaysian opposition. The data are analyzed by applying the critical discourse analysis method. The analysis reveals that meaning inferences embedded in social media narratives are significantly determined by hard power and soft power, and the reproduction of ideology through discourse. The concept of power and discourse is consequential in influencing and affecting public diplomacy. Undoubtedly, it proves that by carefully using meaningful words, creating effective psychosemantics can influence perception and public diplomacy.

Keywords: psychosemantics; psycholinguistics in social media warfare; public diplomacy; digital diplomacy; semantic macrostructure theory; theory of ideological square; cyber warfare

Citation: Shafie, R.S. The Psychosemantic Realm of Social Media Warfare Affecting Public Diplomacy. *Proceedings* **2022**, *82*, 97. https://doi.org/10.3390/proceedings2022082097

Academic Editor: Mohamad Rahimi Mohamad Rosman

Published: 17 October 2022

1. Introduction

Social media is a phenomenon that is able to become the main platform of communication in the digital age [1]. Statista 2021 reported that the number of social media users worldwide is expected to gradually grow from 3.78 billion users in 2021 to 3.96 billion in 2022, with the yearly increase from 2022 to 2025 at a rate of 3.2–4.5% [2]. Social media has become an enabler for internet users to communicate in a borderless environment beyond geographical restrictions [3–5]. The borderless environment opens an opportunity in public diplomacy, where state actors utilize social media to enhance public diplomacy efforts to the extent of digital diplomacy [6–12]. Malaysian social media users have reflected the global phenomenon locally and use social media as a platform to comprehend, construct ideology and perception, construct social cognition, and share the gathered information with other users [13]. A significant number of countries have adopted social media in enhancing their public diplomacy while dispersing ideology and agendas [14–17] as a result of extensive social media usage [18].

As an integral part of communication, language and narratives play an important role in the effort of information dispersion as well as being a tool to persuade and form perception [19]. Language carries significant meaning-making elements known as the semantic macrostructure, which forms comprehension among social media users on a certain discourse [20,21]. As opposed to informal language, state actors often use formal language in online narratives that transmit semantic values constructing psychosemantics, which contributes to social cognition [15,20,21]. Digital diplomacy relies on the effectiveness of online narratives and psychosemantic knowledge to carefully deploy power, either hard power or soft power [7,22,23].

1.1. Background

Social media has evolved from being a social networking platform into a tool for psychological warfare where cyber warfare is deployed. Social media poses invisible threats to any state by offering alternative realities that propel emotional and psychological targeting. Such threats are seen as overwhelming and severely harmful to any state where cultures, social cognition, and unity are at stake [24–27]. These threats are carefully detailed through meaningful narratives, allowing social media users to interpret based on their psychosemantic knowledge [11,28–30]. As these threats pose divisiveness in society, the impact is also seen in international relations where the state's image is tainted with negativity [31]. However, social media itself has the ability to improve a state's performance in digital diplomacy by allowing state actor and non-state actor collaboration to promote the state and enhance the state's image with positive narratives [7,32]. Therefore, social media as an information operation tool is a pronged approach that could either elevate or deteriorate the state's image.

Psychosemantics is a term which denotes how meaning is formed and inferred psychologically [11]. Although the term is used mainly in political psychology, psycholinguistic and semantic elements are the essence for establishing psychosemantics [21]. In this study, psychosemantics is analyzed in terms of its capability to influence perception and public diplomacy. Scholars who have extensively expedited research in psychosemantics are Petrenko [10] and Mitina [11]. Both scholars have explored psychosemantics in the political realm, which includes political psychology, perception and public consciousness, gender issues, political mentality, foreign and national politics, and power in the Russian context. The emergence of this field was explored mainly in the Russian context as opposed to other countries such as China. China is more interested in analyzing soft power and Chinese influence globally [7]. However, psychosemantics in social media is an untapped knowledge that allows the subject to be further explored. The essence of understanding the social media warfare phenomenon lies within the ability to encapsulate the fundamentals of psychosemantic knowledge to disperse power.

Therefore, the study analyzes the Facebook posts of the eighth Malaysian prime minister, YAB Tan Sri Muhyiddin bin Haji Md Yassin, and the Malaysian opposition Facebook page, Pakatan Harapan. Facebook posts from these pages are analyzed in terms of the types of discourse used and types of power used in gaining support or positive perception in the international arena. The sample was taken in March 2020 during the transition period after Tun Mahathir Mohamed, the seventh prime minister, who abated his premiership in late February 2020. The data are analyzed using two of Van Dijk's critical discourse analysis methods, namely the semantic macrostructure theory and the theory of ideological square [21]. The analysis enables an understanding of how meaning dispersed in social media narratives influences public diplomacy.

1.2. Aims and Objectives

The research intends to interpret and comprehend the phenomenon of meaning in the formation of cognition and perception upon meaning interpretation. The diversity of meaning interpretation generated from understanding the semantic macrostructure in social media narratives, as well as its implications for public diplomacy and the use of power and discourse to establish meaning, are investigated. Several objectives that drive the research are to determine psychosemantic knowledge based on discourse and power concepts embedded in online narratives that influence public diplomacy efforts in pandemic situations, and to explore the responses of the WHO on their Facebook page towards online narratives of the eighth Malaysian prime minister, YAB Tan Sri Muhyiddin bin Haji Md Yassin, and the Malaysian opposition Facebook page, Pakatan Harapan. These objectives are able to pave and establish the understanding of how psychosemantics in social media warfare influences public diplomacy. These objectives allow the research to solve the main research question of whether power deployed in online narratives could either enhance or weaken public diplomacy.

1.3. Problem Statement

Effective political communication is integral in ensuring effortless public diplomacy that is established through strategic online narratives [33]. Social media has been the medium of choice for the sole purpose of engagement by political leaders [34] in weaving their ideology, influence, and power through discourse [35] by either spreading fake news, misinformation, or disinformation in influencing and altering public perception [36]. Thus, it has been studied in various disciplines such as communication, applied linguistics, psychology, and functional stylistics [10,11,36,37]. This entails the belief that political leaders are able to shape public perception using social media as a warfare tool.

Hence, exploring online narratives of its psychosemantic elements in the Facebook posts of political leaders is intriguing. Similar studies are scarce globally, suggesting that the subject has not been extensively explored. However, several studies relate to the subject through different angles, including cognitive sciences and political psychology [10,20,38] as well as macro- and micro-psycholinguistic analysis of language and gender [39,40]. Unfortunately, these studies do not explore the knowledge relating to psychosemantics in language and media, which leaves a research gap that needs to be addressed. The analysis will pave a better comprehension of how using psychosemantics in social media warfare influences perception, which will be beneficial to academia as well as the state and state actors.

2. Theoretical Framework

As a multidisciplinary study and in order to ground the research, several theories are applied, respectively. The grand theory of the research is based on a blended theory approach of the behaviorist paradigm and the constructivist paradigm. Both paradigms are seen as significant towards understanding the phenomenon. The social cognitive theory by Bandura [41], representing the behaviorist paradigm, and the social constructivist theory by Vygotsky [42] are predicted to be able to dissect the phenomena accordingly. The theoretical framework is illustrated in Figure 1.

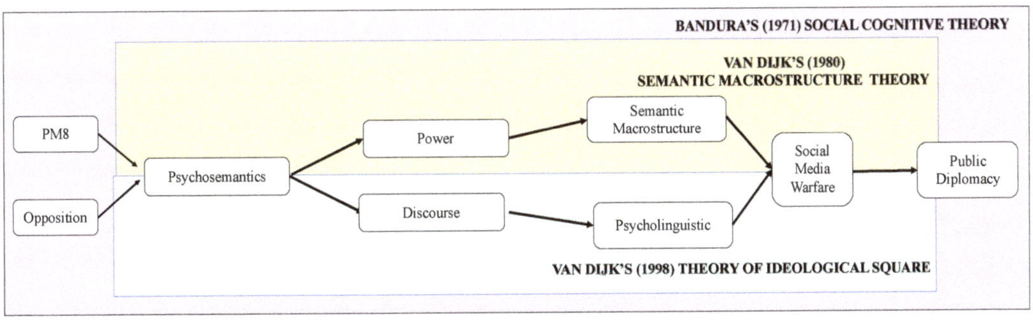

Figure 1. Theoretical framework.

2.1. Psycholinguistics and Semantics (Pyschosemantics) in Psychological Warfare

Every language has its own grammatical structure that generates meaning, and every language interprets meaning differently according to its culture [43–51]. For instance, the English language experiences what is called "World Englishes", a term that describes the emergence of localized, indigenized varieties of English that are influenced by their sociolinguistic context, including history and multicultural settings in various parts and regions of the world [52–55]. This explains why meaning-making or semantics used in other regions might differ from the meaning inferred by native speakers of English [55,56].

Psychological warfare revolves around the use of all levers of state power to influence traditional and non-traditional conflicts of the state [56], which includes nonmilitary strategies involving diplomacy and information in cyberspace. In relation to this, the combination of nonmilitary strategies forms unprecedented forces that lead to effective psychological warfare. Among others, language is core to forming meaning, and the construction of meaning contributes to the comprehension of intended meaning. However, meaning itself can be interpreted differently. It is evident that even command narratives also carry different semantic values to different individuals and different social groups [57]. Psychological warfare using gray zone tactics, including public opinion management, information operations, narrative shaping, and disinformation campaigns, are deployed during peacetime as a form of information or disinformation campaign either for or against the state [58].

2.2. Social Media as a Tool for Psychological Warfare

Cyberspace has become the battleground for contemporary conflicts that range from transnational terror campaigns to state confrontations with other states or nonstate actors, as well as regional insurgencies using online narratives in social media [59]. Social media has evoked and waged emergencies by sparking racism, anarchism, and radicalism globally through its narratives since its emergence in 2001 [6,20,42,43]. It entails the colossal impact of social media warfare that has been used to alter, challenge, and weave ideologies benefitting various parties. The Russian invasion of Ukraine on 24 February 2022 was one of the wars that was covered by the global media, and details of the war were circulated in social media, with numerous reactions from celebrities, politicians, influencers, and even comedians. Mainstream Western media narrated the war to be "an atrocious move", "denying human rights", and claimed that "Ukrainians do not deserve to be refugees as they are Europeans". These narratives created a wave of criticism among social media users, as many feel that these narratives impose white supremacy, racism, and injustice to other war-torn countries, as well as being condescending toward the lives and livelihood of non-Europeans.

These responses revealed a negative impression and perception of social media users towards the crisis. Russia is highly criticized for the unprecedented act that was perceived as an act of violence, and this could tarnish and impact international relations. The severity of the invasion became content for social media users either supporting Russia or against Russia, as well as those either supporting or against Western media narratives. Since the global media is dominated by the Western world [60], the Western-centric media discourse enables Western propaganda to be dispersed through social media, and its narrative opposes the Russian invasion with the intention of establishing a pro-Western psychosemantic knowledge. Hence, it is undeniable that social media is used as a psychological warfare tool to extensively shape the global cognition of the crisis.

2.3. Power and Ideology in Media Discourse

Media discourse is evidently and extensively used in the effort to engage the global public on issues relating to power domination and ideological conflict between different parties. Power and ideology have been at the center of conflicts and wars [58–60], and the media discourse is used extensively to establish power and disperse ideology. The use of power to its extreme extent of either hard or soft power is crucial for establishing a positive or negative perception. This is evident in Russian narratives of the Ukrainian war in which hard power is used, whereas narratives embedded with soft power are often used by China for Chinese diplomacy, referring either to panda diplomacy or the One Belt, One Road diplomacy [7].

2.4. The Social Cognitive Theory as Social Constructivism

The main underlying theory in this research is the social cognitive theory founded by Albert Bandura [41]. However, it is significant that the social cognitive theory is amalgamated with social constructivism. This blended approach combining both social cognitive theory and social constructivism enables the research to look at how society and individual interactions construct psychosemantic knowledge by applying behavioral modeling. In this case, online society is the base of networked learning, in which meaning is learnt and comprehended from discourse. Although meaning interpretation may differ across regions, the basic psychosemantic value is manifested in the discourse, allowing an essential understanding of what is narrated through the social interactions that occur in the communication process. In relation to this, social interactions that are observed during the communication process entail behavioral modeling among target audiences. It is imperative that the research applies a blended approach as it is a progressive form of cognitive learning.

2.5. Van Dijk's Semantic Macrostructure Theory and Theory of Ideological Square

The research also applied two types of linguistic theories in order to understand the psychosemantics embedded in online narratives. These theories are Van Dijk's semantic macrostructure theory [21] and Van Dijk's theory of ideological square [44]. The semantic macrostructure theory is a theory that investigates themes in the form of macropropositions and higher macropropositions in the effort to understand contextual meanings. The theory of ideological square is a theory that examines ideology in discourse that posits positive and negative self-representation as well as positive and negative representation of others, which includes the emphasis on positive representation of "us" and negative representation of "them", as well as de-emphasis on the negative representation of "us" and positive representation of "them".

3. Methodology

The study is a phenomenological research that applies a qualitative method to analyze data. Data are taken from the Facebook posts of the eighth Malaysian prime minister, YAB Tan Sri Muhyiddin bin Haji Md Yassin, and the Malaysian opposition Facebook page, Pakatan Harapan. The sample selected must carry significant semantic value. The research applied two types of critical discourse analysis methods in order to understand the psychosemantics embedded in online narratives, namely Van Dijk's semantic macrostructure theory and Van Dijk's theory of ideological square.

4. Analysis

The analysis involves two data samples from the Facebook posts of the eighth Malaysian Prime Minister, YAB Tan Sri Muhyiddin bin Haji Md Yassin, and the Malaysian opposition Facebook page, Pakatan Harapan. The data are analyzed using the overall meaning of the discourse with reference to the reality at the time the post was made. The data are analyzed and illustrated in Figures 2 and 3, respectively.

Figure 2 is a sample from the Facebook page of YAB Tan Sri Muhyiddin bin Haji Md Yassin. The sample was posted on 25 March 2020. The social cognition at the time of the post was that he was a newly elected PM and was having issues of distrust among MPs, as his cabinet was slandered as a "backdoor government", since he was elected by the Yang diPertuan Agong Al-Sultan Abdullah Ri'ayatuddin Al-Mustafa Billah Shah ibni Almarhum Sultan Haji Ahmad Shah Al-Musta'in Billah. Therefore, he was seen as a PM that did not have the people's vote, and this created issues in his administration. However, he was trying to apply soft power in his narratives by engaging all walks of life. The government was also reproducing an ideology that was ready and capable of providing assistance to all Malaysians equally. This was done by means of emphasizing positivity in its self-representation.

Figure 3 is a sample from the Malaysian opposition Facebook page, Pakatan Harapan, posted on 28 March 2020. This post was a reactive post towards the post made by the PM on 25 March 2020 regarding the Economic Stimulus Package that was launched in order to assist the people. The social cognition that they were trying to impose is that Pakatan Harapan will always be inclusive in their policies and will not allow any discrepancies, as opposed to the eighth PM's government. In addition to that, they were forming a cognition that they care for the well-being of the people. Hence, they applied soft power by using a passive-aggressive manner when they praised the government for such an initiative, while pointing out that the government did not acknowledge and appreciate the Welfare Department's workforce that was also involved directly in enforcing the MCO. Indirectly, they insinuated that the government was careless, while spreading anti-government ideology by applying a negative representation of the government and emphasizing this negativity.

Through the lens of the blended theory, the interaction between individuals online is seen in three different categories. The first is comprised of individuals that wrote the source text, the second involves netizens of the online society, and the third revolves around the individuals or groups that counter-narrate the source text through their posts. In this case, the source text is from the eighth PM, who understood and learned that the nation was at risk from his interactions with the people as well as international organizations such as the WHO. At the same time, the eighth PM modeled behavior according to the response of other nations to battle the pandemic but tailored it to the Malaysian context. On the other hand, the second category involves individuals in the online society interacting and modeling behavior through the Facebook comments section that enables them to interact, learn, and comprehend narratives as well as their intended meanings. In relation to this, the third category consists of individuals and groups that responded and counter-narrated on the source text through new posts and refers directly to the source text, such as the sample taken from Pakatan Harapan's Facebook post. Based on this, it is undeniable that the learning trait in social constructivism, where interactions in society contribute to comprehension, and the behavioral trait in social cognitive theory, where behaviors are modeled and reproduced, are evident and used continuously in meaning comprehension and the formation of psychosemantic knowledge.

Text	Pegawai-pegawai saya di Kementerian Kewangan bertungkus lumus menyediakan satu pakej rancangan ekonomi yang akan memberikan manfaat kepada semua sama ada anda pemandu teksi, pemandu grab, petani, pengusaha restoran, penjusl nasi lemak, penjual goreng pisang, penjual burger, pekerja harian, kerajaan akan berusaha sedaya upaya untuk pastikan anda semua beroleh manfaat. Insya-Allah, no one will be left behind *My officers in the Ministry of Finance are working endlessly to formulate a comprehensive economy stimulus package that will benefit everyone including taxi drivers, grab drivers, farmers, restaurant owners, nasi lemak sellers, goreng pisang sellers, burger sellers, daily workers, the government will ensure that the stimulus package is beneficial to all. Insya-Allah, no one will be left behind*

SEMANTIC MACROSTRUCTURE THEORY	
Theme	Power
Item / Code	Soft Power
Interpretation I / Macroproposition	Comforting the people of the nation by offering solutions
Interpretation II/ Higher Macroproposition	The government is taking aggressive actions in curbing the pandemic and its effects as well as providing assistance to the public

THEORY OF IDEOLOGICAL SQUARE	
Theme	Discourse
Item / Code 1	Reproduction of Ideology
Item / Code 2	Positive Self Representation (postRep-Us)
Item / Code 3	Emphasis on Positive Self (EmpGood)
Interpretation	The government is reproducing ideology by emphasizing on having equal rights of all Malaysians to government assistance.

Overall Interpretation	The government is taking aggressive actions in curbing the pandemic and its effects by providing an inclusive and comprehensive financial assistance to Malaysian citizens.

Figure 2. Data analysis—Facebook post of Tan Sri Muhyiddin Yassin.

Text	
Text	Kenyataan Media Bersama Dato' Seri Dr. Wan Azizah Dr. Wan Ismail (Ahli Parlimen Pandan) dan Hannah Yeoh (Ahli Parlimen Segambut) bertarikh 28 Mac 2020. *Press statement of Dato' Seri Dr. Wan Azizah Dr. Wan Ismail (Pandan MP) and Hannah Yeoh (Segambut MP)* 1. Dalam Pakej Rangsangan Ekonomi Prihatin Rakyat (Prihatin) yang diumum YAB Perdana Menteri petang semalam, kami mengalu-alukan peruntukan elaun khas sebanyak RM200 sebulan kepada anggota tentera, polis, kastam, imigresen, angkatan pertahanan awam serta RELA yang terlibat secara langsung dalam melaksanakan Perintah Kawalan Pergerakan (PKP). *In the recent Economic Stimulus Package announced yesterday by the Prime Minister, we would welcome the allocation for special allowance of RM200 a month to frontliners like the army, police, customs, immigration, civil service defence, and RELA who are involved directly in the enforcement of the Movement Control Order (MCO).* 2. Walau bagaimanapun, kami dapati warga kerja Jabatan Kebajikan Masyarakat (JKM) yang juga terlibat secara langsung dalam pelaksanaan PKP tiada dalam senarai yang diumumkan itu. Mungkin Kerajaan terlepas pandang akan peranan JKM. *However, we found that the Welfare Department's workforce who are also involved directly in enforcing the MCO is not included in the announced list and might have been overlooked by thee government.* 3. Kami berharap warga kerja JKM yang terlibat dalam pelaksanaan PKP ini juga akan menerima elaun khas ini. *We hope that the Welfare Department's workforce involved in the enforcement of the MSO is listed as receivers of the special allowance.* 4. Warga kerja JKM yang menjalankan tugas-tugas penguatkuasaan PKP perlu diiktiraf sebagai petugas barisan hadapan dan dengan itu Kerajaan perlu menyediakan peralatan perlindungan yang sesuai untuk mereka demi melindungi mereka daripada Covid-19 *The Welfare Department's workforce has expedited their duty in enforcing the MCO and should be acknowledged as front liners. Therefore, the government should provide suitable protective equipment accordingly in order to protect them from being infected by Covid-19*

SEMANTIC MACROSTRUCTURE THEORY	
Theme	Power
Item / Code	Soft Power
Interpretation I / Macroproposition	The opposition is negotiating for JKM workers to have the same special allowance as other frontliners
Interpretation II/ Higher Macroproposition	The opposition feels that the financial assistance given was incomprehensive

THEORY OF IDEOLOGICAL SQUARE	
Theme	Discourse
Item / Code 1	Reproduction of Ideology
Item / Code 2	Negative Representation (negRep-Them)
Item / Code 3	Emphasis on Negative Them (EmpBad)
Interpretation	The opposition is reproducing ideology by emphasizing that the government did not include JKM workers in the list of the receivers of the special allowance

Overall Interpretation	The opposition is embedding the notion that they are fighting for the people and for the justice of the people by revealing to the government that there are groups of people that are being left out and are not included in the list of government assistance.

Figure 3. Data analysis—Facebook post of Tan Sri Muhyiddin Yassin.

The blend between social cognitive theory and social constructivism in the construction of social cognition, including the social cognition of international societies, is apparent. It is evident in past literature that argue social constructs influence public diplomacy, and the study proves that soft power and the reproduction of ideology applied in the state's (PM8) online narratives form a positive international social cognition, prompting positive perception. The World Health Organization (WHO) responded positively to the actions taken by most countries, including Malaysia in trying, to sustain their economy by launching economic packages and aid. The WHO acknowledged the difficulties faced by nations, forming a positive perception towards efforts to cope with the pandemic. Therefore, it is apparent that psychosemantic knowledge constructed on the basis of social cognitive theory and social constructivism deployed using social media as a warfare tool succeeded in forming perception in public diplomacy efforts.

5. Conclusions

This research is able to pave and establish a new interdisciplinary study incorporating social media warfare, linguistics, and public diplomacy by exploring meaning-making and its inferences that form the psychosemantic knowledge contributing to perception building. The extent of the research explores new grounds, including the theoretical blended approach by amalgamating the social cognitive theory and social constructivism, in the effort to understand how society and individual interactions construct psychosemantic knowledge by applying behavioral modeling.

The finding enables the study to relate macrostructures, ideology, and overall data interpretation with reality. The study found that macropropositions and higher macropropositions are established in understanding macrostructures leading to the overall interpretation of the text. In relation to that, the emphasis and de-emphasis on either the good or bad, as well as the negative and positive representation of us and them, are used extensively to portray intended meanings. Narratives in the form of passive sentences, as well as having a passive-aggressive style, also influence and connotate meaning. However, it is also found that the perception of reality, as compared to the overall comprehension of current political issues and standings, is vital in understanding meanings beyond the media discourse.

Here, the theoretical blend between the social cognitive theory and social constructivism is found to be evident in the social media environment in both comprehending narratives and reproducing narratives. The obvious and continuous instances of comments responding to posts carrying ideologies came from the behavior modeling of the source of communication or opinion leaders that showed their behavior and opinions through their comments. This suggests the learning and comprehension of meaning through behavior by social interaction. It explains that social behavior and social interactions bring semantic value in social media discourse, forming psychosemantic knowledge. Hence, the study found that meaning-making and its inference-forming psychosemantic knowledge are key in deploying psychological warfare in the effort to influence perception and are key likewise in public diplomacy efforts.

Funding: This research received no external funding.

Institutional Review Board Statement: Not applicable.

Informed Consent Statement: Not applicable.

Data Availability Statement: Not applicable.

Conflicts of Interest: The author declares no conflict of interest.

References

1. Dwivedi, Y.K.; Ismagilova, E.; Hughes, D.L.; Carlson, J.R.; Jacobson, J.; Jain, V.; Karjaluoto, H.; Kefi, H.; Krishen, A. Setting the future of digital and social media marketing research: Perspectives and research propositions. *Int. J. Inf. Mgmt.* **2021**, *59*, 102168. [CrossRef]
2. Social Media—Statistic and Facts. Statista, S. Dixon. 2021. Available online: https://www.statista.com/statistics/715449/social-media-usage-reasons-worldwide/ (accessed on 20 April 2022).
3. Lee, S.; Xenos, M. Social distraction? Social media use and political knowledge in two U.S. Presidential elections. *Comput. Hum. Behav.* **2019**, *90*, 18–25. [CrossRef]
4. Calderaro, A. Social media and politics. In *The SAFE Handbook of Political Sociology: Two Volume Set*; Outhwaite, W., Turner, S., Eds.; Sage Publications: Thousand Oaks, CA, USA, 2018; pp. 781–796.
5. Barnett, G.A.; Xu, W.W.; Chu, J.; Jiang, K.; Huh, C.; Park, J.Y.; Park, H.W. Measuring international relations in social media conversations. *Gov. Inf. Q.* **2017**, *34*, 37–44. [CrossRef]
6. Arceneaux, P. Information intervention: A taxonomy & typology for government communication. *J. Public Dipl.* **2021**, *1*, 5–35.
7. Huang, Z.A.; Wang, R. 'Panda engagement' in China's digital public diplomacy. *Asian J. Comm.* **2020**, *30*, 118–140. [CrossRef]
8. Bjola, C. *Adapting Diplomacy to the Digital Age: Managing the Organisational Culture of Ministries of Foreign Affairs, Working Paper No 1*; Oxford Digital Diplomacy Research Group: Oxford, UK, 2018.
9. Verrekia, B. Digital Diplomacy and Its Effect on International Relations. Available online: https://digitalcollections.sit.edu/cgi/viewcontent.cgi?article=3619&context=isp_collection (accessed on 20 October 2021).
10. Petrenko, V.F.; Mitina, O.V.; Gladkikh, N.Y. Psychosemantics of soft power in geopolitics. *Soc. Stud.* **2018**, *1*, 40–51.
11. Petrenko, V.F.; Mitina, O.V. A psycho-semantic approach to reconstruction of political mentality: Research methods and examples. *Her. Russ. Acad. Sci. T.* **2017**, *87*, 49–62. [CrossRef]
12. Archetti, C. The impact of new media on diplomatic practice: An evolutionary model of change. *Hague J. Dip.* **2012**, *7*, 181–206. [CrossRef]
13. Abdullah, N.H.; Hassan, I.; Ahmad, M.F.; Hassan, N.A.; Ismail, M.M. Social media, youths and political participation in Malaysia: A Review of literature. *Int. J. Acad. Res. Bus. Soc. Sci.* **2021**, *11*, 845–857.

14. Sevin, R.; Ingenhoff, D. Public diplomacy on social media: Analyzing networks and content. *Int. J. Comm.* **2018**, *12*, 3663–3685.
15. Kreis, R. The "Tweet Politics" of President Trump. *J. Lang. Politic.* **2017**, *16*, 607–618. [CrossRef]
16. Hoefte, R.; Veenendaal, W. The challenges of nation building and nation branding in multi-ethnic suriname. *Nat. Ethn. Politics* **2019**, *25*, 173–190. [CrossRef]
17. Jeong, H. Nation Branding and the Representation of a Nation's Identity: The Case of the Study in Sweden Facebook Page. Master's Dissertation, Uppsala Universitet, Uppsala, Sweden, 2018.
18. Costa, V. Shaping public diplomacy through social media networks in the 21st century. *Rom. J. Hist. Int. Stud.* **2017**, *4*, 139–154.
19. Gouws, S.; Metzler, D.; Congxing, C.; Hovy, E. Contextual bearing on linguistic variation in social media. In Proceedings of the Workshop on Language in Social Media (LSM 2011), Portland, OR, USA, 23 June 2011; pp. 20–29.
20. Petrenko, V.F.; Mitina, O.V.; Papovyan, M. Perception of foreign and national political leaders in Russia. *Behav. Sci.* **2020**, *10*, 103. [CrossRef]
21. Van Dijk, T.A. *Macrostructures an Interdisciplinary Study of Global Structures in Discourse, Interaction, and Cognition*; Lawrence Erlbaum Associates: Hillsdale, NJ, USA, 1980; p. 332.
22. Yavuzaslan, K.; Çetin, M. Soft Power Concept and Soft Power Indexes. In *Business Challenges in the Changing Economic Landscape*; Springer: Cham, Switzerland, 2016. [CrossRef]
23. Yecies, B.; Keane, M.; Yu, H.; Zhao, E.J.; Zhong, P.Y.; Leong, S.; Wu, H. The cultural power metric: Toward a reputational analysis of China's soft power in the Asia-Pacific. *Glob. Media China* **2019**, *4*, 203–219. [CrossRef]
24. Johnson, B.D.; Draudt, A.; Brown, J.C.; Ross, R.J. Technical Report Threat Casting Workshop Hosted at Arizona State University Produced by Cyndi Coon. 2019. Available online: https://threatcasting.asu.edu/sites/default/files/2020-07/threatcasting-2020-The%20Future%20of%20Information%20Warfare-WEB_1.pdf (accessed on 22 May 2022).
25. Francois, C.; Lin, L. The strategic surprise of Russian information operations on social media in 2016 in the United States: Mapping a blind spot. *J. Cyber Policy* **2021**, *6*, 9–30. [CrossRef]
26. Andriukaitis, L.; Kalensky, J.; Kargar, S.; Panchulidze, S.; Smętek, J.; Vangeli, A. *The Misuse of Social Media Platforms and other Communication Channels by Authoritarian Regimes*; Lessons Learned; European Union: Brussels, Belgium, 2021.
27. Van Vuuren, R. Information warfare as future weapon of mass-disruption, Africa 2030s scenarios. *J. Futures Stud.* **2018**, *23*, 77–94.
28. Olaniran, B.; Williams, I. Social media effects: Hijacking democracy and civility in Civic Engagement. In *Platforms, Protests, and the Challenge of Networked Democracy Rethoric, Politic and Society*; Jones, J., Trice, M., Eds.; Palgive Macmillan: London, UK, 2020; pp. 77–94.
29. Tucker, J.A.; Guess, a.; Barbera, P.; Vaccari, C.; Siegel, A.; Sanovich, S.; Stukal, D.; Nyhan, B. Social media, political polarization, and political disinformation: A review of the scientific literature. *SSRN Electron. J.* **2018**, *2018*, 1–95. [CrossRef]
30. Campante, F.; Durante, R.; Sobbrio, F. Politics 2.0: The multifaceted effect of broadband internet on political participation. *J. Eu. Econ. Assoc.* **2018**, *16*, 1094–1136. [CrossRef]
31. Ong, J.; Sa'ad, M.F.; Wong, W.L.; Mohd Saudi, N.S.; Singh, I.; Ab Ghani, S. Digital diplomacy: The role of social media. *Solid State Tech.* **2020**, *63*, 7511–7522.
32. Darmastuti, A.; Inayah, A.; Simbolon, K.; Nizar, M. Social media, public participation, and digital diplomacy. *Adv. Soc. Sc. Ed. Humanit. Res.* **2021**, *606*, 38–47.
33. Surowiec, P.; Manor, I. *Public Diplomacy and the Politics of Uncertainty*; Palgrave Macmillan Series in Global Public Diplomacy (GPD); Palgrave Macmillan: London, UK, 2021.
34. Heavey, C.; Simsek, Z.; Kyprianou, C.; Risius, M. How do strategic leaders engage with social media? A theoretical framework for research and practice. *Strat. Manag. J.* **2020**, *41*, 1490–1527. [CrossRef]
35. Surdiasis, F.; Eriyanto, E. Narrative of politics in the era of social media: A multimodal analysis of president Joko Widodo's video blog. *E3S Web Conf.* **2018**, *74*, 10012. [CrossRef]
36. Vowe, G.; Henn, P. *Political Communication in the Online World*; Routledge: London, UK, 2015; p. 310.
37. Sankar, L.V. Discourse and the struggle for power. *Int. J. Acad. Res. Part B* **2013**, *5*, 520–527. [CrossRef]
38. Chalikova, O. Psychosemantic research into the phenomenon of risk. In The Fifth International Luria Memorial Congress. Lurian Approach in International Psychological Science. In Proceedings of the Fifth International Luria Memorial Congress, Yekaterinburg, Russia, 13–16 October 2017.
39. Marinova, I. Comparative psychosemantic distances between the term I and other terms related to responsibility—An exploratory stud. *Contrastive Linguist. XLIV* **2019**, *4*, 57–89.
40. Shevtsov, A.; Hupalovska, V. Psychosemantic content of the concept of sexuality in the language consciousness of adults. *Psycholinguistics* **2020**, *27*, 310–334. [CrossRef]
41. Bandura, A. *Social Learning Theory*; General Learning Press: New York, NY, USA, 1971; p. 247.
42. Vygotsky, L.S. *Mind in Society: The Development of Higher Psychological Processes*; Harvard University Press: Cambridge, UK, 1978; p. 159.
43. Singer, P.W.; Brooking, E.T. *Likewar: The Weaponization of Social Media*; Houghton Mifflin Harcourt: Boston, MA, USA, 2018; p. 432.
44. Van Dijk, T.A. *Ideology: A Multidisciplinary Approach*; Sage Publication: London, UK, 1998; p. 384.
45. Bandura, A.A. A social cognitive theory of personality. In *Handbook of Personality*, 2nd ed.; Pervin, L., John, O., Eds.; Guilford Publications: New York, NY, USA, 1999; pp. 154–196.

46. Van Dijk, T.A. Discourse, Knowledge, And Ideology. In *Communicating Ideologies. Multidisciplinary Perspectives on Language, Discourse, and Social Practice*; Martin, P., JoAnne, N., Van Dijk, T.A., Eds.; Peter Lang: Bern, Switzerland, 2004; pp. 5–38.
47. Van Dijk, T.A. Discourse, context, and cognition. *Discourse Stud.* **2006**, *8*, 159–177. [CrossRef]
48. Finegan, E. *Language, Its Structure, and Use*; Thomson Wadsworth: Belmont, CA, 2008; p. 611.
49. Celce-Murcia, M.; Larsen-Freeman, D. *The Grammar Book: An ESL/EFL Teacher's Course*, 2nd ed.; Heinle & Heinle: Boston, MA, USA, 1999; p. 853.
50. Halliday, M.A.K. On the 'Architecture' of Human Language. In *On Language and Linguistics, Volume 3 in the Collected Works of M.A.K. Halliday*; Webster, J.J., Ed.; Continuum: New York, NY, USA, 2003; p. 29.
51. Fairclough, N. *Language and Power*; Longman: London, UK, 1989; p. 259.
52. Fairclough, N. *Critical Discourse Analysis: The Critical Study of Language*; Longman: London, UK, 1995; p. 268.
53. Fowler, R. *Linguistic Criticism*; University Press: Oxford, UK, 1986; p. 196.
54. Ellis, N.C. Vocabulary acquisition: Word structure, collocation, word-class, and meaning. In *Vocabulary: Description, Acquisition, and Pedagogy*; Schmitt, N., McCarthy, M., Eds.; Cambridge University Press: Cambridge, UK, 1997; pp. 122–139.
55. Corke, S. George Kennan and the inauguration of political warfare. *J. Conf. Stud.* **2016**, *26*, 101–120.
56. Christopher, P.; Matthews, M. The Language of Inform, Influence, and Persuade: Assessment Lexicon and Usage Guide for U.S. European Command Efforts. Santa Monica, CA: RAND Corporation. 2018. Available online: https://www.rand.org/pubs/research_reports/RR2655.html (accessed on 23 May 2022).
57. Harold, S.W.; Beauchamp-Mustafaga, N.; Hornung, J.W. *Chinese Disinformation Efforts on Social Media*; Rand Corporation: Santa Monica, CA, USA, 2021. Available online: https://www.rand.org/pubs/research_reports/RR4373z3.html (accessed on 23 May 2022).
58. Lindsay, G.J. Using Social Media to Understand Narratives in Contemporary Conflicts. Master's Dissertation, Canadian Forces College, Toronto, ON, Canada, 2016; p. 74.
59. Pantserev, K. Manipulative practices in psychological warfare: New challenges in the new age. In Proceedings of the 11th International Scientific and Theoretical Conference—Communicative Strategies of Information Society, Saint Petersburg, Russia, 25–26 October 2019; pp. 281–290. Available online: https://www.researchgate.net/publication/339884604_Manipulative_Practices_In_Psychological_Warfare_New_Challenges_In_The_New_Age (accessed on 23 May 2022).
60. Dimitriu, G. Clausewitz and the politics of war: A contemporary theory. *J. Strateg. Stud.* **2020**, *43*, 645–685. [CrossRef]

proceedings

MDPI

Proceeding Paper

Post-Work Socio-Cultural Life System †

Meida Rachmawati [1,*], Cecep Sukrisna [2] and Mohamad Rahimi Mohamad Rosman [3]

1 Bisnis Manajemen Retail Program, Universitas Ngudi Waluyo, Kabupaten Semarang 50512, Indonesia
2 Universitas Pertahanan, Kabupaten Bogor 16810, Indonesia
3 College of Computing, Informatics, and Media, Universiti Teknologi MARA Kelantan Branch, Machang 18500, Malaysia
* Correspondence: meida_r@unw.ac.id
† Presented at the International Academic Symposium of Social Science 2022, Kota Bharu, Malaysia, 3 July 2022.

Abstract: Many people think that retirement life will turn into a worrying situation. The reality of retirement is not as bad as imagined. On the contrary, many positive things can be done in retirement. Some make retirement a beautiful form of further service in old age, especially with financial problems that are no longer stable. A qualitative descriptive approach was used to find out how the system of post-work socio-cultural life functions. Lifestyle changes render time and financial efficiency a priority along with healthy thinking.

Keywords: life; retirement; social culture

check for updates

Citation: Rachmawati, M.; Sukrisna, C.; Rosman, M.R.M. Post-Work Socio-Cultural Life System. *Proceedings* **2022**, *82*, 98. https://doi.org/10.3390/proceedings2022082098

Academic Editor: Mohd Nasir Ismail

Published: 17 October 2022

1. Introduction

Post-work or retirement, more commonly called retirement, is a necessity for everyone whose routine in fulfilling their life needs involves a job. Many people think that when they retire, life will turn into a worrying situation. It is very reasonable to assume that you can no longer meet with colleagues in activities, no longer wear a uniform with its provisions, lose authority, and be classified as an unproductive person. Additionally, your health condition will probably diminish, which is influenced by the psyche.

Based on the Law of the Republic of Indonesia Number 43 of 1999 concerning Amendments to Law Number 8 of 1974 concerning the Principles of Employment, Article 10, pensions are old age guarantees and compensation for civil servants who have devoted themselves to the state for many years [1]. In essence, everyone must try to guarantee their old age, and for this, every civil servant is obliged to become a participant in a social insurance agency established by the government. Because retirement is not only a guarantee for old age but also a reward, the government makes its contribution to civil servants [2].

The reality of retirement is not as bad as imagined. On the contrary, many positive things can be done in retirement. Some make retirement a beautiful form of further service in old age, especially with financial problems that are no longer stable. This kind of understanding affects a person's way of thinking, which results in being over-sensitive and subjective, so that one becomes sickly in one's retirement. When entering retirement, it is undeniable that there will be changes in daily routines; therefore, the early retirement period is a time to reorganize life patterns. Thus, this paper seeks to answer the following research question: how far has the preparation been undertaken in dealing with the post-work socio-cultural life system?

2. Methodology

The design of this paper uses a qualitative descriptive approach to find out how the post-work socio-cultural life system works. It is done by collecting information from both informants and literature studies and available library materials, mass media, electronic media, electronic books, and other open sources [3]. On the other hand, the inclusion and exclusion criteria are based on previous study of [4–6]. The selection of articles is based on (1) relevance to the context of the study, (2) contribution to a body of knowledge, (3) being based on empirically proven research, and (4) in English only.

To guide the selection of the articles, we utilize several reputable online databases such as Web of Science, Scopus, Emerald, EbschoHost, and IEEE Explore. The search term is "post-work socio-cultural life system". To improve the search results, we omitted non-research articles, patents, and review papers.

3. Finding and Discussion

For most people, the word "retirement" is something to worry about as they become increasingly aware that it is getting closer and closer. Various kinds of feelings and thoughts began to appear, such as feelings of sadness because they would no longer experience work that was loved and colleagues. The feeling of losing a job greatly impacts not only physical health but also psychological health. The activities that have been carried out so far require concentration and the requirement for responsibility along with an emphasis on the ability to think critically, focus, and analyze. These habits always have social aspects and require emotional intelligence. This factor often triggers a decline in mental and physical health psychologically.

According to the Law of the Republic of Indonesia Number 11 of 1992 concerning Pension Funds, pensions are distinguished, among others [7]:

1. Normal Retirement. Pensions are given to employees whose age has reached the retirement period set by the company. In Indonesia, the normal retirement age is usually 55 years.
2. Expedited Retirement. Pension is given due to certain conditions, such as a reduction in the number of employees in the company. This pension allows employees to retire earlier before reaching the predetermined retirement age. Usually, there must be a clear reason for applying for an accelerated retirement.
3. Postponed retirement. Deferred retirement is an employee who stops working before reaching the normal retirement age so that the payment of pension funds is postponed until the employee reaches the normal retirement age.
4. Disability Pension. A pension is given due to an accident such that workers are considered no longer able to be employed at the company. A disability pension is not related to age and will be given if no longer competent or unable to do the job.

As for pension contributions, according to the Human Resources Bureau of the Ministry of Finance, civil servant pension contributions and government contributions are fostered and managed by the social insurance agency for employee pension rights (Law Number: 11 of 1969 article 9); employees who are respectfully dismissed are entitled receive a pension, as follows:

1. Has reached the age of at least 50 years and has a working period before retirement of at least 20 years.
2. Has a working period of at least 4 years and an agency/official appointed by the Ministry of Health has declared them unable to work in any position due to physical or spiritual conditions that are not caused by or because of carrying out their duties based on the regulations on health testing of civil servants.
3. A civil servant who, after carrying out state duties, is not re-employed as a civil servant is entitled to receive an employee pension if he is honorably discharged as a civil servant and is a civil servant at the time of his dismissal, having reached the age of at least 50 years and a period of service before retirement at least 10 years.

The above provisions form a regulation that must be obeyed and carried out by all (state) workers/employees. The next phase is to live a healthy and happy retirement. This has become a phenomenon in itself for some people, considering that not everyone is able or is complacent in preparing for it. Inevitably this condition makes a person depressed. According to the Large Dictionary of Indonesian (KBBI), the meaning of the word depression is a difficult and lethargic state of commerce. Another meaning of depression is a mental disorder in a person characterized by degenerate feelings (such as gloom, sadness, and depression).

Depression is a disorder of the feelings characterized by reduced enthusiasm, low self-esteem, self-blame, sleep disturbances, and eating disorders. People with depression tend to be sad because of a feeling of pressure that almost always involves loss, which can be in the form of losing a close person, losing things, or losing a position. Retirement is the end of an individual's career path; it can be a significant loss event for some people who experience it and have a bad impact on their mental condition, as well as the demand to continue to work and be productive [8].

3.1. Socio-Cultural Life

Retirement is not the end of human life but is the next stage in navigating the phases of life. Yesterday's routine made a beautiful personal life story to remember. Most individuals who have retired still have a strong motivation to return to work for various reasons. However, not all jobs can simply accept pensioners. From the other side, seeing retirees who are still working is a mistake in self-management.

This negative view arises because of the assumption that retirement, which consists of a lot of free time, should be used to gather with family or develop hobbies after dedicating oneself to a job for a dozen or decades. Individuals who work after retirement are also identified as a group that is not well prepared for retirement so their retirement period is considered less prosperous economically and they must continue to work to meet financial needs. Thus, working after retirement is only interpreted by the general public as a form of livelihood strategy to meet economic needs [2].

It is not easy for a person to change his daily routine from a structured pattern of activity to an unstructured one. The role of the family, including the people closest to one, is very influential in providing moral support in living a new life. Therefore, it is necessary to actively socialize to find a new culture of life to create a balance in behavior.

Changing one's lifestyle is a key factor in facing new social and cultural life. Lifestyle is defined as a pattern of work and time as well as a financial pattern. This is important considering that influence and income are already very different. The priority scale for the policy of choosing primary, secondary, and tertiary goods is a necessity. A very big opportunity from the perspective of time, retirement can be used for self-healing with family (wife/husband, children, grandchildren, siblings), whereas previously this was a very rare thing and difficult to obtain because one's time was occupied by activities. If there is still a difference in one's savings, work culture can be continued by opening an independent business as an expressive activity (killing time) including managing hobbies.

3.2. Post-Working Life System

The important thing to do in preparing for retirement is to work positively. There are many ways to prepare yourself for retirement [9].

1. Prepare physically and mentally. Accept with sincerity and with an open feeling that everyone in due time will enter retirement. Talk with your family and inform them that your financial condition may change so that your family can save more or change habits that were previously more consumptive.
2. Apply a healthy lifestyle. Entering retirement, it is hoped that life patterns can change for the better. Diligent health checks, choosing foods that are suitable for body condition, and regular exercise are recommended.
3. Do your favorite hobby. If when you are active, you rarely do your favorite hobby. Now is the time for you to do a hobby that you like and it is even better if that hobby can produce something useful.
4. Plan finances before retirement. If you plan before your retirement arrives, you will not experience financial problems later. Try the financial methods offered by the office where you used to work or the bank where you currently deposit money. If possible,

try to invest. It does not have to be in a big business, but only moderately so that the possibility of loss can be minimized.

5. Keep in touch. Keep in good relations with office friends or with your peers. Sometimes, by maintaining good communication, you can still do activities that you used to often do together or just exchange ideas.
6. Get closer to God. Getting closer to God and performing social and spiritual activities brings peace of mind. Increased gratitude and positive thinking will make it easier for us to live happier days.
7. Positive thoughts and good physical and mental conditions, as well as family support in psychological and financial factors, will make it easier for someone to enter retirement. Enjoy retirement with positive things and be a happy retiree.

A psychologist from North Dakota State University, Jeremy Hamm, and his colleagues were curious. He also undertook research to observe certain personality traits that can predict who will thrive after retirement and who will wither. This is called trait goal disengagement, which refers to the tendency to attack easily when faced with a difficult job or task. Consider a statement like this [10,11]:

1. When my expectations are not met, I lower my expectations.
2. To avoid disappointment, I do not set my goals too high.
3. I feel relieved when I let go of some of my responsibilities.
4. Conclusions

Most people think that retirement may be boring. However, if planned properly, retirement will not hinder productivity or create stress. Retirement is a very big opportunity to realize dreams that have been stored for a long time. One's lifestyle changes to make time and financial efficiency a priority, along with healthy thinking. Retirees need to find ways to stay mentally challenged so that mental health is maintained and moral decline is avoided.

Author Contributions: Conceptualization, M.R. and C.S.; methodology, M.R.; validation, M.R.; formal analysis, C.S.; investigation, C.S.; resources, M.R.; writing—original draft preparation, M.R., C.S. and M.R.M.R.; writing—review and editing, M.R.M.R.; visualization, C.S.; supervision, M.R.M.R. All authors have read and agreed to the published version of the manuscript.

Funding: This research received no external funding.

Institutional Review Board Statement: Not applicable.

Informed Consent Statement: Not applicable.

Data Availability Statement: Not applicable.

Acknowledgments: The authors would like to thank Universitas Ngudi Waluyo, Universitas Pertahanan, Indonesia and Universiti Teknologi MARA Kelantan Branch, Malaysia for supporting the research.

Conflicts of Interest: The authors declare no conflict of interest.

References

1. Undang-Undang Republik Indonesia Nomor 43 Tahun 1999 Tentang Perubahan Atas Undang-Undang Nomor 8 Tahun 1974 Tentang Pokok-Pokok Kepegawaian. Republik Indonesia. Available online: https://peraturan.bpk.go.id/Home/Details/45377/uu-no-43-tahun-1999 (accessed on 10 July 2022).
2. Ariza Marine, P. Makna Kerja Pasca Pensiun Studi Kasus: Purnawirawan TNI AL di Kota Surabaya. Ph.D. Thesis, Universitas Airlangga, Surabaya, Indonesia, 2020.
3. Laher, S.; Kramer, S. *Transforming Research Methods in the Social Sciences: Case Studies from South Africa*; Wits University Press: Johannesburg, South Africa, 2019; p. 456.
4. Webster, J.; Watson, R.T. Analyzing The Past To Prepare For The Future: Writing a Literature Review. *MIS Quarterly* **2002**, *26*, 13–23.
5. Mohamad Rosman, M.R.; Ismail, M.N.; Masrek, M.N. Investigating the predictors of digital library engagement: A structured literature analysis. *Pak. J. Inf. Manag. Libr.* **2021**, *22*, 60–82.

6. Mohamad Rosman, M.R.; Ismail, M.N.; Masrek, M.N. Investigating the determinant and impact of digital library engagement: A conceptual framework. *J. Digit. Inf. Manag.* **2019**, *17*, 215.
7. Undang-Undang Republik Indonesia Nomor 11 tahun 1992 tentang Dana Pensiun. Republik Indonesia. Available online: https://peraturan.bpk.go.id/Home/Details/46603/uu-no-11-tahun-1992 (accessed on 10 July 2022).
8. Tutik Ida Rosanti. Kejadian depresi pada pegawai menjelang pensiun, studi pada kepala desa di lima kecamatan, kabupaten demak. *J. Keperawatan Soedirman* **2010**, *5*, 9–12.
9. Nena Mawar Sari. Bahagia Menghadapi Masa Pensiun. Available online: https://www.rsudwangaya.denpasarkota.go.id/artikel/bahagia-menghadapi-masa-pensiun (accessed on 10 July 2022).
10. Rizal Fadli. Inilah Pengaruh Pensiun Terhadap Kesehatan Mental. 2020. Available online: https://www.halodoc.com/artikel/inilah-pengaruh-pensiun-terhadap-kesehatan-mental (accessed on 10 July 2022).
11. Biro SDM Sekjen Kemkeu RI. Available online: https://www.sdm.kemenkeu.go.id/manajemen.cfm?id=20 (accessed on 10 July 2022).

proceedings

MDPI

Proceeding Paper

Leadership Style and Quality of Work Life among Nurses in Malaysia during the COVID-19 Pandemic Crisis †

Zarina Begum Ebrahim *, Siti Aqilah Hafidzuddin, Muna Kameelah Sauid, Nurul Ain Mustakim and Noorzalyla Mokhtar

Faculty of Business and Management, Universiti Teknologi MARA, Kampus Bandaraya Melaka, Melaka 75350, Malaysia
* Correspondence: zarina148@uitm.edu.my
† Presented at the International Academic Symposium of Social Science 2022, Kota Bharu, Malaysia, 3 July 2022.

Abstract: The aim of this study is to investigate the association between leadership styles and the quality of work life among nurses in Selangor. The researcher used four independent variables, which are transformational, transactional, autocratic, and democratic leadership styles, while the dependent variable is Quality of Work Life (QWL). This study utilized a correlational quantitative design with 202 nurses. The researcher used a simple random sampling method to distribute the questionnaires. The findings revealed that transformational, transactional, autocratic, and democratic leadership styles are correlated with the QWL, indicating positive, moderate, and significant relationships. While regression analysis shows the most influential leadership style on QWL was the Transactional Leadership style, this study is significant to nurses as there are a need to identify the best leadership style that can assist nurses in the healthcare industry to perform their best and to help human resource managers to identify which leadership style is suitable for the management and the nurses. Apart from that, management can use these findings to aid leaders in improving their leadership skills and moulding competent leaders in a way that improves employees' quality of life at work. Empirically, this study is relevant to ensure the 2030 Agenda for Sustainable Development Goal (SDG) is fulfilled in SDG 3 and SDG 8.

Keywords: leadership styles; quality of work life; transactional leadership; democratic leadership; autocratic leadership; transformational leadership

Citation: Ebrahim, Z.B.; Hafidzuddin, S.A.; Sauid, M.K.; Mustakim, N.A.; Mokhtar, N. Leadership Style and Quality of Work Life among Nurses in Malaysia during the COVID-19 Pandemic Crisis. *Proceedings* **2022**, *82*, 99. https://doi.org/10.3390/proceedings2022082099

Academic Editor: Mohamad Rahimi Mohamad Rosman

Published: 17 October 2022

1. Introduction

Nowadays, for every firm to retain and recruit personnel, a good quality of work life (QWL) is essential. It is because, in today's extremely competitive business world, organisations want high-performing employees to achieve their objectives and obtain a competitive advantage [1]. Employee happiness with their work and their healthy life at work have been identified as essential factors of a company's performance [2]. Thus, QWL is an essential factor for improving the working environment. The QWL is a concept based on a set of principles that says that a company's most valuable resource is its people [3]. An et al. said that the excellent relationship between human resources and their work environment that promotes, and boosts job satisfaction is known as "Quality of Work Life" (QWL). Therefore, this study is significant in supporting the 2030 Agenda, which is the sustainable development goal (SDG), specifically in supporting SDG 3: good health and well-being, which focuses on people, and SDG 8: decent work and economic growth, which emphasizes prosperity.

Today, QWL has emerged as a significant concern, with numerous papers published on the subject [1,4–7]. Recently, the emergence of Industrial Era 4.0 and the COVID-19 pandemic cases have affected human lives, specifically in terms of health and QWL. Consumers are placing more demand on health services, both in terms of quantity and quality, as a result of the Industrial Era 4.0 as well as the COVID-19 crisis [1]. Health professionals have been at the forefront of the fight against COVID-19 and have been directly engaged in giving care to COVID-19 patients while the crisis is being managed. In the healthcare industry, nurses have evolved into important frontline participants, and their productivity and contentment are pivotal to the hospital's success. Nursing is a high-demand career that requires an individual's ability to deal with and manage stressful physical and mental dynamics in the workplace [8]. Moreover, due to increasing aging populations in every country, nursing experts and services are in greater demand [9]. Furthermore, nurses, for example, play a critical role because they are on the front lines in responding to the pandemic.

The importance of nursing practise in the care and survival of COVID-19 patients has proven critical. In the nursing profession, QWL is a complicated matter that can have an impact on the quality of care offered to patients [10]. The previous study mentions that in this field, healthcare workers are subjected to established pressure. Available evidence mentions that nurses in most nations have high levels of work stress [11–13]. Moreover, nurses' exhaustion and worry increased as COVID-19 levels continued to increase, resulting in a decline in mental, physical, and other aspects of nurses' quality of life [14]. Furthermore, it has been seen that the balance between professional sacrifice and fear for themselves and their loved ones is worsening. This is crucial to document and manage since a lower QWL among healthcare workers could jeopardise patient care quality [6]. Therefore, organisations must establish an action plan to improve their employees' QWL in order to improve favourable working conditions, with the ultimate goal of accomplishing the organization's objectives.

Nurses also confront numerous obstacles while attempting to master new technology, thus impacting QWL. Various issues influencing nurses' work-life quality, such as an overabundance of workload and inadequate working conditions, are constantly debated, particularly in the present COVID-19 pandemic crisis [5]. In between, nurses who work shifts, such as night shifts or weekends, may face a conflict between work and family life [5]. The COVID-19 pandemic, according to Nakisci-Kavas and Develi, created major challenges, particularly for female health professionals, disrupting the balance between their personal and professional lives. A study by [5] indicates that nurses' QWL is at a moderate level in Sarawak. The findings are also in line with a previous study that reported that the majority of nurses worldwide have a low-to-moderate QWL [7,15]. Nurses are unable to strike a healthy balance between their personal and professional lives. To balance both, the organization must provide QWL for its employees.

Effective leadership from healthcare professionals is required to improve the quality and integration of care [16]. Leadership style is one of the factors that influences the interest and commitment of personnel [17]. By using particular leadership styles and behaviours, a leader would be able to manipulate their employees' QWL. Thus, effective leadership in healthcare is essential in cultivating and increasing healthcare systems' effectiveness and efficiency [18]. Hence, healthcare services management should guarantee the QWL for the nursing profession to develop and improve work performance among employees. Therefore, leadership in health care is vital for keeping up with new developments and responding to changing circumstances [19]. Moreover, the role of leaders in today's businesses has evolved, and each organization's performance is dependent on the leadership styles used by its leaders [17].

Job satisfaction, job involvement, motivation, productivity, health, safety, job stability, competence growth, and a balance between work and non-work life are all connected to QWL [20]. Very few studies have focused on leadership styles on QWL. Even though leadership style has been highlighted as an essential element in healthcare quality in various research [16], it does not attract much attention from researchers to conduct a study between leadership styles and QWL, especially in healthcare settings. In the workplace, different leadership styles exist, each with their own set of benefits and drawbacks.

Moreover, in Malaysia, there is much research conducted on QWL among other professions, but there is little work focused on QWL among nurses. Lots of research has been conducted among academicians, banks, office employees, and factory workers, including public, private, and government organizations, but there has been little research on the nursing profession. To date, very few studies have investigated the association between leadership styles and QWL among nurses in Malaysia. It can be concluded that, up to now, far too little attention has been paid to the role of leadership style and QWL in Malaysia. Thus, it is essential to examine the relationship between leadership styles and the quality of work life among nurses. This study is significant because a better QWL enhances employee development while also advancing the organisation's growth. Furthermore, the poor nurses' working conditions must be improved in order to raise their productivity and job quality and reduce turnover.

2. Literature Review

2.1. Quality of Work Life

The quality of one's work-life balance has become a major concern [5]. Walton was the one who came up with the term "Quality of Work Life" (QWL) to describe the quality of human experience at work. Hermawati et al. explain that employees' comfort level at work is measured by QWL. While [5] referred to QWL as the extent to which personnel are pleased with their personal and work demands while achieving the organization's objectives, employees and management will benefit from QWL programmes because they will work together to address work-related problems, foster cooperation, improve work environments, and manage human resource outcomes and payoffs wisely and equitably [20]. QWL covers a number of areas, such as adequate fair compensation, organization culture, job security, training and development, employees' benefits, alternative work schedules, profit sharing, and workplace participation. According to Nayak, Sahoo and Mohanty, the work environment, professional growth, work stress, social support, remuneration, and rewards are all five dimensions of work-life quality.

Many studies have focused on the QWL of nurses. A study by [1] assessed the effect of QWL on nurse performance via job satisfaction during the COVID-19 pandemic. These findings revealed nurses' job satisfaction has a substantial impact on their performance during pandemics, as does their QWL. Another study by [5] investigated QWL among public hospital nurses in Sarawak General Hospital (SGH) in Kuching, Malaysia. The finding shows 52.9 percent of respondents were dissatisfied with their QWL, while 47.1 percent were satisfied. While a study by [7] reported a moderate QWL among nurses in Jordan, a recent study by [14] found that work satisfaction, anxiety, exhaustion, work-life balance, psychological well-being, and professional self-concept all had a strong link with nursing QWL during the COVID-19 pandemic. With this in mind, finding the right factor that can increase the QWL of nurses is paramount.

2.2. Leadership Style

Leadership is not a new concept in business or in the workplace. In any company, there must be a leader; therefore, the phrase "leadership" is frequently associated with the term "manager" [16]. The ability of a leader to influence and direct subordinates to achieve organisational goals is known as leadership. Solid leadership may assist a business in becoming increasingly representative of trust, fulfilment, and accountability, as well as accommodating people and assets to achieve organisational objectives. Supervisory/management/leadership style, as well as managerial support, have been recognised as significant predictors of work-life quality [2,9,16].

A recent study by [16] stated that in Dumai Public Health, leadership style was found to be positively associated with job satisfaction and nurse performance. Naseem et al. (2018) discovered that leadership styles affected job satisfaction. On the other hand, Sabbah et al. discovered that leadership styles had a positive impact on nurses' well-being in Lebanese hospitals. According to a nurse theorist, the manager's leadership ability is one of the most critical factors in determining job satisfaction. It has been discovered that a leader who cares deeply about her subordinates contributes to greater employee satisfaction because both can sense respect and a friendly relationship [21]. There are several kinds of leadership styles, which are autocratic, democratic, transformational, transactional, laissez-faire, charismatic, authentic, ethical, spiritual, and servant leadership styles. The selection of a leadership style depends upon the objectives of the organisation. In this paper, we study the impact of autocratic, democratic, transformational, and transactional leadership styles on the QWL of nurses.

2.3. *Autocratic*

Autocratic leaders are defined as being directive, controlling, power-oriented, and closed-minded. Autocratic leadership is a type of authoritarian leadership in which the leader makes all decisions without consulting anyone [16]. Because of a lack of consultation with subordinates, opportunities may be overlooked and risks may be underestimated. Igbaekemen [22] stated that an autocratic leader is one who is very conscious and has little faith in their subordinates. However, autocratic leadership should not be shown solely in a negative way only. Autocratic leaders provide precise and concise directions to their subordinates on what to do and how to do it [23]. This will aid in the efficient completion of tasks, the resolution of recognised issues, and the achievement of goals or deadlines. Although employees dislike autocratic leaders, they frequently follow their directions. The advantage of this approach is that it works well in situations where there is little time for discussions, such as crises or chaotic events. This sort of leadership is also particularly effective when crucial business decisions or actions are needed and employees are likely to be instructed exactly what action to take in times of stress or emergency; thus, employees feel appreciated if someone is willing to stand up and take control of what could be a chaotic situation [24]. Especially in the healthcare sector, whereby emergency situations always occur, especially in hospitals, the presence of autocratic leaders will ensure the efficiency of the situation and could reduce unnecessary mistakes by the nurses. Therefore, it was hypothesized that:

H_1. *There is an association between an autocratic leadership style and QWL among nurses.*

2.4. *Democratic*

Democratic leadership, according to Klein, is a process through which individuals discuss their decisions, followed by responsible action. According to Siddique and Harms et al., the democratic leadership style is extensively implemented in most Asian, Middle Eastern, and Western countries. In most leadership circumstances, a democratic leader fosters participation in decision-making, and democratic leadership is associated with improved morale [25]. Meanwhile, according to Kartono, it is a practice that coordinates all members' work and responsibilities by promoting a sense of responsibility and good teamwork. The type of leader who employs this leadership style recognises and respects each employee's worth and regard. Democratic leaders' activities are founded on trust, integrity, equality, honesty, an open mindset, and, last but not least, mutual respect among team members to develop strong connections and relationships through effective listening and understanding. Several authors all found that democratic leadership has a significant impact on major job outcomes, including satisfaction and innovation. Some other recent studies found that it has a good impact on performance. A past study by Elenkov reveals that the democratic leadership style and employee QWL have a positive and significant association. Therefore, it was hypothesized that:

H_2. *There is an association between a democratic leadership style and QWL among nurses.*

2.5. Transformational

Since transformational leaders explain the charismatic vision, followers can obtain information about all other individuals and the company as a whole, which boosts their self-confidence, fosters teamwork, and produces a pleasant work environment. When a leader exhibits transformational behaviours, their followers' self-efficacy to complete a task improves, which in turn improves people's well-being. Several studies have investigated the effects of transformational leadership and discovered that it is favourably related to employee outcomes, such as work engagement, work behaviour, and job performance. For example, Jurado et al. highlighted the value of transformational leadership styles in the healthcare sector for boosting employee wellness. According to previous research, transformational leadership has a favourable impact on the quality of the work life of hospitality employees [2]. IT professionals' work life, job satisfaction, and retention intentions have also been associated with transformational leadership behaviours. Hentschel et al., on the other hand, discovered a positive and negative relationship between transformative leadership and nurse job satisfaction. Therefore, it was hypothesized that:

H$_3$. *There is an association between a transformational leadership style and QWL life among nurses.*

2.6. Transactional

In 1947, Max Weber was the first to describe transactional leadership, followed by Bernard Bass in 1981. Transactional leadership, according to Bass, establishes the foundation for relationships between leaders and followers by defining expectations, clarifying roles, and offering recognition and rewards in order to accomplish the desired results. This type of leadership focuses more on persuading others to join them through a reward or punishment system that honours correct behaviours while eradicating bad ones through corrective measures. Transactional leadership is a style of leadership in which the leader may use both rewards and punishments to entice his followers to follow him [26]. In the same vein, [27] states Transactional leaders can keep staff engaged for the short term by using an award and punishment system. These interactions enable leaders to meet their performance goals, complete required tasks, maintain the current organisational situation, inspire followers through legal arrangements, direct followers' behaviour toward the accomplishment of defined objectives, emphasise extrinsic rewards, prevent unwanted risks, and concentrate on improving organizational efficiency. A past study by Paracha et al. demonstrated that transactional and transformational leadership styles are both significantly correlated with employee performance. However, transactional leadership was shown to be more strongly associated with employee performance than transformational leadership. According to a recent study, using a transactional leadership style will improve the quality of work life. Therefore, it was hypothesized that:

H$_4$. *There is an association between a transactional leadership style and QWL among nurses.*

3. Proposed Research Framework

To conduct this study, a conceptual model was prepared based on previous literature. As a result, this paper suggests an association between leadership style and nurse QWL during the COVID-19 pandemic. It is proposed that there is a positive relationship between leadership styles and QWL. Figure 1 shows the framework of the study. The framework shows leadership styles (autocratic; democratic; transformational; and transactional) as the independent variables used to explain the QWL of nurses as the dependent variable.

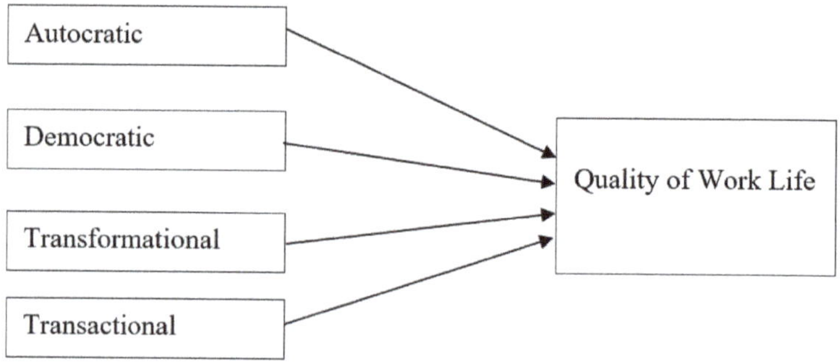

Figure 1. Proposed Framework.

4. Methodology

The research strategy is based on quantitative research. The researcher used correlational research as the research design because this study aimed to investigate the correlation between leadership styles and QWL. The target population of this research is nurses who are working in the healthcare industry in Selangor. The total population for this study was 230 nurses. The researcher used simple random sampling in this study because simple random sampling is a completely random method of picking a sample in which each component and population has an equal chance of being included in the sample. The questionnaire is the primary data-gathering tool for this study. Because of the advantages of low cost and fast speed in transmitting and returning information, an online questionnaire was used as the data collection method. Moreover, it is the best technique to gather data during the pandemic crisis.

4.1. Measures

The instruments are divided into five sections, which consist of Section A, Section B, Section C, Section D, Section E, and Section F. The first section consists of questions on demographics, which include gender, age, years of employment, and department. Next, Section B consists of questions on measuring the independent variables for Democratic Leadership Styles; Section C is focused on Autocratic Leadership Styles; Section D is focused on Transformational Leadership Style; Section E is focused on Transactional Leadership Style; and lastly, Section F is focused on QWL of nurses. To measure the leadership style, the questionnaire was adapted from the Multifactor Leadership Questionnaire while the Quality of Work-Life Scale is adapted from [28]. All item scales were constructed on a 5-point Likert format ranging from 1-strongly disagree to 5-strongly agree.

4.2. Reliability Analysis

Based on Table 1, Democratic Leadership resulted 0.926, Autocratic Leadership resulted in 0.768, Transformational Leadership resulted in 0.900, Transactional Leadership resulted in 0.910, and Quality of Work Life variables resulted in 0.916. According to the description details of Cronbach Alpha's reading, all four variables written are excellent, meanwhile Autocratic Leadership is considered good.

Table 1. Reliability Analysis.

Variables	Cronbach's Alpha	N. of Items	Description
Democratic Leadership	0.926	7	Excellent
Autocratic Leadership	0.768	6	Good
Transformational Leadership	0.900	7	Excellent
Transactional Leadership	0.910	8	Excellent
Quality of Work Life	0.916	6	Excellent

5. Findings

From 230 sets of questionnaires given, only 202 nurses answered the questionnaires and were valid. The results indicated that 87.8 percent of the respondents had participated in answering the questionnaire. Most of the respondents were females (98%) and the highest number of respondents belonged to the age group 30–40 years old (41.6%). Most respondents had worked for more than 10 years (44.1%) of service in the organization, and 41.3% of respondents were from the clinical department.

Table 2 shows Pearson's correlation test results for leadership style and QWL of nurses. The findings reveal all the variables that have a moderate correlation with the QWL of nurses. Transactional leadership style shows the highest correlation (r = 0.688, p = 0.000), followed by Transformational leadership style (r = 0.617, p = 0.000), Democratic leadership style (r = 0.600, p = 0.000), and lastly Autocratic leadership style (r = 0.487, p = 0.000). It can be concluded that all variables have a moderately significant correlation with QWL. It also indicated that applying a transactional leadership style would improve the nurses' quality of life at work.

Table 2. Correlation Analysis of Leadership Style and QWL.

		MEAN_QWL
MEAN_TFL	Pearson Correlation	0.617 **
	Sig. (2-tailed)	0
MEAN_TL	Pearson Correlation	0.688 **
	Sig. (2-tailed)	0
MEAN_A	Pearson Correlation	0.487 **
	Sig. (2-tailed)	0
	Pearson Correlation	0.600 **
MEAN_D	Sig. (2-tailed)	0
	N	202

** $p < 0.01$.

The regression and ANOVA results of the model are presented in Tables 3 and 4. The $R^2 = 0.528$ implies that the dependent variable was influenced by the independent variables, which are transformational leadership, transactional leadership, democratic leadership, and autocratic leadership, by 52.8%. The full model containing all predictors was statistically significant at 5% because the p-value of 0.000 is less than the significance level of 0.05.

Table 3. Model Summary.

Model	R	R Square	Adjusted R Square	Std. Error of the Estimate
1	0.726 [a]	0.528	0.518	0.40083

[a] Predictors: (constant) Democratic Leadership, Autocratic Leadership, Transformational Leadership and Transactional Leadership.

Table 4. ANOVA.

Model	Sum of Squares	df	Mean Square	F	Sig.
1 Regression	35.346	4	8.837	55.000	0.000 [b]
Residual	31.651	197	0.161		
Total	66.997	201			

[b] Predictors: (constant) Democratic Leadership, Autocratic Leadership, Transformational Leadership and Transactional Leadership.

The coefficient results presented in Table 5 indicate that Transactional Leadership Style with a beta of (0.381) and Democratic Leadership Style with a beta of (0.208) are statistically significant and strong predictors of nurses' QWL. The findings clearly show that the transactional leadership style has the greatest impact on nurses' QWL.

Table 5. Coefficients.

Model	Unstandardized Coefficients		Standardized Coefficients	t	Sig.	Collinearity Statistics	
	B	Std. Error	Beta			Tolerance	VIF
1 (Constant)	0.364	0.321		1.135	0.258		
MEAN_DL	0.193	0.063	0.208	3.069	0.002	0.521	1.919
MEAN_AL	0.166	0.087	0.114	1.909	0.058	0.673	1.487
MEAN_TFL	0.166	0.095	0.138	1.737	0.084	0.379	2.637
MEAN_TL	0.420	0.095	0.381	4.436	0.000	0.326	3.069

6. Discussion

Based on the results of Pearson correlation analysis, transformational, transactional, democratic, and autocratic leadership styles have positive and moderate relationships with QWL among nurses. However, through the regression analysis, it can be concluded that the transactional leadership style is the most influential leadership style on QWL among nurses, followed by the democratic leadership style. It is because the beta coefficient for the transactional leadership style is the highest among the four leadership styles being studied, at 0.381 and a significant value of 0.000. It was discovered that using a transactional leadership style would improve the nurses' quality of life at work. The findings are supported by previous findings by [3], who found that transactional leadership style has a strong, positive, and significant relationship with QWL. The respondents also agreed that the manager expresses satisfaction when they meet expectations. Moreover, the manager also guides them in making decisions to establish goals by clarifying roles and requirements. The finding is in line with past findings by Bass presenting that transactional leadership establishes the basis for connections between leaders and followers by defining expectations, clarifying roles, and giving rewards and recognition in order to accomplish the desired results.

Meanwhile, democratic leadership styles also significantly correlate with nurses' QWL. The findings show managers allow nurses to use creativity and ingenuity to solve organizational problems. Subordinates are also more likely to participate in reaching goals, both for the welfare of the organisation and for their own personal development and learning. It was in line with [17], where managers include their subordinates in decision-making processes, taking into consideration their positive and negative viewpoints. Employees become more fully devoted as a result of this. Even though most of the previous findings show transformational leadership is one of the dominating areas that can contribute to employees' work lives, the findings of this study open a new path in leadership and QWL research. Therefore, the current study provides an innovative and valuable perspective through empirical validation by increasing existing knowledge on the quality of work life among nurses through the blending of transactional and democratic leadership styles.

7. Conclusions

Nurses' QWL should be highlighted because their work lives changed during the pandemic as a result of the worry and depression they experienced while treating COVID-19 patients. Therefore, hospitals must be accountable for their QWL, which has an impact on the quality of nursing care offered to patients. The most common element that impacts employee attitudes and behaviours, including QWL, is leadership style. The findings of this study added to the body of knowledge on leadership styles and work-life quality for employees in organisations, especially nurses in the healthcare industry. These findings also indicated that nurse QWL is heavily influenced by leadership, and that leadership can play an important role in defining nurse QWL. As a result, leaders must carefully consider what type of leadership they should adopt if they want to improve staff QWL. The findings yielded Transactional and Democratic leadership styles influence nurses' QWL. Both leadership styles are a mix of different sorts of leadership behaviour and qualities. If a leader needs to focus on close monitoring, finding mistakes and errors, and putting corrective steps in place to address them, a transactional style should be used. On the other hand, when the leader needs to coordinate the activities and responsibilities of all members by emphasising a sense of responsibility and good cooperation, the democratic leadership style will closely fit. Furthermore, democratic leadership is somewhat more productive since it gives employees some autonomy and involves them in decision-making.

7.1. Implication of the Study

This research adds to the current leadership and QWL literature by demonstrating the relationship between leadership styles and nurses' QWL. Therefore, the findings of this study add to the body of knowledge about the impact of transformational, transactional, democratic, and autocratic leadership styles on nurses' QWL in Malaysia. Results from this study can be used to better inform policymakers who want to address issues of QWL. The findings demonstrate that if the leader adopted the right leadership style, the nurses' QWL may improve. Results from this study also yielded that nurses favoured transactional and democratic leadership styles compared to transformational and autocratic leadership styles. It should be known that employees are the most valuable asset in any firm, and employee QWL is heavily influenced by leadership style. Organizations with strong leadership capabilities will increase QWL and performance even more. It was also in line with the Agenda 2030 Sustainable Development Goal (SDG) in supporting SDG 3, ensuring health and well-being at all ages, and SDG 8, promoting decent work and economic growth. Therefore, greater consideration should be given to elevating leadership to a higher priority in order to improve employee QWL. Failure to focus on nurses' QWL will lead to decreased employee retention in the health sector. Thus, paying attention to nurses' QWL will integrate them to produce a more positive work environment and a better work-life balance.

7.2. Recommendation and Limitations

By looking at the results, it was recommended that the organization can apply a transactional leadership approach to improve the QWL of nurses during the pandemic crisis. It shows that there is a significant relationship between transactional leadership and QWL among nurses in Selangor. The leader should be able to skillfully explain about the organization's target to achieve. By applying this, the nurses will realize the significance of self-efficacy when they are achieving the desired goals. The manager should also negotiate often with their subordinates so that the nurses will feel that they also belong to the organization. In addition, managers who want to apply transactional leadership must replace punishments with valuable feedback. If the nurses fail to meet the goal or make careless blunders, respond to their failures with constructive comments instead of anger. Nurses who failed due to communication issues, a lack of skills, or other issues that the boss may resolve through coaching can be reprimanded by engaging them in a dialogue about what went wrong. The manager who applies this leadership approach will intrigue the nurse's loyalty or commitment with a new idea of having solid recognition and rewards for flourishing the initiatives and innovative ideas development, which precisely communicates to the nurses about the program participation of leader's attachments.

In this analysis, the researcher focuses only on four variables that are related to QWL among nurses, which are transformational, transactional, autocratic, and democratic leadership. There are other leadership styles that are not included in this study. Thus, the researcher hopes that future studies can correlate various leadership factors or variables that can increase QWL among nurses such as laissez-faire leadership, resonance leadership, ethical leadership, and spiritual leadership. Future researchers are also suggested to observe for additional factors that can be related to QWL among nurses that can develop a better and greater impact on QWL among nurses. One of the most crucial considerations is that the sample population for this research study is limited to only 202 nurses. This figure does not represent the complete population of Malaysia's working population. As a result, future research must incorporate additional samples from other industries and areas in order to provide a broader perspective and broader conclusion. Since this study proposed a correlational approach, there can be differing results in the relationships between the independent variables and dependent variables when different respondents are used. Future research could also explore the function of mediating and moderating dimensions such as personality traits, psychological empowerment, trust, perceived organisational support, and other relevant aspects in providing new views and adding to the QWL study's body of knowledge.

Author Contributions: Conceptualization, Z.B.E. and S.A.H.; methodology, Z.B.E.; software, M.K.S.; validation, N.A.M., N.M. and Z.B.E.; formal analysis, S.A.H. and M.K.S.; investigation, N.M.; data curation, S.A.H.; writing—original draft preparation, Z.B.E.; writing—review and editing, Z.B.E. and N.M.; visualization, N.A.M.; supervision, Z.B.E. All authors have read and agreed to the published version of the manuscript.

Funding: This research received no external funding.

Institutional Review Board Statement: Not applicable.

Informed Consent Statement: Not applicable.

Data Availability Statement: Not applicable.

Conflicts of Interest: The authors declare no conflict of interest.

References

1. Diana; Eliyana, A.; Mukhtadi; Anwar, A. Creating the path for quality of work life: A study on nurse performance. *Heliyon* **2022**, *8*, e08685. [CrossRef]
2. Adisa, T.A.; Gbadamosi, G. Regional crises and corruption: The eclipse of the quality of working life in Nigeria. *Empl. Relat.* **2019**, *41*, 571–591. [CrossRef]
3. Andrew, L.; Abdul Latiff, S.F.; Burhan, S. Leadership Styles and Quality of Work Life among Administrative Staff at a Public University in Sarawak. *J. Islamic Soc. Econ. Dev. (JISED)* **2019**, *4*, 44–52.
4. Akter, K.M.; Mei Tang, S.; Adnan, Z. Transformational leadership and quality of work life: A mediation model of trust climate. *Probl. Perspect. Manag.* **2021**, *19*, 161–174. [CrossRef]
5. Francis, F.; Sham, F.; Alias, A.; Wahab, S.M.A.; Yusof, S.; Johan, H. Quality of Work Life among Public Hospital Nurses in Sarawak. *Int. J. Serv. Manag. Sustain.* **2021**, *6*, 51–70. [CrossRef]
6. Maqsood, M.B.; Islam, M.A.; Naqvi, A.A.; Al Qarni, A.; Al-karasneh, A.F.; Iffat, W.; Ghori, S.A.; Ishaqui, A.A.; Aljaffan, A.H.; Alghamdi, S.; et al. Assessment of quality of work life (QWL) among healthcare staff of intensive care unit (ICU) and emergency unit during COVID-19 outbreak using WHOQoL-BREF. *Saudi Pharm. J.* **2021**, *29*, 1348–1354. [CrossRef]
7. Suleiman, K.; Hijazi, Z.; Al Kalaldeh, M.; Abu Sharour, L. Quality of nursing work life and related factors among emergency nurses in Jordan. *J. Occup. Health* **2019**, *61*, 398–406. [CrossRef] [PubMed]
8. Permarupan, P.Y.; Al Mamun, A.; Samy, N.K.; Saufi, R.A.; Hayat, N. Predicting Nurses Burnout through Quality of Work Life and Psychological Empowerment: A Study towards Sustainable Healthcare Services in Malaysia. *Sustainability* **2020**, *12*, 388. [CrossRef]
9. Albagawi, B. Leadership Styles of Nurse Managers and Job Satisfaction of Staff Nurses: Correlational Design Study. *Eur. Sci. J. ESJ* **2019**, *15*, 254–275. [CrossRef]

10. Al-maskari, M.A.; Dupo, J.U.; Al-sulaimi, N.K. Quality of Work Life among Nurses A case study from Ad Dakhiliyah Governorate, Oman. *Sultan Qaboos Univ. Med. J.* **2020**, *20*, e304–e311. [CrossRef] [PubMed]
11. Alwani, S.S.; Majeed, M.M.; Ramzan, Z.; Rauf, S.; Syed, M.S.; Shah, S.H.; Hirwani, M.Z.; Hamirani, F.; Alwani, S.S.; Majeed, M.M.; et al. Evaluation of Knowledge, Practices, Attitude, and Anxiety of Nurses towards COVID-19 during the Current Outbreak in Karachi, Pakistan. *Pak. J. Public Health* **2021**, *10*, 82–90. [CrossRef]
12. Luo, M.; Guo, L.; Yu, M.; Jiang, W.; Wang, H. The psychological and mental impact of coronavirus disease 2019 (COVID-19) on medical staff and general public—A systematic review and meta-analysis. *Psychiatry Res.* **2020**, *291*, 113190. [CrossRef]
13. Chui, P.L.; Chong, M.C.; Abdullah, K.L.; Ramoo, V.; Tang, L.Y.; Lee, W.L.; Che, C.C.; Yahaya, N.A.; Rasaiah, K.; Zaini, N.H.; et al. The COVID-19 global pandemic and its impact on the mental health of nurses in malaysia. *Healthcare* **2021**, *9*, 1259. [CrossRef]
14. Agusto, N.W.; Mediawati, A.S.; Komariah, M. Factors Related to Quality of Nursing Work Life (QNWL) during The COVID-19 Pandemic: A Scoping Review. *Malays. J. Med. Health Sci.* **2022**, *18*, 262–266.
15. Kelbiso, L.; Belay, A.; Woldie, M. Determinants of Quality of Work Life among Nurses Working in Hawassa Town Public Health Facilities, South Ethiopia: A Cross-Sectional Study. *Nurs. Res. Pr.* **2017**, *2017*, 5181676. [CrossRef] [PubMed]
16. Sansuwito T bin Prasath, R.; Novera, M.; Sarli, D.; Poddar, S.; Ariani, N. The Effect of Leadership Styles on Nurse Performances and Job Satisfaction Among Nurses in Dumai Public Hospital: Technological Innovation as Mediator. *Malays. J. Med. Health Sci.* **2022**, *18*, 229–234.
17. Veliu, L.; Manxhari, M.; Demiri, V.; Jahaj, L. The Influence of Leadership Styles on Employee's Performance. *Vadyba J. Manag.* **2017**, *31*, 59–69.
18. Ming, A.; Goh, J.; Ang, S.Y.; Della, P.R. Leadership style of nurse managers as perceived by registered nurses: A cross-sectional survey. *Proc. Singap. Heal.* **2018**, *27*, 205–210. [CrossRef]
19. Durmuş, S.C.; Kırca, K. Leadership Styles in Nursing. In *Nursing: New Perspectives*; Durmuş, S.C., Ed.; IntechOpen: London, UK, 2019. [CrossRef]
20. Sekhara Rao, K.S.; Sahyaja, C.; Lahari, S.; Vasavi Pravallika, S. Role of leadership styles in quality of work-life: A study on education sector. *Int. J. Mech. Eng. Technol.* **2018**, *9*, 122–133.
21. Irabor, I.E.; Okolie, U.C. A Review of Employees' Job Satisfaction and Its Affect on Their Retention. *Ann. Spiru Haret Univ. Econ. Ser.* **2019**, *19*, 93–114. [CrossRef]
22. Igbaekemen, G.O. Impact of Leadership Style on Organisation Performance: A Srategic Literature Review. *Public Policy Adm. Res.* **2014**, *4*, 126–136.
23. Dyczkowska, J.; Dyczkowski, T. Democratic or Autocratic Leadership Style? Participative Management and its Links to rewarding Strategies and Job Satisfaction in SMEs. *Athens J. Bus. Econ.* **2018**, *4*, 193–218. [CrossRef]
24. Ping, A.T. The Relationship between Leadership Styles and Employees' Job Satisfaction in Small and Medium Enterprises (SMEs). Ph.D. Thesis, Universiti Tunku Abdul Rahman, Petaling Jaya, Malaysia, 2015.
25. Diana, I.N.; Supriyanto, A.S.; Ekowati, V.M.; Ertanto, A.H. Factor Influencing Employee Performance: The Role of Organizational Culture. *J. Asian Financ. Econ. Bus.* **2021**, *8*, 545–553. [CrossRef]
26. Bhatt, H.C. Leadership styles and quality of work life in small and medium scale enterprises of Kumoun region of Uttarakhand. *J. Strateg. Hum. Resour. Manag.* **2018**, *7*, 23–32.
27. Rathi, N.; Soomro, K.A.; Rehman, F.U. Transformational or Transactional: Leadership Style Preferences during COVID-19 Outbreak. *J. Entrep. Manag. Innov.* **2021**, *3*, 451–473. [CrossRef]
28. Almarsh, S.O. A Measurement Scale for Evaluating Quality of Work Life: Conceptualization and Empirical Validation. *Trends Appl. Sci. Res.* **2015**, *10*, 143–156. [CrossRef]

proceedings

Proceeding Paper

The Impacts of Stress and Resilience on Intentions to Leave the Nursing Profession among Newly Graduated Nurses †

Zamzaliza Abdul Mulud [1,*], Norfidah Mohamad [1], Hidatul Syafizal Zenal Ahmad Rozi [1], Nor Salmie Nadira Ngah [1] and Julinawati Suanda [2]

1 Centre for Nursing Studies, Faculty of Health Sciences, Universiti Teknologi MARA Selangor, Puncak Alam Campus, Bandar Puncak Alam 42300, Selangor, Malaysia
2 School of Business Innovation and Technopreneurship, Universiti Malaysia Perlis, Kangar 01000, Perlis, Malaysia
* Correspondence: zamzaliza@uitm.edu.my

† Presented at the International Academic Symposium of Social Science 2022, Kota Bharu, Malaysia, 3 July 2022.

Abstract: Registered nurses are a professional group that suffer from high levels of stress-related conditions. Thus, this study aimed to identify the impacts of stress and resilience on intentions to leave the nursing profession among newly graduated nurses (n = 165). A descriptive cross-sectional design and purposive sampling were used. Most of the respondents showed a moderate level of resilience and a high level of stress. Of the participants, 12.7% reported an intention to leave the nursing profession. A strong correlation was found between levels of resilience and stress among newly graduated nurses. Effective stress management and interventions to improve resilience in newly graduated nurses are highly recommended.

Keywords: nurses; psychological stress; resilience

check for updates

Citation: Mulud, Z.A.; Mohamad, N.; Rozi, H.S.Z.A.; Ngah, N.S.N.; Suanda, J. The Impacts of Stress and Resilience on Intentions to Leave the Nursing Profession among Newly Graduated Nurses. *Proceedings* **2022**, *82*, 100. https://doi.org/10.3390/proceedings2022082100

Academic Editor: Mohamad Rahimi Mohamad Rosman

Published: 17 October 2022

1. Introduction

Nursing is a stressful job. A study conducted in Hospital Kuala Lumpur among ward nurses reported that approximately 24.6% of nurses experienced job-related stress [1]. Previous research has shown that newly graduated nurses face difficulties in the transition from being a student to becoming a registered nurse [2]. Thus, a 50% turnover rate was reported for the first three years of working [3]. Nurses' time and energy demands have increased over the last ten years. There are many causes of this situation, such as workload levels, lack of support from the management, personal problems, and working hours [4–6]. In addition, uncomfortable working surroundings and the hardship of building a good rapport with co-workers also produce difficulties for newly graduated nurses [5]. Other than these factors, newly graduated nurses, mainly those with bachelor's qualifications, are expected to be skilful within a brief period [2]. These difficulties are likely to affect their psychological well-being, thus increasing the rate of staff turnover and resignations.

The increased nursing-career turnover rate is due to junior nurses' lack of support and high stress levels [7]. Therefore, the stress levels among newly graduated nurses ought to be assessed to remedy the situation and ensure good coping skills to enable nurses to meet the challenges they face and stop turnover. Generally, during their first three months, most newly graduated nurses quit or transfer to other units because they believe they can reduce distress [8]. However, many decide to leave the profession and migrate to other high-income countries for healthier working environments and better career prospects and salaries. A report shows that Malaysia has a low nurse-per-population ratio compared to other developing countries. In 2019, there were 3.5 nurses per 1000 population compared to 11.3 in Japan and 6.2 in Singapore [9]. A high nurse–patient ratio is likely to contribute to poor nursing care quality and jeopardise patient safety [10].

Resilience serves as an adaptive strategy for dealing with adversity. Bouncing back from emotional setbacks and adapting to new circumstances in the face of stress is crucial for a person's health and well-being [11]. Therefore, nurses should have excellent resilience skills to ensure that they can handle their work challenges and be motivated to stay in the profession. Moreover, resilience may be a factor that can facilitate successful transition [12]. Thus, this study aimed to investigate the impacts of stress and resilience on intentions to leave the nursing profession among newly graduated nurses in Malaysia.

2. Methods

2.1. Study Design and Sampling

We used a cross-sectional study design to investigate the impacts of stress and resilience on intentions to leave the nursing profession among newly graduated nurses in Malaysia. The purposive sampling technique was used to recruit the samples in the study. Purposive sampling is also known as judgemental, selective, or subjective sampling. Given the use of this sampling method, the sample included the people who could provide the best information to achieve the study's objectives, were likely to have the information required for the research, and would be willing to share it [13]. The sample size was calculated using the Roasoft, Inc., sample-size calculator. For a population of 280 nursing graduates from 2018 to 2020, with significance at the 5% and a 95% confidence level, 163 samples were required; however, we managed to recruit 165 samples at the end of the data-collection period. The inclusion criteria for this study were that the participants worked as registered nurses in the hospital and had less than three years of work experience. Newly graduated nurses working part-time in the hospital or on study leave were excluded from this study.

2.2. Research Instrument

A self-administered questionnaire that consisted of three sections was used in this study. Section A concerned the subject's demographics and consisted of 6 items: gender, age, academic level, work experience, marital status, and intention to quit nursing. Section B comprised a 10-item Perceived Stress Scale adopted from Cohen, 1994 [14]. Responses were made according to a 5-point Likert-type scale, from 0 = "never" to 4 = "very often" based on the degree to which each statement applied to them over the last month. The values for all of the items scored using the 5-point scale were summed, with a higher score indicating a higher stress level. The total scores for each element indicated the level of stress as being normal (0–7 total score), mild (8–9 total score), moderate (10–12 total score), severe (13–16 total score), or extremely severe (17–40 total score).

Section C included 25 items from the Connor–Davidson Resilience Scale (CD-RISC), adopted from Connor and Davidson [15]. CD-RISC is rated on a five-point scale, from 0, which indicates "not true at all", to (4), which indicates "very often", and higher scores suggest greater resilience. Total scores range from 0 to 100 [15]. Scores ranging from 0–19 are considered to indicate a very low level of resilience, scores from 20–40 a low level of resilience, scores from 41–62 a moderate level, scores from 63-83 a high level of resilience, and scores from 84–100 a very high level of resilience [11]. The original authors granted permission to use the CD-RISC.

A reliability test was conducted to measure the reliability of the questionnaire used in this study. For the first questionnaire, the Perceived Stress Scale, the Cronbach's alpha was 0.75. For the second questionnaire, CD-RISC, the Cronbach's alpha was 0.84. Thus, both scales were reliable and could be used in this study.

2.3. Ethical Considerations

The Ethics Committee, Universiti Teknologi MARA (UiTM), gave ethical approval for this study (600-IRMI (5/1/6)). The participants were assured that their involvement in the study was voluntary and that no personal characteristics would be identifiable from the collected data.

2.4. Data-Collection Process

We obtained a list of alumni who graduated from 2018 to 2020 with a diploma and a bachelor of nursing qualification from the department of student affairs in a public university. The respondents were contacted for permission purposes and to explain the study's objective. After the respondent gave permission, the questionnaires were administered in the form of an online survey; a Google form was sent to the respondents.

2.5. Data Analysis

Statistical Program Package for Social Sciences (SPSS) software version 21.0 (IBM, Armonk, NY, USA) was used to analyse the data. Descriptive statistics were used in frequency (n) and percentage (%) analyses to calculate means (M) and standard deviations (SD). Then, Pearson correlation analyses were performed to test the strength of the relationships between stress and resilience and between stress, resilience and sociodemographic variables (age and work experience). A parametric independent-sample *t*-test was also used to determine the relationships between level of resilience and sociodemographic variables of gender, academic level, and intention to leave the nursing profession.

3. Results

3.1. Sociodemographic Characteristics and Intentions to Leave

Table 1 shows the sociodemographic characteristics of the participants. Most of the respondents, 89.0% (n = 147), were female, with a mean age of 23.9. Of the participants, 58.8% (n = 97) graduated with a degree and 41.2% (n = 68) had a diploma in nursing. The majority of the participants had less than a year of work experience, and 12.7% reported an intention to leave the nursing profession.

Table 1. Descriptive analysis of demographic characteristics and intentions to leave the nursing profession.

Variables	Frequency (n)	Percentage (%)
Gender		
Male	18	10.90
Female	147	89.11
Age (mean, SD)	23.9	2.10
Academic level		
Degree	97	58.8
Diploma	68	41.2
Work experience		
<1 year	128	77.6
$1 \leq 2$ years	30	18.2
$2 \leq 3$ years	7	4.2
Intention to leave		
Yes	21	12.7
No	144	87.3

3.2. Associations between Stress, Resilience, and Intention to Leave

The independent-sample *t*-test was used to measure the differences between the two groups with respect to the means. The test was used to compare the means of sociodemographic characteristics, such as gender, marital status, academic qualifications, and intention to leave the nursing profession, with resilience and perceived stress, respectively. A significant difference in relation to perceived stress could be observed between males and females, with female participants showing higher stress levels than males ($t = -4.20$, $p = 0.001$). Gender differences also existed with respect to resilience, with males scoring higher than females ($t = 2.40$, $p = 0.017$). Table 2 summarises the independent-sample *t*-test results.

Table 2. Summary of the independent-sample *t*-test results comparing sociodemographic characteristics with perceived stress and resilience.

Variables	Perceived Stress			Resilience		
	Mean (SD)	*t*-Stats (df)	*p*-Value	Mean (SD)	*t*-Stats (df)	*p*-Value
Gender						
Male	17.28 (2.74)	−4.20 (163)	0.001 *	51.17 (2.75)	2.40 (163)	0.017 *
Female	20.23 (2.82)			48.07 (5.38)		
Academic qualification						
Degree	19.95 (3.97)	−0.59	0.981	48.47 (5.10)	0.20 (163)	0.315
Diploma	19.99 (3.99)			48.31 (5.47)		
Intention to leave the nursing profession						
Yes	20.71 (3.54)	1.34 (163)	0.182	46.29 (4.89)	−2.00 (163)	0.047 *
No	19.79 (2.86)			48.72 (5.23)		

Note: * Significant at $p < 0.05$.

Pearson correlation coefficients were used to analyse the correlations between age, work experience, resilience, and perceived stress among newly graduated nurses (Table 3). There was a positive correlation between age (r = 0.26, p = 0.01), work experience (r = 0.53, p = 0.01), and resilience. Work experience and perceived stress were moderately correlated (r = 0.33, p = 0.01). Finally, there was a strong negative correlation between resilience and perceived stress, indicating that those with high resilience scored lower for perceived stress (r = −0.99, $p < 0.01$).

Table 3. Pearson correlation coefficient results for age, work experience, resilience, and perceived stress.

Variables	1	2	3	4
1. Age	1			
2. Work experience	0.89 *	1		
3. Resilience	0.26 *	0.53 *	1	
4. Perceived stress	−0.08	0.33 *	−0.99 *	1

Note: * Significant at $p < 0.05$.

4. Discussion

Currently, more than 10% of the registered nurses working in hospitals are new graduates [16]. Previous research has shown that the shift from being a student (graduate) to a professional registered nurse can be demanding for new nurses [16]. Perceived stress is a common factor contributing to nurses' intentions to leave. Resilience is considered one of the psychotherapeutic approaches for dealing with stress among newly graduated nurses. Thus, this study aimed to investigate perceived stress, resilience, and intentions to leave the nursing profession among newly graduated nurses.

This study's findings suggested that most participants reported moderate stress levels. A study by Blomberg et al. in Sweden also found high stress levels among newly graduated nurses [17]. Nurses have some of the highest stress levels among healthcare professionals due to such factors as caring for terminally ill patients, being excluded by co-workers, and having their abilities questioned by their peers. Additionally, shift rotation interrupts their personal lives and may further increase their stress levels [6]. Another study also agreed that shift rotation is another major long-term cause of stress and exhaustion among nurses [4]. Another stressor is night-shift work, with plenty of rest and unfamiliar hospital shifts causing sleep disorders and work–life imbalances [18].

Significant differences between males and females could be observed in terms of stress and resilience levels. Males reported lower levels of stress and higher resilience levels compared to their female counterparts. Many factors could have contributed to these differences. Physiological variations between males and females, such as hereditary susceptibility and hormone and cortisol levels, may manifest themselves emotionally and behaviorally [19]. In addition, females and males adopt different coping skills, with females coping more negatively with adversity. Thus, females reporting high stress and low resilience can be explained.

In this study, there was a relationship between age and level of resilience. This finding indicated that the older the respondents, the higher their resilience levels. Previous studies have also reported similar findings [20,21]. Nearly twice as many nurses between the ages of 27.1 and 40 reported moderate or high levels of resilience compared to those under 27. Compared to nurses under the age of 27, nurses over the age of 40 were three times more likely to indicate moderate or high resilience. [20]. Similarly, nurses older than 36 had a higher resilience level than 18-to-30-year-old nurses [21]. This could be possibly due to individuals acquiring better coping skills as they age. A previous [16] study reported that respondents who worked less than two years had the highest stress levels relative to those with 3–5 years or over five years of work experience, due to a lack of nursing skills [16]. The stress levels among newly graduated nurses with different work experiences are likely to be the same.

This study's findings indicated a strong negative correlation between perceived stress and resilience. The finding is in agreement with [11,22,23], which suggested that low levels of resilience correlated with high levels of stress. Resilience plays a vital role in overcoming stressful challenges in workplaces. Individuals with high levels of resilience can overcome challenges, adapt to surrounding stresses, and adopt positive coping mechanisms.

In this study, 12.4% of all newly graduated nurses intended to leave the nursing profession. A study reported similar findings among newly graduated nurses in China, suggesting that 17.2% had the intention to leave as they entered the nursing profession [6]. Furthermore, the statistic increased to 59.6%, 74.4%, and 71.8% at 4, 8, and 12 months, respectively, due to stress and professionalism. Moreover, according to the research by Liang et al. [8], newly graduated nurses experience stress at the first level of working in the hospital area and stress was still present even though they had acquired ample knowledge and skills during theoretical classes and practical sessions before being employed. Workplace stress puts nurses' physical and mental health at risk, affects their family and social relationships, and contributes to high nursing turnover [4]. As resilience is highly correlated with stress, having a high resilience level provides nurses with a protective factor, thus preventing them from leaving the nursing profession.

5. Conclusions

The findings of this study provide evidence of factors associated with perceived stress and their associations with intentions to leave the profession in newly graduated nurses. In addition, newly graduated nurses reported moderate levels of perceived stress, and this finding was supported by data from previous studies. However, since the current study is a cross-sectional study conducted among newly graduated nurses in only one university, the results can be considered limited. There was also a lack of representativeness for male nurses. However, notwithstanding these limitations, the findings suggest new insights into factors that contribute to perceived stress. Effective stress management and interventions to improve resilience in newly graduated nurses are highly recommended.

Author Contributions: Conceptualization, Z.A.M. and N.M.; methodology, Z.A.M., H.S.Z.A.R., N.S.N.N. and J.S.; software, H.S.Z.A.R. and N.S.N.N.; validation, Z.A.M. and N.M.; formal analysis, Z.A.M., H.S.Z.A.R. and N.S.N.N.; investigation, Z.A.M., H.S.Z.A.R. and N.S.N.N.; writing—original draft preparation, Z.A.M., H.S.Z.A.R. and N.S.N.N.; writing—review and editing, Z.A.M., N.M. and J.S.; supervision, Z.A.M. All authors have read and agreed to the published version of the manuscript.

Funding: This research received no external funding.

Institutional Review Board Statement: The Ethics Committee, Universiti Teknologi MARA (UiTM), gave ethical approval for this study (600-IRMI (5/1/6)).

Informed Consent Statement: Informed consent was obtained from all subjects involved in the study.

Data Availability Statement: Not applicable.

Acknowledgments: We would like to express our gratitude to Universiti Teknologi MARA for the support of this study and all of the individuals who contributed directly or indirectly to this study.

Conflicts of Interest: The authors declare no conflict of interest.

References

1. Zainiyah, S.; Im, A.; Cy, C.; Sara, S. Stress and its associated factors amongst ward nurses in a public hospital Kuala Lumpur. *Malays. J. Health Med.* **2011**, *11*, 78–85.
2. Woo, M.; Newman, S.A. The experience of transition from nursing students to newly graduated registered nurses in Singapore. *Int. J. Nurs. Sci.* **2019**, *7*, 81–90. [CrossRef] [PubMed]
3. Helgesen, K. Stress in New Graduate Nurses: Can They Sweat It? A Correlational Study of Exercise and Stress in New Graduate Nurses. Ph.D. Thesis, California State University, Long Beach, CA, USA, 2017.
4. Lim, J.; Bogossian, F.; Ahern, K. Stress and coping in Australian nurses: A systematic review. *Int. Nurs. Rev.* **2010**, *57*, 22–31. [CrossRef] [PubMed]
5. Lin, Y.Y.; Lee, Y.H.; Chang, S.C.; Lee, D.C.; Lu, K.Y.; Hung, Y.M.; Chang, Y.P. Individual resilience, intention to stay, and work frustration among postgraduate two-year programme nurses. *Collegian* **2019**, *26*, 435–440. [CrossRef]
6. Zhang, A.Y.; Steege, L.M.; Pavek, K.U.; Brown, R.L.; Zhang, Y. Identifying patterns of occupational stress trajectories among newly graduated nurses: A longitudinal study. *Int. J. Nurs. Stud.* **2019**, *99*, 1–21. [CrossRef] [PubMed]
7. Hayes, L.J.; O'Brien-Pallas, L.; Duffield, C.; Shamian, J.; Buchan, J.; Hughes, F.; North, N. Nurse turnover: A literature review—An update. *Int. J. Nurs. Stud.* **2012**, *49*, 887–905. [CrossRef] [PubMed]
8. Liang, H.F.; Lin, C.C.; Wu, K.M. Breaking through the dilemma of whether to continue nursing: Newly graduated nurses' experiences of work challenges. *Nurse Educ. Today* **2018**, *67*, 72–76. [CrossRef] [PubMed]
9. Nurses and Midwives (per 1000 people)—Malaysia. Available online: https://data.worldbank.org/indicator/SH.MED.NUMW.P3?locations=MY (accessed on 1 July 2022).
10. Jarrar, M.; Rahman, H.A.; Shamsudin, A.S. The impact of patient to nurse ratio on quality of care and patient safety in the medical and surgical wards in Malaysian private hospitals: A cross-sectional study. *Asian Soc. Sci.* **2015**, *11*, 326–332. [CrossRef]
11. Elzohary, N.W.; Mekhail, M.N.; Hassan, N.I.; Farag, R.; Menessy, M. Relationship between ego resilience, perceived stress and life satisfaction among faculty nursing students. *IOSR J. Nurs. Health Sci.* **2017**, *6*, 57–70. [CrossRef]
12. Meyer, G.; Shatto, B. Resilience and transition to practice in direct entry nursing graduates. *Nurse Educ. Pr.* **2018**, *28*, 276–279. [CrossRef] [PubMed]
13. Kumar, R. *Research Methodology: A Step-by-Step Guide for Beginners*, 4th ed.; Sage Publications Ltd.: London, UK, 2014.
14. Cohen, S. Perceived Stress Scale by Sheldon Cohen Hosted by Mind Garden. Available online: https://www.mindgarden.com/132-perceived-stress-scale#horizontalTab2 (accessed on 20 August 2021).
15. Connor, K.M.; Davidson, J.R.T. Development of a new resilience scale: The Connor-Davidson Resilience Scale (CD-RISC). *Depress. Anxiety* **2003**, *18*, 76–82. [CrossRef] [PubMed]
16. Wu, T.Y.; Fox, D.P.; Stokes, C.; Adam, C. Work-related stress and intention to quit in newly graduated nurses. *Nurse Educ. Today* **2012**, *32*, 669–674. [CrossRef] [PubMed]
17. Blomberg, K.; Isaksson, A.K.; Allvin, R.; Bisholt, B.; Ewertsson, M.; Kullén Engström, A.; Ohlsson, U.; Sundler Johansson, A.; Gustafsson, M. Work stress among newly graduated nurses in relation to workplace and clinical group supervision. *J. Nurs. Manag.* **2016**, *24*, 80–87. [CrossRef] [PubMed]
18. Hu, Y.; Zhang, Y.; Shen, N.; Wu, J.; Wu, J.; Malmedal, W.K. Stressors of newly graduated nurses in Shanghai paediatric hospital: A qualitative study. *J. Nurs. Manag.* **2017**, *25*, 184–193. [CrossRef] [PubMed]
19. Gao, W.; Ping, S.; Liu, X. Gender differences in depression, anxiety, and stress among college students: A longitudinal study from China. *J. Affect. Disord.* **2020**, *263*, 292–300. [CrossRef] [PubMed]
20. Ang, S.Y.; Uthaman, T.; Ayre, T.C.; Mordiffi, S.Z.; Ang, E.; Lopez, V. Association between demographics and resilience—A cross-sectional study among nurses in Singapore. *Int. Nurs. Rev.* **2018**, *65*, 459–466. [CrossRef] [PubMed]
21. Öksüz, E.; Demiralp, M.; Mersin, S.; Tüzer, H.; Aksu, M.; Sarıkoc, G. Resilience in nurses in terms of perceived social support, job satisfaction and certain variables. *J. Nurs. Manag.* **2019**, *27*, 423–432. [CrossRef] [PubMed]
22. Smith, G.; Yang, F. Stress, resilience and psychological well-being in Chinese undergraduate nursing students. *Nurse Educ. Today* **2017**, *49*, 90–95. [CrossRef] [PubMed]
23. Seyedfatemi, N.; Pourafzal, F.; Inanloo, M.; Haghani, H. Percieved-stress and Resilience in Nursing Students. *Eur. Psychiatry* **2015**, *30*, 1005. [CrossRef]

proceedings

MDPI

Proceeding Paper

The Relationship between Cyberbullying Training Experience, Gender, and Depression among the Malaysian Adolescents during the COVID-19 Pandemic †

Khong Yun Pang, Wen Li Ku, Jaclyn Hui Jie Teng, Ren Wen Lew and Pei Boon Ooi *

School of Medical and Life Sciences, Sunway University, 5, Jalan Universiti, Bandar Sunway, Petaling Jaya 47500, Selangor, Malaysia
* Correspondence: peiboono@sunway.edu.my
† Presented at the International Academic Symposium of Social Science 2022, Kota Bharu, Malaysia, 3 July 2022.

Abstract: The introduction of the online educational system during the COVID-19 pandemic has increased the vulnerability to cyberbullying incidents among adolescents. This study examined the relationship between cyberbullying training and depression among Malaysian adolescents during the COVID-19 pandemic. A total of 1356 Malaysian adolescents participated in the online survey. Results revealed that depression was significantly associated with cyberbullying training. This study concluded that cyberbullying training can protect individuals from depression caused by cyberbullying. Female adolescents were found more vulnerable to depression than males during the COVID-19 pandemic. Therefore, we advocate that cyberbullying training is essential to be enforced in the current Malaysian schools' curriculum.

Keywords: cyberbullying; depression; cyberbullying training; cross-sectional study

Citation: Pang, K.Y.; Ku, W.L.; Teng, J.H.J.; Lew, R.W.; Ooi, P.B. The Relationship between Cyberbullying Training Experience, Gender, and Depression among the Malaysian Adolescents during the COVID-19 Pandemic. *Proceedings* **2022**, *82*, 101. https://doi.org/10.3390/proceedings2022082101

Academic Editor: Mohamad Rahimi Mohamad Rosman

Published: 18 October 2022

1. Introduction

The evolution of technology has created an artificial online world which makes information dissemination and communication convenient and easily accessible. As a result, the Internet has dominated the lives of modern people, especially adolescents [1]. The imposed Movement Control Operation (MCO) rules have made physical interaction difficult, while it has gradually been replaced by online interaction through the existing social networking sites (SNSs) such as WhatsApp, Facebook, Twitter, etc, [2,3]. Hinduja and Patchin [4] identified that Internet deviant behaviors were prevalent among adolescents and directly influenced the well-being of the online population. Accordingly, cyberbullying has risen as an imminent issue amid intensive Internet use [5]. Smith [6] defined cyberbullying as a series of acts that involve the cyber-perpetrators constantly inflicting harm and aggressiveness towards the victim through electronic-assisted technology. Many studies found that there was a growth in cyberbullying cases during the COVID-19 pandemic [7–9]. Wang et al. [10] found that individuals were at a greater risk of being cyberbullied during this pandemic, which can lead to hopelessness and loneliness, which are risk factors for depression. To highlight, Hase et al. [11] and Wright [12] found an association between cyberbullying and depression, while Limone and Toto [13] saw the impact of intensive Internet use during COVID-19, causing depression.

Many cyberbullying interventions and trainings have been designed to prevent cyberbullying perpetration and to countermeasure the negative impact of cyberbullying, for instance, The Media Heroes program [14], Surf-fair [15], Conrad [16], Noncadiamointrappola [17], The Kiva Program [18], and Help-Assert Yourself-Humor-Avoid-Self-Talk-Own [19]. Malaysia also previously launched a cyberbullying campaign (Nethics) which aimed to educate online users on social etiquette on the Internet and prevention strategies for cyberbullying [20]. Gaffney et al. [21] conducted a meta-analysis on 24 cyberbullying interventions and concluded that the current programs showed effectiveness in preventing cyberbullying; typically, they can reduce cyberbullying victimization. These interventional training can help individuals to be prepared encountering cyberbullying issues. In this study, the cyberbullying training experience refers to any of the interventional-based cyberbullying training that had been experienced by the participants which including the school-based cyberbullying training, corporate-based cyberbullying training, online-based cyberbullying training, and especially those mentioned above. Anticipating the growing trend of cyberbullying intervention studies in the future, the research hypotheses were shown below:

1. There was a significant association between the cyberbullying training experience and the likelihood of individuals helping a friend who had been cyberbullied.
2. There was a significant association between cyberbullying training experience and depression.
3. There was a significant association between gender and depression.

2. Methodology

This study received ethical approval from the Sunway Ethics Review Board (SUREC 2018/018). The data collected in this study was part of the overall data collection via a survey initiated by a Malaysian telecom corporate. This survey was launched in Survey Monkey and the participants were recruited through convenience sampling. In this study, we had strictly set the criteria to include only the participants aged 13 to 18 years old, who were also Malaysian and able to provide informed consent. To comply with the rights of the participants, they were briefed through using the online written information sheets and were also required to provide their informed consent on the same webpage before answering the survey. Besides, we had put all questions as compulsory to answer in the survey to address the missing value problem, but the participants were advised to withdraw from the study voluntarily if they were reluctant to answer any of the questions provided.

This questionnaire was made up of the dichotomous scales developed jointly by the telecom corporate and the cyberbullying experts. The questions were designed to understand the cyberbullying experience, cyberbullying training experience, adolescents' habits of assessing the Internet, etc. In addition, the subscale of "Depression" of the Depression, Anxiety, and Stress Scale (DASS-21) [22] was included in the questionnaire to assess the level of depression of the participants. It was made up of a total of 7 items and the scoring of the scale was calculated such that the addition of the 7 items was multiplied by 2. We followed the cut-off score determined originally, as follows: scores of 0 to 9 were indicated as having no depression, a score of 10–13 indicated mild depression, a score of 14 to 20 indicated moderate depression, scores of 21 to 27 indicated severe depression, and scores of 28 and above indicated extremely severe depression.

We used SPSS (Statistical Package for the Social Sciences Version 26, IBM, Armonk, NY, USA) [23] to perform the data analysis in this study. To highlight, a crosstab analysis was used to investigate the significant difference between individuals who had received cyberbullying training previously, and the likeliness of individuals to help a friend who had been cyberbullied. The binary logistic regression was used to measure the relationship between depression in individuals who had received cyberbullying training previously, and gender.

3. Results

3.1. Descriptive Statistics

There was a total of 1356 participants who took part in this study. The mean, standard deviation, skewness, and kurtosis of the variables were shown in Table 1. Table 2 showed the demographics of the participants. Of the participants, 57.2% were male, and 58% of them were individuals aged between 13 and 15 years old, compared to those aged between 16 and 18 years old, while Chinese people made up the majority of the participants (58.7%). Additionally, only 4.5% of the participants reported having experienced cyberbullying in the last 3 months, 44.0% of them reported having underwent a cyberbullying training previously, and 51% of them had displayed at least a mild level of depression.

Table 1. The descriptive statistics information.

Variables	Mean	Std. Deviation	Skewness	Kurtosis
Depression	10.92	10.127	0.775	<0.000

Table 2. Demographics (n = 1356).

Variables	Frequency (n)	Percentage (%)
Gender		
Male	776	57.2
Female	580	42.8
Age		
13-15 years old	786	58.0
16-18 years old	570	42.0
Race		
Malay	337	24.9
Chinese	796	58.7
Indians	200	14.7
Others	23	1.7
Cyberbullying experience		
Yes	61	4.5
No	1121	82.7
Prefer not to say	174	12.8
Previously had been receiving cyberbullying training		
Yes	609	44.0
No	747	55.1
Depression score		
Absent	665	49.0
Mild	96	7.1
Moderate	358	26.4
Severe	131	9.7
Extremely severe	106	7.8

3.2. Hypotheses Testing

In Table 3, we found that there was an association between individuals who had previously received cyberbullying training and the likeliness of individuals to help a friend who had been cyberbullied (X^2 = 63.559, p-value < 0.000). To highlight, people who had received cyberbullying training were more "Very likely" to help a friend who had been cyberbullied (70.4%) compared to their counterparts (51.8%). Similarly, individuals who had not undergone cyberbullying training were more reluctant to help (e.g., "Not likely at all" (4.4%), "Not very likely" (11.2%)).

We excluded those participants who had self-reported not experiencing cyberbullying in the last 3 months, and only 235 participants were included in the logistic regression analysis. The results in Table 4 showed that depression was significantly associated with the female gender (*OR* = 2.117, *C.I.* = 1.084, 4.133) and participants who had previously been receiving cyberbullying training (*OR* = 0.453, *C.I.* = 0.239, 0.859).

Table 3. Chi-square test (n = 1356).

Variables	Previously Had Received Cyberbullying Training		
	No (%)	Yes (%)	Total
How likely would you help a friend who had been cyberbullied			
Not likely at all	33 (4.4)	7 (1.1)	40
Not very likely	84 (11.2)	25 (4.1)	109
I don't know	205 (27.4)	134 (22.0)	339
Somewhat likely	38 (5.1)	14 (2.3)	52
Very likely	387 (51.8)	429 (70.4)	816
Total	747	609	1356
Pearson Chi-square	63.559		
Degree of freedom	4		
p-value	<0.000		

Table 4. Logistic regression (n = 235).

Variable	Depression		Nagelkerke R Square
	OR	95% *C.I.*	
Gender			0.067
Male (Ref)	-	-	
Female	2.117 *	(1.084, 4.133)	
Previously receiving cyberbullying training			
No (Ref)	-	-	
Yes	0.453 *	(0.239, 0.859)	

* $p < 0.05$.

4. Discussion

The result showed that there was a significant association between individuals who had previously received cyberbullying training and the likelihood of individuals helping a friend who had been cyberbullied. Liu et al. [24] found that cyberbullying bystanders will help the cyber-victims based on their perception of the critical level of the cyberbullying situation and the urgency of helping the cyber-victims. Those who had undergone the cyberbullying training were more aware of the negative outcomes of cyberbullying and thus were more likely to see cyberbullying more importantly compared to those who were without the cyberbullying training experience. Additionally, empathy was generally found to be a crucial factor that pushes a bystander to help a victim [25–27]. Hence, most of the cyberbullying interventions were tailor made to boost individuals' empathy levels and equipped them with prevention strategies and healthy coping mechanisms [28]. To highlight, The Media Heroes Project [14] reinforced the empathy and perspective-taking of individuals to approach a potential cyberbullying case. According to the empathy–altruism hypothesis [29], people were more likely to help those for whom they felt empathy. Therefore, the trained "empathy" in the cyberbullying program could encourage individuals to help those who encountered the cyberbullying perpetration.

Our findings showed that females were at a higher risk than males of developing depression. It was consistent with previous studies [30,31]. Kuehner [32] explained that it might be caused by the artefact hypotheses and the biological, psychological, and environmental differences. Artefact hypotheses implied that males were less likely to recognize depression than females, while Eid et al. [33] suggested that females were more vulnerable to mental disorders due to their differential body response to stressors compared to males. Concerning the psychological component, Costa et al. [34] found that females were more susceptible to neuroticism compared to males, which appeared to be a potential risk factor for depression despite the mixed findings in the literature [35,36]. Hankin et al. [37] also found that female adolescents will face more interpersonal environmental stresses such as peer chronic stress compared to male adolescents, which makes them more vulnerable to depression.

We found that individuals who had undergone previous cyberbullying training will be less likely to develop depression compared to their counterparts. This finding was consistent with the literature which stated that cyberbullying intervention can prevent cyber-victimization [21]. The learning outcomes from the cyberbullying training act as a protective factor that help to shield them from the negative psychological influences. Cyberbullying-trained individuals were generally well-versed in cyberbullying issues, while most of them were taught scientific coping strategies against cyberbullying perpetration [28]. As a result, cyberbullying training helped individuals properly cope with the psychological impact of cyberbullying perpetration such as depression.

Implications

The present study showed that cyberbullying training can be a crucial component shielding individuals from the negative psychological impacts derived from the cyberbullying perpetration. Besides, individuals who had undergone cyberbullying trainings were more likely to help their peers who were encountering cyberbullying issues. Therefore, it is important for the policy makers to enforce the cyberbullying training and intervention into the schools' curriculum. We envisioned that the exposure of cyberbullying training towards adolescents can be extremely beneficial for them to face the ever common occurrence of the cyberbullying cases, while better equipping them with the competent skills to deal with the cyberbullying perpetration and victimization.

This study had also revealed that female adolescents were more vulnerable to the negative impact of cyberbullying. Therefore, it was important for parents and teachers to pay a greater attention on the well-being of the female adolescents while providing them with care and empathy, noticing any symptoms of depression.

5. Conclusions

This study found that many individuals were experiencing at least a mild level of depression despite most of them not experiencing cyberbullying in the last 3 months. The findings suggested that cyberbullying training was a protective factor to prevent individuals from experiencing depression that was caused by cyberbullying perpetration and victimization. This was aided by the coping skills that had been taught in most of the cyberbullying intervention programs. Additionally, people who had previous cyberbullying training experience displayed a higher level of willingness to help a friend who was facing a cyberbullying issue. As they were coming in with greater awareness of the cyberbullying issue, their empathy level to offer help was strengthened. We also found that female adolescents were at a greater risk of developing depression compared to male adolescents. Therefore, more attention should be given to them during this difficult period amid the pandemic.

Author Contributions: Conceptualization, K.Y.P. and P.B.O.; methodology, W.L.K.; formal analysis, K.Y.P.; data curation, P.B.O.; visualization, K.Y.P.; validation, P.B.O.; writing—original draft preparation, K.Y.P.; writing—review and editing, W.L.K., J.H.J.T. and R.W.L.; supervision, P.B.O.; project administration, K.Y.P., W.L.K., J.H.J.T., R.W.L. and P.B.O. All authors have read and agreed to the published version of the manuscript.

Funding: This research received no external funding.

Institutional Review Board Statement: The study was conducted approved by the Institutional Review Board of Sunway Ethics Review Board (protocol code SUREC 2018/018 and 2018).

Informed Consent Statement: Informed consent was obtained from all subjects involved in the study.

Data Availability Statement: The data presented in this study are available on request from the corresponding author.

Acknowledgments: We are grateful to all the participants who had given their consent to join this study. We appreciate the effort of Sunway University Review Board to ameliorate our proposal for methodological and ethical concerns. We would also like to thank the telecom corporate company who had kindly shared their data with us.

Conflicts of Interest: The authors declare no conflict of interest.

References

1. López-Bueno, R.; López-Sánchez, G.F.; Casajús, J.A.; Calatayud, J.; Tully, M.A.; Smith, L. Potential health-related behaviors for pre-school and school-aged children during COVID-19 lockdown: A narrative review. *Prev Med.* **2021**, *143*, 106349. [CrossRef] [PubMed]
2. Hussain, W. Role of social media in COVID-19 pandemic. *Int. J. Front. Sci.* **2020**, *4*, 59–60. [CrossRef]
3. Nguyen, L.H.; Drew, D.A.; Graham, M.S.; Joshi, A.D.; Guo, C.-G.; Ma, W.; Mehta, R.S.; Warner, E.T.; Sikavi, D.R.; Lo, C.-H.; et al. Risk of COVID-19 among front-line health-care workers and the general community: A prospective cohort study. *Lancet Public Health* **2020**, *5*, e475–e483. [CrossRef]
4. Hinduja, S.; Patchin, J.W. Bullying, cyberbullying, and suicide. *Arch. Suicide Res.* **2010**, *14*, 206–221. [CrossRef]
5. Chan, N.; Ahrumugam, P.; Scheithauer, H.; Schultze-Krumbholz, A.; Ooi, P.B. A hermeneutic phenomenological study of students' and school counsellors' "lived experiences" of cyberbullying and bullying. *Comput. Educ.* **2020**, *146*, 103755. [CrossRef]
6. Smith, P.K. School bullying. *Sociol. Probl. E Práticas* **2013**, *71*, 81–98. [CrossRef]
7. Alsawalqa, R.O. Cyberbullying, social stigma, and self-esteem: The impact of COVID-19 on students from East and Southeast Asia at the University of Jordan. *Heliyon* **2021**, *7*, e06711. [CrossRef]
8. Fore, H.H. Violence against children in the time of COVID-19: What we have learned, what remains unknown and the opportunities that lie ahead. *Child Abus. Negl.* **2021**, *116*, 104776. [CrossRef] [PubMed]
9. Han, Z.; Wang, Z.; Li, Y. Cyberbullying involvement, resilient coping, and loneliness of adolescents during COVID-19 in rural China. *Front. Psychol.* **2021**, *12*, 664612. [CrossRef]
10. Wang, W.; Xie, X.; Wang, X.; Lei, L.; Hu, Q.; Jiang, S. Cyberbullying and depression among chinese college students: A moderated mediation model of social anxiety and neuroticism. *J. Affect. Disord.* **2019**, *256*, 54–61. [CrossRef]
11. Hase, C.N.; Goldberg, S.B.; Smith, D.; Stuck, A.; Campain, J. Impacts of traditional bullying and cyberbullying on the mental health of the middle school and high school students. *Psychol. Sch.* **2015**, *52*, 607–617. [CrossRef]
12. Wright, M.; Wachs, S.; Harper, B. The moderation of empathy in the longitudinal association between witnessing cyberbullying, depression, and anxiety. *Cyberpsychology: J. Psychosoc. Res. Cyberspace* **2018**, *12*, 6. [CrossRef]
13. Limone, P.; Toto, G.A. Psychological and emotional effects of digital technology on children in COVID-19 pandemic. *Brain Sci.* **2021**, *11*, 1126. [CrossRef] [PubMed]
14. Schultze-Krumbholz, A.; Zagorscak, P.; Scheithauer, H. 11—A school-based cyberbullying preventive intervention approach: The Media Heroes program *. In *Reducing Cyberbullying in Schools*; Campbell, M., Bauman, S., Eds.; Academic Press: Cambridge, MA, USA, 2018; pp. 145–158.
15. Pieschl, S.; Urbasik, S. Does the cyber bullying prevention program surf-fair work? An evaluation study. In *From Cyber Bullying to Cyber Safety: Issues and Approaches in Educational Contexts*; Hanewald, R., Ed.; Nova Science Publishers: Hauppauge, NY, USA, 2013; pp. 205–224.
16. Ortega Ruiz, R.; Rey Alamillo, R.D.; Casas Bolaños, J.A. Knowing, building and living together on internet and social networks: The ConRed cyberbullying prevention program. *Int. J. Confl. Violence* **2012**, *6*, 302–312. [CrossRef]
17. Palladino, B.E.; Nocentini, A.; Menesini, E. Evidence-based intervention against bullying and cyberbullying: Evaluation of the NoTrap! program in two independent trials. *Aggress. Behav.* **2016**, *42*, 194–206. [CrossRef]
18. Salmivalli, C.; Kärnä, A.; Poskiparta, E. Counteracting bullying in Finland: The KiVa program and its effects on different forms of being bullied. *Int. J. Behav. Dev.* **2011**, *35*, 405–411. [CrossRef]
19. Salvatore, A.J. An Anti-Bullying Strategy: Action Research in a 5/6 Intermediate School. Ph.D. Thesis, University of Hartford, West Hartford, CT, USA, 2006.
20. Malaysia Is 2nd in Asia for Youth Cyberbullying 2022. Available online: https://www.thestar.com.my/news/nation/2022/01/14/malaysia-is-2nd-in-asia-for-youth-cyberbullying (accessed on 18 May 2022).
21. Gaffney, H.; Farrington, D.P.; Espelage, D.L.; Ttofi, M.M. Are cyberbullying intervention and prevention programs effective? A systematic and meta-analytical review. *Aggress. Violent Behav.* **2019**, *45*, 134–153. [CrossRef]
22. Lovibond, P.F.; Lovibond, S.H. The structure of negative emotional states: Comparison of the Depression Anxiety Stress Scales (DASS) with the beck depression and anxiety inventories. *Behav. Res. Ther.* **1995**, *33*, 335–343. [CrossRef]
23. IBM Corp. *IBM SPSS Statistics for Windows, Version 26.0*; IBM Corp.: Armonk, NY, USA, 2019.
24. Liu, C.-H.; Yin, X.-R.; Huang, P.-S. Cyberbullying: Effect of emergency perception on the helping tendencies of bystanders. *Telemat. Inform.* **2021**, *62*, 101627. [CrossRef]

25. Haddock, A.D.; Jimerson, S.R. An examination of differences in moral disengagement and empathy among bullying participant groups. *J. Relatsh. Res.* **2017**, *8*, e15. [CrossRef]
26. Lambe, L.J.; Cioppa, V.D.; Hong, I.K.; Craig, W.M. Standing up to bullying: A social ecological review of peer defending in offline and online contexts. *Aggress. Violent Behav.* **2019**, *45*, 51–74. [CrossRef]
27. Zych, I.; Ttofi, M.M.; Farrington, D.P. Empathy and callous–unemotional traits in different bullying roles: A systematic review and meta-analysis. *Trauma Violence Abus.* **2016**, *20*, 3–21. [CrossRef] [PubMed]
28. Hutson, E.; Kelly, S.; Militello, L.K. Systematic review of cyberbullying interventions for youth and parents with implications for evidence-based practice. *Worldviews Evid. Based Nurs.* **2018**, *15*, 72–79. [CrossRef] [PubMed]
29. Toi, M.; Batson, C.D. More evidence that empathy is a source of altruistic motivation. *J. Personal. Soc. Psychol.* **1982**, *43*, 281–292. [CrossRef]
30. Marzo, R.R.; Vinay, V.; Bahari, R.; Chauhan, S.; Ming, D.A.F.; Fernandez, S.F.A.N.; Johnson, C.C.P.; Thivakaran, A.Q.A.; Rahman, M.M.; Goel, S. Depression and anxiety in Malaysian population during third wave of the COVID-19 pandemic. *Clin. Epidemiol. Glob. Heal.* **2021**, *12*, 100868. [CrossRef]
31. Xiong, J.; Lipsitz, O.; Nasri, F.; Lui, L.M.W.; Gill, H.; Phan, L.; Chen-Li, D.; Iacobucci, M.; Ho, R.; Majeed, A.; et al. Impact of COVID-19 pandemic on mental health in the general population: A systematic review. *J. Affect. Disord.* **2020**, *277*, 55–64. [CrossRef]
32. Kuehner, C. Why is depression more common among women than among men? *Lancet Psychiatry* **2017**, *4*, 146–158. [CrossRef]
33. Eid, R.S.; Gobinath, A.R.; Galea, L.A.M. Sex differences in depression: Insights from clinical and preclinical studies. *Prog. Neurobiol.* **2019**, *176*, 86–102. [CrossRef]
34. Costa, P.T.; Terracciano, A.; McCrae, R.R. Gender differences in personality traits across cultures: Robust and surprising findings. *J. Pers. Soc. Psychol.* **2001**, *81*, 322–331. [CrossRef]
35. Hakulinen, C.; Elovainio, M.; Pulkki-Råback, L.; Virtanen, M.; Kivimäki, M.; Jokela, M. Personality and depressive symptoms: Individual participation meta-analysis of 10 Cohort Studies. *Depress. Anxiety* **2015**, *32*, 461–470. [CrossRef]
36. Kendler, K.S.; Gardner, C.O. Sex differences in the pathways to major depression: A Study of opposite-sex twin pairs. *Am. J. Psychiatry* **2014**, *171*, 426–435. [CrossRef] [PubMed]
37. Hankin, B.L.; Young, J.F.; Abela, J.R.Z.; Smolen, A.; Jenness, J.L.; Gulley, L.D.; Technow, J.R.; Gottlieb, A.B.; Cohen, J.R.; Oppenheimer, C.W. Depression from childhood into late adolescence: Influence of gender, development, genetic susceptibility, and peer stress. *J. Abnorm. Psychol.* **2015**, *124*, 803–816. [CrossRef] [PubMed]

proceedings

MDPI

Proceeding Paper

Factors Associated with Adolescent Cyberbullying Perpetration and Victimization in Malaysia †

Pei Boon Ooi [1,2,3,*], Priyadharshini Ahrumugam [4], Phoey Lee Teh [5] and Nee Nee Chan [6]

1 Department of General Studies, School of Interdisciplinary Studies, Sunway University, 5, Jalan Universiti, Bandar Sunway, Petaling Jaya 47500, Malaysia
2 Liveable Cities, Future Cities Research Institute, Sunway University, 5, Jalan Universiti, Bandar Sunway, Petaling Jaya 47500, Malaysia
3 Department of Medical Sciences, School of Medical & Life Sciences, Sunway University, 5, Jalan Universiti, Bandar Sunway, Petaling Jaya 47500, Malaysia
4 Department of Communication, School of Arts, Sunway University, 5, Jalan Universiti, Bandar Sunway, Petaling Jaya 47500, Malaysia
5 Department of Computing and Information Systems, School of Engineering and Technology, Sunway University, 5, Jalan Universiti, Bandar Sunway, Petaling Jaya 47500, Malaysia
6 Faculty of Social Sciences & Liberal Arts, UCSI University, UCSI Heights, 1, Jalan Puncak Menara Gading, Taman Connaught, Kuala Lumpur 56000, Malaysia
* Correspondence: peiboono@sunway.edu.my
† Presented at the International Academic Symposium of Social Science 2022, Kota Bharu, Malaysia, 3 July 2022.

Abstract: Cyberbullying is emerging as a serious concern in schools with the proliferation of digital and communication technologies. This cross-sectional survey study examined the cyberbullying experience, comprised of both cyberbullying perpetration and victimization, of 990 secondary school students in Malaysia, with a mean age of 15.22 years old (S.D. = 1.358). The Partial Least Squares-Structural Equation Modeling (PLS-SEM) approach was used to examine the posited research hypotheses and the results indicated that media exposure ($\beta = 0.364$, $p = 0.00$), environmental factors ($\beta = 0.294$, $p = 0.00$), and individual coping strategies ($\beta = -0.075$, $p = 0.02$) emerged as significant factors and explained 30.9% of cyberbullying perpetration and victimization experiences in Malaysia. Overall, 13.3% of the participants reported having experienced cyberbullying perpetration and victimization in the past 1 year. Educational intervention programs should be built to address and target the factors identified in this work.

Keywords: cyberbullying; prevalence; media exposure; environmental factors; individual coping strategy; cyberbullying perpetration; cyberbullying victimization; Malaysia; youths

Citation: Ooi, P.B.; Ahrumugam, P.; Teh, P.L.; Chan, N.N. Factors Associated with Adolescent Cyberbullying Perpetration and Victimization in Malaysia. *Proceedings* **2022**, *82*, 109. https://doi.org/10.3390/proceedings2022082109

Academic Editor: Mohamad Rahimi Mohamad Rosman

Published: 25 October 2022

1. Introduction

Adolescents today are popular users of the Internet, text messages, and mobile devices such as smartphones and tablets [1] and are the leading consumers of social media platforms such as Instagram and Snapchat [2]. This generation appears to exhibit a growing dependence on technology, mainly due to exposure to the Internet throughout their childhood years and the readily available access to digital media [3]. Such increasing reliance on digital technologies and innovations as well as a shift from face-to-face communication to online communication has led to rising deviant Internet behavior such as cyberbullying, cyber harassment, cyberstalking, and other cyber-related abuses, with studies reporting links to youth suicides in the past decade [4,5].

Cyberbullying is described as "any behaviors performed through electronic or digital media by individuals or groups that repeatedly communicates hostile or aggressive messages intended to inflict harm or discomfort on others" [6] and is associated with systemic abuse of power [5] when a person uses digital technology opportunistically and deliberately to repeatedly harass, humiliate, embarrass, torment, threaten, pick on, or intimidate another person. Cyberbullying victims and perpetrators reported poorer academic performance and greater behavioral problems in school [7] with cyberbullying victims reportedly suffering from lower well-being and mental health status and being the least likely to seek professional help [8].

Interestingly, the factors predicting both cyberbullying perpetration and victimization appeared to be consistent when it comes to media exposure to risky behavior on information and communications technology (ICT) usage—e.g., accepting strangers as online contacts [9] and parental support in monitoring online behaviors [10]. However, research showed mixed findings on individual coping mechanisms' effectiveness [11]. Given the effects and factors are different, for both cyberbullying perpetration and victimization, it would require school counselors to develop, design and implement various intervention programs, based on identified factors, to address them respectively instead of having a program that fixes all.

Cyberbullying incidence rates appear to be rising among Malaysian schoolchildren and university students. It was reported that 37% of school students in Malaysia are encountering or have previously been involved with cyberbullying as internet usage amongst school children in all states in Malaysia is more than 90% on average [12]. One-third of Malaysian young adults confessed that they had cyberbullied someone while two-thirds had witnessed cyberbullying incidents [13], whereby commonly they were bullied either on Facebook or through mobile phones [14]. To achieve the Shared Prosperity Vision (SPV) 2030—i.e., build a progressive society that will make use of modern science and technology by increasing the participants' knowledge, awareness, and behavior towards online user rights and protection—it is imperative for studies to focus on the interaction effects among various factors that contribute to cyberbullying perpetration and victimization as a whole. Hence, this study aims to study the interaction among these factors, based on the research framework below, on cyberbullying experience among adolescents.

The Social Cognitive Theory (SCT) [15] has been widely used to define traditional bullying and cyberbullying, which is a form of individuals' social behavior greatly influenced by personal and environmental factors. Personal factors such as internet self-efficacy, motivation, and cyber victimization experience have contributed to cyberbullying behavior [16]. In addition, the exposure one has to the media also contributes to one's cognitive, behavior, and involvement in cyberbullying [17] according to the Media Effect Model [18]. The Media Effect Model (MEM) postulated that audiences' behaviors and thoughts are affected by the exposure to media, thus, heavy usage or longer time spent using information and communications technology (ICT) were reported to increase individual involvement in cyberbullying [9].

This triadic relationship between personal, environmental, and media exposure was proven statistically significant with Hong Kong university students [16] and is used as a theoretical framework in this current study to identify the factors influencing the cyberbullying perpetration and victimization scenario in Malaysia. They comprised of: (a) Media Exposure, which is comprised of negative online experiences, sexting behaviors, and exposure to harmful sites such as drug, pro-anorexic, or self-harm sites; (b) Environmental Factors, which are comprised of both peer pressure and parent–child relationship; and (c) Personal Factors, which involves individual coping strategies used in dealing with cyberbullying (Figure 1).

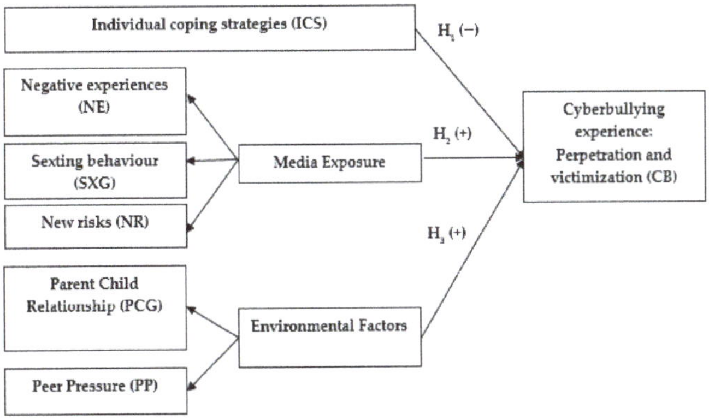

Figure 1. Research model.

2. Materials and Methods

In this cross-sectional study, an anonymous self-report questionnaire was administered to 1200 participants at 12 public secondary schools from the Selangor state in Malaysia, selected randomly via an online randomizer (https://www.random.org/lists/, accessed on 10 June 2019), over 4 months in 2019. All schools share the same setting (i.e., they are public government secondary schools, hosting students aged 13–17 years old). The schools were invited to participate in this study once the approvals from the Ministry of Education Malaysia, Selangor Education Department, and University Institute Ethics Committee were obtained. Consent forms were distributed to all the parents in the schools randomly to allow the students to participate in the study voluntarily. Participants with consent were approached by researchers and recruited to answer a paper-and-pencil questionnaire. No compensation was given and participants could withdraw at any point, without penalty. A total of 1200 copies of questionnaires were randomly distributed to the students with consent, 1009 were returned (84.08%), and 19 outliers were excluded (1.88%), resulting in a total number of 990 participants.

The 48-item questionnaires were part of the Digi 2015 National Survey, derived from the CyberSAFE 2015 Programme, and approval was obtained from Digi Telecommunications for the survey to be used in this study. The survey was submitted to two experts (K.W.L. and M.C.H.) for face and content validations. The questionnaire took 20–30 min to be answered by each participant. Participants were asked to answer on a 5-Likert scale, ranging from "strongly disagree" to "strongly agree". The original questionnaire was made available in both the English Language and translated into Bahasa Malaysia (Malay Language). All schools opted for the Bahasa Malaysia version.

The SmartPLS 3.0 software was used to conduct the Partial Least Squares (PLS) analysis [19] and to examine the association between measured constructs and cyberbullying experience. PLS simultaneously estimated the measurement and the structural parameters involving two stages, which are the measurement model and the structural model. Finally, a bootstrapping method was used to determine the significance of the path coefficients and the factor loadings [20].

3. Results

The participants were 518 (52.3%) females, 417 (42.1%) males, and 55 participants (5.6%) did not indicate their gender. The mean age was 15.22 (S.D. = 1.358) years. The majority of them were Malay (47.2%), followed by Chinese (38.6%), Indian (10.8%), and others (3.4%).

3.1. Prevalence of Cyberbullying Experience

Almost all (91.9%) participants used the Internet in the past year and 82% reported being always online every day (69.9% for always and 12.1% for almost every day). The prevalence of students having experienced cyberbullying was 13.3%—by answering "Yes" to the statement that they have been bullied by others on the Internet in the past 1 year. The majority of the students (64.4%) stated that they were bullied off and online by the same person (Table 1).

Table 1. Descriptive analysis of cyberbullying perpetration and victimization experience (n = 990).

Items	Frequency (Mean)	Percentages (SD)
Have you used the Internet in the past 1 year?		
Yes	910	91.9
No	17	1.7
Missing	63	6.4
Frequency of using the Internet		
Always online every day	692	69.9
Almost every day	120	12.1
3–5 times a week	53	5.3
1–2 times a week	58	5.9
Seldom	0	0
Missing	67	6.8
I have been bullied by others on the Internet in the past 1 year.		
Yes	130	13.13
No	860	86.87
The person who has bullied me on the Internet also bullied me offline (e.g., in school).		
Yes	638	64.4
No	352	35.6

3.2. Assessment of the Measurement Model

In PLS-SEM, the measurement model is assessed for reliability, convergent validity, and discriminant validity. To assess the reliability of measurements in a PLS model, the composite reliability (CR) was used and the threshold value suggested for CR is 0.7 [20,21]. The composite reliability (CR) values of the latent variables were reported to have CR above 0.7, which indicates good reliability of constructs.

The PLS captured the convergent validity of each construct in the scale using the measure of average variance extracted (AVE). AVE provides the overall construct's variance explained by all its indicators. Of note, AVE was considered as established convergent validity if the value was above 0.5, which indicates 50 % of the variance explained [21]. The AVE in the measurement model was above 0.5, therefore it is concluded that the convergent validity of the constructs was established. Lastly, discriminant validity was determined by assessing whether the value of the square root of AVE of a construct was more than the inter-construct correlation between the construct of interest and other constructs assessed in the model [21].

Tables 2 and 3 reported the discriminant validity analysis at the construct level for both the first and second stage model. The inter-construct correlations of all constructs were used to compare with the square root of AVE values of every construct. The square root of AVE values of the constructs are presented as the diagonal entries in Tables 2 and 3 (in bold numbers). The AVE values are greater than any inter-construct correlations [22]. Therefore, in conclusion, the measurement model possessed discriminant validity.

Table 2. Discriminant validity for first stage measurement model.

Variables	CB	ICS	NR	PCG	PP	NE	SXG
CB	**0.738**						
ICS	−0.240	**0.703**					
NR	0.371	−0.223	**0.720**				
PCG	0.389	−0.281	0.321	**0.792**			
PP	0.422	−0.200	0.349	0.402	**0.823**		
NE	0.364	−0.229	0.338	0.377	0.361	**0.708**	
SXG	0.388	−0.173	0.403	0.366	0.383	0.405	**0.843**

Note: CB: Cyberbullying experience; ICS: Individual coping strategies; NR: New risks; PCG: Parent–child relationship; PP: Peer pressure; NE: Negative experience; SXG: Sexting. Bold: The square root of the AVE.

Table 3. Discriminant validity for second stage measurement model.

Variables	Cyberbullying	Environment Factors	Media Exposure	Individual Coping Strategy
Cyberbullying experience	**0.738**			
Environment Factors	0.485	**0.837**		
Media Exposure	0.489	0.560	**0.767**	
Individual coping strategies	−0.240	−0.285	−0.271	**0.703**

The analysis presented above provides additional evidence on the discriminant validity of the measurement model in the first stage and second stage. Bold: The square root of the AVE.

3.3. Analysis of The Structural Model

The principle of a PLS model assessment is based on the latent constructs and the path coefficients' weights on the same lines as regression analysis [23]. The path coefficients, T values, and significance level of the antecedents of cyberbullying post bootstrapping technique were employed with 5,000 sub-samples. The bootstrapping results confirm that all path coefficients report a significant level of 0.05 [22].

The results shown in Table 4 affirmed that environmental factors and media exposure have contributed as positively significant to cyberbullying and individual coping strategy was negatively significant to cyberbullying. Media exposure has the highest contribution ($\beta = 0.364$, $p = 0.00$), followed by environment factors ($\beta = 0.294$, $p = 0.00$), and, finally, individual coping strategies ($\beta = -0.075$, $p = 0.02$), which have very little bearing on cyberbullying experience.

Table 4. Path significance and t values (Bootstrapping results).

	Standard Beta	Standard Deviation	t-Value	Result	R^2
H1: Individual coping strategies → Cyberbullying experience	−0.075	0.035	2.151	Supported	0.309
H2: Media Exposure → Cyberbullying experience	0.364	0.291	8.085	Supported	
H3: Environment Factors → Cyberbullying experience	0.294	0.037	8.319	Supported	

The magnitude of variability of a variable that is shared by another was represented with R^2 measures [24]. R^2 is defined as the squared values of the correlation between a dependent construct and its predicted values, thus it is used to measure a model's predictive accuracy [24]. Then, R^2 also indicates the combined effect of independent constructs on a dependent construct [25]. R^2 values for dependent constructs are considered strong, moderate, or weak if the value is 0.75, 0.50, and 0.25 respectively [22]. In this study, the R^2 or the weight of the endogenous construct of cyberbullying is 0.309, which is a moderate relationship, and this indicated that the environmental factors, media exposure, and individual coping strategies account for 30.9% of the variation in cyberbullying experience.

4. Discussion

The evidence from this study suggests that nine out of ten Malaysian secondary school students go online daily. Despite the high usage, only a small number who took part in this study experienced cyberbullying (i.e., a prevalence rate of 13.13%). The current findings reported lower cyberbullying experiences among secondary school students compared to the other studies conducted in Malaysia [13,14]. There are three main potential reasons for the current findings. Firstly, parental consent was required as part of the research design; thus, the study was unable to establish if the low prevalence rate was reported as a result of the refusal or other concerns when answering. In the past, studies which included parental consent have resulted in lower or no reply or refuse to participate outcomes [26]. Secondly, all the awareness-raising may have sensitized students to the fact that cyberbullying is not well-accepted (by adults) and social desirability may have influenced the reported rates. Lastly, over the years, many awareness campaigns and talks have been exposed to students and family members, which may be associated with greater cyber safety awareness and coping strategies among youths. For example, families were encouraged to have computers placed in common areas of their homes to promote and exercise more rules on cyber safety [12]. In Malaysia with the CyberSAFE campaign, there appeared to be a high degree and increase of digital resilience among the adolescent students surveyed and, generally, positive online usage and positive influence of parents regarding Internet use [27,28].

Interestingly enough, the current study also revealed that 64.4% of the participants reported that the person who has bullied them on the Internet also bullies them offline (i.e., physical bullying) in schools. The relationship between cyberbullying and traditional bullying is inevitable and often cyber victims are also victims of traditional bullying [29]. The power imbalance between bullies and victims may result in the latter being labeled or perceived as weak students, thus, the school counselors or school authorities must consider introducing and implementing cyberbullying prevention and education campaigns that address both traditional bullying and cyberbullying, which frequently overlap, in an integrated manner. Permanent suspension of bullies from school may solve traditional bullying in school but not when the bullies have access to the victims offline. Thus, the cyberbullying prevention and education campaigns should empower the cyberbullying victim/bully victims to recognize, prevent, report, and be free from being the victim.

Among the three factors, media exposure was reported the most significant contributing factor to adolescent cyberbullying experience in Malaysia. Media exposure, especially messaging with sexting and exposure to new risks, is becoming a norm. Proactive actions need to be in place to educate users on users' privacy and digital literacy and not to fall into digital scams. Sexting and exposure to online sexual content is a new element added in the study because of the changeable technology and online environment. It was observed that often Malaysian adolescents endured online harassment and sexually based cyberbullying [30]. Other means of sexual violations are through unwelcomed "exposure to sexual materials such as pornographic pictures, pornographic video clips" and "sexual solicitations"—willingly or unwilling interacting in sexual activity with making the victims feel sexually abused [31]. Despite the dangers implicated, adolescents engage regularly with the Internet and communication technologies to attain knowledge of sexuality and self. This study highlighted the new online bullying behavior that needs immediate attention—a need to call for a healthy relationship and sex education openly in schools and the home, where parents, schools, and students can openly discuss topics such as consent, coercion, prevention, and boundaries in sexting activities.

Lastly, the present result showed there is a statistically negative association between individual coping strategies in dealing with cyberbullying perpetration and victimization experience among Malaysian adolescents. Those who have a higher ability to cope with the cyberbullying experiences and establish a better coping strategy as coping mechanisms reported less cyberbullying experiences. The example of a coping strategy such as changing privacy settings in devices was perceived as a positive step to prevent unwanted viewers and restriction of this access to a limited number of people. Education interventions such as character-building or perspective-changing and empathy workshops could be introduced to build greater digital resilience. With that, youths could enhance their ability to adapt and safeguard their digital privacy when they experience cyberbullying.

5. Conclusions

Cyberbullying experiences are generally associated with individual exposures to risk factors, support systems, and a conducive environment one lives in as well as exposure to unpleasant and potentially unguided and harmful sites. These findings, although provisional, do have practical implications. This information could be used to develop targeted intervention programs for the school setting. The intervention programs should be built to address these factors considering the wide range of stakeholders (parents, schools, school counselors, and authorities) that are involved and contribute indirectly to the cyberbullying experience of the students. In light of the identified risk factors of cyberbullying among adolescents, it is imperative that future research examines potential mediating and moderating processes that might influence the effect of cyberbullying on overall adolescent health and psychological performance. In summary, managing cyberbullying perpetration and victimization would require a holistic approach, targeting not just the environmental factor or media exposure but also strengthening the individual coping strategies.

Author Contributions: Conceptualization, P.B.O., P.A., P.L.T. and N.N.C.; methodology, P.B.O.; software, P.B.O.; validation, P.B.O.; formal analysis, P.B.O.; investigation, P.B.O., P.A. and N.N.C.; resources, P.B.O., P.A., P.L.T. and N.N.C.; data curation, P.B.O.; writing—original draft preparation, P.B.O., P.A., P.L.T. and N.N.C.; writing—review and editing, P.B.O., P.A., P.L.T. and N.N.C.; project administration, P.B.O., P.A. and N.N.C.; funding acquisition, P.B.O. All authors have read and agreed to the published version of the manuscript.

Funding: This research was funded by the Ministry of Higher Education (MOHE) Malaysia, grant number FRGS/1/2020/SS0/SYUC/02/2.

Institutional Review Board Statement: The study was conducted in accordance with the Declaration of Helsinki and approved by the Ethics Committee of Sunway University (SUREC 2018/018 and 6 March 2018).

Informed Consent Statement: Informed consent was obtained from all subjects involved in the study.

Data Availability Statement: The data presented in this study are available on request from the corresponding author. The data are not publicly available due to data being part of the industry partner's property.

Acknowledgments: The authors would like to express their gratitude to all participants and parents who took part in this study as well as Coco Lee and Natasya Azlan for their help with data collection.

Conflicts of Interest: The authors declare no conflict of interest. The funder had no role in the design of the study; in the collection, analyses, or interpretation of data; in the writing of the manuscript; or in the decision to publish the results.

References

1. Livingstone, S. How can children be protected online when the internet has been designed for adults. In *Children's Rights and the Internet: From Guidelines to Practice*; UNICEF: New York, NY, USA, 2016; pp. 12–13.
2. Docu, V. Millennials and anxiety: An exploration into social networking sites as a predisposing factor. *Rom. J. Cogn. Behav. Ther. Hypn.* **2018**, *5*, 1–10.
3. PrakashYadav, G.; Rai, J. The Generation Z and their social media usage: A review and a research outline. *Glob. J. Enterp. Inf. Syst.* **2017**, *9*, 110–116. [CrossRef]
4. Hinduja, S.; Patchin, J. Bullying, Cyberbullying, and Suicide. *Arch. Suicide Res.* **2010**, *14*, 206–221. [CrossRef] [PubMed]
5. Slonje, R.; Smith, P.K.; Frisén, A. The nature of cyberbullying, and strategies for prevention. *Comput. Hum. Behav.* **2013**, *29*, 26–32. [CrossRef]
6. Tokunaga, R.S. Following you home from school: A critical review and synthesis of research on cyberbullying victimization. *Comput. Hum. Behav.* **2010**, *26*, 277–287. [CrossRef]

7. Wright, M.F. Adolescents' cyber aggression perpetration and cyber victimization: The longitudinal associations with school functioning. *Soc. Psychol. Educ.* **2015**, *18*, 653–666. [CrossRef]
8. Spears, B.A.; Taddeo, C.M.; Daly, A.L.; Stretton, A.; Karklins, L.T. Cyberbullying, help-seeking and mental health in young Australians: Implications for public health. *Int. J. Public Heal.* **2015**, *60*, 219–226. [CrossRef] [PubMed]
9. Chen, L.; Ho, S.S.; Lwin, M.O. A meta-analysis of factors predicting cyberbullying perpetration and victimization: From the social cognitive and media effects approach. *New Media Soc.* **2017**, *19*, 1194–1213. [CrossRef]
10. Beyazit, U.; Şimşek, Ş.; Ayhan, A.B. An examination of the predictive factors of cyberbullying in adolescents. *Soc. Behav. Pers. Int. J.* **2017**, *45*, 1511–1522. [CrossRef]
11. Machackova, H.; Cerna, A.; Sevcikova, A.; Dedkova, L.; Daneback, K. Effectiveness of coping strategies for victims of cyberbullying. *Cyberpsychol. J. Psychosoc. Res. Cyberspace* **2013**, *7*, 5. [CrossRef]
12. Digital News Asia. Online Harassment of Schoolkids as high as 70%: Survey. Available online: https://www.digitalnewsasia.com/security/online-harassment-of-schoolkids-as-high-as-70pc-survey (accessed on 30 May 2019).
13. Balakrishnan, V. Cyberbullying among young adults in Malaysia: The roles of gender, age and Internet frequency. *Comput. Hum. Behav.* **2015**, *46*, 149–157. [CrossRef]
14. Lai, C.; Mohamad, M.; Lee, M.; Salleh, K.M.; Sulaiman, N.; Rosli, D.; Chang, W. Prevalence of cyberbullying among students in Malaysian higher learning institutions. *Adv. Sci. Lett.* **2017**, *23*, 781–784. [CrossRef]
15. Bandura, A. Prentice-Hall series in social learning theory. In *Social Foundations of Thought and Action: A Social Cognitive Theory*; Prentice-Hall: Englewood Cliffs, NJ, USA, 1986.
16. Xiao, B.S.; Wong, Y.M. Cyber-bullying among University students: An empirical investigation from the social cognitive perspective. *Int. J. Bus. Inf.* **2013**, *8*, 34.
17. Festl, R.; Scharkow, M.; Quandt, T. Peer influence, internet use and cyberbullying: A comparison of different context effects among German adolescents. *J. Child. Media* **2013**, *7*, 446–462. [CrossRef]
18. Felson, R.B. Mass Media Effects on Violent Behavior. *Annu. Rev. Sociol.* **1996**, *22*, 103–128. [CrossRef]
19. Ringle, C.M.; Wende, S.; Becker, J.-M. SmartPLS 3. In *SmartPLS GmbH*; SmartPLS GmbH: Bönningstedt, Germany, 2015.
20. Hair, J.F.; Black, W.C.; Ralph, A.; Babin, B.J. *Multivariate Data Analysis*, 8th ed.; Cengage Learning EMEA: London, UK, 2018.
21. Fornell, C.; Larcker, D.F. Evaluating Structural Equation Models with Unobservable Variables and Measurement Error. *J. Mark. Res.* **1981**, *18*, 39–50. [CrossRef]
22. Hair, J.F., Jr.; Hult, G.T.M.; Ringle, C.; Sarstedt, M. *A Primer on Partial Least Squares Structural Equation Modeling (PLS-SEM)*; Sage Publications: Newbury Park, CA, USA, 2016.
23. Chin, W.W. *Handbook of Partial Least Squares*; Springer: Berlin/Heidelberg, Germany, 2010; Volume 201.
24. Field, A. *Discovering Statistics Using IBM SPSS Statistics*; Sage Publications: Newbury Park, CA, USA, 2013.
25. Pallant, J. *SPSS Survival Manual*, 6th ed.; Open University Press: London, UK, 2013.
26. Garett, R.; Lord, L.; Young, S. Associations between social media and cyberbullying: A review of the literature. *mHealth* **2016**, *2*, 46. [CrossRef] [PubMed]
27. Digi. Sustainability Report. Available online: http://www.digi.com.my/sustainability/sustainability-reports/download/digi-sustainability-report-2015.pdf (accessed on 30 May 2019).
28. Digi. Digi Yellow Heart Cyberbullying and YOUTH Disposition Survey 2018. Available online: http://www.digi.com.my/sustainability/sustainability-reports/download/yh_cyberbully-disposition-survey-2018.pdf (accessed on 30 May 2019).
29. Von Marées, N.; Petermann, F. Cyberbullying: An increasing challenge for schools. *Sch. Psychol. Int.* **2012**, *33*, 467–476. [CrossRef]
30. Bakar, H.S.A. Investigating the emergence themes of cyberbullying phenomenon: A grounded theory approach. In Proceedings of the 2013 IEEE 63rd Annual Conference International Council for Education Media (ICEM), Singapore, 1–4 October 2013.
31. Yusuf, S.; Hassan, M.S.; Ibrahim, A.M. Cyberbullying among Malaysian children based on research evidence. In *Advanced Methodologies and Technologies in System Security, Information Privacy, and Forensics*; IGI Global: Hershey, PA, USA, 2019; pp. 115–137.

proceedings

MDPI

Proceeding Paper

The Phases Before, During, and After Becoming a Drug Abuser: Are They Interrelated? †

Nadhira Ahmad Afip [1,*], Iskandar Hasan Tan Abdullah [2] and Tuan Nooriani Tuan Ismail [1]

1 Faculty of Administrative Science & Policy Studies, Universiti Teknologi Mara (UiTM) Selangor, Shah Alam 40450, Malaysia
2 Faculty of Administrative Science & Policy Studies, Universiti Teknologi Mara (UiTM) Cawangan Kelantan, Bukit Ilmu, Machang 18500, Malaysia
* Correspondence: nadhira.afip96@gmail.com
† Presented at the International Academic Symposium of Social Science 2022, Kota Bharu, Malaysia, 3 July 2022.

Abstract: This paper focuses on the issue of women's involvement with drug abuse in Malaysia. The involvement with drug abuse is much more common among men; however, understanding women's participation in drug use is critical, as it is increasing year-on-year. In this study, the researchers aim to understand the factors that contribute towards the involvement of women in drugs and the effectiveness of treatments provided by drug rehabilitation centers in order to help clients to stay sober. It is vital that this issue be curbed, as it will create a butterfly effect. This is because it not only impacts the individuals themselves, but also the community. Additionally, the situation becomes more critical as the opportunity for women drug abusers to be treated is quite limited compared to the opportunities given to men. Hence, this research finds that it is important to understand the phases before, during, and after becoming a drug abuser as it will also help to determine the effectiveness of the treatment and mitigate the tendency of women drug users to relapse.

Keywords: women and drugs; rehabilitation center; drug treatments

Citation: Afip, N.A.; Abdullah, I.H.T.; Ismail, T.N.T. The Phases Before, During, and After Becoming a Drug Abuser: Are They Interrelated? *Proceedings* **2022**, *82*, 110. https://doi.org/10.3390/proceedings2022082110

Academic Editor: Mohamad Rahimi Mohamad Rosman

Published: 25 October 2022

1. Introduction

The issue of drug abuse has become a worldwide issue regardless of whether a country is a developed or developing country. It is critical that drug abuse be addressed as it may influence the stability of the country in term of safety, society, and health. However, drugs are important for medical purposes, yet they are often misused by some people. Hence, if a drug is not being used as prescribed, it can lead to an addiction. Addiction may cause negative changes toward an individual mentally, emotionally, and physically.

It is important to curb drug misuse as it can create a butterfly effect within the country. This means that the existence of drug abuse will not only impact the individual (drug abuser), but will also have an impact on the family and community. For instance, in America, the issue of drug abuse costs more than USD 700 billion a year, as there is an increment in the cost of health care, crime, and loss of productivity [1]. As stated above, drug abuse may cause addiction, which is a chronic disease that can disrupt the functions of the brain and body, and if left untreated, it can lead to harmful effects or death [1]; hence, this can create social issues such as prostitution, gambling, and chronic disease.

In Malaysia, since 2018, the number of women participating in drug abuse has increased year-by-year. Referring to Table 1, below, it can be seen that the number of males involved in drug abuse is higher compared to women; however, the percentage of women becoming involved from 2018 to 2019 is much higher compared to men, with the percentages of 19.0% and 2.73%, respectively. However, in 2020, the number of involvements between the two sexes has declined. This is due to the COVID-19 pandemic and the government of Malaysia enforcing a Movement Control Order (MCO), which has indirectly helped the government to curb the issue of drugs. In 2021, the National Anti-Drug Agency reported that the number of women involved in drug abuse started to rise again.

Table 1. Women's involvement in drug abuse from 2018–2021.

Year	Males		Females	
2018	24,025		947	
		2.73%		19.0%
2019	24,682		1127	
		−21.59%		−14.73%
2020	19,352		961	
2021	117,822		5317	

2. Need for This Study

This research will be relevant in understanding the relationship between the phases before, during, and after being a drug abuser. Examining these phases will help to obtain a better understanding of what triggers an individual to become involved with drugs, the effectiveness of treatments at the rehabilitation center, and how the treatments have helped clients to survive after they have been released from the rehabilitation center. In this research, the researcher selected a group of Malay women who are currently serving their time at a rehabilitation center, clients who have been released from the center (alumni), and officers in charge of handling the treatment sessions. By referring to the information given by the selected group of informants, the researcher was able to determine what triggers the involvement of Malay women in drugs and the effectiveness of the treatments provided to them. This proves that this study is able to fill the gap relating to the issue regarding the involvement of women in drugs and how the treatments provided are able to prevent the client from relapsing.

3. Overview of the Drug Abusers' Life

3.1. *Before*

During this first phase, the researcher sought to identify the causes that trigger women to be involved with drugs. Generally, there are many reasons that can cause women to participate in drug use. According to [2], one of the reasons that may cause women to be involved with drugs is the influence of their intimate partner. This is because intimate partners play a huge role in an individual's life, especially women, hence they are more likely to be introduced to drugs by their partner. Furthermore, the association of drugs among women may also be caused by physical violence. When a partner has great control and power over them, the tendency for a woman to be influenced by drugs, either voluntarily or involuntarily, is high. This is because drug abuse is often related to emotional disorders such as stress, anxiety, and depression [3]. In addition, drug abuse can affect learning achievement, cause addiction, and increase mental health issues [4]. Moreover, intimate partners usually share the same equipment to inject drugs, and with that, they are vulnerable to contracting the human immunodeficiency virus (HIV). According to [5], peer influence, a problematic family environment, and sensation-seeking are also contributing factors that cause women to be involved with drugs.

3.2. *During*

Access to a rehabilitation center plays an important role in treating people who abuse drugs. This is because the rehabilitation center will act as a tool to monitor and guide the individual to recover. However, according to past research, up until now, rehabilitation centers are unable to guarantee the recovery of the clients [6]. Studies have found that this is because clients still have the urge to continue consuming drugs even after they have been released from the center [7]. In ensuring the effectiveness of the treatment toward the clients, early-stage identification needs to be made so that a suitable treatment is able to be given. This proves that it is crucial to provide a treatment that can tackle the root cause of the issue by educating the client about the recovery process and encouraging good practices and activities in their daily life, which then will help them to achieve lifelong recovery. To support the previous statement, [8] suggested that future studies need to evaluate the effectiveness of the current treatment program and whether it needs to be adjusted in order to tailor the program toward the desired goal.

3.3. *After*

In this phase, the post-treatment process is crucial to be looked into as it will determine whether the client is able to avoid relapse. According to [5], the client tends to start becoming involved with drugs again or relapsing as they face challenges such as having trouble obtaining and retaining jobs. This is because, during their time at the center, most of their basic needs are being taken care of by the center; thus, the client tends to be in a comfort zone. With that, it is important to ensure that the client is able to be independent. Hence, past researchers suggest that a well-established method should work together with the current treatment program so that the root cause of relapse can be addressed [9]. This is because the treatment will determine the quality of life of the clients after they are released from the center [10].

4. Research Questions

4.1. *What Are the Factors That Trigger Malay Women to Be Involved with Drugs?*

As for this research question, it will assist the researchers in order to identify the factors that has triggered Malay Women to be involved with drugs. Identifying the factors that triggers women to involve with drug is important as it will help the researcher to classify whether the client are willing or being forced to be associating with drugs.

4.2. *How Effective Are the Programs Provided by the Rehabilitation Center?*

From this research question, the researcher able to understand which programs gave more positive impact toward the client in supporting their journey to be free from drug dependency. The researcher also able to provide modifications to improvise the current treatment program.

5. Methodology

5.1. *Research Design*

This research is a qualitative study that adopts a case study design. It involves an intensive examination of a real-life setting. By adopting the case study design, interview sessions and focus group discussions will be used in order to gather information from the informants.

5.2. *Unit of Analysis*

The study involves Malay women from PUSPEN Bachok who have undergone treatment for more than 3 months, previous clients (alumni) who have been released from the center, and officers in charge who were involved during the treatment process. PUSPEN Bachok was chosen as it is a pioneering drug treatment center that only focuses on women.

5.3. *Sample Size*

Rule of thumb will be applied in this study as it will involve 10–20 knowledgeable people in order to obtain sufficient understanding and help to define the case studied [11].

5.4. Sampling Technique

As for the sampling technique, this study uses purposive sampling. This is because it is a specific issue; hence, the sample will only be among a group of informants that possess the experience and information pertaining the issue studied. Moreover, by applying purposive sampling, it will help the researcher to capitulate useful information from the informants.

5.5. Instrumentation

A set of interview questions will be posed to the informants (female Malay clients, clients who have been released (alumni), and officers in charge at the center). The questions are presented in both Bahasa and English.

5.6. Data Analysis

The analysis of this study will be completed using NVivo version 12 to analyze the information that is obtained from the informants. Other than that, the analysis will include import, explore, code, query, reflect, visualize, and memo [12].

6. Discussion and Conclusions

Figure 1 represents the whole focus of the study. The researcher determines whether the phases before, during, and after becoming a female drug user are interrelated. Hence, this study attempts to determine the factors affecting the tendency for drug abuse among Malay women. Furthermore, this study will also examine the category of admission among women clients, which will determine the intention and willingness to enter into and participate with treatment at the rehabilitation center. However, past studies revealed that the effectiveness of the treatment cannot rely on the program alone, but rather requires the willingness of the client to ensure the effectiveness of the treatment. According to [13], the willingness of the clients plays a crucial role as it will determine their level of participation during the recovery process. This is because, if the client voluntarily participates, they will have a better understanding of their addiction and may be more determined to be recover. The study conducted by [6] found that clients who voluntarily admit themselves into the center are more participative during the treatment process. Voluntary clients are more open to accept and participate during the treatment process, which will help them to reduce the tendency to relapse. One of the informants in the study also mentioned that the rehabilitation center can only provide programs for the clients, and at the end of the day, it fully depends on the individual to be ready and willing to make the changes. This proves that ensuring the client is willing to participate in treatment process is crucial. If the client is being forced to enroll at the center, they will have trouble adapting to the situation and this will affect the smoothness of the treatment process [14]. Moreover, [13] also stated that if a client is not willing to participate, they will lose the opportunity to understand their current situation and what the right treatment is that they need in order to recover. Finally, the effectiveness of the treatment program will depend on the client acknowledging how the program is able to help them to improve their productivity and quality of life after being released. This is because it is important to see how the program is able to help the clients to progress and survive in the real world.

To conclude, this study is important to fill the research gap on women and drug abuse, which is an area that is very understudied, especially focusing on Malay women. It is hoped that this study will be able to help develop a better understanding of the importance of addressing this problem and to plan an effective action plan to reduce the number of relapse cases among women.

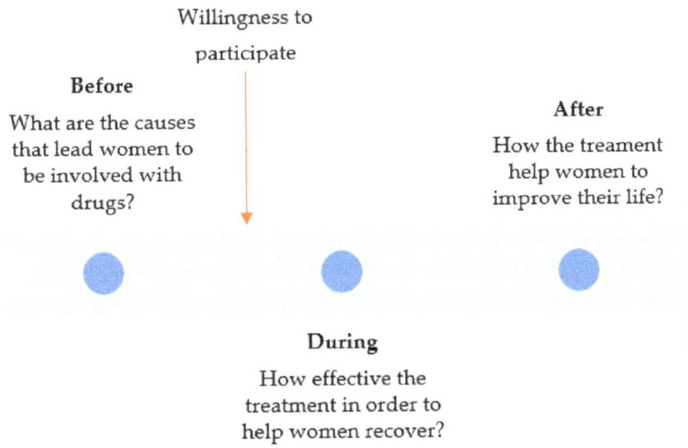

Figure 1. Representation Phases of women involvement in drug abuse and their journey to recover.

Author Contributions: Conceptualization, N.A.A., I.H.T.A. and T.N.T.I.; methodology, N.A.A., I.H.T.A. and T.N.T.I.; software, N.A.A., I.H.T.A. and T.N.T.I.; validation, N.A.A., I.H.T.A. and T.N.T.I.; formal analysis, N.A.A., I.H.T.A. and T.N.T.I.; investigation, N.A.A., I.H.T.A. and T.N.T.I.; resources, N.A.A., I.H.T.A. and T.N.T.I.; data curation, N.A.A., I.H.T.A. and T.N.T.I.; writing—original draft preparation, N.A.A.; writing—review and editing, N.A.A., I.H.T.A. and T.N.T.I.; visualization, N.A.A., I.H.T.A. and T.N.T.I.; supervision, I.H.T.A. and T.N.T.I.; project administration, N.A.A., I.H.T.A. and T.N.T.I.; funding acquisition, N.A.A., I.H.T.A. and T.N.T.I. All authors have read and agreed to the published version of the manuscript.

Funding: This research received no external funding.

Institutional Review Board Statement: Not applicable.

Informed Consent Statement: Not applicable.

Data Availability Statement: Not applicable.

Acknowledgments: Paper presented during a panel organized by the Faculty of Information Management in the context of the Social Science and Information System conference, organized and hosted by the Faculty of Information Management, during the International Academic Symposium of Social Science 2022 (IASSC 2022).

Conflicts of Interest: The authors declare no conflict of interest.

References

1. National Institute on Drug Abuse (NIDA). *National Institute on Drug Abuse*; U.S. Department of Health and Human Services: Washington, DC, USA, 2020.
2. Wickersham, J.A.; Loeliger, K.B.; Marcus, R.; Pillai, V.; Kamarulzaman, A.; Altice, F.L. Patterns of substance use and correlates of lifetime and active injection drug use among women in Malaysia. *Am. J. Drug Alcohol Abus.* **2016**, *421*, 98–110. [CrossRef] [PubMed]
3. Skóra, M.N.; Pattij, T.; Beroun, A.; Kogias, G.; Mielenz, D.; Vries, T.D.; Müller, C.P. Personality driven alcohol and drug abuse: New mechanisms revealed. *Neurosci. Biobehav. Rev.* **2020**, *116*, 64–73. [CrossRef]
4. Guttmannova, K.; Kosterman, R.; White, H.R.; Bailey, J.A.; Lee, J.O.; Epstein, M.; Hawkins, J.D. The association between regular marijuana use and adult mental health outcomes. *Drug Alcohol Depend.* **2017**, *179*, 109–116. [CrossRef]
5. Nasir, N.C.; Kamaluddin, M.R.; Jasni, M.A. Risk factors of drugs relapse behavior among female clients in Cure and Care Rehabilitation Centre (CRCC). *J. Psikol. Malays.* **2020**, *34*, 53–66.
6. Deuraseh, N.; Murad, M.N. Keberkesanan rawatan dan pemulihan dadah dari perspektif residen di Rumah Pengasih, Bukit Tunku, Kuala Lumpur Malaysia. *South-East Asian J. Youth Sports Health Educ.* **2018**, *4*, 95–124.
7. Mohd Rashid, S.; Osman, N.; Mohamad, L.Z. Penilaian tahap kepulihan penagihan dadah dalam kalangan penagih lelaki yang sedang menjalani rawatan dan pemulihan. *Malays. J. Soc. Sci. Humanit. (MJSSH)* **2021**, *6*, 15–27. [CrossRef]

8. Kabisa, E.; Biracyaza, E.; Habagusenga, J.D.; Umubyeyi, A. Determinants and prevalence of relapse among patients with substance use disorders: Case of Icyizere Psychotherapeutic Centre. *Subst. Abus. Treat. Prev. Policy* **2021**, *16*, 13. [CrossRef] [PubMed]
9. Amat, M.A.; Ahmad, J.; Jailani, O.; Jaafar, W.M.; Zaremohzzabieh, Z. Relapse among drug addicts in east coast Malaysia: A qualitative study of risk factors. *Int. J. Acad. Res. Bus. Soc. Sci.* **2020**, *10*, 432–447. [CrossRef]
10. Danquah-Amoah, A.; Charan, A.A. The relevance of rehabilitation centers in our communities. *Int. J. Sci. Res.* **2017**, *6*, 1393–1395.
11. Bernard, H.R. *Social Research Methods: Qualitative and Quantitative Approaches*, 2nd ed.; Sage: Thousand Oaks, CA, USA, 2013.
12. Jackson, K.; Bazeley, P. *Qualitative Data Analysis with NVIVO*, 3rd ed.; Sage Publication: London, UK, 2019.
13. Choo, P.C.; Ismail, H.; Tan, K.S.; Lye, C.T.; Yuen, Y.Y.; Lai, K.P. A framework for enhancing intention to recover and credibility of the rehabilitation centre among older adults in Malaysia. *J. Tour. Hosp. Environ. Manag.* **2020**, *5*, 13–39. [CrossRef]
14. Greenfield, S.; Kaplan, S.; Ware, J.E. Expanding patient involvement in care: Effects on patient outcomes. *Ann. Intern. Med.* **1985**, *102*, 520–528. [CrossRef] [PubMed]

MDPI
St. Alban-Anlage 66
4052 Basel
Switzerland
Tel. +41 61 683 77 34
Fax +41 61 302 89 18
www.mdpi.com

Engineering Proceedings Editorial Office
E-mail: engproc@mdpi.com
www.mdpi.com/journal/engproc

www.ingramcontent.com/pod-product-compliance
Lightning Source LLC
Chambersburg PA
CBHW061543220926
51850CB00005B/233
* 9 7 8 3 0 3 6 5 6 0 9 5 3 *